Lecture Notes in Networks and Systems

Volume 581

The series "Lecture Notes in Networks and Systems" publishes the latest developments in Networks and Systems—quickly, informally and with high quality. Original research reported in proceedings and post-proceedings represents the core of LNNS.

Volumes published in LNNS embrace all aspects and subfields of, as well as new challenges in, Networks and Systems.

The series contains proceedings and edited volumes in systems and networks, spanning the areas of Cyber-Physical Systems, Autonomous Systems, Sensor Networks, Control Systems, Energy Systems, Automotive Systems, Biological Systems, Vehicular Networking and Connected Vehicles, Aerospace Systems, Automation, Manufacturing, Smart Grids, Nonlinear Systems, Power Systems, Robotics, Social Systems, Economic Systems and other. Of particular value to both the contributors and the readership are the short publication timeframe and the world-wide distribution and exposure which enable both a wide and rapid dissemination of research output.

The series covers the theory, applications, and perspectives on the state of the art and future developments relevant to systems and networks, decision making, control, complex processes and related areas, as embedded in the fields of interdisciplinary and applied sciences, engineering, computer science, physics, economics, social, and life sciences, as well as the paradigms and methodologies behind them.

Indexed by SCOPUS, INSPEC, WTI Frankfurt eG, zbMATH, SCImago.

All books published in the series are submitted for consideration in Web of Science.

For proposals from Asia please contact Aninda Bose (aninda.bose@springer.com).

David Guralnick · Michael E. Auer ·
Antonella Poce

Editors

Innovative Approaches to Technology-Enhanced Learning for the Workplace and Higher Education

Proceedings of 'The Learning Ideas Conference' 2022

 Springer

Editors
David Guralnick
Kaleidoscope Learning
International E-Learning Association
New York, NY, USA

Antonella Poce
Università degli Studi di
Modena e Reggio Emilia
Modena, Italy

Michael E. Auer
CTI Global
Frankfurt, Germany

Carinthia University of Applied Sciences
Villach, Austria

ISSN 2367-3370   ISSN 2367-3389 (electronic)
Lecture Notes in Networks and Systems
ISBN 978-3-031-21568-1   ISBN 978-3-031-21569-8 (eBook)
https://doi.org/10.1007/978-3-031-21569-8

This Springer imprint is published by the registered company Springer Nature Switzerland AG
The registered company address is: Gewerbestrasse 11, 6330 Cham, Switzerland

Committees

Conference Chair

David Guralnick, Ph.D.   Kaleidoscope Learning and Columbia University, New York, New York, USA

Executive Committee Chairs

Michael E. Auer, Ph.D.   CTI, Frankfurt, Germany
Antonella Poce, Ph.D.   Università degli Studi di Modena e Reggio Emilia, Rome, Italy

Publication Chair

Lara Ramsey   Kaleidoscope Learning, New York, New York, USA

Executive Committee

Mohammed Ali Akour, Ph.D.   A'Sharqiyah University, Ibra, Oman
Kostas Apostolou, Ph.D.   McMaster University, Hamilton, Ontario, Canada
Sharon Bailin, Ph.D.   Simon Fraser University, Vancouver, British Columbia, Canada
Ryan Baker, Ph.D.   University of Pennsylvania, Philadelphia, Pennsylvania, USA
Patricia Behar, Ph.D.   Federal University of Rio Grande do Sul, Brazil

John Black, Ph.D.	Teachers College, Columbia University, New York, New York, USA
Patrick Blum, Ph.D.	Blum Consulting, Aachen, Germany
Santi Caballé, Ph.D.	Open University of Catalonia, Barcelona, Spain
Nicola Capuano, Ph.D.	University of Basilicata, Potenza, Italy
Imogen Casebourne	University of Oxford, Oxford, UK
Manuel Castro, Ph.D.	Universidad Nacional de Educacion a Distancia (UNED), Madrid, Spain
Veronica Chehtman	AySA Water and Sanitation Argentina, Buenos Aires, Argentina
Hal Christensen	QuickCompetence, New York, New York, USA
Samir El-Seoud, Ph.D.	The British University in Egypt (BUE), Egypt
Kai Erenli	University of Applied Sciences bfi Vienna, Vienna, Austria
Matthias Gottlieb, Ph.D.	Technical University of Munich, Munich, Germany
Christian Guetl, Ph.D.	Graz University of Technology, Graz, Austria
Alexander Kist, Ph.D.	University of Southern Queensland, Queensland, Australia
Gila Kurtz, Ph.D.	Holon Institute of Technology, Holon, Israel
Mark J. W. Lee, Ph.D.	Charles Sturt University, Bathurst, Australia
Christy Levy	Kaleidoscope Learning, Chicago, Illinois, USA
Matthea Marquart	Columbia University, New York, New York, USA
Bruce McLaren, Ph.D.	Carnegie Mellon University, Pittsburgh, Pennsylvania, USA
Jorge Membrillo Hernández, Ph.D.	Tecnológico de Monterrey, México
Dominik May, Ph.D.	University of Georgia, Athens, Georgia, USA
Christina Merl, Ph.D.	TalkShop/2CG®, Vienna, Austria
Gary Natriello, Ph.D.	Teachers College, Columbia University, New York, New York, USA
Barbara Oakley, Ph.D.	Oakland University, Oakland, Michigan, USA
Andreas Pester, Ph.D.	Carinthia Tech Institute, Villach, Austria
Robert Pucher, Ph.D.	University of Applied Sciences Technikum Wien, Vienna, Austria
Teresa Restivo, Ph.D.	University of Porto, Porto, Portugal
Fernando Salvetti, Ph.D.	Logosnet, Turin, Italy
Alicia Sanchez, Ph.D.	Czarina Games, Alexandria, Virginia, USA
Sabine Seufert, Ph.D.	Universität St.Gallen (HSG), St. Gallen, Switzerland
Thrasyvoulos Tsiatsos, Ph.D.	Aristotle University of Thessaloniki, Thessaloniki, Greece
James Uhomoibhi, Ph.D.	Ulster University, Newtownabbey, UK
Matthias Utesch, Ph.D.	Technical University of Munich, Munich, Germany

Ellen Wagner, Ph.D.	North Coast EduVisory Services, Sonoma, California; and University of Central Florida, Orlando, Florida, USA
Sarah Wang, Ph.D.	Xi'an Jiaotong-Liverpool University, Suzhou, Jiangsu, China
Xiao-Guang Yue, Ph.D.	European University Cyprus, Nicosia, Cyprus

Program Committee

Carme Anguera Iglesias	Open University of Catalonia, Barcelona, Spain
Fahriye Altinay Aksal, Ph.D.	Near East University, Nicosia, Cyprus
Zehra Altinay Gazi, Ph.D.	Near East University, Nicosia, Cyprus
Sarah Appleby	Online Learning International, New York, New York, USA
Anabel Bugallo	ADP, New York, New York, USA
Martha Burkle, Ph.D.	Southern Alberta Institute of Technology, Calgary, Canada
Mihai Caramihai, Ph.D.	University Politehnica Bucharest, Bucharest, Romania
Nunzio Casalino, Ph.D.	Guglielmo Marconi University and LUISS Business School, Rome, Italy
Mark Cassetta	Pfizer, New York, New York, USA
Gary J. Dickelman	EPSSCentral, Annandale, Virginia, USA
David Foster, Ph.D.	ExecOnline, Chapel Hill, North Carolina, USA
Sarah Frame	University of East London, London, England
Marga Franco i Casamitjana	Open University of Catalonia, Barcelona, Spain
Genevieve Gallant, Ph.D.	GG Consultants Limited, St. John's, Newfoundland, Canada
Abel Henry	United Nations Development Programme, Copenhagen, Denmark
Manir Abdullahi Kamba, Ph.D.	Bayero University, Kano, Nigeria
Okba Kazar, Ph.D.	Biskra University, Biskra, Algeria
J. C. Kinnamon, Ph.D.	Practising Law Institute, New York, New York, USA
Adamantios Koumpis, Ph.D.	Bern University of Applied Sciences, Bern, Switzerland
Molly Koenen	Pioneer Management Consulting, Minneapolis, Minnesota, USA
Maria Lambrou, Ph.D.	University of the Aegean, Greece
Stacy Lindenberg	Talent Seed Consulting, Columbia, South Carolina, USA
Allison Littlejohn, Ph.D.	Glasgow Caledonian University, Glasgow, Scotland

Luis Ochoa Siguencia, Ph.D.	Jerzy Kukuczka Academy of Physical Education, Katowice, Poland
Grace O'Malley, Ph.D.	National College of Ireland, Dublin, Ireland
Rikke Orngreen, Ph.D.	Aalborg University, Copenhagen, Denmark
Michael Paraskevas, Ph.D.	Technological Educational Institute of Western Greece, Antirrio, Greece
Iina Paarma	United Nations Development Programme, Copenhagen, Denmark
Kinga Petrovai, Ph.D.	The Art and Science of Learning, Ottawa, Ontario, Canada
Stefanie Quade	Berlin School of Economics and Law, Berlin, Germany
Maria Rosaria Re	Roma Tre University, Rome, Italy
Laura Ricci	MIA Digital University, Barcelona, Spain
Gina Ann Richter, Ph.D.	St. Charles Consulting Group, New York, New York, USA
Andree Roy, Ph.D.	University of Moncton, Moncton, Canada
John Sandler	Telstra Corporation, Melbourne, Australia
Steven Schmidt	East Carolina University, Greenville, North Carolina, USA
Barbara Schwartz-Bechet, Ph.D.	Misericordia University, Elkins Park, Pennsylvania, USA
Julie-Ann Sime, Ph.D.	Lancaster University, Lancashire, UK
Amando P. Singun, Jr.	Higher College of Technology, Muscat Sultanate of Oman
Anelise Spyer	Docta, São Paulo, Brazil
Christian Stracke, Ph.D.	eLC Institute for eLearning, Bonn, Germany
Kyla L. Tennin, Ph.D.	University of Phoenix, Tempe, Arizona, USA
Terrie Lynn Thompson	University of Alberta, Edmonton, Alberta, Canada
Caryn Tilton	MyPlacetoLearn, Welches, Oregon, USA
Leyla Y. Tokman, Ph.D.	Anadolu University, Eskisehir, Turkey
Christos Troussas, Ph.D.	University of West Attica, Athens, Greece
Chris Turner, Ph.D.	University of Winchester, Winchester, UK
Karin Tweddell Levinsen, Ph.D.	Danish University of Education, Denmark
Maggie M. Wang, Ph.D.	The University of Hong Kong, Hong Kong
Steve Wheeler	University of Plymouth, Plymouth, UK
Annika Wiklund-Engblom, Ph.D.	Abo Akademi University, Vasa, Finland
Rusen Yamacli, Ph.D.	Anadolu University, Eskisehir, Turkey

ALICE Special Track Co-chairs

Santi Cabellé, Ph.D.	Open University of Catalonia, Barcelona, Spain
Nicola Capuano, Ph.D.	University of Basilicata, Potenza, Italy

ALICE Special Track Committee

Joan Casas, Ph.D.	Open University of Catalonia, Spain
Jordi Conesa, Ph.D.	Open University of Catalonia, Spain
Thanasis Daradoumis, Ph.D.	University of the Aegean, Greece
Sara De Freitas, Ph.D.	Coventry University, UK
Christian Gütl, Ph.D.	Graz University of Technology, Austria
Giuseppina Rita Mangione, Ph.D.	Institute of Educational Documentation, Innovation and Research, Italy
Agathe Merceron, Ph.D.	Beuth University of Applied Sciences Berlin, Germany
Anna Pierri, Ph.D.	University of Salerno, Italy
Antonio Sarasa, Ph.D.	Universidad Complutense De Madrid, Spain
Marco Temperini, Ph.D.	Sapienza University of Rome, Italy
Daniele Toti, Ph.D.	Catholic University of the Sacred Heart, Italy

Inclusive Learning Special Track Co-chairs

Fahriye Altinay Aksal, Ph.D.	Near East University, Nicosia, Cyprus
Zehra Altinay Gazi, Ph.D.	Near East University, Nicosia, Cyprus

Inclusive Learning Special Track Committee

Umut Akçil	Near East University, Nicosia, Cyprus
Meryem Bastas	University of Kyrenia, Kyrenia, Cyprus
Muhammet Berigel	Karadeniz Technical University, Trabzon, Turkey
Huseyin Bicen	Near East University, Nicosia, Cyprus
Gokmen Dagli	University of Kyrenia, Kyrenia, Cyprus
Ipek Danju	Near East University, Nicosia, Cyprus
Didem Islek	Near East University, Nicosia, Cyprus
Ceren Karaatmaca	University of Kyrenia, Kyrenia, Cyprus
Nesrin Menemenci	Near East University, Nicosia, Cyprus

Preface

The Learning Ideas Conference, with the subtitle "Innovations in Learning and Technology for the Workplace and Higher Education," looks to bring together people from around the world to help reimagine what learning can be, particularly using, and inventing, new technologies.

The Learning Ideas Conference 2022 was held as a hybrid event, with significant participation from in-person and virtual participants. Our goal was for conference participation to be easily accessible to all, and for in-person and virtual participants to have the opportunity to interact, thus creating an experience that was truly hybrid. The conference featured 3 wonderful keynote speakers:

- Dr. Ryan Baker, Associate Professor, Graduate School of Education and Director, Penn Center for Learning Analytics at the University of Pennsylvania, Philadelphia, Pennsylvania, USA. *"Algorithmic Bias in Education."*
- Dr. Ian Bogost, Professor, Computer Science and Engineering and Professor and Director, Program in Film & Media Studies, Washington University in St. Louis, St. Louis, Missouri, USA. *"How to Learn Playfully."*
- Dr. Antonella Poce, Full Professor in Experimental Pedagogy, University of Modena and Reggio Emilia, Rome, Italy. *"Creating Cultural Assets to Foster Social Inclusion and Development."*

The conference featured two panel discussions, one on "Learning, Technology, and Society: The Present and the Future" and the other on "Approaches to Workplace Learning and Performance in Today's Hybrid World," and a total of over 100 sessions. All papers were double-blind peer reviewed.

I very much appreciate all of the work it took to make The Learning Ideas Conference 2022 a success, from our keynotes, our Executive Committee and Program Committee members, our reviewers, and of course our conference organizing team.

I am looking forward to The Learning Ideas Conference 2023, to be held again as a hybrid event, both in New York and online.

New York, USA   David Guralnick, Ph.D.
                                                                                                          Conference Chair
                                                                                     The Learning Ideas Conference 2022

Contents

Main Conference

CyEd: A Cyberinfrastructure for Computer Education

Sherif Abdelhamid(✉) ⓘ, Tanner Mallari ⓘ, and Tristen Stower ⓘ

Virginia Military Institute, Lexington, VA 24450, USA
`abdelhamidse@vmi.edu`

Abstract. Technology-supported learning (TSL) is a promising discipline within computer science, concerned with designing and building tools and services that improve the students' learning environment. TLS can foster group-based interactions, collaborations, and accessibility to learning material, computing resources, and data. This work is situated within the TLS research area. It aims to provide educators and students with an easy-to-use cyberinfrastructure (CI) accessible from their desktops and integrated into their daily study activities. CIs could impact how STEM research is conducted. Many agencies are supporting this technology as a tool for scientific discoveries. However, fewer applications are observed in utilizing CIs in the classroom as a scientific educational tool. There is an urgent need to increase the use of these cyber environments in STEM and computer education. In this work, we present CyEd, a CI for computer education. A key goal is to form a community of practice around the provided learning resources, to enable the sharing of resources and insights, and to enhance the students' involvement and engagement. In a community of practice, people exchange knowledge regularly to deepen their expertise and understanding of a topic. According to Wenger's theory of "community of practice," such communities act as "social learning systems" where students collaborate with educators to solve problems, share ideas, build tools, and build relationships. CyEd will be designed for scalability, usability, extensibility, and continuity. Users will gain access to the many resources within and can also contribute new learning resources, data, and methods in an innovative way that advances Computer Science education. This model will facilitate the improvement and expansion of the system and increase the self-sustainability of CyEd, allowing it to meet the changing needs of its community.

Keywords: Computer education · Cyberinfrastructure · Collaborative learning · Cooperative learning · Student motivation · Student engagement

1 Introduction

Computer Education Research (CER) is a field of inquiry that evolved over the last three decades, focusing on the teaching scholarship and aims to define, inform, and improve the students' computer science education and address various research topics such as assessment, pedagogy, and diversity in Computer Science. In addition to these topics, technology-supported learning (TSL) is another promising discipline within CER

© The Author(s), under exclusive license to Springer Nature Switzerland AG 2023
D. Guralnick et al. (Eds.): TLIC 2022, LNNS 581, pp. 3–14, 2023.
https://doi.org/10.1007/978-3-031-21569-8_1

concerned with designing and building tools and services that improve the students' learning environment through fostering group-based interactions, collaborations, and accessibility to learning material, computing resources, and data. These systems can be valuable sources for collecting students' learning data to monitor students' academic progress, identify key challenges, and promote factors leading to success.

1.1 Motivation

Our motivation is to provide educators and students interested in Computer Science with an easy-to-use Cyberinfrastructure (CI) or technology-supported learning environment that is accessible from their desktop and integrates into their daily study activities. According to [11], Indiana University provided a comprehensive definition of CIs:

Cyberinfrastructure consists of computational systems, data and information management, advanced instruments, visualization environments, and people, all linked together by software and advanced networks to improve scholarly productivity and enable knowledge breakthroughs and discoveries not otherwise possible.

A key goal for the CI is to form a community of practice around the CI resources to enable sharing resources, insights and enhance students' involvement. Communities of practice are groups of people who share a concern or passion about a topic and who interact regularly in order to gain a deeper understanding of that topic [13]. Recent reports show that CIs can impact how Science, Technology, Engineering, and Mathematics (STEM) research is conducted. Many agencies, including the National Science Foundation (NSF), Department of Defense (DOD), Department of Energy (DOE), and Central Intelligence Agency (CIA), are supporting this technology as a tool for research and scientific discoveries. However, fewer applications are noticed in utilizing CI resources in the classroom as a scientific educational tool. There is an urgent need to increase the use of CIs in STEM.

1.2 Innovation and Approach

CyEd will enable educators to teach Computer Science to students spanning various academic levels and disciplines. It is designed for scalability, usability, extensibility, and continuity. The implemented system (CyEd) is built to be pervasive, flexible, and extensible. Each user gains access to the many resources within and can also contribute new learning resources, data, and methods in an integrative way that advances computer education. This will facilitate the improvement and expansion of the system and increase the self-sustainability of CyEd, allowing it to meet the changing needs of its community.

1.3 Research Objectives and Goal

The researchers behind CyEd launched a continuous comprehensive education and outreach plan composed of short courses, workshops, and focused-user group meetings to provide a path towards adopting the new CI and recommending user-guided improvements. Selected CyEd tools were used as part of the CyberSmart 2021 workshop organized by Virginia Military Institute for high-school students and will be used again during the upcoming 2022 event. The undergraduate educational plan will start with students from Computer Science and other related departments, students from minority and under-represented groups, and high school students in Virginia state. The researchers behind this work plan to use CyEd next fall as part of several undergraduate courses within the computer and information sciences and the mathematics departments. By achieving these objectives, we will achieve our goal of forming a collaborative community of students, educators, and researchers around CyEd and creating an inclusive and engaging learning environment.

2 Related Work

Our related work review used a thematic approach where sources were grouped and discussed according to similar themes, concepts, or topics. This organizational method is often considered efficient because it helps analyze and synthesize sources into a well-organized, coherent review. In addition, we utilized a multi-method sequential approach. Researchers behind CyEd conducted a literature review followed by a bibliometric analysis to observe the evolution of the scientific literature, identify specific characteristics of the related knowledge domain, and identify main research themes and trends related to the field of CIs and online learning.

2.1 Literature Review

Two main research themes emerged during the related work review on CIs and online education. The first theme focuses on online teaching pedagogies, methods, and techniques employed within the learning environment, while the second theme is about the different factors affecting online learning effectiveness.

Numerous research papers have empirically tested and demonstrated the positive impact of online cooperative learning [1, 8, 9, 14]. One research work [4] revealed that cooperative learning is the most beneficial strategy in online education, with a positive impact in 71% of 17 papers surveyed. Another work [8] explained that collaboration could increase exposure to different views, opinions, reiterations of course material, which is consistent with positive results in other studies, including references [1, 14]. Another work [4] revealed a concern about cooperative learning that students who participated in cooperative learning were less engaged than those who did not. Other sources [7, 9] identified a prominent online education strategy which is the use of interactive multimedia. Of the 31 papers evaluated, 64% had positive empirical results of using interactive multimedia methods for content delivery. Learners preferred interactive tutorials to basic video tutorials. Results showed that multimodal teaching format could lead to better outcomes than just using text. The benefits of interactive multimedia are discussed in various papers. Several factors need to be considered for an adequate educational CI. A research work [8] summarized the factors influencing online learning effectiveness based on four sources: the "technology, format, instructor, and learner." The paper prioritized and reviewed 79 out of 317 identified publications and investigated each source's role in learner effectiveness.

According to this work, technology capabilities can trigger a learner's motivation. In addition, they can positively impact the learning experience with passive components, such as videos, and active ingredients, such as quizzes, adequately utilized in the learning process. However, other works noted that technical complications could reduce the learner and lecturer motivation [6]. Numerous papers show the benefit and preference for interactive multimedia [4, 7, 9]. Frequently identified factors for the instructor include the instructor's attitude towards the technology, their role in a technology-supported environment as facilitators, and instructor feedback. An instructor's attitude towards technology "affects learners' attitudes toward the format and technology" [8] and therefore predicts perceived learning. The prior experiences and preferences significantly impact the instructor's attitude. Another research work [6] surveyed 26 randomly sampled university lecturers in Indonesia and found that lecturers heavily preferred publicly available learning management systems, such as Google Classroom and Edmodo, to video conferencing and university-developed e-learning platforms. Instructors who adopt an interactive teaching style as a guide to the platform were found to have a "positive impact on learners' involvement and participation, cognitive engagement, and attitudes toward format and technology." More identified factors for the learner include student motivation, prior technological experience, and prior academic achievements. Based on existing literature [1, 8, 9], the most impactful factor in learning is the student's motivation. Previous academic achievements, especially those related to online or blended learning environments, can also impact the learning experience, positively impacting their view on the format of the technology.

2.2 Bibliographic Analysis

The bibliometric mapping of publications can identify key research themes and which articles have had the most significant impact (in which fields). According to [12], two main characteristics of bibliometric mapping are the creation process of bibliometric maps and the visual representation. Additionally, the visual representations offer various perspectives on the same data from various angles, enhancing a researcher's understanding of the underlying maps and their ability to recognize underlying patterns and themes. In this work, we conducted a bibliometric mapping using the Microsoft Academics publication database [10]. We specifically aimed to study the major research themes and patterns related to CIs and online learning. From the retrieved publications (relevant to our topic of interest), we extracted terms (with minimum occurrences of 10) from the abstract and title, resulting in 476 key terms. Based on the co-occurrences of the terms within the same title or abstract, we constructed the term co-occurrence network, which consists of 16,279 edges/links and 476 nodes. Each node represents a term, and each edge represents a co-occurrence relationship. Finally, we applied a network clustering algorithm, which identified four main research clusters (see Fig. 1).

Cluster 1 (in red) contains key terms related to psychology, social science, social relations, metacognition, social cognition, and related topics. This cluster includes research on teaching pedagogies, one of the main themes identified in the literature review. It reveals multidisciplinary research that spans psychology, social science, and education research. Another cluster, number 2 (in blue), contains key terms like educational technology, digital learning, experiential learning, and related topics.

Cluster 2 is consistent with the second theme identified in the literature review and reflects the cohort of research that focuses on the uses of technology for learning. The bibliographic analysis revealed two more clusters of research that did not show up in the literature review.

Cluster 3 (in green) reflects the research relevant to the computer science field and includes subtopics like human-computer interaction, natural language processing, and artificial intelligence. This group mainly researches the technical aspects of designing and building CIs, learning environments, and other technologies, including pedagogical chatbots. Cluster 3 concentrates on delivering the needed infrastructure for online collaborative learning environments.

Another new cluster, cluster 4 (in yellow), arose during the analysis. It reflects the research on collaborative knowledge, knowledge management, knowledge engineering, procedural knowledge, and knowledge-based systems.

In addition to the key term analysis, we extracted both author and institution information from each publication. Collected data reveal the most active researchers based on publications and collaborations with other researchers (Fig. 2). Moreover, we collected institutional data, and results showed top institutions and research centers based on the number of publications produced and external collaborations with other institutions (see Fig. 3).

3   System Features and Overview

3.1   Student Development Support

CyEd supports students and educators by providing features and tools that address various aspects of students' development, including technical, cognitive, metacognitive, and social skills. Additionally, CyEd aims to engage students in various activities and promote extrinsic and intrinsic motivation through digital badges, instructor feedback, and peer support. We envision four classes of users: (a) undergraduate and high school students who are interested in studying various topical areas related to computer and information science; (b) instructors at the above levels; (c) researchers interested in using the educational data analytics tools; and (d) CI researchers who can experiment with novel user interfaces, integrate new learning tools, or add computational resources and services (science-as-service). CyEd is planned for integration in undergraduate courses at various departments. Faculty members will provide learning resources, and students can provide support through tool development, feedback, peer support, or participation in online activities.

3.2   Integrated Tools

CyEd is designed to be pervasive, flexible, and extensible, and new learning supporting tools will be introduced over time. An example of a current tool integrated within CyEd is vizLab, which uses the "Block-based" programming approach. The tool intends to introduce first-year students to programming and reduce the cognitive load of learning programming constructs and syntax. Students will develop algorithms in a visual programming language (Blockly), where graphical icons represent the language's essential

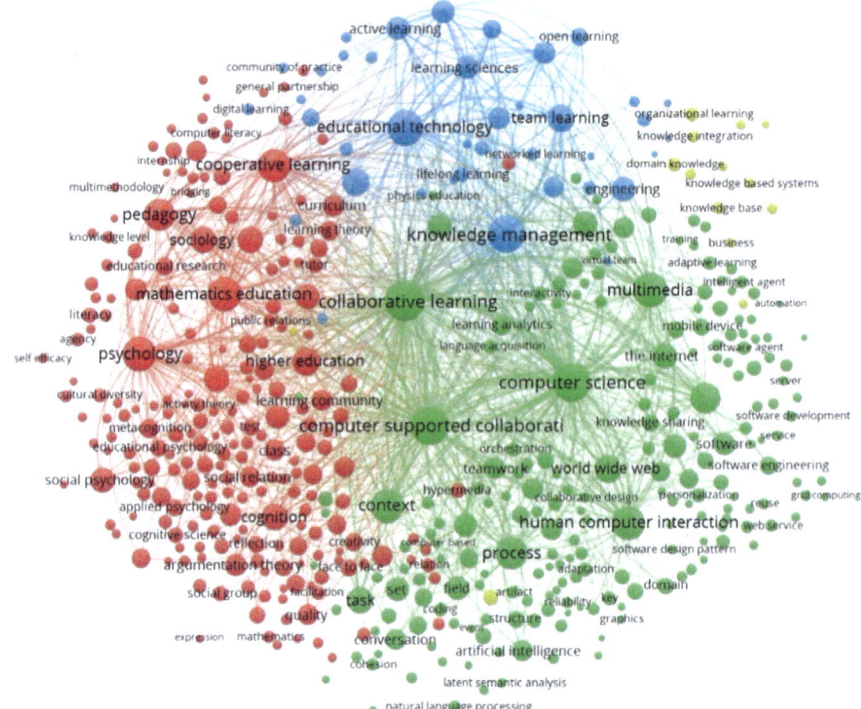

Fig. 1. The term co-occurrence network where each node represents a term, and each edge represents a co-occurrence relationship. In addition, nodes within the same cluster are assigned the same color. A total of nine clusters were identified, and each color can represent a research theme, collective efforts, or trend related to CIs and online education.

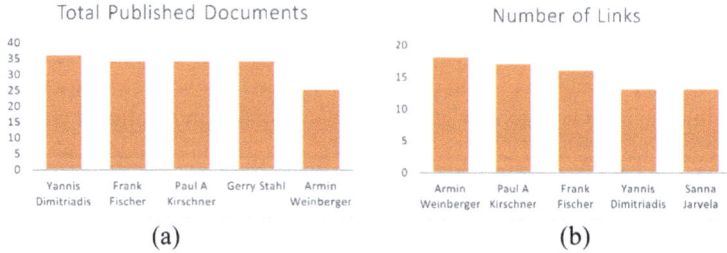

Fig. 2. Top authors based on the number of publications produced and the total links/collaborations with other authors. The number of links was calculated based on each publication's authorship data.

elements. The graphical icons can be selected, copied, and moved around in a workspace to create a complete program. The blocks can be directly executed and translated into the code of standard, text-based programming languages (Python). Students will be able to see the translation of the blocks into Python in real-time and create programming

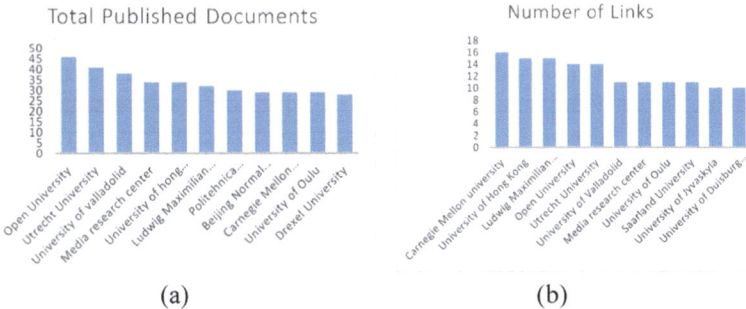

(a)   (b)

Fig. 3. Top institutions and research centers based on the number of publications produced and the total external collaborations with other institutions. The number of links was calculated based on each author's affiliation data.

elements, including (e.g., action sequence, decision, iteration, and functions). Additionally, CyEd stores the created blocks in the database and can be retrieved later. Students can share and collaborate with peers on various projects, as shown in the right menu in Fig. 4. Students can build new programs on top of existing programs created by their peers.

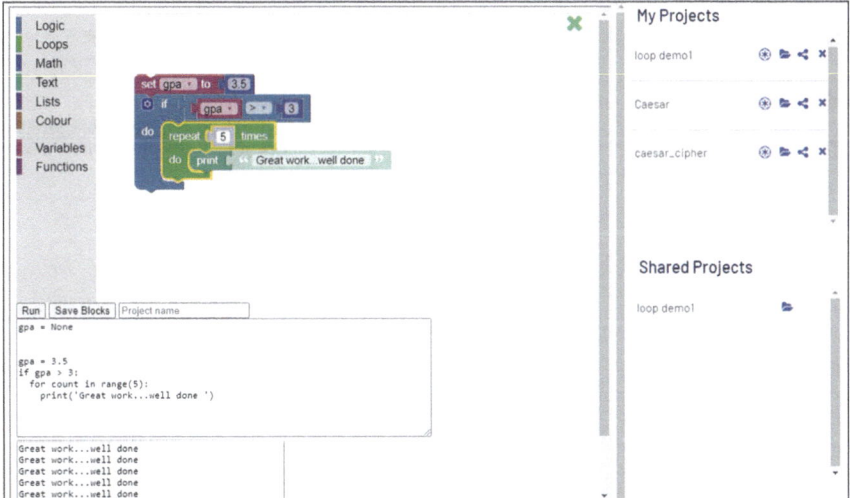

Fig. 4. Block-based learning of programming languages using VizLab. Blocks are directly translated into Python code, and students can execute the code online. Students can create and share block-based programs with their peers and instructors. In addition, students can submit their code, as blocks, to a learning management system (Canvas) for evaluation and grading by instructors. This approach facilitates team-based learning and collaboration amongst students.

Another example of a learning tool integrated into the CI is "Algorithm Visualizer," which uses the work by [5] to help teachers and students write Python programs in the browser, step forwards and backward through execution to view the run-time state of data structures and share their program visualizations. Currently, CyEd provides students and faculty with the ability to create project repositories containing learning resources and data (see Fig. 5a). In addition, users can share their projects with peers and form teams as shown in Fig. 5b. Each project or learning resource can include data, text, multimedia files. CyEd provides a chat-like instant messaging feature for each project team. This tool helps improve communication and collaboration between team members (Fig. 6). NoteBoard is another tool within CyEd that promotes collaboration amongst students within the learning environment. NoteBoard focuses on crowdsourcing the note-taking process and allows students to add new notes or modify existing ones simultaneously in a shared online space (see Fig. 7).

CyEd provides the capability for students and instructors to link their user accounts with external learning management systems, including Canvas. As shown in Fig. 8a, users can provide their canvas access token and can link multiple Canvas accounts at the same time. Once the integration is completed, all the users' courses will be displayed with their roles, as in Fig. 8b. In-addition to these learning tools, pre-existing tools were integrated into CyEd, including Colab [15], an online enclosed environment with educational video content created by the community of students and moderated by instructors, this tool aims to make the classes more motivational through peer-based learning and can significantly reduce face-to-face students' lecturing.

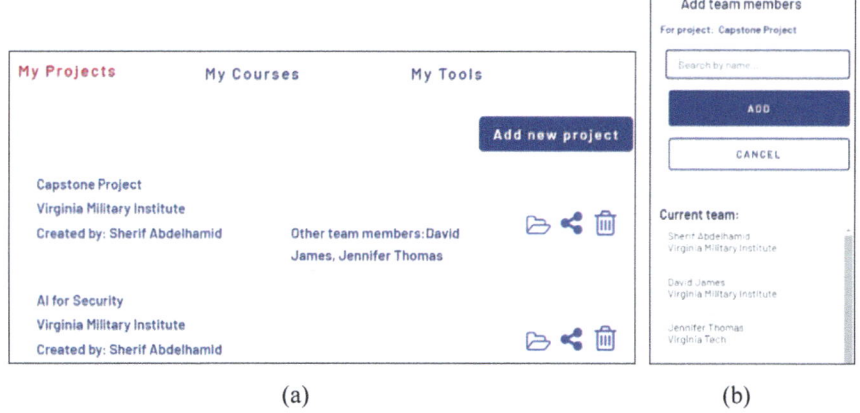

Fig. 5. Students can create project repositories containing learning resources and data. Project creators can manage teams and invite team members for collaboration. All team members can have access to each project's resources and data.

3.3   System Architecture

The implemented system utilizes a coordinating architecture that serves as a multiplexer for various components (see Fig. 9). It consists of a web-based user interface, a database

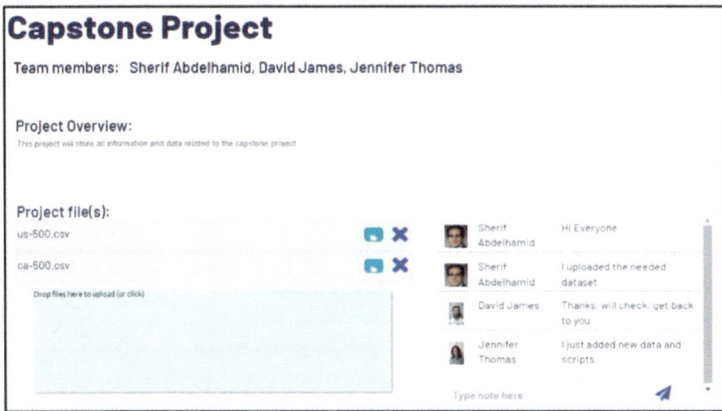

Fig. 6. Each project has a dedicated, dynamic page. All team members can upload to the cloud and share various resources and data. In addition, CyEd provides an instant messaging capability for team members to communicate in real-time.

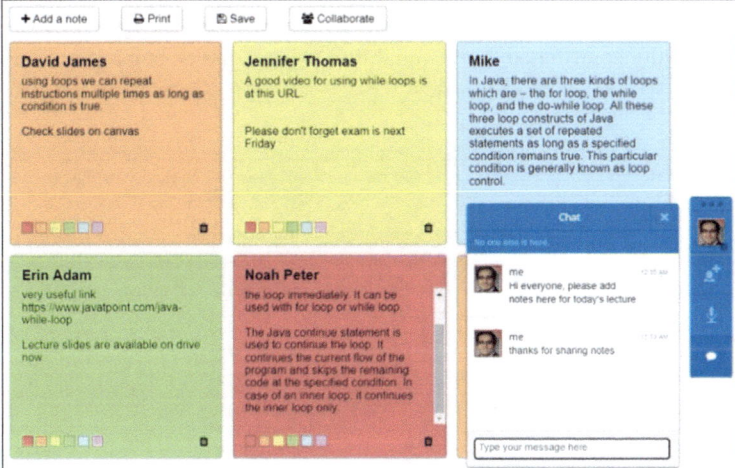

Fig. 7. Crowd-sourcing the note-taking process can be done using NoteBoard. Students can simultaneously add, remove, or group notes within a shared note repository per each lecture. In addition, students can communicate using a real-time chat service.

(contains user account information, educational videos, students' interactions, block-based codes, projects' information, and learning resources), learning analytics tools, and a resource manager (connects the system to different computing resources including Amazon and Google web services and learning management systems including Canvas). The vision behind this architecture is to make the system self-sustainable, where students and educators can contribute new computing and learning resources. In addition, the design helps the system be self-manageable as it hides the complexities of resource allocation, scheduling, cross-platform interactions, and other low-level concerns from

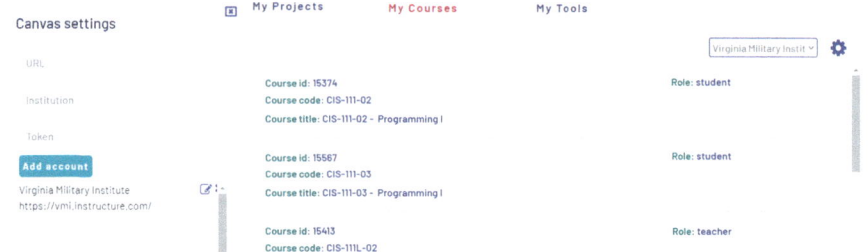

Fig. 8. Integration of CyEd with an external learning management system, "Canvas." Student and instructors can link their Canvas accounts with CyEd.

end-users. We follow a spiral development approach [3], informed by modern Agile development practices [2]. We are working in short iterations (2–3 months), with each iteration, new tools are added and used internally by faculty and students. Updates will then be made available to all users. Since this will be a web-based system, releases and interim bug fixes will be easy to deploy with minimal impact. To evaluate the software, we are collecting metrics along several dimensions: engagement (number of users and their interactions), usefulness (number of tools and amount of data in the system), extensibility (number of faculty contributed tools), reliability (system faults, bug reports, and feature requests).

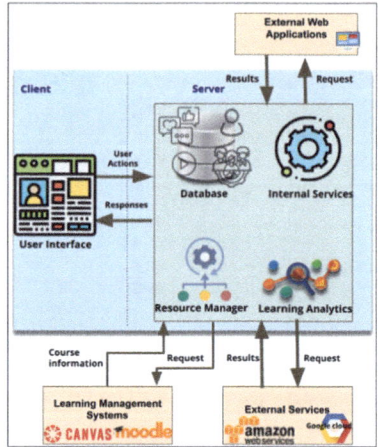

Fig. 9. CyEd CI multi-tier architecture.

4   Conclusion and Future Work

Digitization of information, communication, and data has affected every aspect of our lives and is replacing and extending traditional science and engineering education efforts.

Even conventional research approaches, whether theoretical or experimental, are being transformed into in silico simulations and experiments to take advantage of the new possibilities at new spatial and temporal levels. CyEd main research contribution is to enable people, learning tools, and knowledge to be linked in ways that reduce barriers of location, time, institution, and discipline. It will be designed for scalability, usability, extensibility, and continuity. Our future work will include new learning tools and learning analytics services. We plan to integrate CyEd within undergraduate courses and use it during an upcoming summer workshop for high school students. Periodical usability assessments will be conducted on the user interface to make changes or add new features.

Acknowledgments. This work was supported in part by the Commonwealth Cyber Initiative, an investment in the advancement of cyber R&D, innovation, and workforce development. For more information about CCI, visit https://cyberinitiative.org/.

References

1. Abuhassna, H., Al-Rahmi, W.M., Yahya, N., Zakaria, M.A.Z.M., Kosnin, A.B., Darwish, M., et al.: Development of a new model on utilizing online learning plat-forms to improve students' academic achievements and satisfaction. Int. J. Educ. Technol. High. Educ. **17**(1), 1–23 (2020)
2. Beck, K.: Embracing change with extreme programming. Computer **32**(10), 70–77 (1999)
3. Boehm, B.: A spiral model of software development and enhancement. ACM SIG-SOFT Softwa. Eng. Notes **11**(4), 14–24 (1986)
4. Davis, D., Chen, G., Hauff, C., Houben, G.J.: Activating learning at scale: a review of innovations in online learning strategies. Comput. Educ. **125**, 327–344 (2018)
5. Guo, P.J.: Online python tutor: embeddable web-based program visualization for cs education. In: Proceeding of the 44th ACM Technical Symposium on Computer Science Education, pp. 579–584 (2013)
6. Irfan, M., Kusumaningrum, B., Yulia, Y., Widodo, S.A.: Challenges during the pandemic: use of e-learning in mathematics learning in higher education. Infin. J. **9**(2), 147–158 (2020)
7. Lin, C.H., Wu, W.H., Lee, T.N.: Using an online learning platform to show students' achievements and attention in the video lecture and online practice learning environments. Educ. Technol. Soc. **25**(1), 155–165 (2022)
8. Müller, F.A., Wulf, T.: Technology-supported management education: a systematic review of antecedents of learning effectiveness. Int. J. Educ. Technol. High. Educ. **17**(1), 1–33 (2020). https://doi.org/10.1186/s41239-020-00226-x
9. Nurhayati, A., Bandung, Y.: Elearning platform for adult learner based on mobile instant messaging for collaborative learning. In: 2020 International Conference on Information Technology Systems and Innovation (ICITSI), pp. 38–43. IEEE (2020)
10. Sinha, A., Shen, Z., Song, Y., Ma, H., Eide, D., Hsu, B.J., Wang, K.: An overview of microsoft academic service (mas) and applications. In: Proceedings of the 24th International Conference on World Wide Web, pp. 243–246 (2015)
11. Stewart, C.A., Simms, S., Plale, B., Link, M., Hancock, D.Y., Fox, G.C.: What is cyberinfrastructure. In: Proceedings of the 38th Annual ACM SIGUCCS Fall Conference: Navigation and Discovery, pp. 37–44 (2010)
12. Van Eck, N., Waltman, L.: Software survey: Vosviewer, a computer program for bibliometric mapping. Scientometrics **84**(2), 523–538 (2010)

13. Wenger, E.: Communities of practice: learning, meaning, and identity. Learning in doing: social, cognitive and computational perspectives. Cambridge University Press (1998). https://doi.org/10.1017/CBO9780511803932

14. Yadegaridehkordi, E., Shuib, L., Nilashi, M., Asadi, S.: Decision to adopt online collaborative learning tools in higher education: a case of top malaysian universities. Educ. Inf. Technol. **24**(1), 79–102 (2019)

15. Abdelhamid, S., Stower, T.: Use of online educational videos for concept oriented peer-based learning. In: Filodiritto Editore, 11th International Conference New Perspectives in Science Education, pp. 155–160 (2022)

Socio-affective Profiles in Virtual Learning Environments: Using Learning Analytics

Jacqueline Mayumi Akazaki$^{(\boxtimes)}$ (ID), Leticia Rocha Machado (ID), and Patricia Alejandra Behar (ID)

Federal University of Rio Grande do Sul, Porto Alegre, Brazil

Abstract. This article aims to apply Learning Analytics (LA) in a Virtual Learning Environment (VLE) considering the social and affective aspects. The Reference Model based on four dimensions, which are: what, why, how, and who. In addition, the General Process of LA was used, consisting of three steps—namely, data collection and preprocessing, analysis and action, and postprocessing. The methodology of this work has a qualitative and quantitative approach based on 13 case studies. Data collection from the 285 Brazilian students took place from the interaction and records in the VLE Cooperative Learning Network (in Portuguese: ROODA). As a result, 38 Socio-affective Scenarios were mapped, discovered from the Reference Model and the General Process of LA. These are important because, through them, the professor can visualize the social and affective aspects of his/her students, being able to intervene, if desired, in order to try to avoid drop outs in response to absent, discouraged, feelings of separation from the class.

Keywords: Learning analytics · Distance education · Virtual learning environment · Social and affective aspects

1 Introduction

The term Learning Analytics (LA) was coined in 2011, from the development of the LA and Knowledge conference (https://tekri.athabascau.ca/analytics), which explicitly emphasized the role of LA as a bridge between Computer Science and the Sociology/Psychology of learning. At this meeting, which took place in Canada and attracted more than one hundred participants, the LA assumed that "the technical, pedagogical and social domains must be placed in dialogue with each other to ensure that interventions and organizational systems meet the needs of all parties' stakeholders" [1].

Subsequently, in the same year, the Society for LA Research (solaresearch.org) was formed to support the conference, develop and advance a research agenda in LA, as well as advocate and elucidate its use. George Siemens was one of the first authors to start studying LA, and in 2012 he defined it as:

> [...] the measurement, collection, analysis and reporting of data about students and their contexts for the purposes of understanding and optimizing learning and the environments in which it takes place. Learning Analytics refers to using data

D. Guralnick et al. (Eds.): TLIC 2022, LNNS 581, pp. 15–26, 2023.
https://doi.org/10.1007/978-3-031-21569-8_2

produced by students, building analysis models to discover information and social connections, predict progress and advise on learning [2, pp. 4].

LA is one of the areas that has been attracting the most attention within the scientific community that works with technologies aimed at education. Its objective is to promote more effective ways for subjects who work in Distance Education (DE), in relation to monitoring the performance and engagement of students in the Virtual Learning Environment (VLE), anticipating the identification of possible problems, which can lead the student to drop out [3]. The VLE can serve as a tool for in-person class, hybrid, and DE. In general, it has several features that provide opportunities for exchanges, such as Chat, Forum, places for posting work personal, or collective writing, among others [4].

Johnson et al. [5] identified LA as an important future trend for education, as it makes possible the individualized analysis of the student. In this scenario, in DE, one of its main challenges is to consider the social and affective aspects of each individual when working in the VLE.

Thus, although LA is an emerging field, it is not new and is linked to several areas, involving machine learning, artificial intelligence, information retrieval, statistics, visualization and research models. Thus, LA can be understood as multidisciplinary, and "the novelty in relation to other approaches is in human intervention throughout the process" [6, pp. 83].

In this sense, LA is a way for institutions to use their information to gain a better understanding of what they are doing and, in this way, achieve a more personalized teaching strategy for students [7]. Universities, according to Filatro [6], are producing and storing large amounts of data, but there is still a lack of systems that offer fast, predictive, and specific information. This scenario opened space for the development, over the years, of actions related to the use of data in the educational field; thus, the area of LA emerged.

Therefore, the analysis of social and affective data can be converted into useful information to favor the automatic discovery of knowledge related to educational practice. The activity of collecting, analyzing and interpreting this information is usually complex, time-consuming, and exhausting, and it is often left out in the pedagogical routine. Professors signal the lack of functionalities to support their decision making. A careful investigation of the data generated in the interactions carried out in these virtual spaces can help the professor accompany the student in order to engage him/her in learning activities [8].

Thus, the objective of the research is to apply LA in a VLE considering the social and affective aspects. Thus, based on this information, the professor can intervene according to the profile, context and demand of the participants.

The article is divided into five sections. In Sect. 2, the General Process of LA, its model and the LA in VLEs are presented. In Sect. 3, the research methodology is described. In Sect. 4, the results are listed. In Sect. 5, the conclusions are exemplified.

2   General Process of Learning Analytics

LA consists of collecting and analyzing student data, and the results are used to improve some aspect of teaching and learning through human interventions [9]. The General Process of LA proposed by Chatti et al. [10] comprises: (1) Data collection and Preprocessing; (2) Analysis and Action; and (3) Postprocessing, shown in Fig. 1.

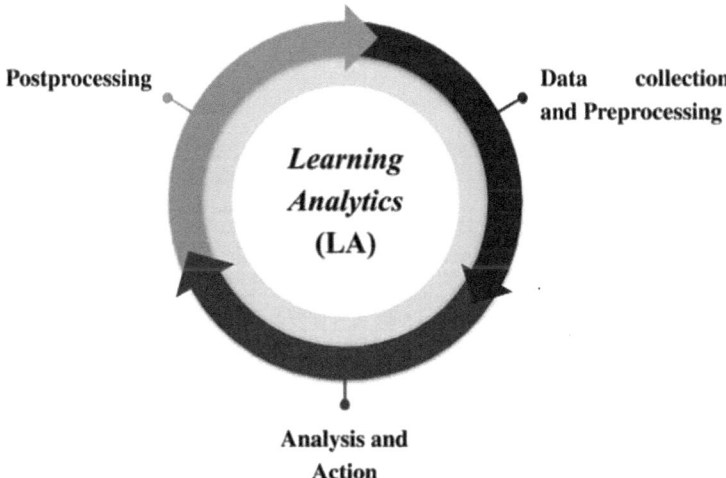

Fig. 1. General process of learning analytics. Source: created by the authors (2022) based on Chatti et al. [10].

Thus, as shown in Fig. 1, Data collection gathers information from educational environments and systems, and Preprocessing aims to eliminate irrelevant attributes. Therefore, at the end of this step, the data is available in a format that can be used as input to an LA method. Subsequently, the objective of the Analysis and Action step is to apply different techniques to the preprocessed data in order to discover useful hidden patterns. The main purpose of any analytical process is decision making, which incorporates adaptation, analysis, evaluation, feedback, intervention, mentoring/tutoring, monitoring, personalization, prediction, recommendation, and reflection. Finally, Postprocessing takes place, in which data are compiled and refined from additional sources, establishment of attributes for iterations, identification of indicators/metrics and modification of analysis variables.

In this context, the author Filatro [6] points out the need for advanced studies on the use of educational data, in order to expand the development of processes and functionalities aimed at improving teaching and learning, both for students and to support professors. Thus, it is understood that it is essential to develop models that address these characteristics and possibilities, as discussed below.

2.1 Learning Analytics Reference Model

The best-known LA model in the literature is the "Learning Analytics Reference Model" [10]. It provides a systematic overview based on four dimensions, which are: what, why, how, and who, discussed in Fig. 2.

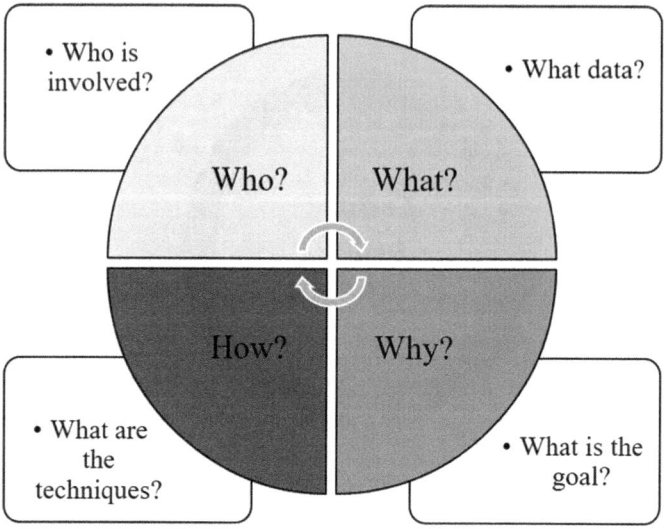

Fig. 2. Learning analytics reference model. Source: created by the authors (2022) based on Chatti et al. [10].

Thus, in Fig. 2, within each dimension it is possible to identify elements necessary to build systems from the LA approach, which are:

What? Refers to the types of data collected. These can come from VLEs, instructional sources, social networks, among others.

Why? Related to the goals and results of the analysis carried out, which may be:

- **Monitoring and Analysis**: dealing with the monitoring of activities carried out by students aggregated to reports and enabling changes in the instructional design;
- **Prediction and Intervention**: outlining a possible model to try to predict the student's performance situation in the future, in order to envision additional personalized and effective aids to improve their performance;
- **Mentoring/Tutoring**: supervising students during the teaching process to achieve goals;
- **Assessment and Feedback**: support self-assessment and obtain intelligent feedback for students and faculty;
- **Adaptation**: informing students about the next steps, organizing resources in an adaptive way;
- **Personalization and Recommendation**: helping students to decide how to achieve their goals, recommending changes in behavior and content;

- **Reflection**: helping actors to self-reflect on their progress, based on comparisons of previous data.

How? Linked to the different techniques that can be used to detect patterns contained in the data, such as information visualization, descriptive statistics, social network analysis, among others.

Who? Aimed at the public involved, which may be students, professors, educational institutions, researchers, system designers, among others.

The Learning Analytics Reference Model is currently considered the most complete because, in addition to pointing out the dimensions, it presents the elements and application situations. Thus, from the analysis of the models found, it was possible to observe that no study considered the Socio-affective aspects. Moissa, Gasparini and Kemczinski [9] point out that it is necessary to research the profiles of students and model computational features that allow supporting qualified teaching activities, with regard to pedagogical interventions during the teaching process and learning.

The works located contemplate the deepening of the individual differences of the students, the interaction and the behavior indicators, but they do not explore the social and affective part together. Therefore, based on the relevance of considering these aspects, so that it is possible to understand the student as a whole and due to the absence of studies found, there is a need to research on the subject. Thus, in this study, the understanding of the Learning Analytics Reference Model proposed by Chatti et al. [10]. The definition of the LA Model is a figurative system that reproduces reality in a more abstract way and that serves as a reference in which it is composed of the analyzes of the different types of profiles of the students in the VLE.

Thus, for a better understanding of the subject, it is important to discuss the possibilities in the VLEs, explained below.

2.2   Learning Analytics in Virtual Learning Environment

In Brazil, studies applying LA in the context of VLEs are still recent and scarce, with few works in the area and representing only 5% in relation to the world. Evidence shows that its use is still in the early stages, as only 20 national institutions were found that are involved with LA [11]. In this context, research by Filvà et al. [12] and Brito et al. [13] contributed to investigate the theme of LA in VLEs in the Brazilian panorama.

Filva et al. [12] analyzed and visualized the student's interaction with the VLE Moodle. The generated data were displayed to support the professor's interpretation from the learning point of view, with the inclusion of textual explanations. Most professors found the view that summarized the interactions of the students and the student with the rest of the group useful. This study is relevant because it was possible to observe that the most important data in the VLE was the interaction, but that there is still no consensus on which aspects are essential for the learning effect.

Brito et al. [13] analyzed the contribution of social, cognitive and behavioral indicators of student learning, created based on VLE Moodle data, to help professors, tutors, and coordinators of online courses in the identification of students at risk of drop out. The survey and selection of indicators were displayed in the form of reports in a plugin, through a data visualization tool, carried out in a literature review on the subject. This analysis allowed monitoring the sections that the student visualizes and the use of available resources, in which it was possible to monitor the performance, interaction, and path of each student within the environment. However, there was no practical mapping of indicators, but a study of those that already exist.

In the international scenario, there are few significant studies on the subject. Beheshitha et al. [14] examined three different ways of visualizing student participation in online discussions in a course setting with LA. This analysis revealed that, for better effects of views on the quantities and qualities of messages posted, the individual differences of students should be considered.

The article by Azevedo et al. [15] explored the estimated time students spent on each topic and activity included in VLE Moodle. As findings of this study, students between 35 and 44 years old spend more time in sessions than younger ones, the months with the highest numbers of access are November and January, and the materials were considered by the public as very important in the learning process.

Therefore, it was found that there are few works that address the topic of LA in VLE. This analysis of the national and international scenario made it possible to find gaps that research points out, such as the importance of: interactions, personalization, student monitoring, social, cognitive and behavioral indicators, and individual differences. Therefore, the research methodology adopted for this work is presented and detailed below.

3   Methodology

The research aims to apply LA in a VLE considering the social and affective aspects. The methodology of this study has a qualitative and quantitative approach based on 13 case studies. In this way, disciplines and courses that used the VLE of the Cooperative Learning Network (in Portuguese: ROODA) were monitored. The choice of ROODA (http://ead.ufrgs.br/rooda) is due to the fact that it is one of the official platforms of the Federal University of Rio Grande do Sul (Brazil), and it is possible to collect analysis data on the performance of users through social and affective aspects.

Thus, the LA model was used, based on Chatti et al. [10], which provides a systematic overview based on four dimensions, which are: what, why, how, and who. In this work, in the "What?" dimension, Socio-affective data from the VLE ROODA were collected. In the "Why?" dimension, the objective and main results are in the following classes: monitoring and analysis; prediction and intervention; and personalization and recommendation. In the "How?" dimension, filtering techniques and graph theory are used. By dimension "Who?", the target audience is the student. Figure 3 summarizes the LA Reference Model used in this research.

After the application of the LA Reference Model, the General Process of LA was used, which is an interactive cycle. For this, it was necessary to carry out three sub-steps:

- **Data collection and Preprocessing**: the collection was performed manually in six graduate, two undergraduate and five extension courses that used ROODA between the years 2019 and 2021, resulting in a total of 285 students. The Preprocessing was carried out by mapping the data for each mood, which are: satisfied, animated,

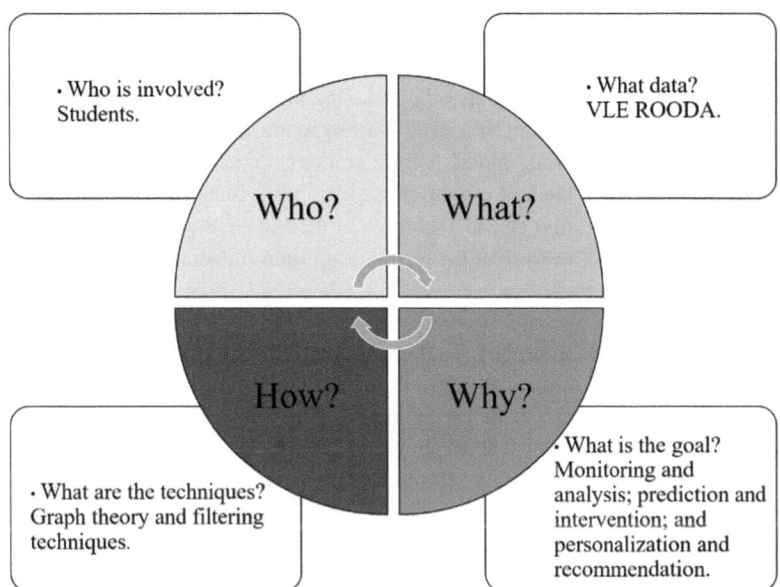

Fig. 3. Learning analytics reference model used in this work. Source: created by the authors (2022) based on Chatti et al. [10].

discouraged, and dissatisfied with social indicators (absence, collaboration, feelings of separation from the class, drop out, informal groups and popularity). As a result of this step, the first social and affective intersections were inferred in an online spreadsheet, generating called the Socio-affective Scenarios.

- **Analysis and Action**: analysis was applied to social indicators and moods in order to discover useful hidden patterns in the data.
- **Postprocessing**: in this sub-step, there was the compilation, refinement of data, identification of indicators and modification of analysis variables.

In this way, the result of the crossing between social and affective indicators was organized, presented in the next section.

4   Results

The present research aimed to apply LA in VLE to analyze the social and affective aspects. Thus, the Reference Model by Chatti et al. [10] in the process. The data indicate that the participants have a profile that varies between adults and the elderly, as their ages correspond to the age group of 18 to 75 years. It is important to point out that in the graduate subjects there were students who had just left high school and entered the University; on the other hand, there were courses that were aimed at the elderly public. In order to carry out the research, the social interactions and affectivity of the students in ROODA were inferred. Therefore, formal evaluations were not carried out expressed in notes in the posts, but only the amount and writing in ROODA.

In Table 1, the Socio-affective Scenarios composed by the crossing between the affective and social indicators are shown. Thus, these Scenarios were analyzed weekly per student, and this process was carried out during all the weeks of the disciplines and courses. In the first column, from the combinations made, are the Scenarios created. The second corresponds to the four moods of the students. The third column is composed of social indicators, but a student can be present in more than one because the number of interactions is counted, that is, the exchange of messages, the frequency, sending and sharing of files. Thus, if the student performs more than one action in ROODA, he/she can have more than one indicator.

Based on Table 1, S1 Scenario is composed of the student who showed to be animated and absence. It is possible to identify that the subject, despite being animated, uses little of the communication features of the environment. One way to increase your interaction is to encourage you to engage in social exchanges with colleagues through synchronous communication using the Chat feature. For this, a topic related to the content could be selected and students could be invited to participate in a Chat, for example. This strategy will allow the sharing of interests, confidence and anxieties of this subject with their peers, in addition to the possibility for the professor to analyze the reasons that led to the absence of the student in the VLE.

It is important to note that, in some cases, there is more than one social indicator linked to the affective one, as is the case, for example, in the S11 Scenario, which corresponds to the affective indicator "animated" plus the social indicators "collaboration", "informal groups" and "popularity". An example of a pedagogical strategy for this Scenario is: the student has shown to be animated, collaborative, popular and participatory in one or more informal groups. The characteristics point to a greater commitment of this subject in relation to the class, and he values the interaction and sharing of materials. This type of student can be an ally during the teaching and learning process in the VLE, as they can assume greater responsibilities. The Chat functionality can be used by this subject as an instantaneous means that allows to speed up the communication process and the creation of contact networks with those colleagues who are distant, absent and discouraged. Thus, you can instigate that student to create their own Chat rooms aimed at both doubts, exchanges of experiences, or even sharing links with the rest of the class. These encounters can be held more than once by the subject, if necessary. Don't forget to encourage and congratulate this student's collaboration, this can be done through the contacts tool or the general questions Forum. It is also important that a self-assessment is carried out in order to verify that the application of this strategy is not demanding too much from the student and causing an overload of responsibility.

In this sense, it is important to analyze from Table 1, as Threats to Validity, which in this study are carried out according to Wohlin et al. [16] composed of four, as follows:

(1) **Internal Validity**: it is focused on validating the current study. In this case, all the students of the disciplines and courses agreed to participate in the research. However, it is worth noting that the Scenarios found can be repeated on other occasions, and the pedagogical practices, the features adopted and the profile of the students were adopted.

Table 1. Mapping of socio-affective scenarios.

Socio-affective scenarios	Affective indicator	Social indicator		
S1	Animated	Absence	–	–
S2	Animated	Collaboration	–	–
S3	Animated	Feelings of separation from the class	–	–
S4	Animated	Informal groups	–	–
S5	Animated	Popularity	–	–
S6	Animated	Absence	Collaboration	–
S7	Animated	Collaboration	Feelings of separation from the class	–
S8	Animated	Collaboration	Informal groups	–
S9	Animated	Collaboration	Popularity	–
S10	Animated	Informal groups	Popularity	–
S11	Animated	Collaboration	Informal groups	Popularity
S12	Discouraged	Absence	–	–
S13	Discouraged	Collaboration	–	–
S14	Discouraged	Feelings of separation from the class	–	–
S15	Discouraged	Popularity	–	–
S16	Discouraged	Absence	Collaboration	–
S17	Discouraged	Collaboration	Feelings of separation from the class	–
S18	Discouraged	Collaboration	Informal groups	–
S19	Discouraged	Collaboration	Popularity	–
S20	Discouraged	Informal groups	Popularity	–
S21	Discouraged	Collaboration	Informal groups	Popularity
S22	Dissatisfied	Absence	–	–
S23	Dissatisfied	Collaboration	–	–
S24	Dissatisfied	Absence	Collaboration	–

(continued)

Table 1. (*continued*)

Socio-affective scenarios	Affective indicator	Social indicator		
S25	Dissatisfied	Collaboration	Feelings of separation from the class	–
S26	Dissatisfied	Collaboration	Popularity	–
S27	Dissatisfied	Informal groups	Popularity	–
S28	Dissatisfied	Collaboration	Informal groups	Popularity
S29	Satisfied	Absence	–	–
S30	Satisfied	Collaboration	–	–
S31	Satisfied	Feelings of separation from the class	–	–
S32	Satisfied	Informal groups	–	–
S33	Satisfied	Popularity	–	–
S34	Satisfied	Absence	Collaboration	–
S35	Satisfied	Collaboration	Feelings of separation from the class	–
S36	Satisfied	Collaboration	Popularity	–
S37	Satisfied	Informal groups	Popularity	–
S38	satisfied	Collaboration	Informal groups	Popularity

(2) **External Validity**: it is associated with the ability to generalize the results. It is very likely that similar Scenarios can fulfill LA based Socio-affective responses results from other may disciplinary or generalized courses.

(3) **Reliability**: concerns the extent to which data and analysis are dependent on researchers. This was done through two functionalities, the Social Map (social indicators) and the Affective Map (moods), with no human intervention. At the time that LA was inserted in the researcher's analysis, however, all the data tabulated in the spreadsheet were selected by two other specialists in the area. Thus, it is believed that they were minimized as reliably reliable.

4) **Construction Validity**: focuses on the generalization of results in relation to the theory or concept of the experiment. In this context, it was explained at the beginning of the disciplines or courses that the participation of the students would not be burdened and the identification of the subject (sensitive information) was omitted.

Therefore, the professor, based on the Socio-affective aspects, can intervene pedagogically, if he wants, in a possible drop out or discouragement of the student.

5   Conclusion

Educational institutions are increasingly producing significant amounts of data. These have information that, when extracted from DE platforms and subsequently analyzed, can indicate possible drop out Scenarios from students to professors. Thus, new functionalities need to be developed to enable the identification of social and affective aspects, in case the professor wants to intervene pedagogically in support of the student.

As contributions to this work, the 38 Socio-affective Scenarios mapped by the Reference Model and General Process of LA are available. These can improve the professors practice in teaching.

As limitations of the present study, some Socio-affective Scenarios were identified, but it is possible that there are others. The variation can happen when different functionalities are used in different situations.

The possibility of future research is related to the application of Socio-affective Scenarios in disciplines and extension courses to verify their validity.

Acknowledgments. This study was financed in part by the Coordenação de Aperfeiçoamento de Pessoal de Nível Superior – Brasil (CAPES) – Finance Code 001.

References

1. Siemens, G., de Baker, R.S.J.: Learning analytics and educational data mining: towards communication and collaboration. In: Proceedings of the 2nd International Conference on Learning Analytics and Knowledge, pp. 252–254. ACM, Canadá (2012). https://doi.org/10.1145/2330601.2330661
2. Siemens, G.: Learning analytics: envisioning a research discipline and a domain of practice. In: Proceedings of the 2nd International Conference on Learning Analytics and Knowledge, pp. 4–8. ACM, Canadá (2012). https://doi.org/10.1145/2330601.2330605
3. Einhardt, L., Tavares, T.A., Cechinel, C.: Moodle analytics dashboard: a learning analytics tool to visualize users interactions in Moodle. In: Proceedings of XI Latin American Conference on Learning Objects and Technology, pp. 1–6. IEEE, Costa Rica (2016). https://doi.org/10.1109/LACLO.2016.7751805
4. Ribeiro, A.C.R., Behar, P.A.: Pedagogical work based on affective aspects in the virtual learning environment ROODA. In: Proceedings of the XIII Brazilian Congress of Distance Higher Education, II International Congress of Distance Higher Education, pp. 1–10. ESUD/CIESUD, Brazil (2016)
5. Johnson, L., et al.: NMC horizon report: 2016 higher education edition, 1st edn. The New Media Consortium, United States of America (2016)
6. Filatro, A.: Data Science in education: in person class, distance and corporate, 1st edn. Saraiva Education, Brazil (2021)
7. Pineda, A.F., Cadavid, J.M.: A systematic literature review in learning analytics. In: Proceedings of the VII Workshops of the Brazilian Congress of Informatics in Education, pp. 429–438. CBIE, Brazil (2018). https://doi.org/10.5753/cbie.wcbie.2018.429
8. Dutt, A., Ismail, M.A., Herawan, T.: A systematic review on educational data mining. Inst. Electr. Electron. Eng. **5**(1), 15991–16005 (2017). https://doi.org/10.1109/ACCESS.2017.2654247

9. Moissa, B., Gasparini, I., Kemczinski, A.: Educational data mining versus learning analytics: are we reinventing the wheel? A systematic mapping. In: Proceedings of the XXVI Brazilian Symposium on Informatics in Education, pp. 1167–1176. SBIE, Brazil (2015). https://doi.org/10.5753/cbie.sbie.2015.1167

10. Chatti, M.A., et al.: A reference model for learning analytics. Int. J. Technol. Enhanc. Learn. **4**(5–6), 318–331 (2013). https://doi.org/10.1504/IJTEL.2012.051815

11. Brasil, P.C., Medeiros, T.J., Nunes, I.D.: A systematic review on the use of learning analytics in Brazilian virtual learning environments. In: Proceedings of the III Congress on Technologies in Education, pp. 371–380 (Ctrl + E), Brazil (2018)

12. Filvà, D.A. et al.: A learning analytics tool with hybrid graphical and textual interpretation generation. In: Proceedings of the Fourth International Conference on Technological Ecosystems for Enhancing Multiculturality, pp. 327–333. ACM, Espanha (2016). https://doi.org/10.1145/3012430.3012536

13. de Souza Brito, M.T., et al.: Contributions of a report-type plugin for the identification of evasion risk in VLE Moodle based on data visualization. Braz. J. Inform. Educ. **28**(1), 1–29 (2020). https://doi.org/10.5753/rbie.2020.28.0.01

14. Beheshitha, S.S., et al.: The role of achievement goal orientations when studying effect of learning analytics visualizations. In: Proceedings of the 1st International Conference on Learning Analytics and Knowledge, pp. 54–63. ACM, Canadá (2016). https://doi.org/10.1145/2883851.2883904

15. Azevedo, J.M. et al.: Learning analytics: a way to monitoring and improving students' learning. In: Proceedings of the International Conference on Computer Supported Education, pp. 641–648. ACM, Portugal (2017). https://doi.org/10.5220/0006390106410648

16. Wohlin, C., et al.: Experimentation in Software Engineering, 1st edn. Springer Science & Business Media, United States of America (2012)

Adaptive Scaffolding Toward Transdisciplinary Collaboration: Reflective Polyvocal Self-study

Mara Alagic[✉], Maria Sclafani, Nathan Filbert, Glyn Rimmington, Zelalem Demissie, Atri Dutta, Aaron Bowen, Ethan Lindsay, Meghann Kuhlmann, Ajita Rattani, and Atul Rai

Wichita State University, Wichita, KS, USA
mara.alagic@wichita.edu

Abstract. Contemporary global challenges require experts from various disciplines to work together. Since every field of knowledge has its unique language and discipline-based culture, collaborative inquiry presents an additional challenge during such collaboration. Ideally, collaborators from each discipline can transcend their respective linguistic and cultural boundaries to achieve transdisciplinarity, where this includes sharing and taking perspectives, active listening; and adaptive, relational metacognitive scaffolding. Within such a framework, the merging of ideas, theories, research design, and methodologies can allow technological applications from each discipline to be achieved through active collaborative, sense-making, and sustained constructivist relations. Within the context of the Disaster Resilience Analytics Center (DRAC) research team, we developed a model of adaptive scaffolding via self-consistent, iterative refinement. This convergence project focused on socio-economic aspects, outreach, and STEAM education, along with postgraduate education. The research team comprised researchers from STEAM disciplines in physical sciences, mathematics, computer sciences, social sciences, humanities, education, and library science. It proved essential to occasionally step away from the research topic and to critically co-reflect on the initial and ongoing challenges in the convergence path. This resulted in more constructive integration and transcendence of disciplines, leading to the development of an adaptive scaffolding framework. We present this framework and additional reflective insights and limitations related to its potential application in different contexts.

Keywords: Adaptive Scaffolding Framework · Convergence Science · Disaster Science

1 Introduction

Figure 1 serves as an outline of this paper and visually captures our discovery of an adaptive scaffolding model that might be applied in collaborative explorations when appropriately contextualized. Briefly, we will first introduce our work within the Disaster Resilience Analytics Center (DRAC), the significance of both disaster science (DS) and its transdisciplinary nature, and how that came about to be at the center of our team's

© The Author(s), under exclusive license to Springer Nature Switzerland AG 2023
D. Guralnick et al. (Eds.): TLIC 2022, LNNS 581, pp. 27–40, 2023.
https://doi.org/10.1007/978-3-031-21569-8_3

attention. We will further describe the challenges we faced due to the team's composition related to the expected collaborative work, the discovery of potential improvements through reflective thinking, and the consequent conclusion to investigate this experience as a reflective polyvocal self-study. The purpose of this polyvocal self-study was to capture lessons learned during the team's engagement in research within the DRAC [54]. Furthermore, we describe a potentially transferable framework of collaborative inquiry that emerged from our work and which contributes further to the reflective nature of this paper.

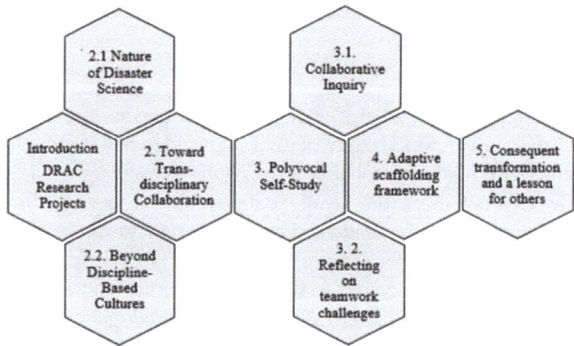

Fig. 1. Outline: Visualizing the process.

The DRAC project and its multidisciplinary team were one of a number of successful proposals under a convergence initiative by the university. It provided an opportunity for a diverse group of faculty members to engage in research for improving the prediction of natural disasters and the resilience of communities to such disasters, and to explore, investigate and provide community outreach.

Fig. 2. Visual proposal for the competitive Convergence Initiative [54].

More precisely, our project had the following goal:

The ultimate goal of the research cluster is to create a digital platform for the Great Plains area that can help the region with all four phases—mitigation, preparedness response, and recovery—by enabling the prediction of the various type of disasters as well as the potential impact of the disasters on the economic, social, ecosystem, human capital, and other aspects. We propose to create a test-bed digital platform for a small geographic region of the Wichita area that allows us to predict/analyze/simulate various what-if scenarios so that the community is more resilient to disaster focusing on Mitigation and Preparedness [54].

2 Toward Transdisciplinary Collaboration

2.1 Nature of Disaster Science

Transdisciplinarity is perhaps above all a new way of thinking about and engaging in inquiry. [31, p. ix].

It appears that the *transdisciplinarity concept* was first mentioned by the Swiss psychologist Jean Piaget at the seminar on interdisciplinarity in universities sponsored by the Organization of Economic Cooperation and Development and the French Ministry of Education in 1970. Piaget [35] mentioned transdisciplinarity as placing, "… relationships within a total system without any firm boundaries between disciplines…" (p. 138). The same year, Jack Lee Mahan [28] wrote in his doctoral dissertation, "Transdisciplinary inquiry would be characterized by a common orientation to transcend disciplinary boundaries and an attempt to bring continuity to inquiry and knowledge." Nicolescu elaborated, "Transdisciplinarity concerns that which is at once between the disciplines, across the different disciplines, and beyond all disciplines. Its goal is the understanding of the present world, of which one of the imperatives is the unity of knowledge," [32, p. 19]. That seems to be a foundational statement although there are many other contextual elaborations about how the transdisciplinary way of thinking challenges discipline-based thinking and seeks to create new knowledge through inquiry into problems that cannot be solved within a single discipline [5]. The history of the development of the modern concept of transdisciplinary research has been well-documented and represented. For example, Mittelstrass [29] explains this as an issue of the "Order of Knowledge," which he discusses as being "From Disciplinarity to Transdisciplinarity and Back" (p. 68), while most historical reviews present a reversal of this depiction. From the *socially responsible science perspective*, current complex social and environmental problems, from climate change disasters to a sustainable future, lend themselves to transdisciplinary inquiry. These complex, interconnected, so-called "wicked problems" [7] need creative solutions, between and beyond disciplines, traversing knowledge bases and value systems. The problems inherent to the prediction of disasters, disaster risk reduction, and resilience are those kinds of problems—problems requiring a transdisciplinary way of thinking, understanding, and knowledge creation.

DS as a field of study has become established through the various initiatives focused on the research of disasters and Disaster Risk Reduction (DRR) and by the creation of the journals *Journal of Disaster Science* in 2016, *Progress in Disaster Science* in 2019. The actual term has been used since the early nineties [3] and comes with a recognition that an essential change in scientific approaches to disaster risk reduction is needed by shifting the emphasis from individual hazard and risk assessment to a system analysis with transdisciplinary action-oriented research co-facilitated by researchers and stakeholders [24]. Many authors argue for the value of the transdisciplinary approach to research involving scientists from different disciplines to comprehensively study problems relevant to DRR. Although they do not offer a unique roadmap, there is a general agreement about the need to move beyond discipline-specific approaches to problem-related contextualization of disciplinary ontologies, and phenomenological, theoretical, and methodological innovations (e.g., [34]). Van Niekerk [50, p.13] suggested, "Transdisciplinarity provides a binding paradigm for disaster risk reduction." In our context, the three dominant features of this binding paradigm [24], extend to disaster education and are conceptualized as Collaborative Inquiry, Applied Problems, and Value to Multiple Disciplines:

- *Collaborative Inquiry:* Agreeing on the research problem and how team members from different disciplines are going to communicate and collaborate.
- *Applied Problem:* Creating a consensus that the goal of the study is the amelioration of the effects of well-specified problems through solution-focused, transferable knowledge.
- *Value to multiple disciplines:* Creating the knowledge in ways that are of value from a practical as well as a scientific standpoint to individual disciplines.

DRR education in both formal and informal educational settings [41] is a crucial component of all efforts to predict and recover from disasters, especially considering the increased frequency and severity of natural and manmade disasters. In this context of education, interdisciplinary STEAM learning offers the potential for disrupting traditional disciplinary boundaries and collaboratively engaging with topics and ideas in a critical, and interconnected manner [15]. STEAM education should be designed to not only transcend disciplines but also impact individuals and communities outside of the classroom [8]. This is particularly important with disaster education, as recent large-scale disasters have revealed the inadequacy of disciplinary approaches to disaster education.

DS Challenges: Global and Local Disasters and Hazards. The increased frequency and severity of natural disasters observed in recent times have been linked to anthropogenically driven greenhouse gas accumulation, which consequently has raised the global average sea-level air temperature by more than 1 °C compared to the preindustrial era [1]. The global average air temperature reflects the overall energy content of the atmosphere. However, this energy is not homogeneously distributed due to the dynamic nature of atmospheric circulation and its interaction with oceanic circulation and the topography and albedo of landmasses. Thus, it is not unusual to observe simultaneous events at opposite extremes within the same continent or region [6, 11, 25], for example with record floods in Eastern Australia at the same time as heatwaves and wildfires in Western Australia, during January 2022. This increased frequency and severity of weather-related disasters is a matter of some concern to insurance companies, governments, and humanitarian organizations [4, 12, 16].

The consequences of natural disasters go beyond economic costs and insurance claims to include environmental damage and human lives and well-being. As stated above, the increased severity and frequency of these natural disasters can be attributed, in part, to climate change and in part to a lack of disaster resilience for known, recurrent disasters. Climate change could be the insurance industry's primary concern in the coming decades [42]. For example, in Britain, "…more than 570,000 new homes had been built since 2016 that would not be resilient to future high temperatures and more than 70,000 had been built since 2009 on flood plains" [45, p1]. Furthermore, flood mapping data showed that 19% of British properties were susceptible to surface water flooding.

High fuel loads in grasslands coupled with abnormally windy, hot and dry conditions in Western and Central Kansas [39] attributable, in part to climate change [20, 23] helped the December 2021 wildfires to spread quickly, burning more than 350,000 acres in parts of Western and Central Kansas, leaving two people dead and three injured, and more than 42 structures destroyed [37] at a cost of more than $2.3 m [33]. Additional consequences of these wildfires included disruption to grazing and cropping enterprises, and power outages, due to downed power lines. In the case of US Western wildfires [26], it is likely, insurance companies will not survive unless they manage the resilience of communities located in potential wildfire locations. Urgent action is needed from policymakers, developers, and insurers to protect homes and businesses from flooding and wildfires. Added to this challenge is the problem that the communities most vulnerable to flooding or wildfires are the least able to change or adapt. Social stratification affords the higher SES communities the least vulnerable areas while penalizing low SES communities who are most often living in vulnerable areas. Low-lying areas in Houston, during Hurricane Harvey in 2017 and New Orleans during Hurricane Katrina in 2005, illustrate this point.

DS Opportunities: Remote-sensing, deep learning, and GIS. The science and application of remote-sensing technologies have advanced rapidly during the past five decades. An early example was its application to crop hail damage [48]. While the insurance industry has been slow to adopt remote sensing, it has been applied to fire [44], hail [58] and drought [40]. High-resolution (0.25–2 m) multi-spectral or Synthetic Aperture Radar (SAR) imagery of before and after disasters can be used to detect changes [47] such as damaged houses, blockages to transportation routes, spills of toxic materials, or disruption to utilities; usually within a couple of days. Such information is helpful for disaster recovery efforts. In the longer term, it can also identify where to invest resources to improve community resilience for future disasters. Remotely sensed differences, before and after a disaster, have the potential as a tool for insurance companies [10] to either re-calculate premiums for higher-risk levels or to provide incentives to improve resilience. Since local governments have access to georeferenced data for buildings and infrastructure, costs, and values, they can rapidly estimate losses due to a disaster and in the process compare the costs of disasters versus the cost of improved resilience [22, 38, 43]. The tedium of this type of work is being overcome with the application of deep learning and object recognition to satellite imagery [57].

...development of efficient risk mitigation and adaptation strategies is impossible without joint interdisciplinary efforts among actuaries, statisticians, atmospheric scientists, civil engineers, and policymakers; such truly interdisciplinary initiatives are still relatively rare. [27, p. 13]

The ultimate solution to this problem calls for participation by team members from different disciplinary backgrounds-science, engineering, economics, sociology, education. For this to succeed, adaptive scaffolding will be needed to allow the development of a common language and shared perspectives among participants.

2.2   Beyond Discipline-based Cultures

"Disciplines are the result of artificial fragmentation of knowledge. Real-world problems are rarely confined to the artificial boundaries of academic disciplines" [9, p. 357). The recognition that one perspective is often not enough to address a challenge is not new, nor is the desire to seek out perspectives other than that held by a person or group of people within a field to address that challenge. One famous literary example of this practice may be traced back to ancient Athens. Plato's recorded version of Socrates' *Apology* [36] to the Athenian court contains a story in which Socrates recounts his being told that no one was wiser than he. Questioning the truth of that assertion, he proceeds to consult with others around Athens attempting to find a person or group of people wiser than him. There are reasons to view this story as peripheral to any contemporary discussion of disciplinary culture, or any attempt to foster transdisciplinary dialogue (for example, Socrates concludes that, as of his *Apology*, he had yet to find anyone wiser than him, and had begun questioning the value of wisdom anyway). That said, the story represents an ancient articulation of an idea still present in contemporary scientific inquiry—outside perspectives are essential to address a challenge perceptible to a researcher or group of researchers within a field. It is from this broad perspective that any endeavor to build a transdisciplinary model of thinking and apply it to a specified challenge begins [52].

As an examination of global and local disaster-related challenges highlighted, knowledge, practices, and technologies developed by multiple fields may be brought to bear on both preparing a community to respond to a disaster as well as responding in the wake of a disaster. Certainly, all voices in that conversation possess a shared desire to address the challenge posed to a community by a disaster. While this shared desire makes an excellent starting point, it cannot by itself facilitate the full conversation. Many models exist to consider aspects of communication, such as intercultural aspects [53] or sustainable aspects. That said, articulating a broad model of transdisciplinary communication remains a challenge and this challenge, in turn, hampers actionable efforts to construct transdisciplinary teams to analyze issues of disaster resilience. Any such model must account for the fact that every field of knowledge has its unique ontology, vocabulary, and disciplinary culture(s). It must provide a framework to build upon those field-specific elements as a means of drawing all relevant language, concepts, and technologies into the model to help a community prepare for a disaster or respond to one in a meaningful way. *Thus, an essential criterion for judging the success of transdisciplinary research is the extent to which they promote the development of novel conceptual models and empirical investigations that integrate and extend the concepts, theories, and methods of particular fields* [46, p. 67].

3 Polyvocal Self-Study

The polyvocal aspect of this participatory self-study is dialogic [17], and it has been co-constructed through intentional reflections and co-reflections while evolving and unpacking an understanding of multiple layers comprising the DRAC team's collaborative inquiry as well as necessary relational understandings for keeping up with that inquiry.

3.1 Collaborative Inquiry

There is a notable uniqueness along the continua of collaborative inquiry to which convergence initiatives strive. Attempting to converge on a real-life problem-inspired model that transcends multiple disciplinary boundaries and is engaged with all societal stakeholders reveals many collaborative tensions and obstacles [13]. In this context, principles of collaborative inquiry are concerned with the development of mutually beneficial relationships, which is achieved not only as a component of research but also as a way of presenting an alternative formulation of what counts as knowledge [49]. The convergence of these models and methods toward complex research endeavors and the collaboration of diverse and often disparate project partnerships points to the opportunities for knowledge innovation. Gibbons [14], have conceptualized this goal as Mode 2 in the production of knowledge:

> *Mode 2 knowledge production is transdisciplinary. It is characterized by a constant flow back and forth between the fundamental and the applied, between the theoretical and the practical. Typically, discovery occurs in contexts where knowledge is developed for and put to use, while results—which would have been traditionally characterized as applied—fuel further theoretical advances... Knowledge is always produced under an aspect of continuous negotiation, and it will not be produced unless and until the interests of the various actors are included.*

This kind of collaborative inquiry goes beyond *collaboration* in integrating and applying contributing knowledge and practices toward *convergence* via co-creating novel concepts, methodologies, theories, and models for genuinely innovative approaches relative to 'wicked' or increasingly complex global socio-environmental and community-based problem-solving strategies.

3.2 Reflections: What Did We Learn Through This Inquiry Process?

One of the biggest challenges was having *in-depth conversations with scholars in other disciplines* since almost everyone was looking at DS from their discipline-based perspective. It might be worth mentioning that the COVID-19 pandemic resulted in having only virtual meetings and so, side conversations that might have been helpful were almost nonexistent, and not easy to facilitate. After all, each member was an expert in a particular discipline but not nearly as well versed in others. At times team discussions took significant time and energy for a scholar in Sociology, for example, to communicate with scholars in Economics and the Geosciences. One member noted that the team needed to foster more dialogue, between those with expertise in machine learning and those with expertise in meteorology early in the project when some students were studying how machine learning can be used to predict hurricane trajectories, for instance. From such a dialogue, the machine learning specialists could learn more about meteorological processes underlying hurricane formation and trajectories, while the meteorologist could learn more about how machine learning works. Purposeful dialog, supported by appropriate scaffolding, often needed to follow in which the gaps in knowledge and understanding were filled. Listening carefully and formulating the questions to fill gaps in knowledge, required a sustained and substantial investment of time. In fact, on some occasions, technical vocabulary needed to be explained before team conversations could proceed. In addition, team members did not always share the same assumptions about the research process, such as what questions to ask about natural disasters, and how to proceed in researching natural disasters. These cross-disciplinary challenges often appeared during the first year of the DRAC project.

Another challenge was the *bifurcation of the research team into two smaller teams titled* Prediction and Resilience. This division enabled highly technical research to proceed within the Prediction team and its weekly group meetings. Disappointingly, this was without the benefit of insights from other disciplines, such as sociology, economics, or education. During this same period, the Resilience team held its weekly group meetings and made advances in studying disaster risk reduction and community-focused research as well as theoretical discussions about the ideal of achieving true convergence per the initial proposal leading to conversations about scaffolding, understanding each other's perspective and as mentioned in a reflection, *learning to collaborate in a transdisciplinary way of thinking.*

Members of the entire DRAC team, from both Prediction and Resilience, met monthly and shared progress. However, the initial division within the overall team, especially when combined with the corresponding separate meeting times, created some unforeseen problems. For example, it was difficult to participate in whole-team conversations or to enjoy the fruits of truly convergent scientific research. On a more basic level, it was hard to stay abreast of the work of the other team. It became clear that the entire DRAC team needed to meet and integrate all activities. The DRAC team has settled into this pattern and is showing signs of helping members to evolve toward disciplinary transcendence. What we were learning along the way was that for this evolution a better-structured awareness of adaptive scaffolding is required.

4   Towards a Collaborative Framework: Adaptive Scaffolding

The analogy of scaffolding construction for a building to cognitive scaffolding in the construction of knowledge [56] combined with the Zone of Proximal Development [51] is usually considered a teacher education concept. The Zone of Proximal Development (ZPD) is the "…distance between the actual developmental level…and the level of potential development…" (p. 86). It may be limited to small distances without scaffolding, or larger when higher developmental levels are desired and addition of appropriate scaffolding is required. Scaffolding may be defined as elements that help bridge the gap between one's initial developmental level and a higher, desired developmental level, by providing the appropriate amount of support. Thus, it implies the important role of the professional educator or teacher in providing the scaffolding. Careful consideration of these concepts leads to the recognition that the goal of a teacher (expert in education and learning theories) is to equip the learners with the ability to eventually scaffold for themselves as well as for other learners. This leads to an expanded concept of scaffolding—*reciprocal/relational scaffolding* [18] where team members learn from each other's knowledge with scaffolding fluctuating constantly as the DRAC team collaborates toward shared goals and collaborative inquiry into DS. The scaffolding metaphor has been extended by considering the scaffolding *domain* (conceptual and heuristic), *agency* (expert, reciprocal, and self-scaffolding), and self-scaffolding as *metacognition*. Furthermore, based on a combination of domain and agency there are six potential zones (Fig. 2) of scaffolding activity distinguished by the relative positioning of the team members in the associated scaffolding: (a) Conceptual/Heuristic by an expert; (b) in a reciprocal situation; or (c) by a self-scaffolder [18]. These zones are inherently related to the four functions of scaffolding: (a) to navigate inquiry; (b) to structure tasks; (c) to support communication; and (d) to foster reflection [19]. Each needs to be implicitly integrated into DRAC teamwork.

Creating an environment that fosters successful team culture requires, in addition to a shared goal, the attention of all team members to dialogic interactions. That kind of *relational scaffolding* with attention to metacognitive self-scaffolding is necessary for successful teamwork. Some consideration has been given to similar concepts in the literature. For example, conversational scaffolding—"how organizational members engaged with various media alone and in combination to accomplish both individual and concurrent conversations" [55, p.1]. Building from conversational scaffolding, three dimensions—content, relationships, and technology—of relational scaffolding have been considered in computer-mediated communication [30]. Furthermore, in this context, *adaptive scaffolding* refers to regulated learning and understanding by activated prior knowledge, monitored emerging understanding, and engaged seeking of a reciprocal understanding. In terms of scaffolding categories, attention has been given to *conceptual* (conceptual understanding and knowledge development) and *procedural scaffolding* related to processes and supporting structures. *Strategic scaffolding* that promotes analysis, planning, and decision-making during the study, was carried through collaboration, and, via the DRAC team's reflections, it was learned along the way. *Metacognitive scaffolding* (Fig. 4) concerns how to deal with thoughts during the learning in the self-thinking process, as mentioned in DRAC team reflections but not discussed before the design of Fig. 3, which captures our thoughts about the adaptive scaffolding framework that should be considered before any convergence initiative.

Dialogic sharing and construction of meaning is one type of relational scaffolding that may help in a project like ours. Consider the word, "field." For library scientists, it may have a particular meaning, but it may have many meanings for people in other disciplines. Physicists discuss gravitational and electromagnetic fields. In abstract algebra,

ADAPTIVE SCAFFOLDING		AGENCY		
		expert	peer/ reciprocal	self
DOMAIN	conceptual	zone1	zone3	zone5
	heuristic	zone2	zone4	zone6
	relational	**PSPT**	reciprocal/ relational	**METACOGNITION**

Fig. 3. Adaptive scaffolding framework.

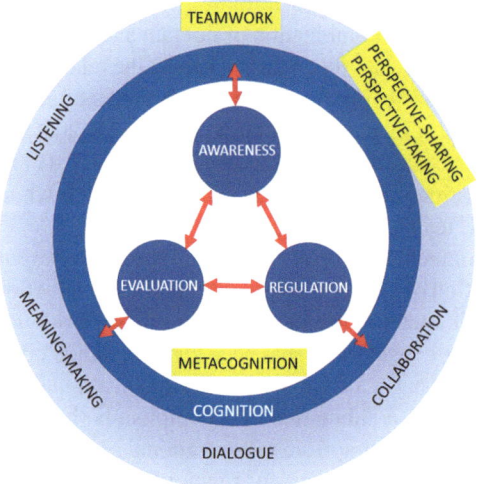

Fig. 4. Role of metacognition within the adaptive scaffolding framework.

the field has another. In the graphic design of graphical user interfaces and programming, it has yet another. If it is easy in a conversation to start forming a concept in your mind based on your discipline's meaning for a word, but then become confused when the person in the other discipline elaborates and starts making use of their discipline's meaning, which makes no sense to you. So in our dialogs, we tried to be more deliberate about questioning each other, to make sure we are constructing the same meaning. It is quite a process of understanding each other's perspective, *perspective sharing and perspective-taking*, PSPT [2]. The Japanese Enryo-Sasshi concept [21], teases out stages of the process of communicating in which we are trying to either articulate our intended meaning in such a way that the other will arrive at the same meaning, and the interpretation of the messages from the other, to try to construct what we think they meant. Enryo-Sasshi occurs in cycles, with pauses or thinking time between utterances.

5   Teamwork Transformation & Lessons for Others

In this paper, we first introduced the DRAC research project in which faculty members and graduate students from different disciplines collaborate on prediction of natural disasters and improvement of community resilience to such disasters. We considered the prevalence of natural disasters, particularly under the effects of global climate change and the need to improve both prediction and community resilience. We discovered early in the project that the success of this multidisciplinary team and its collaboration would hinge on thinking beyond the topic and to giving some consideration to the processes of transdisciplinary collaboration. As a team, we reflected on the first two years of the project, through a polyvocal, reflective self-study. We considered characteristics and relative success of the collaborative inquiry and the challenges experienced as we worked in a team. These reflections gave rise to an adaptive scaffolding framework, which incorporated a *Relational* domain with PSPT, reciprocity in mutual, metacognitive scaffolding, continuous dialog for meaning making, and active listening in which we aimed to practice Enryo-Sasshi (Figs. 3, 4). This framework carries with it additional tasks and investment of time that may be seen to take away from time that should be spent on the research topic of disaster resilience. On the other hand, we could argue that progress with the research topic was enhanced by our application of the framework. These findings are preliminary and more time will be needed to fully assess the degree of success due to the adoption of a framework for adaptive scaffolding. One could argue that the most successful multidisciplinary project team would be one that comprises members, who are already well-versed in the application of adaptive scaffolding. However, in reality, most teams will need to be both working on the research topic and putting into practice the adaptive scaffolding framework. There is also always scope for improvement in the framework itself in a variety of contexts. Each multidisciplinary context, with its unique characteristics, will require appropriate application of the adaptive scaffolding framework. Ideally, having an understanding of this framework, through formal/informal education and/or professional development, promises a fast track to successful collaborative teamwork.

References

1. Abram, N.J., et al.: Connections of climate change and variability to large and extreme forest fires in southeast Australia. Commun. Earth & Environ. **2**(1), 1–17 (2021). https://doi.org/10.1038/s43247-020-00065-8
2. Alagic, M., Orel, T.: Third place learning environments: Perspective sharing and perspective-taking. J. Adv. Corp. Learn. (iJAC) **2**(4), 4–8 (2009)
3. Aguirre, B.E., El-Tawil, S.: The emergence of transdisciplinary research and disaster science. Am. Behav. Sci. **64**(8), 1162–1178 (2020)
4. Albek-Ripka, L.: One year on from horrific fires, Australians struggle to rebuild. National Geographic. Retrieved from: https://www.nationalgeographic.com/environment/article/one-year-on-horrific-fires-australians-struggle-to-rebuild on Mar 5, 2022 (Dec 30, 2020)
5. Bernstein, J.H.: Transdisciplinarity: a review of its origins, development, and current issues. J. Res. Pract. **11**(1), 1–20 (2015)
6. Betts, R.A.: Heed blame for extreme weather. Nature **589**(7843), 493–494 (2021)

7. Brown, V. A., Deane, P. M., Harris, J. A., & Russell, J. Y.: Towards a just and sustainable future. In V. A. Brown, J. A. Harris, & J. Y. Russell (Eds.), Tackling wicked problems: Through the transdisciplinary imagination (pp. 3–15). Abingdon, Oxon, UK: Earthscan (2010)

8. Burnard, P., Colucci-Gray, L., Sinha, P.: Transdisciplinarity: Letting arts and science teach together. Curric. Perspect. **41**(1), 113–118 (2021)

9. Choi, B., & Pak, A.: Multidisciplinarity, interdisciplinarity and transdisciplinarity in health research, services, education and policy: 1. Definitions, objectives, and evidence of effectiveness. Clin. Investig. Med., **29**(6), 351–364 (2006)

10. De Leeuw, Jan, Anton Vrieling, Apurba Shee, Clement Atzberger, Kiros M. Hadgu, Chandrashekhar M. Biradar, Humphrey Keah, Calum Turvey.: The potential and uptake of remote sensing in insurance: A review. Remote. Sens. **6**(11) (2014): 10888–10912

11. Dolan, D.A.: Multiple partial couplings in the multiple streams framework: The case of extreme weather and climate change adaptation. Policy Stud. J. **49**(1), 164–189 (2021)

12. Doyle, R.: Natural disaster H1 insured losses hit 10-year high-Aon. Reuters. Retrieved from https://www.reuters.com/business/finance/natural-disaster-h1-insured-losses-hit-10-year-high-aon-2021-07-21/ on Mar 5, 2022) (Jul 21, 2021)

13. Fam, D., Neuhauser, L., & Gibbs, P. (Eds.).: Transdisciplinary theory, practice and education: the art of collaborative research and collective learning (1st ed. 2018 edition). Springer (2018)

14. Gibbons, M.: Mode 2 society and the emergence of context-sensitive science. Sci. Public Policy **27**(3), 159–163 (2000)

15. Guyotte, K.W.: Toward a philosophy of STEAM in the anthropocene. Educ. Philos. Theory **52**(7), 769–779 (2020)

16. Henriques-Gomes, L.: Planning experts call for state governments to buy back land from people in most bushfire-prone areas. Guardian. Retrieved from: https://www.theguardian.com/australia-news/2020/jan/19/bushfire-destroyed-homes-should-not-be-rebuilt-in-riskiest-areas-experts-say on Mar 5, 2022 (Jan 18, 2020)

17. Holquist M.: Glossary for Bakhtin M. The dialogic imagination: Four essays (Emerson, C., Holquist, M., Transl Holquist, M, ed). Austin: University of Texas Press (1981).

18. Holton, D., Clarke, D.: Scaffolding and metacognition. Int. J. Math. Educ. Sci. Technol. **37**(2), 127–143 (2006)

19. Hsu, Y.S., Lai, T.L., Hsu, W.H.: A design model of distributed scaffolding for inquiry-based learning. Res. Sci. Educ. **45**(2), 241–273 (2015)

20. Huber, D. G., Gulledge, J.: Extreme weather and climate change: Understanding the link, managing the risk. Arlington: Pew Center on Global Climate Change (2011)

21. Ishii, S.: Enryo-sasshi communication: A key to understanding Japanese interpersonal relations. Cross Currents **11**(1), 49–58 (1984)

22. Irfan, U.: The $5 trillion insurance industry faces a reckoning. Blame climate change. Insurers are getting rocked by climate disasters. They're also shaping how we prepare for the next one. Vox (Oct 15, 2021)

23. Irfan, U. and Jones, B.: Why Hurricane Ida has been so devastating to Louisiana and the Gulf Coast, rising seas, a warming world, and Covid-19 are shaping the impact of "one of the strongest storms to ever hit Louisiana. Vox. Retrieved from: https://www.vox.com/22648189/hurricane-ida-new-orleans-louisiana-flood-climate-change-covid on Mar 5, 2022 (Aug 30, 2021)

24. Ismail-Zadeh, A.T., Cutter, S.L., Takeuchi, K., Paton, D.: Forging a paradigm shift in disaster science. Nat. Hazards **86**(2), 969–988 (2016). https://doi.org/10.1007/s11069-016-2726-x

25. Kozoil, M., O'Malley, N., Cormack, L. & Grieve, C.: 'Face reality': Don't build in flood prone areas, resilience boss says. Sydney Morning Herald. Retrieved from: https://www.smh.com.au/national/nsw/face-reality-don-t-build-in-flood-prone-areas-resilience-boss-says-20220303-p5a1iw.html on Mar 5, 2022 (Mar 4, 2022).

26. Lavietes, M.: Western U.S. wildfires cost insurers up to $13 billion in 2020. Reuters. Retrieved from: https://www.reuters.com/article/us-usa-wildfires-insured-losses-trfn/western-u-s-wildfires-cost-insurers-up-to-13-billion-in-2020-idUSKBN28P2NQ on Mar 5, 2022 (Dec 10, 2020)
27. Lyubchich, V., Newlands, N.K., Ghahari, A., Mahdi, T., Gel, Y.R.: Insurance risk assessment in the face of climate change: Integrating data science and statistics. Wiley Interdiscip. Rev.: Comput. Stat. **11**(4), e1462 (2019)
28. Mahan, J. L., Jr.: Toward transdisciplinary inquiry in the humane sciences. Doctoral dissertation, United States International University. UMI No. 702145. Retrieved from ProQuest Dissertations & Theses Global (1970)
29. Mittelstrass, J.: The Order of Knowledge: From Disciplinarity to Transdisciplinarity and Back. European Review **26**(S2), S68–S75 (2018). https://doi.org/10.1017/S1062798718000273
30. Meissner, J. O., Tuckermann, H. A.: Relational Scaffolding Model of Hybrid. In Communities and Technologies 2007: Proceedings of the Third Communities and Technologies Conference, Michigan State University 2007 (p. 479). Springer Science & Business Media (2010, May)
31. Montuori, A.: Gregory Bateson and the promise of transdisciplinarity. Cybern. & Hum. Knowing **12**(1–2), 147–158 (2005)
32. Nicolescu, B.: Methodology of transdisciplinarity. World Futures **70**(3–4), 186–199 (2014)
33. NIFC National Large Incident Year to Date Report 12/30/2021 (PDF) National Interagency Fire Center (pp. 27–90) retrieved from https://gacc.nifc.gov/sacc/predictive/intelligence/NationalLargeIncidentYTDReport.pdf on 4/10/2022 (2021).
34. Peek, L., Champeau, H., Austin, J., Matthews, M., Wu, H.: What methods do social scientists use to study disasters? An analysis of the Social Science Extreme (2020)
35. Piaget, J.: The epistemology of interdisciplinary relationships. In Centre for Educational (1972)
36. Plato. Apology. In The Trial and Death of Socrates, tr. G.M.A. Grube.: Hackett Publishing Company, Inc. (1975)
37. Perkins, O.: Estimated 350,000 acres continue to burn in central Kansas. The Hutchinson News. Retrieved from https://www.hutchnews.com/story/news/2021/12/16/kansas-weather-wildfires-350000-acres-burning-forest-service-wind-gusts/8925332002/ on 4/10/2022 (2021)
38. Prevatt, D. O., Coulbourne, W., Graettinger, A. J., Pei, S., Gupta, R., Grau, D.: Joplin, Missouri, tornado of May 22, 2011: Structural damage survey and case for tornado-resilient building codes. Am. Soc. Civ. Eng. (2012)
39. Pugh, B. & Rippey, B.: Kansas drought map, released 12/30/2021. U.S. Drought monitor. NOAA/CPC. retrieved from https://web.archive.org/web/20211231025959, https://droughtmonitor.unl.edu/CurrentMap/StateDroughtMonitor.aspx?KS on 4 October 2022 (2021)
40. Rojas, O., Vrieling, A., Rembold, F.: Assessing drought probability for agricultural areas in Africa with coarse resolution remote sensing imagery. Remote Sens. Environ. **115**, 343–352 (2011)
41. Sandlin, J., Schultz, B., Burdick, J. (eds.): Understanding, mapping, and exploring the terrain of public pedagogy. Handb. Public Pedagog., Routledge (2009)
42. Scism, S. S.: Underlying Conditions: The COVID-19 Pandemic and Xenophobia Trends in Costa Rica (Doctoral dissertation, Tulane University, Biomedical Sciences) (2021)
43. Simmons, K.M., Kovacs, P., Kopp, G.A.: Tornado damage mitigation: Benefit–cost analysis of enhanced building codes in Oklahoma. Weather, climate, and society **7**(2), 169–178 (2015)
44. Smith, J. R.: Industrial and commercial geography. H. Holt (1913)
45. Smith, L.: British homes, businesses unprepared for climate change, Aviva says. Reuters. Retrieved from: https://www.reuters.com/world/uk/british-homes-businesses-unprepared-climate-change-aviva-says-2021-07-20/ on Mar 5, 2022 (Jan 22, 2021)

46. Stokols, D., Harvey, R., Gress, J., Fuqua, J., Phillips, K.: In vivo studies of transdisciplinary scientific collaboration lessons learned and implications for active living research. Am. J. Prev. Med. **28**(2 Suppl 2), 202–213 (2005)

47. Towery, N.G., Eyton, J.R., Changnon, S.A., Dailey, C.L.: Remote sensing of crop hail damage, illinois state water survey, Champaign, IL, USA (1975)

48. Towery, N.G., Morgan, G.M., Changnon, S.A.: Examples of the wind factor in crop-hail damage. J. Appl. Meteorology.**15**(10): 1116–1120 (1976)

49. Townsend, A.: Collaborative action research. In the SAGE encyclopedia of action research (Vol. 1–2, pp. 117–119). SAGE Publications Ltd. (2014) https://doi.org/10.4135/978144629 4406

50. Van Niekerk, D.: Transdisciplinary: The binding paradigm for disaster risk reduction (Inaugural address of the Scientific Contributions Series 254). (2012) https://dspace.nwu.ac.za/bit stream/handle/10394/8572/Van_Niekerk_D.pdf?sequence=1

51. Vygotsky, L. S., Cole, M. Mind in society: Development of higher psychological processes. Harvard university press (1978)

52. Walter, A.I., Helgenberger, S., Wiek, A., Scholz, R.W.: Measuring societal effects of transdisciplinary research projects: design and application of an evaluation method. Eval. Program Plann. **30**(4), 325–338 (2007)

53. Wang, J., Aenis, T., Siew, T.F.: Communication processes in intercultural transdisciplinary research: framework from a group perspective. Sustain. Sci. **14**(6), 1673–1684 (2019). https://doi.org/10.1007/s11625-019-00661-4

54. WSU DRAC Wichita State University, Disaster Resilience Center Proposal. (2020). https://www.wichita.edu/research/drac/

55. Woerner, S. L., Orlikowski, W. J., & Yates, J. (2005, July). Scaffolding conversations: engaging multiple media in organizational communication. In *21st EGOS Colloquium, Berlin*

56. Wood, D., Bruner, J.S., Ross, G.: The role of tutoring in problem solving. J. Child Psychol. Psychiatry **17**(2), 89–100 (1976)

57. Yang, L., Cervone, G.: Analysis of remote sensing imagery for disaster assessment using deep learning: a case study of flooding event. Soft. Comput. **23**(24), 13393–13408 (2019). https://doi.org/10.1007/s00500-019-03878-8

58. Young, F.; Chandler, O.; Apan, A.: Crop hail damage: insurance loss assessment using remote sensing. In Proceedings of the Remote Sensing and Photogrammetry Society Conference, Aberdeen, U.K., 7–10 September 2004 (2004)

Emergency Remote Teaching: A Case Study

Macedonio Alanis[✉]

Tecnologico de Monterrey, Monterrey, Mexico
alanix@tec.mx

Abstract. During the first quarter of 2020, nearly every country declared a COVID-19 pandemic emergency, and schools suddenly had to migrate to remote teaching modalities. As a result, 1.29 billion learners in 151 countries changed their learning models. Every school reacted differently, with varying degrees of success. One success story is the Tecnologico de Monterrey university system in Mexico. The school's previous extensive experience with e-learning and a recent emergency remote teaching event (ERT) enabled Tecnologico de Monterrey to preempt the emergency declaration by going online two weeks before any other school in Mexico. More than 90,000 students attend 55,000 class sessions per week on its 33 campuses throughout Mexico. Students' acceptance rate of the models implemented was high. The case study results of the university's success highlight four factors that may simplify a school's transition to ERT: the school's previous experience with emergencies, experience with distance learning and course design, availability of technology, and commitment from the students, teachers, school administrators, and parents. The analysis also identifies five distinct stages in an ERT event: before the ERT event, the ERT declaration, the first week of ERT, the following weeks of ERT, and the return to normal operations. Despite the tragic events of the COVID-19 pandemic, we must still try to learn something from it and work to build a better world. The lessons learned in the last two years might help bring a better response in a (hopefully distant) future emergency.

Keywords: Emergency remote teaching · Distance learning · COVID-19 · Distance education · Online learning · Educational innovation · Professional education · Higher education

1 Introduction

During the first quarter of 2020, nearly every country declared a COVID-19 pandemic emergency. Schools around the world closed their doors and massively migrated to distance education. At its peak, according to UNESCO [1]:

- 81.8% of the world's student population was affected by school closures.
- 1.29 billion learners were out of school.
- 151 countries were affected by school closures.

D. Guralnick et al. (Eds.): TLIC 2022, LNNS 581, pp. 41–52, 2023.
https://doi.org/10.1007/978-3-031-21569-8_4

Emergency Remote Teaching (ERT) refers to "a temporary shift of instructional delivery to an alternate delivery model due to crisis circumstances. It involves the use of fully remote teaching solutions for instruction or education that would otherwise be delivered face-to-face or as blended or hybrid courses" [2].

Different events might trigger an ERT situation. Sometimes, these events are life-threatening or represent a risk of direct physical harm. In those cases, it is clear that the dangers at hand must be addressed before dealing with the impact on the educational process. The United States Government has a web page (www.ready.gov) with advice on planning for and reacting to different emergencies.

This paper concentrates only on the aspects of an emergency that disrupts the educational process for a medium to a long time.

In early 2020, when schools closed due to the COVID-19 emergency confinements, it was unclear when they would reopen. There was no turning back, no plan B. Schools had to come up with ideas to finish the school year.

Every school reacted differently, with varying degrees of success. One success story is the Tecnologico de Monterrey university system in Mexico. The school's previous extensive experience with e-learning and a recent ERT event enabled Tecnologico de Monterrey to preempt the emergency declaration by going online two weeks before any other school in Mexico. More than 90,000 students attend 55,000 class sessions per week on its 33 campuses throughout Mexico; a high student acceptance of online learning already existed.

The research reported in this paper aimed to analyze the lessons learned from the Tecnologico de Monterrey experience and other schools worldwide to identify the ERT steps and the factors that can positively influence the outcome.

This exploratory research tries to identify relevant variables and constructs that can serve as a basis for future qualitative and in-depth studies. The challenges faced during the COVID-19 emergency should lead to advances in distance learning and emergency response in the education environment.

2 Previous Experiences Around the World with Emergencies in Education

The 2020 COVID-19 crisis was not the first time schools suddenly closed and tried alternative teaching methods. It had happened before, although not at the same level.

In 2003, the Severe Acute Respiratory Syndrome (SARS) epidemic occurred mainly in China and Hong Kong. It also spread to Vietnam, Singapore, Taiwan, the Philippines, Mongolia, Canada, and other countries. The SARS epidemic forced the closure of schools in China and Hong Kong for almost four weeks and then in Beijing for four more weeks. In that period, the educational systems resorted to TV programs and a website for online learning and teacher guidance [3].

The Ebola Virus Disease epidemic of 2014 affected mainly Liberia and Sierra Leone. The crisis required specialized training for teachers and distance learning options for schools. The schools had to use accelerated learning programs to complete the school year for children who had no access to technology [4].

The lessons learned from the SARS crisis helped early discussions and planning and put countries in a better position to face the COVID emergency [5, 6]. The Ebola epidemic brought forward the legal aspects of containment measures and the need for specific actions to tackle inequities in the global health systems and the responses to help vulnerable populations [7, 8].

3 Emergency Remote Teaching at Tecnologico De Monterrey

Tecnologico de Monterrey is a university system based in Monterrey, Mexico, with 33 university and high school campuses in 20 Mexican states, with programs extending to Central and South America. It is accredited by the Southern Association of Colleges and Schools (SACS), headquartered in the USA.

For the past 30 years, Tecnologico de Monterrey has been testing different distance learning techniques for graduate, undergraduate, and high school courses. Depending on the type of program, the delivery methods tested have been via:

- Synchronous satellite (or web) transmissions.
- Semi-synchronous web-based classes.
- Asynchronous online classes.

Some courses required a combination of models, and some included a face-to-face component.

Before 2020, many teachers at Tecnologico de Monterrey had experience teaching online courses, and all students were required to take at least two online courses before graduating. The technology and technical support were widely available. The school used Blackboard or Canvas, two of the most popular learning management systems (LMS), as platforms for all its courses.

3.1 Previous Experiences with ERT at Tecnologico De Monterrey

An example of a school's response to emergencies before 2020 occurred in Mexico in 2017. A 7.1 magnitude earthquake hit Central Mexico, severely affecting certain types of structures across Mexico City. An area badly hit was the Mexico City location of Tecnologico de Monterrey.

The earthquake caused severe damage to the campus facilities, requiring a complete reconstruction of many buildings. It took about six months to complete the construction project. All classes on that campus had to go online. All the resources of the Tecnologico de Monterrey system supported its campus in Mexico City. Educators and support staff from all 33 campus locations collaborated to assist online education design and delivery for the students in Mexico City.

The earthquake in Mexico City provided valuable experience for many teachers, but, most importantly, for school management, who learned how to respond quickly to emergencies.

3.2 Tecnologico De MONTERrey's Response to the COVID-19 Emergency

Based on the experience with ERT of 2017 and e-learning, the university decided to start ERT preventively. On March 12, 2020, the school declared a voluntary closing and resorted to ERT two-and-a-half weeks before the Mexican Government officially called for social distancing and school closings. Many other universities and businesses followed Tecnologico de Monterrey's lead and voluntarily closed before it was required.

The model was implemented without serious incidents at all 33 campuses of Tecnologico de Monterrey.

The effort involved:

- 90,000 + students.
- 10,000 + teachers.
- 55,000 + class sessions per week.

Week 1 of ERT had 44,404 sessions online, with 1,000 concurrent sessions and some learning blocks had 1,400 sessions. In total, 95% of the classes operated uneventfully. A total of 5% had technical or connectivity problems, or the teacher did not attend.

In the first student satisfaction survey, 64% indicated satisfaction, 26% neutral, and 10% indicated dissatisfaction.

At the start of May 2020, the Government decided to extend the COVID-19 emergency measures. What had started as a four-week exercise was now the format for all classes for the remainder of that semester. The school announced that the semester would continue in that format until at least the end of June.

The fully online model continued through the fall semester of 2020 and the spring of 2021. By the fall of 2021, the school was ready to start hybrid models, where some students could attend the classroom while others would attend online. The transition back to "normality" would continue through the spring of 2022, but the educational models would adapt to take advantage of widely available new tools and techniques.

4 Four Factors that Determine the Readiness for ERT

Some schools are in a better position than others to face ERT events. Better-prepared schools can easily migrate to e-learning models and consider starting distance education as a preventive measure before it becomes the only alternative. An advantage of an early start is that schools can test the technology and refine the methodologies before starting a quarantine period or before the campus becomes inaccessible.

Among the factors that may simplify the transition to ERT are:

1. School's experience with emergencies:

    a. Availability of a previous continuity-of-operations plan.
    b. Previous experience with ERT.

2. Experience with distance learning and course design.

   a. School's involvement with distance learning.
   b. Access to instructional designers.
   c. Use of a LMS.

3. Availability of technology

   a. Student's and teacher's access to technology.
   b. Access to technical support.
   c. Access to a commonly accepted communication channel.

4. Commitment

   a. From management, teachers, students, and parents (Fig. 1).

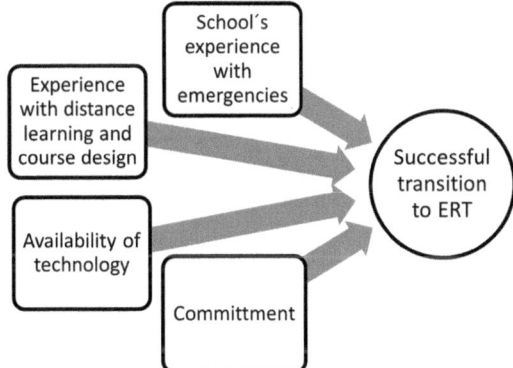

Fig. 1. Factors that simplify the transition to ERT

4.1 SCHool's Experience with Emergencies

Loyola University in New Orleans has a complete continuity-of-operations plan because they are on the Gulf of Mexico coast and must prepare for hurricane season every year. Their current program primarily addresses contingencies for suspension of on-campus activities for periods of up to two weeks, but there are some strategies for managing more extended evacuation periods [9].

Other schools have faced ERT situations of different magnitudes. Previous experience can help management determine the school's preparedness for any such future event and may prove valuable in handling new emergencies.

4.2 Experience with Distance Learning and Course Design

If the students and teachers have had experience teaching and attending distance learning classes, the switch to ERT might not be very stressful. Before 2020, many schools offered distance learning courses along with their face-to-face curricula. By 2016, 80% of all higher education institutions had online courses [10]. Even if not all the faculty were involved in distance learning, experienced instructors could tutor others with less experience to help them succeed.

Another advantage of having distance learning courses in place is that this generally requires access to instructional designers familiar with the differences between face-to-face and distance methodologies. As in the previous case, experienced designers can quickly train less experienced ones, and the group can support the teachers in their efforts to redesign their courses for new media.

The third factor in this category is the availability of a LMS. Generally, an LMS is a software application for the documentation, administration, reporting, tracking, and delivery of educational courses for learners [11]. Some examples of educational platforms are:

- Blackboard (https://www.blackboard.com/).
- Canvas (https://www.instructure.com/).
- Moodle (https://www.moodle.com/).

Having an LMS in place would simplify the communication with the students, the distribution of materials, and even the application of exams during an ERT event.

4.3 Availability of Technology

The technology available to faculty and students (computer equipment, bandwidth, and software) determines the type of course offered and the technology used for distance learning. Sometimes, the school might open spaces for the production or delivery of lectures. However, if the school facilities must close, the alternative would be to lend the equipment to teachers and students who might need them. Schools can go as far as to provide access points or cellular connections to faculty and students in need.

Just as important as the equipment is the technical support. Things break, and the software does not always work. It is vital to have a call number ("hotline") to answer technical questions and keep the school's network and resources operational throughout the emergency.

Another critical technology is communication. Having a commonly accepted technology (e.g., public e-mail, institutional e-mail, message boards, Facebook, Skype) will simplify sending messages and avoid miscommunications. It is essential to be consistent with the channel. For example, if students expect to receive news via e-mail, do not send messages through Facebook, where no one would be looking.

4.4 Commitment

The last but most important factor that is key to the success of any ERT initiative is the commitment from the faculty, students, administration, and society. ERT places much stress on the faculty by demanding extra work to adapt courses to foreign media. It also stresses the students by requiring better time management skills, reducing contact with classmates and teachers, and emphasizing self-instruction. School administrators also have more stress from extra work and a higher demand for resources.

In a challenging situation like ERT, commitment, love for the craft, respect for the institution, and a culture of cooperation among students, faculty, and staff will ultimately determine the success or failure of the initiative.

5 Factors to Consider During an Emergency Remote Teaching Event

There are five distinct stages in the ERT event:

1. Before the ERT event occurs.
2. The ERT declaration.
3. The first week of ERT.
4. The following weeks of ERT.
5. The return to normal operations after the ERT event is over.

Important factors change depending on the stage of the process. The following sections discuss the details of each phase (Fig. 2).

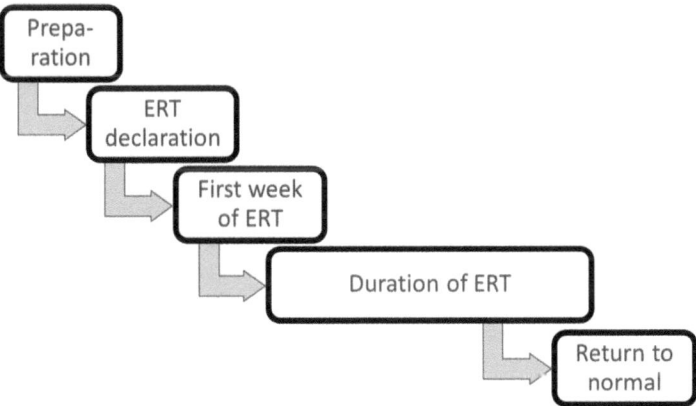

Fig. 2. Five stages of an ERT event

5.1 Before an ERT Event Occurs

The question to answer before an ERT event occurs is, "Are we prepared?".

Sometimes you can see an emergency coming—for example, when a hurricane is heading towards your area. Other times, there is no warning—for example, earthquakes or volcanic eruptions. In any case, the timing of the emergency declaration may make the difference between chaos or control.

Even before there is a need for ERT, some procedures should be in place. At the very least, these are:

- a communication channel and a backup,
- a chain of command, and
- a safe emergency meeting point in case people become disbanded and need to regroup.

Five steps to assess readiness and prepare for an ERT event are:

1. Secure a communication channel and plan for a backup. Stick to one media for communication. People would not know whom to listen to if instructions came from multiple sources. However, an alternative must be ready in case the primary communication channel gets disrupted. The basic media could be e-mail or the school's web page. Alternative channels can include mass media (TV, radio, or newspapers), cell phone messages (SMS or WhatsApp), or even a network of contacts to spread news by word of mouth.
2. Define a chain of command. It must be clear who has the authority to declare and terminate an emergency. It must be clear which type of communication goes directly to everyone in the school and which goes through channels (from the school administration to the department heads, the teachers, and the students).
3. If there is a continuity-of-operations plan, check if it applies to the current situation or needs a revision. If the program is ready, initiate it.
4. Assess the readiness factors (see the discussion of readiness factors in the previous section). If there is a plan that needs revision or no plan, the strengths and weaknesses should be assessed in this step. This exercise should bring out training needs and the types of technologies required during emergency teaching.
5. Draft a plan to include time to regroup and train the faculty; define the new learning interactions, the initiation of the online classes, the continuing maintenance of the online courses, and the process for the termination of the emergency.

5.2 ERT Declaration

In the "ERT declaration" phase, the question is, "Are we ready to start with distance education?".

The process starts when the appropriate school authority declares the emergency and informs that the school is going to the ERT mode. Make sure everyone receives the message. Be prepared to activate (or quickly design) the contingency plans for the situation at hand.

Reserve some time (maybe one week) between the moment you make the ERT declaration (and suspend face-to-face classes) and the time when remote teaching starts. Generally, a one-week delay will not change educational programs much. The delay will give the school time to make sure the hardware and software required are in place, and the faculty have the tools and training they need to start the process.

During the preparation period, complete the following steps:

1. Communicate with students.

    a. How classes will continue.
    b. Where they will find information about their courses.
    c. How to communicate with their teachers.

2. Create support groups in at least three different areas:

    a. A technical support group will be in charge of deciding the technology to use, securing the hardware and software required, operating the communications channels, solving technical problems in personal equipment, and training the faculty to achieve the minimum technical skills required.
    b. A course design group, composed mainly of instructional designers or people with experience with remote teaching, will define the course work's direction. The group must decide if there will be synchronous or asynchronous teaching, the size of the groups, and the way to evaluate the learning results. This group is also responsible for training the faculty on their minimum course redesign process.
    c. A teacher's online resources group is a group of peers who support each other, share best practices, and revise each other's redesigns and online teaching skills. This group should also identify further training requirements and measure the results of the distance learning efforts.

3. Assess and solve hardware and software needs so teachers can start online classes, whether from home or an office at a specific location, and students can receive the instruction needed. If the teachers do not have computers with the necessary power to work at home, consider lending them some equipment or, if the situation permits, setting up some space in the school or at a different location to host the live transmissions or the recording of educational capsules. In some cases, schools went as far as letting faculty borrow office chairs to take home.
4. Train the teachers on the basics of hardware and software and distance learning. This phase only covers general training and assessments of needs. Based on the needs identified, the school can set up a training program for the following weeks to make incremental improvements in parallel with the remote teaching operations.
5. Redesign the courses to adapt to the new environment.

5.3 The First Week of ERT (Make Sure the Materials and Environments Are Ready)

The question to answer in the first week of the ERT phase is, "Is everything working?".

When remote teaching starts, the initial focus is on getting everything to work. Ensure that all the students are connected, receive the messages, and know what to do.

1. Start by distributing the educational materials and instructions, and then initiate remote courses.

2. If you have live sessions, hold the first class. Try to spend some time talking to students to assess their concerns and answer their questions. Get them to participate and lose their fear of the technology.
3. If you are doing work asynchronously, contact everyone by e-mail to ensure the program starts correctly, answer questions, and address student concerns.
4. Make sure everyone knows the rules of interaction and how to manage the software.
5. Evaluate the initial results and be ready to adjust your programs.
6. Most of the time, the novelty effect will result in people attending and participating in classes this first week.

5.4 The Following Weeks of ERT (Revise and Improve)

During the following weeks of ERT, the vital question is, "Are the students learning?".

By the second week, the process is no longer new. The focus shifts from technology to education, and the aim is continuous improvement. Different players assume more clearly their roles:

1. The administration:

   a. Holds periodic meetings to keep everyone informed, shares the good news and examples.
   b. Works on continuously enhancing the teacher's delivery skills and improving the available materials and the learning experience.
   c. Measures what is working and the expectations and feelings of faculty and students.

2. The teachers:

   a. Are ready to modify activities, assignments, or technology that do not work as expected.
   b. Are flexible. People can get sick; technology can fail. Assignments can be late or misplaced by people not familiar with the technology.
   c. Most importantly, give feedback to the students. Grade their assignments quickly and hold question and answer sessions.

3. The support groups:

   a. The technical and educational design groups reduce their activity as people learn how to work things out.
   b. The teacher's group becomes critical because teachers can attend other teachers' sessions to give feedback, support, and share best practices.

5.5 Returning to Normal Operations (After the ERT Event is Over)

There are two questions to answer when returning to normal operations after the ERT event is over: "Are we ready to go back? What did we learn?".

While the decision to go into ERT is generally not optional and external forces define the dates, once the emergency subsides, the decision to return to normal operations can wait until everything is ready and in order. UNESCO has several publications dealing with the details of returning to normal activities [4].

The priority is safety. Ensure the emergency that caused the ERT decision is over and conditions are safe to return to the school. If the school has been vacant for some time, verify the safety of the facilities and equipment. Take advantage of this opportunity to do maintenance on the facilities.

If there were problems with remote teaching, you might want to return to normal operations as soon as possible. However, you could desire to keep some distance learning aspects if they contributed to the educational program's objectives.

One alternative is to complete an entire period with remote instruction. This option permits more time to prepare the facilities and wastes no time transitioning back to face-to-face instruction. The return to normal operations can take place in the following school period.

As at the start of the process, good communication at the end is also essential. People must have one official communication channel with instructions on how to return to standard operations.

The return can be:

- All at once: setting a date for everyone to return to school and resume operations.
- By school year: starting with those students who might need more support.
- By subject: leaving some courses in the distance learning mode while others return to face-to-face instruction or trying hybrid approaches, where some students attend face-to-face while others continue online.

Just as at the beginning of the emergency, changing learning modalities (back to face-to-face instruction at the end, for example) is a transition. It is advisable to talk to students to assess their concerns about it and answer their questions.

The last and most critical step of returning to normal operations is learning from the experience.

- Administrators can learn how to respond better to unusual situations.
- Teachers can bring new tools (video capsules, offline assignments, best-practice sharing) to their face-to-face classes. They might also choose to continue some activities remotely, developing new educational models that enhance face-to-face learning.
- The school and society will hopefully enjoy a new sense of community.

6 Conclusions

Despite the tragic events of the COVID-19 pandemic, we must still try to learn something from it and work to build a better world.

This paper reports exploratory research. The objective is to use qualitative techniques to identify the relevant variables and constructs that can serve as a basis for future quantitative and in-depth studies. Documenting the events, decisions, and outcomes of some participants can define the starting points in designing better emergency procedures for schools and aid in the formal transition to distance learning, or hybrid programs, by institutions that follow more traditional face-to-face conventions.

Acknowledgments. The author acknowledges the technical support of Writing Lab, Institute for the Future of Education, Tecnologico de Monterrey, Mexico, in the production of this work.

References

1. UNESCO: Global monitoring of school closures. https://en.unesco.org/covid19/educationres ponse#schoolclosures. Last accessed 14 March 2022
2. Hodges, C., Moore, S., Lockee, B., Trust, T., Bond, A.: The difference between emergency remote teaching and online learning. Educ. Rev. (2020). https://er.educause.edu/articles/2020/3/the-difference-between-emergency-remote-teaching-and-online-learning. Last accessed 14 March 2022
3. World Health Organization—Western Pacific Region: SARS: How a Global Epidemic was Stopped. WHO Press, Geneva (2006)
4. UNESCO: Preparing the Reopening of Schools (2020/5/5) https://unesdoc.unesco.org/ark:/48223/pf0000373401. Last accessed 14 March 2022
5. Nicholas, D., Gearing, R., Koller, D., Salter, R., Selkirk, E.: Pediatric epidemic crisis: lessons for policy and practice development. Health Policy **88**(2/3), 200–208 (2008)
6. Webster, P.: Canada and COVID-19: learning from SARS. Lancet **395**(10228), 936–937 (2020)
7. McCollum, R., Taegtmeyer, M.: Let's not make the same mistake again: a political economy analysis of Sierra Leone's Cholera and Ebola epidemic responses. Global Health Governance **11**(2), 71–83 (2017)
8. Jobe, K.: The constitutionality of quarantine and isolation orders in an Ebola epidemic and beyond. Wake Forest Law Rev. **51**(1), 165–188 (2016)
9. Loyola University New Orleans, Continuity of Operations Plan. http://academicaffairs.loyno.edu/continuity-operations-plans. Last accessed 15 May 2020
10. Goodman, J., Melkers, A, Pallais, J.: Can online delivery increase access to education? J. Labor Econ. **37**(1) (2019)
11. Kushwaha, R.C., Singhal, A.: Online learning: an emergence of new model of education. Int. J. Recent Technol. Eng. (IJRTE) **7**(6), 75–78 (2019)

Cascades of Concepts of Virtual Time Travel Games for the Training of Industrial Accident Prevention

Oksana Arnold[1]([⊠]), Ronny Franke[2], Klaus P. Jantke[3], and Hans-Holger Wache[4]

[1] Fachhochschule Erfurt, Altonaer Str. 25, 99085 Erfurt, Germany
oksana.arnold@fh-erfurt.de
[2] Fraunhofer Institut für Fabrikbetrieb und -automatisierung IFF, Sandtorstr. 22, 39106 Magdeburg, Germany
[3] ADICOM Software, Frauentorstr. 11, 99423 Weimar, Germany
[4] Berufsgenossenschaft Rohstoffe und chemische Industrie, Präventionszentrum Berlin, Innsbrucker Str. 26/27, 10825 Berlin, Germany

Abstract. Time travel games are a recent form of edutainment media having high potential in areas such as environmental education and prevention training. At TLIC 2021, the present authors delivered a contribution advocating planning of human training experiences as dynamically as driving a disturbed technical system back into a normal mode of operation. A trainee's inappropriate actions may be possibly ruining a—fortunately only virtual—technical installation. Time travel backwards offers opportunities to influence the fate to make good of the damage. This requires concepts of Artificial Intelligence to plan and execute time travel adaptively according to a trainee's training history, to a trainee's strengths and weaknesses, and to possibly changing environmental conditions. There is introduced a cascade of gradually more intricate categories of time travel games. With every step from one category to the next, the deployed AI gets more powerful and effective in providing adaptive guidance of trainees. The most advanced time travel games are those that allow for the dynamic modification of events experienced in the virtual past. In this way, the game system evolves over time and adapts to the needs of human trainees with emphasis on guidance for trainees who fail repeatedly. Consequently, every trainee gets an individual training experience. The AI takes care that all trainees experience their individual success as a result of their own efforts—i.e., as a gratification for mastery. All concepts and their respective representations are illustrated through examples from a novel application. This includes the way in which concepts are used and the impact they have on affective and effective training.

Keywords: Time travel exploratory game · Time travel prevention game · Dynamic time travel prevention game · Accident prevention training · Game design · Didactic design · Design of experience · Adaptivity · Didactic principles

1   Introduction

This work expands on references [2, 3, 7], the two first reflecting the authors' contribution to The Learning Ideas Conference 2021. The present work is based on a novel time travel prevention game aiming at accident prevention in the paint and coatings industry where paint, stain, varnish, and the like are produced. Several screenshots from the application are used to provide a touch and feel of training with digital time travel prevention games and, more importantly, to illustrate the authors' concepts, methodology, and effectiveness.

1.1   The Application: Prevention Training for the Paint and Coatings Industry

The first figure is intended to provide an impression of the authors' underlying novel time travel prevention game showing the virtual factory as a whole and in some detail.

Fig. 1. The virtual factory, receiving store, intermediate bearing, and basket mill workplace

The fourth screenshot in the lower right corner of Fig. 1 shows the workplace where trainees stir the mixture to produce the desired stain or varnish. In contrast to the case study of references [2, 3] where inaccurate trainee behavior may result in an explosion and fire, the undesired events here are more subtle. Humans may be exposed to varying noxious substances of a certain intensity and for a varying time.

1.2   Storyboarding—The Technology of Designing Trainee Experience

As already put in reference [2], it makes a difference whether a human learner or trainee has experienced an accident, especially a self-induced one. This, however, shall not be misinterpreted as a call for a didactic principle of deliberately damaging industrial installations or injuring humans at their workplace for more affective and effective learning about accident prevention. And it makes a difference whether a human learner or trainee has exposed herself to a lethal concentration of vapor.

What we currently have at our fingertips are virtual training environments in which humans can act and interact without fear and without the danger of real damages and injuries. The present authors' approach is to offer affecting experience to learn from, preferably human experience so touching and exciting that learners and trainees find it worth telling.

Following reference [14], storyboarding is the organization of experience. For more details and in-depth discussions with emphasis on complex applications, interested readers are directed to publications such as references [4, 8, 15, 16]. The above-quoted sources [2, 3, 7] may do, as well. Storyboard interpretation technology [9] might be of a particular interest. Seen in the right light, the approach to digital storyboarding adopted and adapted from reference [14] is an approach to dynamic plan generation derived from reference [1]; see also references [5, 6]. Consequently, one might say that storyboarding as used here means the planning of forthcoming experiences of game play and training that takes the dynamic needs and desires of varying trainees into account.

Throughout this contribution, the authors confine themselves to an introduction of the essentials by example.

Storyboards are hierarchically structured families of finite directed graphs, so-called pin graphs [10], describing potentially forthcoming experience of game play and training. A first storyboard graph is on display in Fig. 2.

There are two types of nodes called *episodes* and *scenes*. The nodes in a graph that are smaller in height than other nodes represent scenes. The scenes have a certain operational meaning in the domain. In contrast, episodes are placeholders for more complex (inter-) actions. They are subject to substitution by other graphs residing in the storyboard.

What a node means, be it an episode or a scene, is left open as long as possible. Dynamics is guaranteed by postponing decisions until execution time—one of the key concepts adopted from reference [1]. In the authors' application domain, the term execution time means the time playing the game or, in other words, the training time. For this purpose, graphs to be substituted and operational meanings to be assigned are equipped with logical conditions. They depend on variables that refer to the history of interaction, to the user/player/learner model, and to some environmental conditions. These conditions are called substitution conditions and assignment conditions, respectively.

1.3   Storyboarding—The Dynamics

According to the present focus, emphasis is put on storyboarding the experience of virtual time travel. Subsequently, this is used to demonstrate adaptive dynamics.

At the end of a training session for accident prevention in the paint and coatings industry, the trainee arrives at a scene named Exit Menu. The operational meaning of

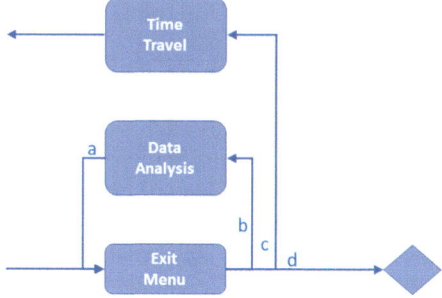

Fig. 2. Storyboard graph surveying essentials of embedding time travel into game play

the scene Exit Menu is an interface that enables the trainee to decide about whether to exit the training/game session or not. There are two alternative variants of the interface as on display in Fig. 3 having different logical assignment conditions.

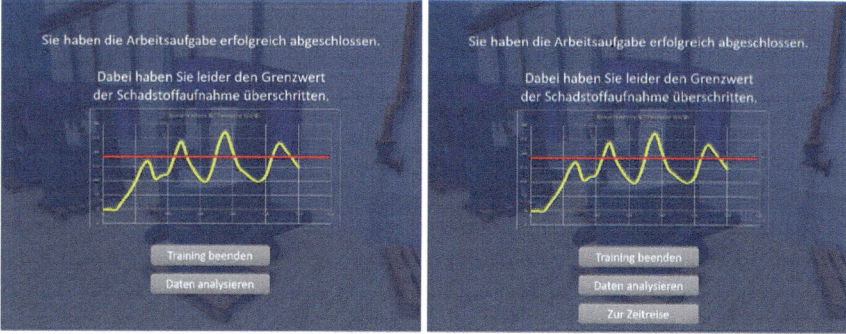

Fig. 3. Two alternative meanings assigned to the scene Exit Menu dynamically at play time

In case the trainee arrives at this scene for the first time, the left interface on display in Fig. 3 is offered. It is possible either to quit training or to inspect the data about intensity and duration of exposure to noxious substances. After seeing the data, the trainee gets offered an opportunity of virtual time travel as shown on the right of Fig. 3. Its assignment condition is generally true after the first arrival at Exit Menu.

Alternative edges leaving a node of a storyboard graph, such as the edges labeled by b, c, and d, in Fig. 3 carry execution conditions, as well. This is another concept allowing for dynamically adapting to different trainees. By way of illustration, the execution condition of edge c is that Zur Zeitreise has been selected.

2   Cascades of Time Travel Concepts

The authors distinguish a cascade of concepts, so to speak, on a macroscopic level and another one on a microscopic level of description. The further down the cascade, the more adaptive and, thus, potentially more affective and effective is the game play. This section is aimed at an introduction of these concepts, whereas the remaining part of the paper discusses applications and, in the process, provides several illustrations.

2.1 A Hierarchy of Time Travel Game Concepts

The authors distinguish three principal categories of educational games relying on the peculiarity of and drawing benefit from virtual time travel:

- Time Travel Exploratory Games
- Time Travel Prevention Games
- Dynamic Time Travel Prevention Games

The latter is a novelty explicitly presented here for the first time, to our knowledge. It will be detailed in Sect. 2.2. Before doing so, a few words about the first two concepts shall do.

Reference [7] introduces a time travel exploratory game for the purpose of environmental education. The problem in focus is global ocean warming. The players/learners may travel back in time and make thermal pictures of different areas. They bring their findings back to the present time. After individual explorations, teams of players may compare and discuss their findings varying in location and time of the origin. This leads to a collaborative phase of game play resulting in a common representation of insights such as a video visualizing the ocean warming worldwide.

Beyond the limits of time travel exploratory games, time travel prevention games—the term was coined on the German Prevention Day (Deutscher Präventionstag) 2015 [13]—enable the human players to change the virtual past and, in this way, to impact on later experiences of game play. Erroneous behavior and inaccurate or possibly incomplete actions may be corrected such that undesirable events disappear from the story experienced. This is a typical case of nonmonotonicity of story space evolution as introduced and discussed in reference [12]. The key concepts of time travel prevention games, their touch and feel including a complex application for the purpose of accident prevention in the industry, are introduced and discussed in some detail by the conference contribution [2] and by the related journal article [3].

In reference [3], the authors discuss approaches to an adaptive guidance of trainees who fail repeatedly. The ultimate goal is to enable trainees to master the problem under consideration. Trainees must experience the success of finally solving problems by themselves. To be effective, training must be affective. Frustration must be avoided. Being on the same journey backwards in time repeatedly is intolerable.

The solution is dynamically changing time travel in such a way that the trainee's success is becoming more and more likely. As a side-effect, different human players experience slightly varying stories of game play adaptive to their individual needs and desires. This makes training by means of dynamic time travel prevention games even more worth telling and, thus, sustainable.

2.2 A Hierarchy of Dynamic Time Travel Prevention Concepts

The step toward *dynamic* time travel prevention games is bringing those games closer to pieces of artwork in the area of interactive digital storytelling. According to Stern, "the first wish that most players, developers and researchers originally feel when first encountering and considering interactive story, is the implicit promise to the player to be able to directly affect the plot of the story, taking it in whatever direction they wish" [18]. Human players in an industrial training context should experience the freedom to drive the story in the direction they wish—mastery and successful completion of the task. This includes the opportunity of traveling backwards in time. Being back at an earlier point of time, one may fix inaccurate actions of the virtual past aiming at better results the next time.

As already demonstrated in reference [3], players/trainees get access to a time tunnel as visualized by means of Fig. 4. The screenshots from the left to the right illustrate the trainee's diving back into the past virtually. Every object on display is the iconic representation of an elapsed episode. Objects deeper in the tunnel—i.e., visible slightly more to the left—refer to earlier points of time. Selecting an object by clicking the above center button brings the trainee back to the related episode.

Fig. 4. The appearance of the time tunnel when training the production of a certain varnish similar to the original time tunnel introduced in references [2, 3]

Thinking about what it means to arrive at an episode a player/trainee has already experienced earlier, several alternatives arise. The authors distinguish the following options of making time travel prevention games considerably more dynamic going quite far beyond the idea of modifying the destinations offered in the time tunnel.

- **Modifying the destination episode** to make inaccurate actions less likely. By way of illustration, one may remove objects that shall not be used.
- **Reducing the history of play** following the destination time for more rigor. By way of illustration, one may skip episodes and implement their results.
- **Expanding the destination episode** to provide information and guidance. By way of illustration, one may visualize hazards such as effused vapor.

Needless to say that all these approaches may be combined for more adaptivity. Furthermore, they may be generalized, applying principles not only to the destination episode, but to certain subsequent episodes, as well.

3   The Design, Implementation and Application of Dynamics in Time Travel Prevention Games

This section is intended to communicate the authors' concepts of dynamics in time travel prevention games by means of details from a training session including some information about the overall game design. Due to the complexity of training processes, this section's presentation remains unavoidably incomplete.

Every training session begins with task formulation as specified in Fig. 5 below.

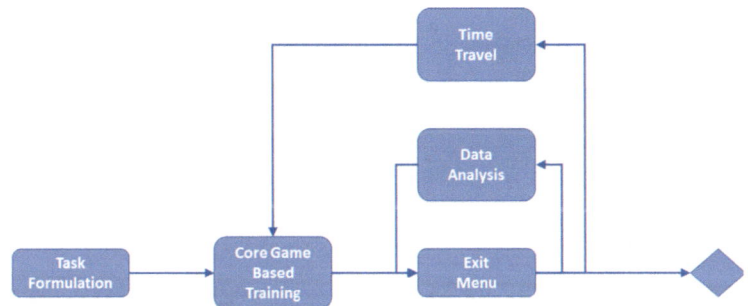

Fig. 5. Top-level storyboard graph of a time travel prevention game with result analysis

In time travel prevention games with the peculiarity that the result of game play is either successful or simply a uniquely defined failure such as the explosion and fire investigated in references [2, 3], there is no need for an explicit analysis of this result. As said earlier, the present case study is more subtle. This leads to a top-level design as visualized in Fig. 5.

Issues concerning the varying description levels of game play and storyboarding are beyond the limits of this contribution. Interested readers are directed to what the authors call layered languages of ludology [11, 17] (Fig. 6).

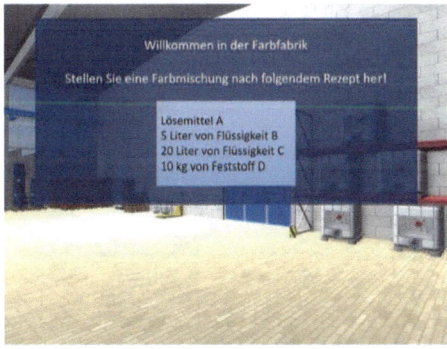

Fig. 6. Setting the goal in the task formulation episode—the trainee shall mix some paint from solvent A, 5 L of the liquid B, 20 L of the liquid C, and 10 kg of the original substance D

Subsequent steps of interaction may take place during the main episode denoted by Core Game Based Training in Fig. 5. There are branching points that allow for actions in varying order and for varying choices of objects such as containers used to carry liquids. The following figures are intended to provide an impression.

Fig. 7. Filling in the solvent and placing the barrel with the solvent in one of the basket mills

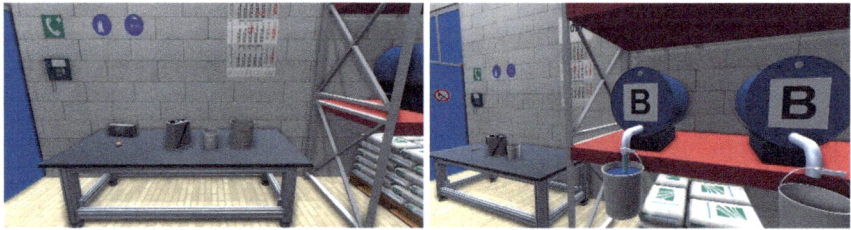

Fig. 8. Selection of a bucket from the available containers and filling in the needed liquid B

The four screenshots on display in the Figs. 7 and 8 may be seen an illustration of the human trainee's activities when progressing from one episode to the next one. Storyboard graphs prepared for expansion of the node Core Game Based Training determine which actions are available and in which order they may be executed (Fig. 9).

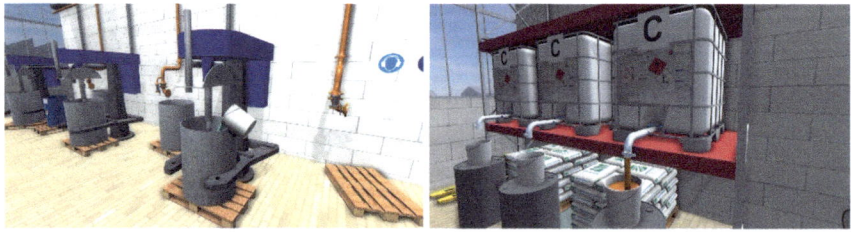

Fig. 9. Filling in liquid B and getting liquid C to be carried to the basket mill in the next step

There is neither the need nor the space to show every detail. The authors confine themselves to a few more screenshots (Figs. 10 and 11).

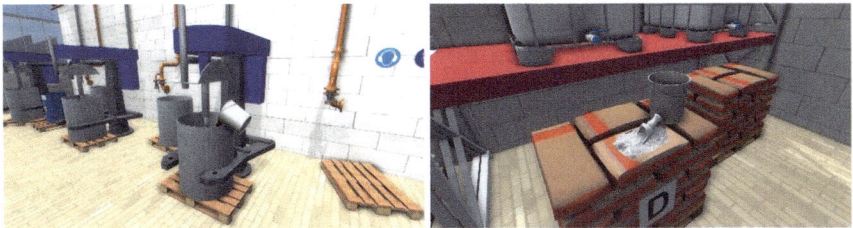

Fig. 10. Filling in liquid C and going to the storage place for getting the solid component D

Fig. 11. After filling in the solid component, the mill may be turned on. The main episode of Core Game Based Training is completed. Specified by the storyboard, the Exit Menu follows

The exit menu informs the trainee that the threshold for exposure to noxious vapor has been exceeded. Those trainees who take this information serious and who are interested in their achievements will not exit immediately, but they will inspect the data (Fig. 12).

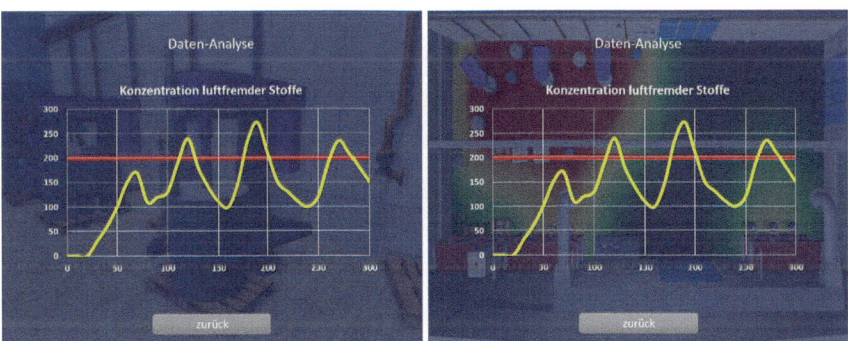

Fig. 12. Two alternative variants of data analysis provided; the one on the rights shows in the background a bird's eye view of the workplace with a heatmap of noxious vapor concentration

After leaving the data analysis scene, trainees come back to the exit menu that is now offering to travel back in time. Who travels back for the first time may freely choose the destination in the time tunnel. Later on, dynamic modifications according to ideas systematized in Sect. 2.2 may be applied for the sake of better adaptivity.

The selection of buckets is not really optimal for the transport of fluids, and they are likely to effuse noxious vapor; instead, they may be fixed by using a jerrycan. If trainees fail to do so, they are likely to arrive at similarly bad—i.e., unhealthy—results the next time. Adaptivity comes into play by applying dynamical changes of game experience.

Fig. 13. Two modifications of the destination episode providing extra information to the trainee

After returning once more to the workplace, the location has changed slightly as on display in Fig. 13 on the left or it is changing drastically during interaction as shown in Fig. 13 on the right—instances of "Modifying the destination episode."

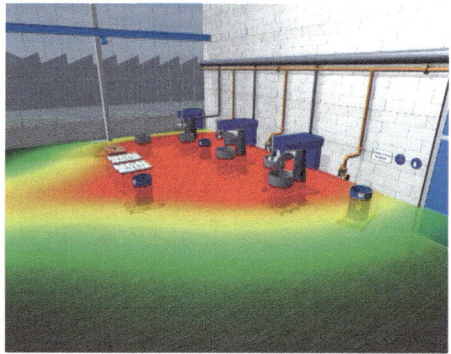

Fig. 14. A very explicit modification of the destination episode on a bit more abstract level to provide information to the trainee about the distribution of noxious substances at the workplace

The heatmap overlay shown in Fig. 14 is explicit because of being artificial. This is obviously no decoration, but a technical information. Such a change of the virtual world is a cause of concern. It provokes the human trainee to interpret what is on display. And it is intended to motivate pondering ways out.

In response to unsatisfactory training results, the digital game system is adaptively moving to a meta level of communication, so to speak. The change of the system's communication strategy in learning and training, however exciting, is unfortunately beyond the limits of the present contribution.

4 Conclusion

Virtual *time travel for educational purposes* is the authors preferred didactic concept. In *time travel exploratory games*, traveling back in time allows for exciting and, sometimes, surprising opportunities of gaining information and, possibly, bringing virtual artifacts to the present time. In *time travel prevention games*, the authors go even further offering options to impact the fate. The present time may be changed in response to changes in the past.

Dynamic time travel prevention games introduced in the present contribution open up unimagined potential—*the past is no longer what it used to be.* Not only that the human trainee has permission to change the past, the past is dynamically changed by the game system to provide better guidance aiming at the trainee's ultimate success and, in this way, at affective and effective training.

References

1. Arnold, O.: Die Therapiesteuerungskomponente einer wissensbasierten Systemarchitektur für Aufgaben der Prozeßführung, vol. 130 of DISKI, infix, St. Augustin, Germany (1996)
2. Arnold, O., Franke, R., Jantke, K.P. Wache, H.-H.: Dynamic plan generation and digital storyboarding for the professional training of accident prevention with time travel games. In: Guralnick, D., Auer M.E., Poce, A. (eds.) Innovations in Learning and Technology for the Workplace and Higher Education: Proceedings of 'The Learning Ideas Conference' 2021, vol. 349 of Lecture Notes in Networks and Systems, pp. 3–18, Springer (2021)
3. Arnold, O., Franke, R., Jantke, K.P., Wache, H.-H.: Professional training for industrial accident prevention with time travel games. Int. J. Adv. Corp. Learn. **15**(1), 20–34 (2022)
4. Arnold, S., Fujima, J., Jantke, K.P.: Storyboarding serious games for large-scale training applications. In: Foley, O., Restivo, M., Uhomoibhi, J., Helfert, M. (eds.) Proceedings of the 5th International Conference on Computer Supported Education, CSEDU 2013, Aachen, Germany, pp. 651–655, SciTePress (2013)
5. Arnold, O., Jantke, K.P.: Therapy plan generation in complex dynamic environments. In: ICSI Report TR-94-054. International Computer Science Institute, Berkeley (1994a)
6. Arnold, O., Jantke, K.P.: Therapy plans as hierarchically structured graphs. In: Fifth International Workshop on Graph Grammars and their Application to Computer Science, Williamsburg, VA, USA (1994b)
7. Arnold, O., Jantke, K.P.: AI planning for unique learning experiences: The time travel exploratory games approach. In: Csapó, B., Uhomoibhi, J. (eds.) Proceedings of the 13th International Conference on Computer Supported Education, CSEDU 2021, vol. 1, pp. 124–132. SciTePress (2021)
8. Arnold, O., Jantke, K.P., Spundflasch, S.: Hierarchies of pervasive games by storyboarding. In: Proceedings of 5th International of Games Innovation Conference, IGIC 2013, Vancouver, Canada, pp. 8–15. IEEE Consumer Electronics Society (2013)
9. Fujima, J., Jantke, K.P., Arnold, S.: Digital game playing as storyboard interpretation. In: Proceedings of the 5th International Games Innovation Conference, Vancouver, BC, Canada. pp. 64–71. IEEE Consumer Electronics Society (2013)
10. Höfting, F., Lengauer, T., Wanke, E.: Processing of hierarchically defined graphs and graph families. In: Monien, B., Th., Ottmann (eds.) Data structures and efficient algorithms. LNCS, vol. 594, pp. 44–69. Springer, Heidelberg (1992). https://doi.org/10.1007/3-540-55488-2_21

11. Jantke, K.P.: Layered languages of ludology: the core approach. Diskussionsbeiträge 25, Inst. Medien- und Kommunikationswissenschaften, TU Ilmenau (2006)

12. Jantke, K.P.: The evolution of story spaces of digital games beyond the limits of linearity and monotonicity. In: Iurgel, I.A., Zagalo, N., Petta, P. (eds.) Proceedings of the 2nd International Conference on Digital Storytelling, ICIDS 2009, vol. 5915 of Lecture Notes Computer Science, pp. 308–311, Springer (2009)

13. Jantke, K.P. (2015). https://www.praeventionstag.de/nano.cms/vortraege/begriff/Time-Travel-Prevention-Games?sb=Time+Travel+Prevention+Games. Last accessed 15 March 2021

14. Jantke, K.P., Knauf, R.: Didactic design through storyboarding: Standard concepts for standard tools. In: Baltes, B.R., et al. (eds.) First International Workshop on E-Learning Technologies and Applications, Cape Town, South Africa, pp. 20–25, Computer Science Press, Trinity College, Dublin, Ireland (2005)

15. Jantke, K.P., Spundflasch, S.: Storyboarding pervasive learning games. In: Tan, D. (ed.) Proceedings of International Conference Advanced Information and Communication Technology for Education, ICAICTE 2013, Hainan, China, pp. 42–53, Atlantis Press (2013)

16. Krebs, J., Jantke, K.P.: Methods and technologies for wrapping educational theory into serious games. In: Zvacek, S., Restivo, M.T., Uhomoibhi, J., Helfert, M. (eds.) Proceedings of 6th International Conference Computer Supported Education, CSEDU 2014, Barcelona, Spain, May 2014, pp. 497–502, SciTePress (2014)

17. Lenerz, C.: Layered languages of ludology. In: Beyer, A., Kreuzberger, G. (eds.) Digitale Spiele – Herausforderung und Chance, pp. 39–52. VWH, Boitzenburg, Germany (2009)

18. Stern, A.: Embracing the Combinatorial Explosion: A Brief Prescription for Interactive Story R&D. In: Spierling, U., Szilas, N. (eds.) ICIDS 2008. LNCS, vol. 5334, pp. 1–5. Springer, Heidelberg (2008). https://doi.org/10.1007/978-3-540-89454-4_1

Emotional Intelligence: A Journey Inside the Emotional Life Within an Immersive Interactive Setting

Barbara Bertagni[1,2,3](✉), Roxane Gardner[4,5], Rebecca Minehart[4,6], and Fernando Salvetti[1,2,3]

[1] Centro Studi Logos, e-REAL Labs, 10143 Turin, Italy
`bertagni@logosnet.org`
[2] Logosnet, e-REAL Labs, 6900 Lugano, Switzerland
[3] Houston, TX 77008, USA
[4] Center for Medical Simulation, Boston, MA 02129, USA
[5] Brigham and Women's Hospital/Children's Hospital/Massachusetts General Hospital and Harvard Medical School, Boston, MA 02114, USA
[6] Massachusetts General Hospital and Harvard Medical School, Boston, MA 02114, USA

Abstract. Emotions are an essential part of being human. Whether we experience them deeply, overcontrol them, or let them overwhelm us, emotions have a big influence on our experiences, our performance, and our quality of life. Being emotional intelligent helps us make better decisions, forge stronger relationships, and overcome challenging situations. To improve emotional intelligence, we designed an immersive setting where participants interact with a variety of situations digitally displayed with fully cognitive and emotional involvement practicing empathy, self-awareness, and self-regulation. We work at the intersection of thinking and feeling, increasing participants' awareness about the way the mind works, experimenting with the intensity of the emotions in the body and the impact on behavior. Scientific evidence from neuroscience shows that, as human beings, we are not passive receivers of sensory input, but our brain constructs meaning actively from sensory input, beliefs, past experiences, and culture. In our brains, thoughts and feelings are highly interconnected; we feel what our brain believes, so reflection on emotion is crucial to make sense of our experiences and regulate our feelings. It's not about controlling emotions, but rather about connecting with our emotional life.

Keywords: Emotional awareness · Empathy · Immersive experience

1 Emotional Intelligence

Emotional Intelligence refers to the ability to recognize and regulate our own emotions, as well as our ability to understand other people's emotions and to be able to deal with them empathetically. Being emotionally intelligent doesn't mean that we control and suppress emotions, but rather that we are connected with our emotional life (Fig. 1).

© The Author(s), under exclusive license to Springer Nature Switzerland AG 2023
D. Guralnick et al. (Eds.): TLIC 2022, LNNS 581, pp. 65–68, 2023.
https://doi.org/10.1007/978-3-031-21569-8_6

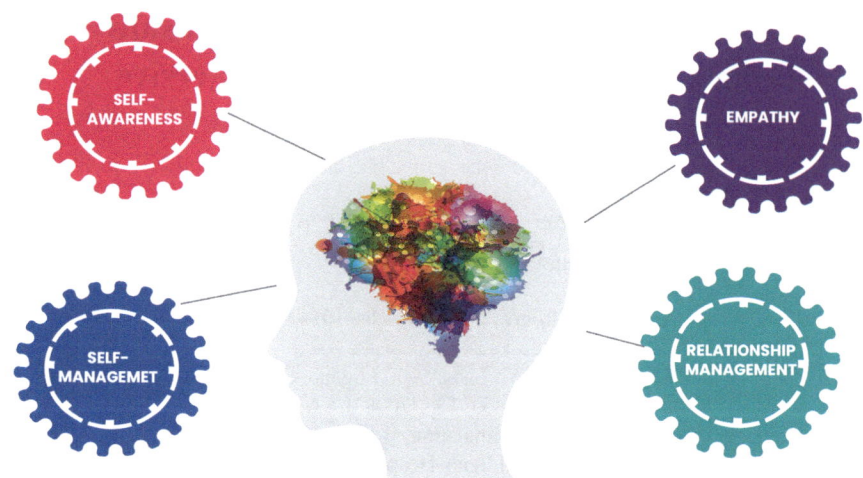

Fig. 1. The four components of emotional intelligence.

Emotions are an essential part of being human, and they have a big influence on our experiences, our performance, and our quality of life. Improving our EI starts by noticing how we feel in different situations. Neuroscientist Lisa Feldman Barrett [1] did interesting research asking over 700 test subjects to keep track of their emotional experiences for weeks or months. The results of this research showed that people vary tremendously in how they differentiate their emotional experiences. In a lot of cases, words like angry, sad, anxious, and afraid were used interchangeably to express an unpleasant feeling, without distinguishing the different shadows of these emotions. The same can be said for pleasant emotions like happiness, calmness, and pride. On the contrary, other people were able to distinguish among different feelings using appropriate words to express them [2–6].

This ability to describe fine-grained emotional experiences is called Emotional Granularity, and it is what makes us emotion experts, able to go beyond general expressions like "awesome" and "bad," using all the shades of emotions: happy, content, thrilled, relaxed, joyful, hopeful, inspired, prideful, angry, afraid, envious, melancholy, etc. At the other end of the spectrum are people who haven't developed an emotional vocabulary and use "sad" and "mad" interchangeably. If we knew only a few emotional concepts, whenever we experience an emotion or perceive someone else as emotional, we'd categorize only with this broad brush.

A collection of scientific studies shows that people with higher emotional granularity are more flexible when regulating their emotions, less likely to drink excessively under stress, less likely to react aggressively against someone who has hurt them, and likely to have a more satisfying life.

2   Immersive Experience to Develop Emotional Granularity

Being aware of our emotions and being able to describe them with their differences and shadows play a strategic role in managing emotions. The question is, how can we improve this competence?

We developed a learning experience in the e-REAL immersive lab where learners are actively involved facing engaging scenarios digitally displayed in a glasses-free mode and being involved in difficult conversations with avatars that engage them both at a cognitive and at an emotional level (Fig. 2).

Fig. 2. The Parkour Extreme scenario.

After each experience, we ask participants to describe the emotions they felt, and the answers are usually very different. Someone might say that there was no emotion in the experience, while others can identify multiple emotions. During the debriefing of this experience, one of the most interesting findings is that the same situation makes people feel very different emotions. For instance, in the Parkour Extreme scenario, one participant says that this situation made him/her scared or disappointed, and at the same time, other participants feel a positive excitement like enthusiasm or amusement. Scenario after scenario, people become more aware of their emotions, and it becomes clear that it's not the situation, but the meaning that we give to the situation that triggers the emotion. This opens an interesting space of introspection, and the debriefing helps with improving the emotional awareness, along with the extension of the vocabulary available to talk about emotions.

How many emotions can you name? Being able to recognize and name our emotional state accurately is the first step to boosting our emotional intelligence. With this goal, another exercise that we do is exploring the way emotions are expressed and named in different cultures. Have you ever heard about abhiman, awumbuk, or gezelligheid? These are only a few examples of words used to express a specific feeling in a specific culture. For instance, the Dutch call gezelligheid when the rain is muzzling, and the damp rises from the canals. It describes the feeling to be comforted in a homey place surrounded by good friends. The more we enlarge our vocabulary, exploring emotions, listening to the ways other people describe their emotions and trying to put in words what seems unspeakable, the more we gain in emotional intelligence. We are not passive receivers of sensory input, but active constructors of our emotions. We build meaning from sensory inputs, past experiences, shared values, and culture. Learning to name the emotions more accurately helps us be more aware of our emotional life and more effective in our behavior.

3 Practicing Empathy

Empathy is the key competence to connecting with other people. It's the capacity to understand the moods of people around us, in an immediate way, without any criticism or judgment. It involves the ability to identify with other people's emotions and situations, even if we are not in agreement with them.

With e-REAL, we help learners practicing empathy by interacting with different recorded or virtual people in the immersive room. We begin working only with the audio, listening to different voices telling different stories; then, we add the video and, in the end, the interaction. Step by step, we give everyone time to really feel the situation and deal with it. The results are astonishing: the opportunity to explore the emotional world, immersing oneself in engaging experiences, really improves the self-awareness and the capacity to deal with challenging situations and people.

Now, because of the pandemic, we have developed the same program online, in our virtual immersive platform, and we are happy about the results. The feedback that we receive most often is that working with this methodology on the emotions is transformational and life changing.

References

1. Feldman Barrett, L.: How Emotions Are Made. The Secret Life of the Brain. Pan Macmillan, London (2017)
2. Watt Smith, T.: The Book of Human Emotions. Profile Books, London (2015)
3. Goleman, D.: Emotional Intelligence. Why It Can Matter more than IQ. Bantam Books, New York (1995)
4. Bachoud-Lévi, A.C., et al.: L'Empathie. Odile Jacob, Paris (2004)
5. Salvetti, F., Bertagni, B. (eds.): Learning 4.0. Advanced Simulation, Immersive Experiences and Artificial Intelligence, Flipped Classrooms, Mentoring and Coaching. Franco Angeli, Milan (2018)
6. Hoffman, H., Vu, D.: Virtual reality: teaching tool of the 21st century? Acad Med. **72**, 1076–1081 (1997)

Skill Scanner: Connecting and Supporting Employers, Job Seekers and Educational Institutions with an AI-Based Recommendation System

Koen Bothmer and Tim Schlippe[(⊠)]

IU International University of Applied Sciences, Erfurt, Germany
`tim.schlippe@iu.org`

Abstract. Usually employers, job seekers and educational institutions use AI in isolation from one another. However, skills are the common ground between these three parties which can be analyzed with the help of AI. Employers want to automatically check which of their required skills are covered by applicants' CVs and know which courses their employees can take to acquire missing skills. Job seekers want to know which skills from job postings are missing in their CV and which study programs they can take to acquire missing skills. In addition, educational institutions want to make sure that skills required in job postings are covered in their curricula, and they want to recommend study programs. Consequently, we investigated several natural language processing techniques to extract, vectorize, cluster and compare skills, thereby connecting and supporting employers, job seekers and educational institutions. Our application *Skill Scanner* uses our best algorithms and outputs statistics and recommendations for all groups. The results of our survey demonstrate that the majority finds that with the help of *Skill Scanner*, processes related to skills are carried out more effectively, faster, fairer, more explainably, and in a more supported manner. In total, 89% of all participants are not averse to apply our recommendation system for their tasks, and 67% of job seekers would certainly use it.

Keywords: Artificial intelligence in education · Upskilling · Recommender systems · Clustering · Natural language processing

1 Introduction

Access to education is one of people's most important assets, and ensuring inclusive and equitable quality education is goal 4 of United Nations' Sustainable Development Goals. This goal should not only refer to general education, but also to specific education in the professional environment. If people have the right education for the professional environment, they have a better chance to get jobs that allow them to have a good life. Unfortunately, there are often still gaps between the skills that are needed in the job market, the skills that job seekers have and the skills that are taught in educational

© The Author(s), under exclusive license to Springer Nature Switzerland AG 2023
D. Guralnick et al. (Eds.): TLIC 2022, LNNS 581, pp. 69–80, 2023.
https://doi.org/10.1007/978-3-031-21569-8_7

institutions like schools, universities, online platforms and massive open online courses (MOOCs) [1].

To solve this problem, all three players—employers, job seekers[1] and educational institutions—need to be aligned. There are already natural language processing (NLP) approaches to extract text data from job seekers' CVs (curriculum vitae), employers' job postings or educational institutions' learning curricula and give recommendations to one of these players. However, this way all three parties use AI in isolation from one another. For example, [2] present a Word2Vec-based [3] system that informs employers how well job seekers' CVs fit job postings. LinkedIn has a system that recommends employers' jobs to job seekers based on their personal profile [4]. [5] investigate how AI-based recommendations help job seekers find study programs based on their profile. [6] also use a combination of knowledge graph and BERT [7] for helping employers find suitable candidates in a corpus of CVs.

Our approach leverages similar NLP methods [8], but it benefits not only one, but all three players involved. Connecting and supporting them all allows the greatest possible exchange of information and satisfies their needs as illustrated in Fig. 1:

(1) Employers want to automatically check which of their required skills are covered by applicants' CVs (*Find and Select*) and know which courses their employees can take to acquire missing skills (*Upskill Workforce*).
(2) Job seekers want to know which skills from job postings are missing in their CVs (*Fit to Demand*) and which study programs they can take to acquire missing skills (*Find Program*).
(3) Educational institutions want to ensure that the skills required in job postings are covered in their curricula (*Fit to Demand*), recommend study programs and advise students (*Advise*).

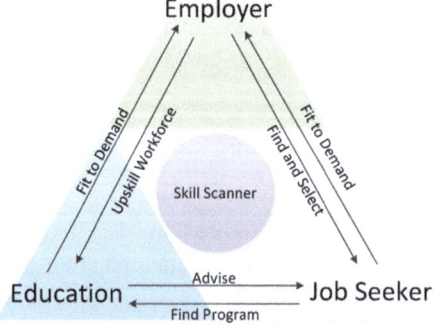

Fig. 1. Connecting and supporting employers, job seekers and educational institutions.

---

[1] The term "job seeker" refers to current applicants and individuals who wish to advance towards a position.

Since skills are the common ground between these three players, we developed the application *Skill Scanner*[2], which combines NLP techniques to extract, vectorize, cluster and compare skills in a pipeline and outputs statistics and recommendations for all three players in form of reports. Our goal was to help employers, job seekers and educational institutions adapt to the job market's needs. Consequently, we used job postings, which represent the job market's needs as reference. Our recommendation system determines which skills in the job market's job postings are covered and which skills are missing. These representative skills, which we draw from a large set of job postings, are referred to as "*market skills*" in this paper.

Since companies hiring data scientists state that it is increasingly difficult to find a so-called "unicorn data scientist" [9], we conducted our analyses using companies' job postings for a data scientist position, job seekers' CVs for that position and a curriculum from a master's program in data science. However, our investigated methods and our final recommendation system can be applied to other job positions, as well.

Finally, we present our detailed analysis of the feedback from employers, job seekers and educational institutions on the reports generated with *Skill Scanner*, demonstrating the potential benefits of finding covered and missing skills with the help of our cluster-based algorithms to all three parties.

2 Related Work

In this section, we will present the latest approaches of recommendation systems for employers, job seekers, and educational institutions.

2.1 Recommendation Systems for Employers

Automatically ranking CVs is a valuable tool for employers. For example, [10] rank candidates for a job based on semantic matching of skills from LinkedIn profiles and skills from their job description, relying on a taxonomy of skills. They determine the semantic similarity of the skills reported in an applicant's LinkedIn profile to the skills required for a job using the node distance—i.e., the distance to the lowest common ancestor in the taxonomy of skills. Recent NLP techniques offer opportunities to improve these methods: [2] use word embeddings from Word2Vec [3] to match CVs to jobs. [6] combine a knowledge graph and BERT to find suitable candidates in a corpus of CVs.

Our system also works with embeddings—with sentence embeddings—to vectorize the skills. In addition, we use a cluster approach to find synonymous skills.

2.2 Recommendation Systems for Job Seekers

Recommendation systems for job seekers have been investigated by [11–13]. As in the systems for employers, text data from social media profiles such as LinkedIn or Facebook is usually processed [5, 14]. Researchers at LinkedIn [15] have built a taxonomy of 35k standardized skills and use semantic matching to measure the similarity in job descriptions and job seekers' profiles.

---

[2] https://github.com/KoenBothmer/SkillScanner.

The benefit of our clustering approach compared to a taxonomy is that our model can pick up new skills without the need to update a taxonomy.

2.3 Recommendation Systems for Educational Institutions

[16] give a systematic review of recent publications on course recommendation. Most related work focuses on recommending courses to potential students. They report a growing popularity of data mining techniques in those systems. To cope with the challenges of different levels of abstraction and synonyms in the course materials and students' documents, some researchers first cluster the content, which they can then compare. K-means is usually used for clustering. To help employers recommend appropriate courses for their employees, [17] suggest a framework called "Demand-aware Collaborative Bayesian Variational Network (DCBVN)."

Compared to the related work, we propose courses for students and employees based on K-means clustering extended with additional steps to detect outliers in the clusters. While the job market is not considered in the recommendation process of other approaches, we use information from employers' job postings—denoted as *market skills* in this paper—as valuable information to enhance our recommendations.

3   Extracting, Vectorizing, Clustering and Comparing Skills

Our goal was to help employers, job seekers, and educational institutions adapt to the *market skills*. Our recommendation system *Skill Scanner* determines which skills in the job postings are covered and which skills are missing. In this section, we will describe our pipeline to extract, vectorize, cluster and compare skills.

3.1 Our Pipeline to Extract, Vectorize, Cluster and Compare Skills

For a certain job position, (1) *Skill Scanner* takes a CV, a job posting or a learning curriculum as input, (2) extracts the skills of the provided document, (3) compares the document's extracted skills to a skill set which represents the market's needs (*market skills*), and (4) returns information of which *market skills* are covered or missing in the provided document compared to the market's needs [8].

To be able to compare the skills in the provided document to the *market skills*, we need to cope with the challenges of different levels of abstraction and synonyms among the skills in the uploaded document and the *market skills*. Consequently, we apply the following 4 steps when we gather the *market skills* and when we upload a document to be analyzed which are visualized in Fig. 2:

1. *Skill Extraction*: Extract skill requirements.
2. *Vectorization*: Map skill requirements to a semantic vector space, where skills with similar meanings are closer together and skills with different meanings are farther apart.
3. *Clustering*: Cluster skill requirements to cope with the challenges of different levels of abstraction and synonyms.

4. *Comparison and Analysis*: Compute intersections among the skill sets of the provided documents and the *market skills* and visualize recommendations based on covered and missing *market skills*.

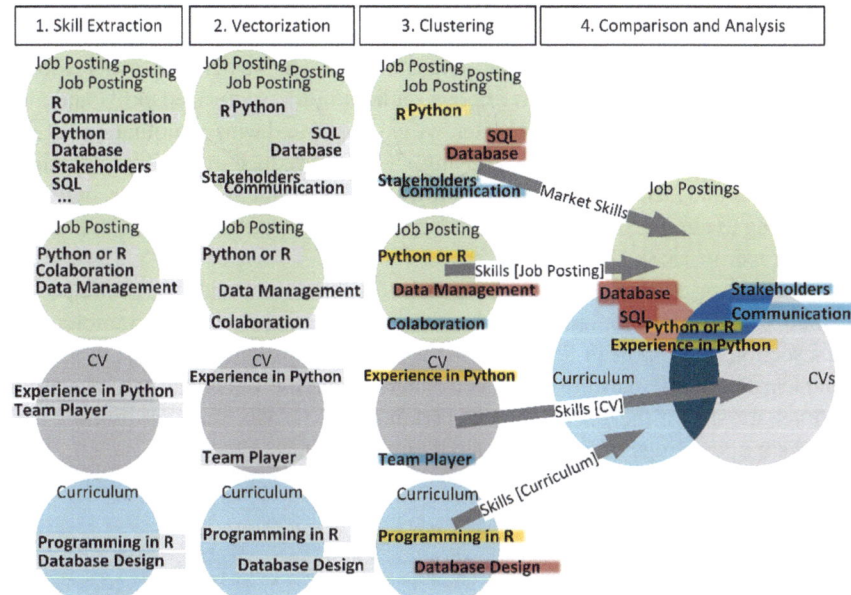

Fig. 2. Our pipeline to extract, vectorize, cluster and compare skills.

3.2 Implementation

In job postings, CVs and learning curricula, skills are usually expressed in bullet points. Therefore, in step 1 *Skill Extraction*, we developed keyword- and rule-based techniques to extract bullet points from these sources. For job postings, we used the BeautifulSoup package [18] to gather and extract 21.5k bullet points from 2,633 job postings for data scientists in English from Indeed.com and Kaggle.com. In this work, we refer to this representative set of skills as the *market skills*. Since some bullet points in a job posting are not skill requirements, we analyzed methods to deal with outliers that are not skill requirements in step 3 *Clustering*.

Like in [2, 6], we experimented with word embeddings to vectorize the skills in step 2 *Vectorization*. To represent the skills which usually consist of several words, we investigated stacking and averaging the word embeddings in a skill after they were produced with Word2Vec [3] and GloVe [19]. In addition, we explored sentence embeddings. As Bidirectional Encoder Representations from Transformers (BERT) [7] models are successful in NLP tasks, we also experimented with Sentence-BERT [20], a modification of the pre-trained BERT transformers. Sentence-BERT (44.2%) outperformed word embedding like GloVe (39.5%) by 12% in Silhouette score [21] at the end of our pipeline.

The benefit of our clustering approach compared to a taxonomy like in [17] is that our model can pick up new skills without the need to update a taxonomy. While hierarchical clustering approaches have not proven to be robust against outliers [22], K-means clustering has been successfully used in clustering word embeddings [23] and is adaptable and scalable [24]. Consequently, we used K-means to cluster our 768-dimensional vectors with the cosine distance as the distance metric. K was chosen as 31 with the highest Silhouette score of 44%. To remove outliers in the vectorized skills and allow our clustering techniques to perform better, we experimented with combinations of PCA [25], UMAP [26] and DBSCAN [27]. Using UMAP to reduce the vectorized skills to two dimensions and DBSCAN to remove outliers in the 2-dimensional space, performed best according to our manual checks and reduced the 21.5k potential skills retrieved with our web scraper to 18.8k skills.

After retrieving clusters and vectors representing the skill of each cluster, we perform mathematical operations to find covered and missing *market skills*. For example, Fig. 3 shows a section of a report in *Skill Scanner* where overlaps between skills in job postings and learning curricula were calculated. There, using the visualizations, educational institutions are shown the importance of certain skills in data scientist profiles along with the lack of coverage of those skills in their curricula.

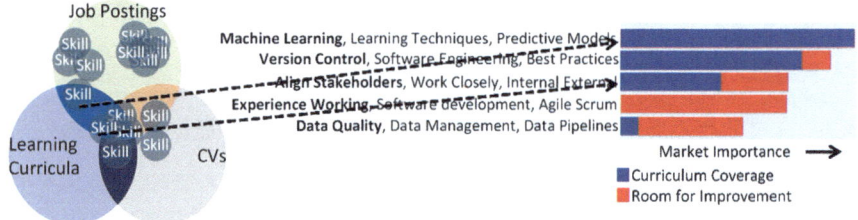

Fig. 3. Skill comparison and analysis between job postings and learning curricula.

4   Reports for Employers, Job Seekers and Educational Institutions

In this section, we will present reports for employers, job seekers and educational institutions which *Skill Scanner* outputs after analyzing the skills of the provided document and the *market skills*.

4.1   CV-Market Report

Figure 4 demonstrates an excerpt of the *CV-Market Report* which is generated if *Skill Scanner* receives a CV as input given the *market skills*. Alternatively, the report can be generated when a CV and a single job posting are provided and then compared. The *CV-Market Report* visualizes the coverage and importance of the skills in the CV which supports (1) job seekers to find out which skills are still missing and which are already covered in their CV to be used for the application for a desired position, and supports (2) employers to find out which skills the applicant is still missing and which are already covered when applying for an advertised position.

In case of a single provided job posting, the bars show exclusively the skills specified in this provided job posting. Otherwise, the bars show all *market skills*. In both cases, the skills are sorted by importance in the *market skills*, which is represented by the bar lengths. Each skill is described by the 3 most frequent bigrams used in the *market skills*. The blue part in each bar demonstrates how well the skills in the provided CV match, whereas the red part shows how much is missing indicating the room for improvement. This representation of the skills with the bigrams, the coverage and the room for improvement is used consistently in all reports.

Fig. 4. CV-Market Report.

Technically, the bigrams at each bar on the y-axis are the most common bigrams that are located in a *market skill* cluster gained in step 3 *Clustering* of *Skill Scanner*'s pipeline (see Fig. 2). How well a skill in the provided CV matches a skill in the job posting or in the *market skills* is determined by the distance of the skill's vector specified in the CV to the centroid of the cluster.

4.2 CV-Curriculum Report

Figure 5 demonstrates an excerpt of the *CV-Curriculum Report*, which displays learning modules that best cover the skill gaps of an input CV given the *market skills* and a set of learning modules. The *CV-Curriculum Report* supports (1) job seekers to find a study module for the targeted expansion of skills with regard to a desired job position, (2) employers to find study modules for the targeted upskilling of employees, and (3) educational institutions to attract and advise students.

4.3 Curriculum-Market Reports

Figure 6 shows an excerpt of the *Curriculum-Market Report* which is generated if *Skill Scanner* receives a learning curriculum as input given the *market skills*. The *Curriculum-Market Report* displays the coverage and importance of the skills in the curriculum and the *market skills*, which helps educational institutions adapt the taught content with regard to the skills required in the job market.

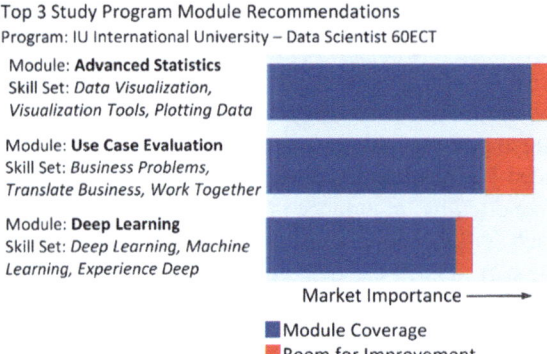

Fig. 5. CV-Curriculum Report.

Fig. 6. Curriculum-Market Report.

4.4 CV-CVs Report

Figure 7 demonstrates an excerpt of the *CV-CVs Report*, which is generated if *Skill Scanner* receives a CV given other CVs, a job posting, and the *market skills*. The *CV-CVs Report* visualizes a comparison of skill coverage in one CV to skills in other CVs, which supports employers to select the best candidate from a group of applicants by processing each CV and comparing the scores.

How well a skill in the provided CV matches a skill in the job posting is determined by the distance of the skill's vector specified in the CV to the centroid of the skill cluster gained in step 3 *Clustering* of *Skill Scanner*'s pipeline (see Fig. 2). The average of these distances determines the applicant's score. The score is computed for each CV, then the CVs are ranked by their scores.

Fig. 7. CV-CVs Report.

5 Feedback from Employers, Job Seekers and Educational Institutions

In this section, we describe the design and results of our survey, in which we asked for feedback on our reports.

Table 1. Overview of which reports were shown in which questionnaire.

Job seekers	CV-Curriculum Report	CV-Market Report
Employers	CV-Curriculum Report	CV-CVs Report
Educational institutions	CV-Curriculum Report	Curriculum-Market Report

5.1 Experimental Setup

As described, *Skill Scanner* receives a set of skills from a document, compares it to the job market's demands and returns reports based on the input document. To figure out if with the help of these reports processes related to skills are carried out more effectively, faster, fairer, more explainably, and in a more supported manner, we analyzed the feedback on the reports with 3 questionnaires —1 questionnaire for representatives of job seekers, 1 questionnaire for representatives of employers and 1 questionnaire for representatives of educational institutions. Table 1 gives an overview of which reports were shown in which questionnaire.

In each questionnaire, we asked questions about the reports presented. The participants evaluated most questions with a score. The score range follows the rules of a forced choice Likert scale, which ranges from (1) *strongly disagree* to (5) *strongly agree*. Each questionnaire was designed in English and translated to Dutch and German. In total, 108 participants (54 female, 54 male) filled out our questionnaires. Most of them live either in the Netherlands (68.52%) or in Germany (28.71%). Participants were evenly distributed among the three user groups: 33 stated that they were job seekers (30.56%), 36 reported working in the human resources department of employers (33.33%) and 39 reported working for an educational institution (36.11%).

5.2 Effectiveness

First, we asked if the participants agree that *Skill Scanner* would help performing their skill-related tasks more effectively. Figure 8a illustrates the feedback of our 36 representatives of *employers*, 33 representatives of *job seekers* and 39 representatives of educational institutions (*education*). While *effectiveness* was rated best with 4.09 (*agree*) on average by *job seekers*, it was rated with 3.56 (between *neutral* and *agree*) by *education* and 3.42 (between *neutral* and *agree*) by *employers*. The feedback from the *job seekers* is 20% better than from *employers* and 15% better than from *education*.

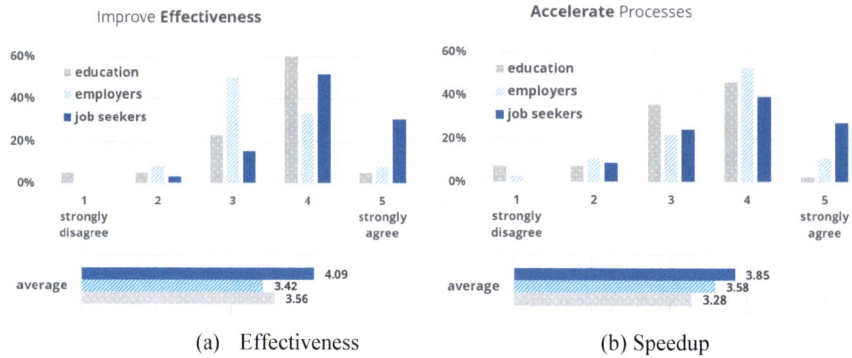

(a) Effectiveness   (b) Speedup

Fig. 8. Feedback on *effectiveness* and *speedup*.

5.3 Speedup

Figure 8b shows our results in relation to the potential speedup of skill-related tasks with *Skill Scanner*. The feedback from *job seekers* is again the best with 3.85 on average (*agree*). This time the feedback from *employers*, with an average of 3.58 (between *neural* and *agree*), is better than that from *education* with 3.28 (better than *neutral*). The feedback from *job seekers* is 8% better than from *employers* and 17% better than from *education*.

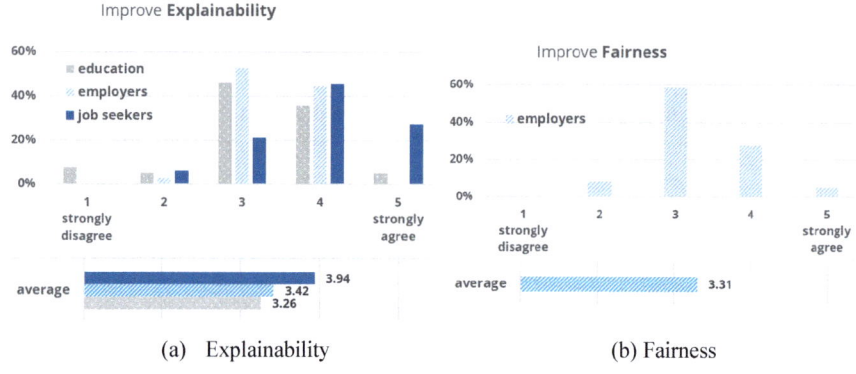

(a) Explainability   (b) Fairness

Fig. 9. Feedback on *explainability* and *fairness*.

5.4 Explainability

Then we asked the participants if they agree that *Skill Scanner* helps explain strengths and weaknesses in CVs and curricula. The results are demonstrated in Fig. 9a. While the feedback from *job seekers* is again rated highest with 3.94 (*agree*) on average, *employers* rated with 4.42 (between *neutral* and *agree*) and *education* with 3.26 (*neutral*) on average. The feedback from *job seekers* is 15% better than from *employers* and 21% better than from *education*.

5.5 Fairness

To find out if *Skill Scanner* can contribute to a fairer selection of applicants based on the analyzed CVs, we asked only the employers if they agree, since their HR departments look through the CVs. The feedback is visualized in Fig. 9b and is a little better than *neutral* with an average of 3.31. *Skill Scanner*'s impact on fairness is not as great as for the other aspects we asked about, since sympathy and soft skills also play a role in hiring an applicant fairly.

5.6 Usage

In total, 89% of all participants are not averse to apply our recommendation system. As with all other questions, job seekers were the most agreeable respondents when asked about usage. 67% of job seekers would certainly use *Skill Scanner*. This might be explained by the ease with which job seekers could adopt *Skill Scanner* in an application process. For employers and educational institutions, the introduction of our recommendation system would mean that they would have to change many processes, which is why they are likely to be more critical.

6 Conclusion and Future Work

The labor market dictates what job seekers should learn, and educational institutions should teach. Therefore, *Skill Scanner* processes skills in job postings, CVs and curricula and outputs recommendations for employers, job seekers, and educational institutions based on present and missing skills and their importance to employers. *Skill Scanner*'s reports were shown to 108 representatives of our 3 parties in a survey. The majority finds that with our system, skill-related processes more effective, faster, fairer, more explainable, more autonomous and performed in a more supported manner. After these initial estimates of *Skill Scanner's* potential, further analysis could include measuring time and cost savings. Future work may also be to apply our pipeline to other job positions and expand it to other domains. In addition, we would like to extend *Skill Scanner* with further reports, e.g., based on a comparison of job posting and *market skills*, which is easily possible with our clustering pipeline.

References

1. Palmer, R.: Jobs and skills mismatch in the informal economy. 978–92–2–131613–8 (2017)
2. Fernández-Reyes, F.C., Shinde, S.: CV Retrieval system based on job description matching using hybrid word embeddings, Computer Speech & Language, **56** (2019)
3. Mikolov, T., Chen, K., Corrado, G., Dean, J.: Efficient estimation of word representations in vector space. ICLR (Workshop Poster) (2013)
4. Geyik, S.C., Guo, Q., Hu, B., Ozcaglar, C., Thakkar, K., Wu, X., Kenthapadi, K.: Talent search and recommendation systems at LinkedIn: Practical challenges and lessons learned. SIGIR (2018)
5. Guruge, D.B., Kadel, R., Halder, S.J.: The state of the art in methodologies of course recommender systems—a review of recent research data, **6**(2), 18 (2021)

6. Wang, Y., Allouache, Y., Joubert, C.: Analysing CV corpus for finding suitable candidates using knowledge graph and BERT. DBKDA (2021)
7. Devlin, J., Chang, M.-W., Lee, K., Toutanova, K.: BERT: Pre-training of deep bidirectional transformers for language understanding (2019)
8. Bothmer, K., Schlippe, T.: Investigating natural language processing techniques for a recommendation system to support employers, job seekers and educational institutions. In: The 23rd International Conference on Artificial Intelligence in Education (AIED) (2022)
9. Baškarada, S., Koronios, A.: Unicorn data scientist: The rarest of breeds, Program: Electronic Library and Information Systems, **51**(1), pp. 65–74. (2017)
10. Faliagka, E., et al.: On-line consistent ranking on e-recruitment: Seeking the truth behind a Well-Formed CV. Artif. Intell. Rev. **42**, 515–528 (2014)
11. Si-ting, Z., Wenxing, H., Ning, Z., Fan, Yang: Job recommender systems: A survey. ICCSE (2012)
12. Hong, W., Zheng, S., Wang, H., Shi, J.: A job recommender system based on user clustering. J. Comput. **8**, 1960–1967 (2013)
13. Alotaibi, S: A survey of job recommender systems. Int. J. Phys. Sci. (2012)
14. Diaby, M., Viennet, E., Launay, T.: Toward the next generation of recruitment tools: An online social network-based job recommender system. ASONAM (2013)
15. Li, J., Arya, D., Ha-Thuc, V., Sinha, S.: How to get them a dream job? Entity-aware features for personalized job search ranking. SIGKDD (2016)
16. Deepani B. Guruge, Rajan Kadel, Sharly J. Halder: The state of the art in methodologies of course recommender systems—a review of recent research. data **6**(2): 18. (2021)
17. Wang, C., Zhu, H., Wang, P., Zhu, C., Zhang, X., Chen, E., Xiong, H.: Personalized and explainable employee training course recommendations: A bayesian variational approach. ACM Trans. Inf. Syst. (2021)
18. Hajba, G.L.: Using beautiful soup. In: Website Scraping with Python. Apress (2018)
19. Pennington, J., Socher, R., Manning, C.D.: GloVe: Global vectors for word representation. EMNLP (2014)
20. Reimers, N., Gurevych, I.: Sentence-BERT: Sentence embeddings using siamese BERT-networks, EMNLP-IJCNLP (2019)
21. Rousseeuw, P.J.: Silhouettes: A graphical aid to the interpretation and validation of cluster analysis. Comput. Appl. Math. **20**, 53–65 (1987)
22. Rani, Y., Rohil, H.: A study of hierarchical clustering algorithm. Int. J. Inf. Comput. Technol. (Vol. 3, Issue 10) (2013)
23. Zhang, Y., et al.: Does deep learning help topic extraction? A kernel K-means clustering method with word embedding. J. Informet. **12**(4), 1099–1117 (2018)
24. Lloyd, S.P.: Least squares quantization in PCM. Techn. Report RR-5497, Bell Lab (1957)
25. Pearson, K.: On lines and planes of closest fit to systems of points in space. Phil. Mag. **2**(11), 559–572 (1901)
26. McInnes, L., Healy J.: UMAP: Uniform manifold approximation and projection for dimension reduction. ArXiv, abs/1802.03426 (2018)
27. Ester, M., Kriegel, H.P., Sander, J., Xu, X.: A Density-based algorithm for discovering clusters in large spatial databases with noise. KDD. AAAI Press, 226–231 (1996)

Creating Affective Collaborative Adult Teams and Groups Guided by Spiral Dynamic Theory

Lisa R. Brown[✉] [ID], Pamela McCray, and Jeffery Neal

University of the Incarnate Word, San Antonio, TX 78209, USA
lrbrown5@uiwtx.edu

Abstract. The field of Adult and Continuing Education caters its teaching and learning to adults who are 25 years of age and older. This group brings to the higher education environment a unique set of skills and life experiences that require pedagogical delivery that is innovative and motivating. For example, older adults (who are often technology adverse) enter the higher education space as graduate students with reservations due to perceptions of disconnected and impersonal learning (e.g., online learning management systems (LMS) platforms). This proposal offers Spiral Dynamic Theory (SDT) as an instrument of course design scaffoldi2, ng for contemporary graduate-level courses that integrate technology, embodied learning, and memetic ways of knowing. Social constructivist worldviews aid in delivering this innovative learning that facilitates adult development through cultural diversity, student group collaborations, and team-based cooperation learning strategies. SDT is a theoretical framework of evolving psychosocial adult development using a color-coded mnemonic of hierarchical paradigms and worldview constructs identifying similarities and differences in human thinking. SDT helps adults recognize the deep-value systems at play within the group dynamics such that distributive leadership and interpersonal effectiveness for meeting collective goals are optimized in both academic and work environments. Ideally, adult learning progression moves from a simplistic to more complex neuropsychology and problem-solving capacity. Memes (i.e., units of culture) are negotiated among group members who pursue collaborative team goals and achievements. However, there is the potential for progression, entrenchment, and/or regression of thinking with the open-ended SDT framework serving as an interpretive guide to advance innovation.

Keywords: Adult development · Affective adult pedagogy · Collaborative teams · Experiential learning · Spiral dynamic theory

1 Introduction

Learning environments and concepts are integrally connected to how adults engage and problem-solve within social contexts and organizations. Marsick and Watkins [43] emphasize the importance of innovation-and-organizational knowledge perspectives that give attention to how adults store, retrieve, and manage knowledge through "finding ways to harvest the tacit knowledge embedded in routines and processes" (p. 62). Hence,

© The Author(s), under exclusive license to Springer Nature Switzerland AG 2023
D. Guralnick et al. (Eds.): TLIC 2022, LNNS 581, pp. 81–96, 2023.
https://doi.org/10.1007/978-3-031-21569-8_8

learning organizations made up of individuals are "socially created," these collections of people work together, playing a primary role in changing how they respond to challenges within and outside of any institution [42, p. 18]. Learning for adults also includes the cultural perspectives that position them to utilize organic processes to make meaning as they co-create knowledge. By introducing the concept of culture, we can more deeply examine how it informs, facilitates, or impedes learning and team building. Deep analysis of technology's role in higher education and the convergence of changing demographics require optimal approaches and strategies for adult learners and their psychosocial development [8].

1.1 Reimagining Concepts in Adult Learning

This conference presentation and article explores and deconstructs each presenter's interpretative experiences relative to the role of traditional adult learning philosophical perspectives and theories juxtaposing them to more contemporary emergent theoretical frameworks (e.g., Spiral Dynamic Theory). We engage with the concept of complexity and diversity as we operationalize *affective*[1] collaboration strategies for teams of adult learners, taking into consideration the developmental change that occurs in the process. We privilege the social context of blended online learning in this discussion, focusing on teaching and learning experiences in graduate education [37]. The examinations of how adult educators use innovation in their delivery of pedagogy and praxis while considering how autonomy (self-directed learning) and collaboration are essential methods for team and group work [23, 32] among adults is discussed.

Additionally, we recognize that the inclusion of *life experience* and embodied ways of knowing have increasingly become critical aspects of how adults want to learn in innovative ways [59]. Due to excessive technology, disconnections in adult education are an essential gap in the literature that merits further investigation and deconstruction. The COVID pandemic has been a mitigating variable requiring that in-person learning—commonplace in adult education—be modulated downward as remote instruction becomes more pronounced. Consequently, many adults found themselves thrust into a learning space, possessing low to no digital literacy (including some college and university instructors). For example, the use of LMSs can be met with technology aversion among older lifelong learner groups [44, 52]. The sudden use of technology became the primary mode of pedagogy delivery within a group (i.e., adults ages 25 and over) who have traditionally been averse to remote learning. The LMS digital learning modality—absent an in-person facilitator—combined with diminished opportunities for the face-to-face qualitative collaborative experiences that in-person instruction allows [50] are highlighted in our presentation.

Crentsil, Gschwandtner, and Wahhaj [15] provide valuable insight into how technology aversion among adults might be overcome when examining its impact on small-scale farmers who were more influenced by issues of ambiguity associated with risk-taking

---

[1] *Affective* collaboration and strategies, for our purposes, involve moods, feelings, and attitudes that can emerge during team/group student work due to the diverse SDT ontological worldview(s) of the adult learner(s) and the social context that locates the graduate-level education.

in their use of technology. The ambiguity relative to new farming processes became a primary variable that inhibited the Ghanian farmers from adopting innovative technologies. This suggests that cognitively meeting adults at their level of complexity relative to problem-solving is crucial for integrating technology into their daily life and work.

Piaget [49] developed a stage theory to provide models for cognitive development—primarily among children—finding that as one matures, new capacities for problem-solving and knowledge acquisition allow for more complex and accurate real worldviews. However, the final stage of his theory termed formal operational thinking served as the beginning of adult thinking. Neo-Piagetian paradigms [51] built upon the concept of formal operational thinking, leading to a more emergent understanding of adults' thinking and developmental problem solving over time. In making a case for what Piaget believed, von Glasersfeld [60] describes the theorist's model of human existence as that of *Radical Constructivism*. The neo-Piagetian held that Piaget's theory of knowledge required a reorganization of ideas about knowledge stating:

> It is not a question of merely adjusting a definition here and there or rearranging familiar concepts in a somewhat novel fashion. The change that is required is of a far more drastic nature. It involves the demolition of our everyday conception of reality and, thus, of everything explicitly or implicitly based on naïve realism; it shakes the very foundations on which 19th-century science and most 20th-century psychology have been built [58, p. 95].

Adult learning and development scholarship identify linkages between how well adults can navigate complexity and ambiguity using models of hierarchical cognitive thinking [6, 8, 9, 11, 25]. Moreover, the evidence suggests that dynamic changes in the ability to problem-solve among adults require a forward change in their psychosocial realities [3, 26]. This change directly results from meeting particular conditions (Table 1) that lead to higher-order thinking and consolidation of change in SDT worldview(s).

1.2 Spiral Dynamic Theory

The theoretical framework guiding our conference proposal and this conceptual article is based upon the idea of expanding levels of human existence—termed Spiral Dynamic Theory (SDT)—that direct the thinking and problem-solving capabilities among individual adults, organizations, and larger societies [3]. We use this knowledge to inform our delivery of adult learning pedagogy to groups of graduate students in higher education.

SDT has its origins in the scholarship of Clare W. Graves [25, 26, 28]—a psychologist and Professor Emeritus—who retired in 1978 from Union College in New York. The theory advanced by Graves was initially described as his Emergent.

Cyclical Level of Existence Theory (ECLET), grounded in his interdisciplinary research that bridged the adult biopsychosocial systems disciplines at Union College. Graves contended that adult learning and development was an open-ended system of emergent thinking and evolving worldview constructs [9, 25, 28]. However, he was initially discouraged about the ECLET due to witnessing a colleague (Abraham Maslow) being *torn to pieces* about his theory on the hierarchy of human needs due to underdeveloped data at an APA seminar in the 1950s [28]. Graves vowed he would never subject

Table 1. Six stages are required for SDT worldview level movement.

The chronological SDT change states	Conditions needed for problem resolution
1. Change potential (open/closed)	The change potential for the adult must be met through the acquiring of new insights. Some adults will remain resistant to change, becoming static in their ability to problem-solve or advance in ontological reality
2. Solutions	The problem is recognized cognitively, and an understanding of a need to address the problem is made clear
3. Dissonance	The disturbance generated by the solution stage subsequently triggers a regressive movement in how to solve the problem
4. Insight	New clarity and insight are introduced into the thinking about the problem that halts the regression movement as strong enlightenment points that facilitate change are realized
5. Barriers are removed or neutralized	Non-interference for the change to occur or properly timed aid is provided to the adult leading to assistance with overcoming the problem
6. Consummation of change	The adult experiences an SDT worldview level jump to consolidate the change in thinking and resolve the problem

Note During the consummation of change (Step 6), quantum leaps can occur in the SDT worldview typologies stages, disrupting the normal progression of change and problem-solving ability. However, such leaps are very unstable and typically result in a regression downward to a more optimal SDT worldview level change stage [7, 28] .

himself to such humiliation and have his research be strongly maligned in the field due to weak empiricism. Hence, he retained several incomplete manuscripts until he died in 1986. Therefore, Grave's scholarship was not as highly published and implemented until it was taken up and shepherded by his protégés in the late 20th century [3]. Nevertheless, Graves [28] used the rejection experienced by Maslow to fine-tune his theory. He later goes on to state that:

Maslow came around to my point of view. If you look at some of his later writings, you will see that he accepted both the cyclical idea that there are more than one kind of expressive system and more than one kind of belonging system and that the systems were open-ended [27, p. 52].

As Grave's protégés Don Beck and Chris Cowan continued to advance his scholarship—the latter of whom mentored and trained this author in the use and application of SDT—Brown [6, 9] has refined several of the worldview constructs. The resulting evolution of the prior frameworks helps make the application of the emergent model (offered in this presentation) useful in examining graduate students' thinking and values-based responses in connection to cultural diversity and even more specifically civic engagement.

Memes the Transferable Units of Culture. Although Clare W. Graves [26] never explicitly used the term *meme*, a unit of culture that transfers from person to person through non-genetic human imitation [16], his conceptualization of the term was present in his research on how thinking and problem-solving capacities evolved from one stage (worldview construct) to the next on the ECLET spiraling framework. His protégés later advanced the theory of the movement up the framework as a dynamic spiral of value memes (i.e., ^{V}MEME^2) of increasing cognitive complexity [3]. Additionally, movement along the framework is induced by problem resolution, as shown in Table 1.

Hierarchical and Evolving Worldview Constructs. Brown [6, 8] provides some descriptive details about the unique worldview constructs that exist in the framework. The diagram (Fig. 1) uses a mnemonically color-coded system to distinguish each worldview construct that begins with low-order simplistic thinking and then progresses upward in a zig-zag fashion toward more complex taxonomies of higher-order thinking and problem-solving abilities.

The graduate students are provided multiple in-class lessons to become familiar with Spiral Dynamic Theory and its taxonomy. Then they are asked to self-assess their positionality on the SDT framework using case study scenario exercises to guide their evaluations. Throughout the semester, the instructor makes continuous nonformal assessments by observing the student group collaborations. These interpretations will assist the course facilitator to identify optimal project team recommendations that are concomitant with the emergent SDT model. The horizontal worldview constructs (Fig. 1) experience the most conflict while the thinking among the immediately vertical neighboring system tends to maintain the most harmonious collaborations. The following descriptions are associated with each of the color-coded ^{V}MEME with conceptions of teal and coral currently at the hypothesis stage of investigation.

- *Basic Instinctive*. The Beige-colored A/N coded *value meme* system is representative of a precultural existence. Wherein the case of the individual, he would possess low self-awareness. The thinking is driven by physiological imperatives and the simple purpose of staying alive [4, 9]. The locus of control is internal, and the thematic orientation is individualistic.
- *Magical Mystical World*. The Purple-colored B/O coded *value meme* system is concerned with safety and security through kinship ties. There exists in this worldview system the belief in obeying the desires of magical-mystical spiritual beings and divine authority figures (e.g., priests, Shamans, tribal elders). The purple-coded is marked by fantastical thinking, tribalism, and traditionalism. The locus of control is external, and the thematic orientation is collectivist.

---

^{2} ^{V}MEME(s) represent ontological Spiral Dynamics organizing principles and values embedded within a taxonomy of emergent worldviews and hierarchical cognitive problem-solving abilities. The ^{V}MEME is mimetically influenced and therefore functions as a type of meta-meme within the SDT framework [9].

SDT Constructivist Theoretical Framework Model

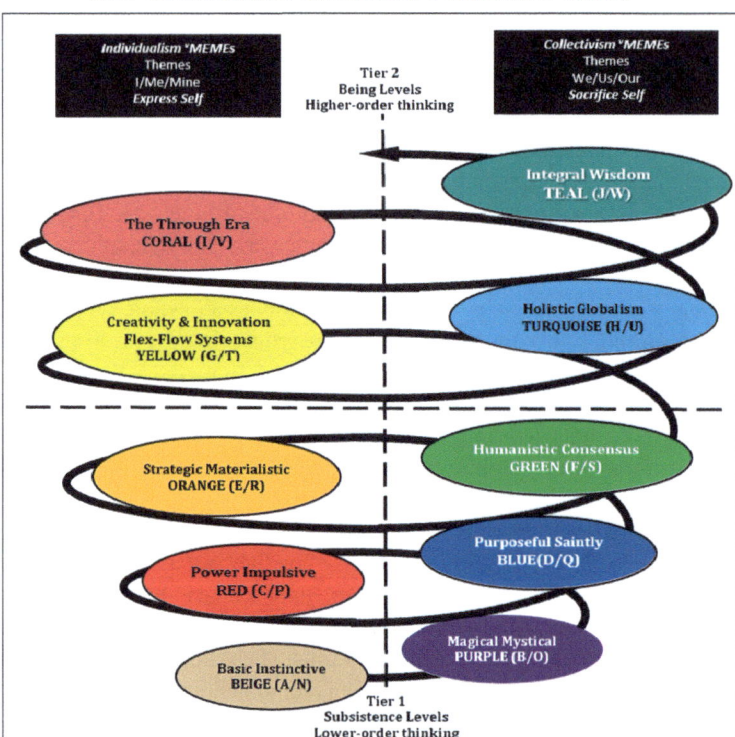

Fig. 1. This figure displays the oscillating zig-zag movement up the spiral representing the changes in SDT worldview constructs, thinking, and problem-solving abilities (reproduced with permission [6]).

- *Power Impulsive.* The Red-colored C/P coded *value meme* system is the egocentric memetic worldview often marked by imperatives of domination by force and intimidation. It is represented by thinking and actions that are impulsive. The perception in thinking holds that life is a jungle where there exist the haves and the have-nots. The driving force is to be among the haves. The locus of control is internal and individualistic.
- *Purposeful and Saintly Living.* The Blue-colored D/Q coded *value meme* system represents honor and a good versus evil memetic worldview. This thinking believes that we are assigned to a specific place in life and must accept our lots in life as predestined. This thinking can be marked by dogmatic absolutism where calls for sacrifice are urged to bring order and stability to a chaotic space. The thinking can be paternalistic and authoritarian with goals for obtaining the betterment of everyone. Rules are to be followed and are often non-negotiable. The locus of control is external and collectivist.
- *Strategic Materialism.* The Orange-colored E/R coded *value meme* system is marked by manipulation and a memetic worldview that values autonomy and independence. Winning and competition are prevailing values of this system. Using scheming tactics

and cunning strategies to obtain desired outcomes is a driving force. Gaining high achievements that brings praise and material possessions is viewed as the most optimal reward. The thinking is grounded in logic and a reasonable calculated certainty for success. The locus of control is internal and individualistic.

- *Humanistic Consensus.* The Green-colored F/S coded *value meme* system F/S is the relativistic memetic worldview that believes in human dignity and consensus building in contrast to religious edicts. There is an effort to explore the personal *inner self* in cooperation with the inner self-discovery also being made by others. The thinking in this value meme system prioritizes the life of community, unity, and harmony, as the group seeks to share societal resources for the benefit of all. The locus of control is external and collectivist.

- *Creativity and Innovation/Flex-Flow Systems.* The Yellow-colored G/T coded *value meme* system represents the first of the more complexity-oriented constructs (i.e., Tier 2 worldviews). In the literature, the yellow value meme represents a departure from the preoccupations of subsistence living found with the Tier 1 constructed thinking. Yellow marks the first worldview of Tier 2 *being levels* where the problems of subsistence living are clearly understood even though they are not necessarily under control [9] The SDT yellow-coded value meme thinking welcomes paradoxes and uncertainty. It accepts a flex-flow [9] perspective that appreciates the layered dimensions of human nature and societies. The locus of control is internal and individualistic.

- *Holistic Globalism Dichotomies.* The Turquoise-colored H/U coded *value meme* thinking represents the second of the Tier 2 constructs. This system reflects the globalism memetic worldview marked by its ability to more easily negotiate complexity and recognize patterns more immediately than seen operating under the prior lower six, Tier 1 ᵛMEMEs. Its priorities include the pursuit of the good for all living things, and the thinking views the world as a single dynamic organism possessing its own type of independent human energy of *mind.* Turquoise thinking is more driven by purpose versus achieving harmony with other persons. The locus of control is external and collectivist.

- *The Through Era.* The Coral-colored I/V coded *value meme* system reflects the worldview that privileges the use of perspectives that merge human biology and technology, thinking it to be the most optimal approach to adult existence. At this level, adults have advanced in their problem-solving abilities as they are no longer burdened by subsistence living. As a result, there is a freeing up of the capacity for higher-order thinking to engage more complex problems of the human condition. This construct is among the last two value meme constructs currently included in the Brown [7] SDT model and is in developing formulation. The research limitation for the Coral and Teal constructs is constrained due to the impediments for the author to secure access to a sufficient sample of individuals holding to these types of adult thinking patterns. Coral thinking represents "a secular vision of unlimited technoscientific progress" where notions of God and faith are integral [63, p. 67]. The locus of control is internal and individualistic.

Integral Wisdom and Human Insight. Like the prior Coral-colored I/V coded *value meme* system, the Teal-colored J/W coded value meme system represents a construction based upon theoretical propositions by the author using interpretive observational pilot research and adult development literature [19, 20]. Note that at the Tier 2 level of the SDT framework, the differences between collectivist orientations juxtaposition individualistic worldview orientations begin to become less pronounced. The tensions and disagreements between the horizontal conflict patterns among the adults engaged in Tier 1 level thinking led to more significant stagnation concerning problem-solving between diametrically oppositional perspectives [3, 9]. However, the Tier 2 teal-colored J/W system departs more gently from the prior Coral-colored systems of adult thinking introducing a more tempered reliance on technology and mechanization to resolve complex human problems. The locus of control is external and collectivist.

2 Review of the Literature

Constructivism theory views learning as constructing new knowledge that includes social, cultural, and experiential components with early roots dating back to Socrates and his teaching method of allowing his followers to challenge their ideas [2, 42]. Jean Piaget is attributed with developing the modern theory of Constructivism; however, many well-known contributors, such as John Dewey, Lev Vygotsky, and Seymour Papert, have also informed the theory's development. Constructivism purports that learners interpret new information through personal, contextual lenses and build upon existing knowledge and previous experiences to construct a new understanding of their world [10, 12]. Constructivism centers on the learner's characteristics of self-direction and experience to create new knowledge. It reflects real-life problem solving and supports the general principles of adult education and lifelong learning [13, 47].

2.1 Constructivism Informs Student Engagement

Constructivism involves student engagement in knowledge construction activities such as project-based learning, group discourse, peer learning, and team collaboration to support the application and transfer of knowledge [10, 13, 44, 47]. Along with the cognitive processes for new knowledge creation, teacher-led activities and student-centered learning are critical theory components [2, 10, 13, 44]. Vygotsky's socio-culture theories also contribute to social and cultural influences on learning.

Constructivism views learning as an organic process, constantly changing based on new information or experiences, complementing aspects of Spiral Dynamic Theory. In educational research, technology use has been stressed for many years, especially in problem-based learning to help build higher-order thinking and apply the new skills and knowledge to real-world situations [12, 38, 40]. The remote learning constraints imposed by the COVID-19 pandemic helped fuel technology innovation in adult education. Constructivist pedagogies were incorporated into remote adult education using digital platforms. Progressive instructors introduced new applications, such as *Instant Messaging* and the *Flipped Classroom* environment, to enhance student and instructor interaction [12].

The current socio-political environment of the United States has sparked debate on the efficacy of one's experience in constructing new knowledge in Constructivism. The impact of the media and the cry of fake news challenge the meaning given to reality and one's experience [57]. Hence, Constructivism has been criticized for not recognizing the influence of the social and cultural context on one's perceptions and experiences, which impact the subsequent creation of new knowledge. For example, Goings's [22] Black Male Adult Learner Success Theory provides a robust example of the complex and unique experiences of Black male adult learners in higher education and the impact of the environment on academic success. The reality perspective advanced by Goings [22] is often missed or diminished in the realm of meaning-making and construction in higher education policy.

In summary, Constructivism centers on the learner's characteristics of self-direction and experience to create new knowledge to support real-life problem solving, and promote lifelong learning [10, 12, 13, 44, 47, 57]. Applying the SDT framework, we propose, supports the benefits of applying innovative constructivist approaches, primarily through a lens recognizing social and cultural differences and the advantages of embedded technology.

2.2 Experiential Learning

As mentioned, Constructivism involves a synthesis of multiple theorists from Dewey [17] to Piaget [49] and Vygotsky [61], who profoundly affected Kolb's [36] Experiential Learning theory. The latter theory emphasizes the significance of experience in the adult learning process, and it views learning as a more holistic and integrated process that combines experience, perception, cognition, and behavior [1, 5]. Experiential Learning is the construction of knowledge and meaning from real-life incidents or simply learning by doing and undergirded by individual reflection [5, 14, 45].

Experiential Learning fosters new neurological connections in the brain, thus helping adults learn new things [5, 18]. Brown [9], as an element of the SDT framework, advances the concept of transferable *mental units* of culture termed *memes*. She goes on to recognize that in the field of adult education, there is a "Growing body of knowledge on adult development and how cognition—that can be dynamic and evolving—contributes to the adults' problem-solving capacities and their abilities to negotiate environmentally complex social contexts and thinking" (p. 206). Therefore, the role of experience and neuroplasticity[3] can operate as key variables for examining how adults learn and develop. SDT offers an interpretive model for how students in higher education as adult learners engage in adaptive complex critical thinking in new and innovative ways based on the connections formed (or uncoupling) that occur during group collaborations. Nevertheless, Schenck and Cruickshank [53] offer criticism of the Kolb [36] model due to its cyclical nature and its inability to adapt to newer research. We advance that SDT introduces a contemporary lens of epistemology that bridges the need for a newer way to examine adult learning using an integral view of the body, mind, and spiritual ways of knowing within dynamically changing social contexts [6, 8].

---

[3] A general umbrella term that refers to the brain's ability to modify, change, and adapt both structure and function throughout life and in response to experience.

3   Innovation for Building Affective Teams in Graduate Education

Innovative pedagogy and instructor facilitation can develop a participatory form of learning that supports an active and immersive environment. Individuals possess cognitive references from early childhood that expand throughout their life, primarily through contextual and social interactions with people, places, situations, and backgrounds [1, 5, 45, 54]. These contextual and social interactions influence the success of adult learners [46]. Vygotsky refers to cultural activities as *tools of intellectual adaptation* that include memories, mnemonics, and mind-maps, that help individuals acquire knowledge (as cited in [5]). Therefore, experiential learning requires acknowledging social interaction as a pivotal factor in students' psychological mindsets and knowledge interpretation. For example, Bell and Bell [5] discuss an experiential entrepreneurship educational framework that goes beyond merely teaching about entrepreneurship or theory, but to teaching for entrepreneurship, which develops learners' entrepreneurial skills, competencies, and experiences for employment [5]. The roles of both the educator and the adult learner must complement learner accountability and educator guidance and support. Active learning pedagogies, including experiential learning, play a crucial role in developing adult learners' higher-order thinking and cognitive skills.

Hence, the adult educator must create a supportive, low-risk learning environment to encourage adult students to test new ideas, ask questions, and challenge assumptions and past experiences [1]. Reflection and team discussion help adult learners think about strategies and outcomes and adjust. Incorporating simulation games, field trips, and industry experts provide authentic educational opportunities that represent real-life dimensions and complexity [54]. A constructivist scaffolding approach allows adult learners to engage in complex tasks which would otherwise be beyond their abilities [5].

4   Ways of Knowing and Pragmatic Collaboration

Adult Learning is expansive having a plethora of theories, modalities of pedagogy, and innovations for teaching and instruction in higher education. Hence, we hold that learning among graduate students tends to be a very organic process due in part to the role of life experience and that learning, generally, is far more developed in humans than in any other living species [31]. Thus, the goal to sustain *affective student* teams in graduate education entails the coordination of human phenomena (e.g., moods, feelings, and attitudes) as we consider embodied learning in online spaces [44].

Gallagher [21] describes an approach to meaning-making termed *enactivist* that involves mixing the brain and bodily processes. He held that cognition is not just in our minds but also in our bodies related to environmental factors. Therefore, enactivist interventions are designed to show that the mind is not exclusively in the head, nor would our responses to problem-solving. SDT offers, per our proposal, an aperture to the mind-body domain to the complexity of adult learning.

Opposite the approach of enactivist is pragmatism which holds that the *truth* of beliefs, theories, or meaning-making happens via the practical success of their applications. SDT holds that beliefs are a part of our conscious and unconscious value systems. Thus, a practical application of a known truth (i.e., pragmatism) requires some level of predictability but does not necessarily offer prescriptive pathways for the adult learner when their practical truths are met by uncertainty or chaos. We suggest that the incongruence of knowledge and experience among graduate student collaborative teams could be mitigated by properly blending like-minded and diverse thinkers among its members [58].

Illeris [31] contends that four notable styles are associated with specific languages and geographical regions. They include the Gestalt view, American behaviorism, Russian cultural-historical theory, and Piaget's Constructivism. As technology has expanded from the late 20th century until the present, team dynamics have needed to evolve. Firstly, one must understand what elements comprise *affective* teams and group collaborations. According to Johnson and Johnson [32], a team is a kluge of interpersonal communications and actions molded to contend for an established goal. In graduate education, students are challenged to form various connections and schemes that lead to high-quality participation among team members. We hold that clear and informed application of the SDT levels, particularly those found in Tier 1, provides for conditions where one can see the scaffolding necessary to engage in problem-solving [9] that facilitates higher-order thinking among individual students and groups.

4.1 Managing Diversity in Online Communities of Practice

The literature shows that *formal learning* is the pedagogical model that adult learners find most helpful in establishing policy assessments about access, curriculum preparation, resource management, and improvement of learning capabilities [48]. Moreover, students in formal and nonformal settings tend to identify via socioeconomic, cultural, and lived experiences. Schugurensky and Myers [55] argue that lifelong learning is often centered on the normative and ontological components. Hence, the potential for human disconnection via the overuse of innovative learning technologies, such as exclusively asynchronous learning, for the delivery of adult education can present emotional anxiety for instructors and graduate students [39]. Conversely, in a virtual learning setting, some team members are more expressive, blunt, and free with their responses on discussion boards and chat rooms during lectures [32]. We suggest that an adequate preliminary evaluation and assessment of students' diverse worldviews can foster optimal innovation and *affective* team building using SDT to identify and foster sensory attributes [9] that may match well between graduate student collaborators as they pursue mutually accepted goals.

5 Conclusion

The COVID-19 pandemic forced changes in adult education for students, faculty, and administrators. Many adult students quickly adapted their learning behavior to a virtual, digital environment [33, 34, 37]. Instructors had to reinvent their teaching approaches to fit this environment as administrators had to wrestle with practical support and retaining the quality of education delivery. An online learning environment requires new strategies, skills, and pedagogical changes [33, 34, 37].

Although online technology has been used in adult education for many years, the COVID-19 pandemic unmasked the gaps in adult digital literacy, the effective use of technology to create meaningful learning experiences, and the broader social challenges and inequities in digital access [56]. Wang et al. [62] contend distinctions are needed in adult education between the convenience of using technology to access course material and the use of technology as a strategic learning tool embedded into the course design. Affective reflectivity provides adult learners with an awareness of their perceptions, thoughts, and actions [6, 62]. Spiral Dynamic Theory (SDT) interprets the individual, organization, and societal domains into an adult learning framework that demonstrates how behavior oscillates between dynamically changing worldviews [6]. Since technology is a crucial factor in each domain understanding its role in the lives of adult learners is important.

In the past, many technology-based learning programs emerged from a history of computer-based training aligned with behaviorism [62]. In today's adult learning environment, technology plays a crucial role in providing a learning framework for deeper personal learning while enabling real-world learning experiences through virtual environments [62]. Student collaboration and digital social communities impact learning both within and outside the classroom. Additionally, the educator's epistemological view influences the use of technology in the teaching process [62]. Technology can support an adult learning environment. However, it requires reflection and planning by instructors and course designers on how to best implement strategies to impact positive innovations within adult learning processes [41, 56, 62].

Introducing any form of innovation and technology and/or Blended Synchronous [or asynchronous] Learning (BSL) approaches, we contend that a proper SDT assessment is necessary as a precursor for *affective* adult learning and development (diagnostic) or collaborative group formations. Assumptions that all learners' preferences for the delivery of pedagogy, technology, and organization/logistics [39] are valid concerns in adult education. There has arguably been a memetic assumption leading to the privileging of LMS technology as an all-encompassing panacea for the delivery of learning in higher education. It would be unreasonable to dispute how the use of innovative technology allows flexibility of attendance (including mode of access) for adult learners. However, the reduced face-to-face engagement for students may ultimately prove to be an insufficient and misdirected strategy relative to traditional adult education approaches. It remains to be seen. Nonformal and the emerging informal learning in local communities and online social media spaces—particularly during the COVID pandemic—have experienced our highest adult participant increases [44]. That is especially relevant when we consider adult education where participants are inhibited by a digital divide due to limits of economic access that may result in adult learning being the exclusive purview of the elite and privileged dominant groups.

In conclusion, measures should be avoided to constrict the BSL environment where the instructor takes total charge of selecting course content and who might be required to engage online on a particular day—hence assuming that a rotating or flexible schedule is, in fact, a form of educational democracy. Adult learning principles support the assumptions that adults learn best when they are allotted choices about their learning

[35]. Our supposition and hypothesis in this presentation hold that being able to collaborate well within teams and groups, for optimal adult learning to occur (especially when using innovative technologies), we intentionally include the *affective* dimensions of intercultural relations among diverse graduate students engaged in higher education.

References

1. Abderrahim, L., Gutiérrez-Colón Plana, M.: A theoretical journey from social Constructivism to digital storytelling. EUROCALL Rev. **29**(1), 38–49 (2021). https://files.eric.ed.gov/fulltext/EJ1305303.pdf
2. Alt, D.: Contemporary constructivist practices in higher education settings and academic motivational factors. Aust. J. Adult Learn. **56**(3), 375–399 (2016). https://uiwtx.idm.oclc.org/login?url=https://www-proquest-com.uiwtx.idm.oclc.org/scholarly-journals/contemporary-constructivist-practices-higher/docview/1851582676/se-2?accountid=7139
3. Beck, D.E., Cowan, C.C.: *Spiral Dynamics: Mastering Values, Leadership, and Change.* Blackwell Publishing (2006)
4. Beck, D., Herbo-Larsen, T., Solonin, S., Vilijoen, R., Thomas, J.Q.: Spiral Dynamics in Action: Humanity's Master Code. Wiley (2018)
5. Bell, R., Bell, H.: Applying educational theory to develop a framework to support the delivery of experiential entrepreneurship education. J. Small Bus. Enterp. Dev. **27**(6), 987–1004 (2020). https://doi.org/10.1108/JSBED-01-2020-0012
6. Brown, L.R.: Reparations and adult education: civic & community engagement for lifelong learners. In: Darity, W.A. Hubbard, L. (eds.) The Black Reparations Project: Report of the Reparations Planning Committee. UC Press (projected 2022)
7. Brown, L.R.: Spiral dynamic theory—part 3: memetic worldviews and ways of thinking (lecture presentation), The University of the Incarnate Word, San Antonio, TX, December 2, 2021) (2021)
8. Brown, L.R.: Comparing graduate student civic engagement outcomes among for-profit and public university adult learners in Chile. J. High. Educ. Outreach Engagem. **22**(4), 81–112 (2018)
9. Brown, L.R.: Civic engagement activities and outcomes in Chilean private for-profit and public graduate education. [Electronic Dissertation]. The University of Georgia, Athens, GA (2016)
10. Burrows, J., Brown, J.: Creating a tool to evaluate teaching materials for older beginner piano students through the lens of constructivism. Aust. J. Music. Educ. **52**(2), 33–45 (2019)
11. Butters, A.M.: A brief history of spiral dynamics. Approaching Relig. **5**(2), 67–78 (2015). https://doi.org/10.30664/ar.67574
12. Charania, A., Bakshani, U., Paltiwale, S., Kaur, I., Nasrin, N.: Constructivist teaching and learning with technologies in the COVID-19 lockdown in eastern India. Br. J. Educ. Technol. **52**(4), 1478–1493 (2021). https://doi.org/10.1111/bjet.13111
13. Chuang, S.: The applications of constructivist learning theory and social learning theory on adult continuous development. Perform. Improv. **60**(3), 6–14 (2021). https://doi.org/10.1002/pfi.21963
14. Correia, R., Klea, L., Campbell, G., Costa, A.: Fostering intergenerational education: an experiential learning program for medical students and older adults. Can. Med. Educ. J. **11**(5), e74–e77 (2020). https://doi.org/10.36834/cmej.69327
15. Crentsil, C., Gschwandtner, A., Wahhaj, Z.: The effects of risk and ambiguity aversion on technology adoption: evidence from aquaculture in Ghana. J. Econ. Behav. Organ. **179**, 46–68 (2020). https://doi.org/10.1016/j.jebo.2020.07.035

16. Dawkins, R.: The Selfish Gene. Oxford University Press (1976)
17. Dewey, J.: Experience and Education. Collier Books (1938)
18. Dopico, E., Ardura, A., Borrell, Y.J., Miralles, L., García-Vázquez, E.: Boosting adults' scientific literacy with experiential learning practices. Eur. J. Res. Educ. Learn. Adults **12**(2), 223–238 (2021). https://doi.org/10.3384/rela.2000-7426.ojs1717
19. Erickson, D.M.: A developmental re-forming of the phases of meaning in transformational learning. Adult Educ. Q. **58**(1), 61–80 (2007)
20. Erikson, E.H.: Identity and the life cycle: selected papers. Psychol. Issues **1**(1), 5–165 (1959)
21. Gallagher, S.: Enactivist Interventions: Rethinking the Mind. Oxford University Press (2017)
22. Goings, R.B.: Introduce the Black male adult learner success theory. Adult Educ. Q. **71**(2), 128–147 (2021). https://doi.org/10.1177/0741713620959603
23. Gordon, J., Smith, M.: You Win in the Locker Room First: The 7 C's to Build a Winning Team, in Business, sports, and Life. Wiley (2015)
24. Graves, C.W.: Deterioration of work standards. Harv. Bus. Rev. **44**(5), 117–126 (1966)
25. Graves, C.W.: Levels of existence: an open system theory of values. J. Hum. Psychol. **10**(2), 131–155 (1970)
26. Graves, C.W.: Human nature prepares for a momentous leap. The Futurist, pp. 72–87 (1974)
27. Graves, C.W.: In: Lee, W.R., Cowan, C.C., Todorovic, N. (eds.) Levels of Human Existence. ECLET Publishing, Santa Barbara, CA (2002)
28. Graves, C.W.: In: Cowan, C.C., Todorovic, N. (eds.) The Never Ending Quest. ECLET Publishing, Santa Barbara, CA (2005)
29. Goldman, J., Kuper, A., Wong, B.: How theory can inform our understanding of experiential learning in quality improvement education. Acad. Med. **93**(12), 1784–1790 (2018). https://doi.org/10.1097/ACM.0000000000002329
30. Hendricks, N., Aploon-Zokufa, K.: A 'curriculum moment' for adult and community education and training: acknowledging the voices and experiential knowledge of lecturers and students at community learning sites. J. Vocat. Adult Continuing Educ. Train. **4**(1), 15 (2021). https://doi.org/10.14426/jovacet.v4i1.181
31. Illeris, K.: An overview of the history of learning theory. Eur. J. Educ. **53**(1), 86–101 (2018)
32. Johnson, D.W., Johnson, F.P.: *Joining Together: Group Theory and Group Skills*, 12th edn. Pearson (2017)
33. Karakaya, K.: Design considerations in emergency remote teaching during the COVID-19 pandemic: a human-centered approach. Educ. Tech. Res. Dev. **69**(1), 295–299 (2020). https://doi.org/10.1007/s11423-020-09884-0
34. Kabir, M.R.: Impact of faculty and student readiness on virtual learning adoption amid COVID-19. Rev. Internacional de Educ. para la Justicia Soc. **9**(3e), 387–414 (2020). https://doi.org/10.15366/riejs2020.9.3.021
35. Knowles, M.: Self-Directed Learning: A Guide for Learners and Teachers. Follett Publishing Company, Chicago (1975)
36. Kolb, D.A.: Experiential Learning: Experience as the Source of Learning and Development. Prentice-Hall (1984)
37. Korkmaz, G., Toraman, C.: Are we ready for post-COVID-19 educational practice? An investigation into what educators think as to online learning. Int. J. Technol. Educ. Sci. **4**(4), 293–300 (2020)
38. Kozińska, K.: TED talks as resources for the development of listening, speaking, and interaction skills in teaching EFL to university students. Neofilolog **56**, 201–221 (2021). https://doi.org/10.14746/n.2021.56.2.4
39. Lakhal, S., Mukamurera, J., Bédard, M., Heilporn, G., Chauret, M.: Students and instructors perspective on blended synchronous learning in a Canadian graduate program. J. Comput. Assist. Learn. **37**(5), 1383–1396 (2021). https://doi.org/10.1111/jcal.12578

40. Larison Karen, D.: On beyond constructivism. Sci. Educ. **31**(1), 213–239 (2022). https://doi.org/10.1007/s11191-021-00237-8
41. Li, W., Ornstein, K.A., Li, Y., Liu, B.: Barriers to learning a new technology to go online among older adults during the COVID-19 pandemic. J. Am. Geriatr. Soc. (JAGS) **69**(11), 3051–3057 (2021). https://doi.org/10.1111/jgs.17433
42. Marsick, V.J., Watkins, K.E.: Adult educators and the challenge of the learning organization. Adult Learn. **7**(4) (March 1996), 18 (1996)
43. Marsick, V.J., Watkins, K.E.: Group and organizational learning. In: Kasworm, C., Rose, A., Ross-Gordon, J. (eds.) Handbook of Adult and Continuing Education, pp. 59–68. Sage, Thousand Oaks, CA (2010)
44. Merriam, S.B., Bierema, L.L.: Adult Learning: Linking Theory and Practice. Jossey-Bass (2014)
45. Michelson, E.: Truthiness, alternative facts, and experiential learning. New Dir. Adult Continuing Educ. **2020**(165), 103–114 (2020). https://doi.org/10.1002/ace.20371
46. Mohammadi, A., Grosskopf, K., Killingsworth, J.: An experiential online training approach for underrepresented engineering and technology students. Educ. Sci. **10**(3), 46 (2020). https://doi.org/10.3390/educsci10030046
47. Mohammed, S., Kinyo, L.: Constructivist theory as a foundation for the utilization of digital technology in the lifelong learning process. Turk. Online J. Distance Educ. **21**(4), 1–20 (2020)
48. Owusu-Agyeman, Y.: An analysis of theoretical perspectives that define adult learners for effective and inclusive adult education policies. Int. Rev. Educ. **65**(6), 929–953 (2019). https://doi.org/10.1007/s11159-019-09811-3
49. Piaget, J.: Six Psychological Studies. Random House, New York (1967)
50. Ransom, A., LaGrant, B., Spiteri, A., Kushnir, T., Anderson, A.K., De Rosa, E.: Face-to-face learning enhances the social transmission of information. PLoS ONE **18**(2), 1–17 (2022). https://doi.org/10.1371/journal.pone.0264250
51. Robinson, O.: Development Through Adulthood: An Integrative Sourcebook, 2nd edn. Palgrave Macmillan/Springer Nature (2020). https://doi.org/10.1007/978-1-137-29121-9
52. Romero-Hall, E., Jaramillo Cherrez, N.: Teaching in times of disruption: faculty digital literacy in higher education during the COVID-19 pandemic. Innov. Educ. Teach. Int. **59**(1), 1–11 (2022). https://doi.org/10.1080/14703297.2022.2030782
53. Schenck, J., Cruickshank, J.: Evolving Kolb: experiential education in the age of neuroscience. J. Experiential Educ. **38**(1), 73–95 (2015). https://doi.org/10.1177/1053825914547153
54. Schrimp, F., Davis, S., Regan, S., Goodnow, K., Gausvik, C., Pallerla, H., Schlaudecker, J.D.: Tell me your story: Experiential learning using in-home interviews of healthy older adults. J. Am. Geriatr. Soc. (JAGS), **69**(12), 3608–3616 (2021). https://doi.org/10.1111/jgs.17483
55. Schugurensky, D., Myers, J.P.: A framework to explore lifelong learning: the case of the civic education of civics teachers. Int. J. Lifelong Educ. **22**(4), 325–352 (2003)
56. Smith, M.C., Bohonos, J., Patterson, M.: Adult and continuing education and human resource development: responses to the COVID-19 pandemic. New Horiz. Adult Educ. Hum. Res. Dev. **33**(2), 1–3 (2021). https://doi.org/10.1002/nha3.20310
57. Stern, D.B.: Constructivism in the age of Trump: truth, lies, and knowing the difference. Psychoanal. Dialogues **29**(2), 189–196 (2019). https://doi.org/10.1080/10481885.2019.1587996
58. Stoller, A.: The flipped curriculum: Dewey's pragmatic university. Stud. Philos. Educ. **37**(5), 451–465 (2018). https://doi.org/10.1007/s11217-017-95921
59. Tisdell, E.J.: Exploring Spirituality and Culture in Adult and Higher Education. Jossey-Bass (2003)
60. von Glasersfeld, E.: Piaget and the radical constructivist epistemology. Constructivism **1**, 94–107 (2014)

61. Vygotsky, L.S.: Mind in Society: The Development of Higher Psychological Processes. Harvard University Press (1978)
62. Wang, V., Torrisi-Steele, G., Reinsfield, E.: Transformative Learning, Epistemology and Technology in Adult Education. SAGE Publications (1978). https://doi.org/10.1177/147797142 0918602
63. Winyard Sr., D.C.: Transhumanism: Christian destiny or distraction? Perspect. Sci. Christ. Faith **72**(2), 67–82 (2020)

Using Social Media to Teach Advocacy to Students

Angela N. Bullock[1]([⊠]), Alex D. Colvin[2], and M. Sebrena Jackson[3]

[1] Social Work Program, University of the District of Columbia, Washington, D.C., USA
angela.bullock@udc.edu
[2] Department of Social Work, Texas Woman's University, Denton, USA
[3] School of Social Work, The University of Alabama, Tuscaloosa, USA

Abstract. With the advent of new social media platforms, students are now expected to engage with technology. This has resulted in educators integrating social media into the classroom as a means of communication for active learning. Social media, specifically platforms like Facebook and Twitter, has created opportunities for individuals to make a greater impact addressing social problems and social justice issues. The article's purpose is to examine the practical usage of technological innovations through the exploration of the experiential learning theory by illustrating how social media platforms can be used as tools for engaging in electronic advocacy. Additionally, recommendations on the support students can receive to effectively use the tools for advocacy efforts are made. Suggested recommendations include educators integrating practical guides into their course materials that provide information on the use of social media platforms and universities providing financial support to help secure licenses for technology that incur a cost among others.

Keywords: Electronic advocacy · Experiential learning · Higher education · Social media

1 First Section

Research shows that most college students use some type of social media on a daily basis for personal or "school-related" purposes; according to Head and Eisenberg [1], more than half of them use social networks for "everyday life research" (p. 16). Today's students expect to be engaged technologically as well as intellectually, and as a result, university professors have been turning to social media as a communication avenue for active learning or immediate application of knowledge through engagement [2, 3]. Whether education is delivered in traditional, online or hybrid classrooms, students and professors are taking advantage of the latest advances in technology to transform the learning environment [4]. In an era when Facebook, Twitter, and YouTube have a pervasive presence, it is not surprising that university faculty regard social media as a potentially powerful pedagogical tool. With that in mind, it can be assumed that technology is no longer a separate or distinct field within education; but has become a

D. Guralnick et al. (Eds.): TLIC 2022, LNNS 581, pp. 97–103, 2023.
https://doi.org/10.1007/978-3-031-21569-8_9

fundamental part of the way we interact socially, inside the classroom as well as outside of it [4].

Educators now have an opportunity to help students engage in electronic advocacy, also known as e-advocacy using these media tools to affect political, economic, social, and environmental systems. Networking sites such as Facebook, LinkedIn, Pinterest, Instagram, and Twitter [5], provide rich forums that focus on social change and create a way to connect individuals to current issues related to advocacy [6]. These platforms have changed the way individuals advocate, organize, and mobilize support for causes, campaigns, and coalition actions [7]. Guo and Saxton [8] have identified media advocacy as a specific tactic, the ultimate goal of which is to mobilize supporters. This sense of connectedness and community, in combination with their familiarity and facility with digital technology, create opportunities for educators to help students hone their skills to effectively impact marginalized populations. The purpose of this article is to explore the practical usage of technological innovations through the exploration of the experiential learning theory, specifically showing how social media platforms can be used as tools for engaging in electronic advocacy.

2 Background

Anyone who has been in a classroom in recent years, either as a teacher or a student, knows that technology has changed the nature of education [9]. Laptops, smart boards, and iPads have become as ubiquitous as blackboards and chalk once were. Today's students expect to be engaged both technologically as well as intellectually. Technology use gives students more freedom to actively think about information, make choices, and execute skills than is typical in teacher-led lessons. Moreover, when technology is used as a tool to support students in performing authentic tasks, the students are in the position of defining their goals, making design decisions, and evaluating their progress. In a technology lead classroom, the teacher's role changes, as well. The teacher is no longer the center of attention as the dispenser of information but, rather, is using technological platforms to build communities and extend discussion beyond the classroom. This includes the way that educators prepare students to engage in advocacy.

Teaching e-advocacy can effect change within political, economic, social, and environmental systems. Advocacy, according to the National Association of Social Workers (NASW) [10] is "the act of arguing on behalf of a particular issue's idea or person". According to Bowen [11], "effective advocacy increases the power of people to make institutions more responsive to human needs, and it influences public policy and decisions regarding the allocation of resources" (p. 53). Obar et al. [12] reported that U.S. advocacy groups believed that social media could facilitate civic engagement and collective action by strengthening outreach efforts, enabling "engaging feedback loops" (p. 15), increasing speed of communication, and being cost-effective. The benefits of social media for civic engagement include the ability of individuals to interact and collaborate in real time [8] and to build a necessary sense of community that can facilitate collective action. In their study of social media use in the nonprofit sector, Lovejoy and Saxton [13], identified "action," encompassing participation in advocacy campaigns, as a communicative function of social media (p. 342).

By bringing social media into assignments and classroom discussions, educators can provide students with the opportunity to demonstrate knowledge, skills, and value acquisition, notably learning to respond to contexts that shape practice and advance social and economic justice [14]. It can also be an effective way to help students comprehend global issues. In addition to information-sharing, social media can be used as a tool of advocacy to promote program initiatives and bring awareness of social problems to others.

As innovation continues to expand and means of educating students consistently change, it is imperative that educators use new innovations to advance social and economic justice. This includes educators cultivating their own social media literacies to assist students in developing the skills they need to be informed professionals [15, 16]. By making social media an essential component of curricula, educators can prepare students for a climate in which advocacy, information-sharing, and community organizing occur on networks that are collaborative and publicly available.

3 Theoretical Framework

An essential component in education includes teachers' ability to convey course content and make it applicable for students. Experiential learning, which is an approach to applied learning, calls for students to draw from various experiences within and beyond the classroom [17]. Developed in 1938 by philosopher, psychologist and educator John Dewey, experiential learning emphasizes "the continuity of experience," which entails the constant nurturing of student curiosity through engaging in learning experiences so that students ultimately extend beyond the boundaries of their current knowledge [17]. Furthermore, experiential learning offers students a smooth approach to expand their intellectual perspectives because they come to realize that their experiences are as critical as any assigned textbooks or classroom lecture [17]. In essence, "students learn by observing and engaging in actions, internalizing those actions, applying the actions to theory and ideas, interpreting others' viewpoints and actions, and then revising their own viewpoints and actions in response" [18] (p. 196).

The experiential learning theory (ELT) is a multi-dimensional model of adult development [19]. According to Kolb [20], learning is "the process whereby knowledge is created through the transformation of experiences. Knowledge results from the combination of grasping and transforming experience" (p. 41). "Grasping experience" refers to the process of receiving information, and "transforming experience" is how learners understand information and act on that information [21]. Specifically, learners grasp experience through concrete experience and abstract conceptualization, and learners transform experience through reflective observation and active experimentation [21, 22]. Subsequently, experiential learning promotes the creation of knowledge through a spiral-form learning cycle that entails experiencing, reflecting, observing, and acting within the context of the learning environment and specific topic [21].

4 Opportunities

Educators across various disciplines have utilized social media as a tool to infuse experiential learning into the classroom. For instance, Madden and colleagues [18] completed a cross-institutional assignment incorporating social media into a course project. To help college students develop skills in social media, educators developed a cross-institutional group project that required students to create an instructional video about a social media topic. Students across institutions also engaged in Twitter chats related to the course assignment [18]. Upon completion of this project, researchers found that student engagement was higher than in traditional lectures and discussions, students drew on content learned in their current and previous courses to complete the project, and students were able to make the connection between the class project and real world social media situation.

Social work educators have also integrated social media in the classroom and added an e-advocacy component to class projects. Hitchcock [23] created an online toolkit and created useful course material and assignments that teach students to advance social justice and demonstrate advocacy using social media. Hitchcock et al. [24] also offer a comprehensive guide for educators who want to integrate technology into the classroom. Assignments that encourage the use of social media as an avenue for advocacy include: (1) connecting with peers and professionals using Twitter (any course); (2) using infographics to promote advocacy and brokering skills (any course); (3) engaging in YouTube electronic advocacy for LGBTQ Issues (practice courses); (4) engaging in social work practice advocacy, including policy critique exercise (practice courses); (5) implementing Twitter chats (practice courses); and (6) utilizing social media and technology use policy assignment (practice courses).

Integrating social media in the classroom offers several benefits. One key benefit includes engagement of students as active learners. Students who are actively learning are actively engaged. Whether solving a problem, debating an issue or researching a concept, they are processing ideas and forging deeper understanding. Active learning also increases critical thinking skills. Active learning shifts the focus of learning from passively (and possibly unquestioningly) digesting information to being accountable for actively engaging with sources and perspectives. The skills developed through various social media assignments and projects promote students in becoming competent in utilizing social media to engage in social action [10].

5 Recommendations for Practice

Social advocacy continues to be connected with technology tools and Internet-based outreach such as social media [25]. Integrating social media into education provides opportunities for students to engage on a different level and ensures that students are equipped to teach their future clients how to leverage social media [26].

Students can be encouraged to create a personal social media policy to guide their use of social media in practice. This can encompass educators integrating handouts, tutorials, and media simulations into their course materials that provide information on the use of social media platforms [18]. Doing this helps aid the students in making the connection of how to effectively use the tools when teaching advocacy effects themselves. Additionally, institutions of higher education (IHE) can support these efforts by offering financial support to help with securing licenses for tools that require a cost. Still more, instructors must be uniquely positioned to facilitate training and provide guidelines for students to keep them informed with the current state of practice. This includes embedding seminar and workshop activities and discussions in the curriculum that emphasize the use of social media in education [27].

Furthermore, to help students understand the impact and effectiveness of social media, educators can look for examples of e-advocacy methods in action [28, 29]. Belluomini [30] offers four steps students can apply when choosing self-advocacy methods. Step 1 involves the evaluation of technology literacy of those they are trying to empower individuals, and assistance of digitally literate individuals to understand how to use the technology to effectively engage in self-advocacy. Step 2 calls for the identification of the areas of concern that warrant advocacy, as well as the identification of appropriate e-advocacy tools. Step 3 involves practicing using the e-advocacy tools that were identified. Lastly, step 4 entails evaluating the effectiveness of the selected e-advocacy tool, identifying what is and what is not effective, then finally determining why certain aspects are not effective. Once mastery of the use of the e-advocacy tool has been achieved, self-reflection of the process occurs and the skills learned are used to move forward in advocacy.

6 Conclusion

Social advocacy continues to be connected with technology tools and internet-based outreach such as social media [25]. Students can be encouraged to create a personal social media policy to guide their use of social media in practice. As Madden and colleagues [18] suggest, handouts, tutorials, and orientations that provide information on the social media platforms should be developed to ensure that all students know how to effectively use the tools for class and in practice. Teaching students to engage in e-advocacy can effect change within political, economic, social, and environmental systems. E-advocacy assignments that integrate social media in the classroom will prepare students to use social media as a tool to "advance community organizing and policy advocacy and to support community mobilization, particularly in times of personal or community crisis" [31] (pp. 8–9). By focusing on these efforts, educators can inform a new generation of students who will engage in e-advocacy.

References

1. Head, A.J., Eisenberg, M.B.: How college students seek information in the digital age. Project Information Literacy Progress Report, p. 7 (2009)
2. Kassens-Noor, E.: Twitter as a teaching practice to enhance active and informal learning in higher education: the case of sustainable tweets. Act. Learn. High. Educ. **13**(1), 9–21 (2012). https://doi.org/10.1177/1469787411429190
3. Prescott, J.: Teaching style and attitudes towards Facebook as an educational tool. Act. Learn. High. Educ. **15**(2), 117–128 (2014). https://doi.org/10.1177/1469787414527392

4. Reamer, F.G.: Ethical standards for social workers' use of technology: emerging consensus. J. Soc. Values Ethics **15**(2), 71–80 (2018)
5. Duggan, M., Ellison, N.B., Lampe, C., Lenhart, A., Madden, M.: Social media update 2014. Pew Res. Center **19**, 1–2 (2015)
6. Maben, S., Helvie-Mason, L.: When Twitter meets advocacy: a multicultural undergraduate research project from a first-year seminar. Int. J. Teach. Learn. High. Educ. **29**(1), 162–176 (2017)
7. Bullock, A.N., Colvin, A.D.: # SocialWorkAdvocacy. J. Soc. Work Glob. Commun. **3**(1), 2 (2018)
8. Guo, C., Saxton, G.D.: Tweeting social change: how social media are changing nonprofit advocacy. Nonprofit Volunt. Sect. Q. **43**(1), 57–79 (2014)
9. Nguyen, A.: Technol. Soc. Work Educ. (2013). https://socialworklicensemap.com/blog/technology-and-social-work-education/
10. National Association of Social Workers, Association of Social Work Boards, Council on Social Work Education, Clinical Social Work Association: Standards for technology in social work practice (2017). https://www.socialworkers.org/includes/newincludes/homepage/PRA-BRO-33617.TechStandards_FINAL_POSTING.pdf
11. Bowen, G.A.: Promoting social change through service-learning in the curriculum. J. Effect. Teach. **14**(1), 51–62 (2014)
12. Obar, J.A., Zube, P., Lampe, C.: Advocacy 2.0: an analysis of how advocacy groups in the United States perceive and use social media as tools for facilitating civic engagement and collective action. J. Inf. Policy **2**, 1–25 (2012)
13. Lovejoy, K., Saxton, G.D.: Information, community, and action: how nonprofit organizations use social media. J. Comput.-Mediat. Commun. **17**(3), 337–353 (2012)
14. Hitchcock, L.I., Battista, A.: Social media for professional practice: integrating Twitter with social work pedagogy. J. Bac. Soc. Work. **18**(Special Issue), 33–45 (2013). https://doi.org/10.18084/basw.18.suppl-1.3751j3g390xx3g56
15. Mishna, F., Bogo, M., Root, J., Sawyer, J.L., Khoury-Kassabri, M.: "It just crept in": the digital age and implications for social work practice. Clin. Soc. Work J. **40**(3), 277–286 (2012)
16. Mukherjee, D., Clark, J.: Students' participation in social networking sites: implications for social work education. J. Teach. Soc. Work. **32**(2), 161–173 (2012)
17. Weinstein, N.: Experiential Learning. Salem Press Encyclopedia (2021)
18. Madden, S., Briones Winkler, R., Fraustino, J.D., Janoske, M.: Teaching, tweeting and teleworking: experiential and cross-institutional learning through social media. Commun. Teach. **30**(4), 195–205 (2016). https://doi.org/10.1080/17404622.2016.1219040
19. Kolb, A.Y., Kolb, D.A., Passarelli, A., Sharma, G.: On becoming an experiential educator: the educator role profile. Simul. Gaming **45**(2), 204–234 (2014). https://doi.org/10.1177/1046878114534383
20. Kolb, D.A.: Experiential Learning: Experience as the Source of Learning and Development. Prentice-Hall (1984)
21. Passarelli, A., Kolb, D.A.: Using experiential learning theory to promote student learning and development. Working Paper No.11–03 (2011). https://weatherhead.case.edu/departments/organizational-behavior/workingPapers/WP-11-03.pdf
22. Fraustino, J.D., Briones, R., Janoske, M.: Can every class be a Twitter chat?: cross-institutional and experiential learning in the social media classroom. J. Public Relat. Educ. **1**(1), 1–18 (2015)
23. Hitchcock, L.: Second edition of the social media toolkit for social work educators. Teach. Learn. Soc. Work. (2018). https://laureliversonhitchcock.org/2018/11/05/revised-social-media-toolkit/
24. Hitchcock, L., Sage, M., Smyth, N.: Teaching Social Work with Digital Technology (2019). CSWE Press

25. Goucher, K., Wolf, L., Goldkind, L.: Social media in agency settings. In: Goldkind, L., Wolf, L., Freddolino, P. (eds.) Digital Social Work: Tools for Practice with Individuals, Organizations, and Communities, pp. 168–182. Oxford University Press (2019)

26. Berzin, S., Singer, J., Chan, C.: Practice innovation through technology in the digital age: a grand challenge for social work. Working Paper No. 12. American Academy of Social Work and Social Welfare (2015). https://grandchallengesforsocialwork.org/wp-content/upl oads/2015/12/WP12-with-cover.pdf

27. Fang, L., Al-Raes, M., Zhang, V.: "How to connect the two": social media in field education. Field Educ. **9**(2), 1–24 (2019)

28. McNutt, J.: Advocacy, social change and activism: perspectives on traditional and electronic practice in a digital world. In: McNutt, J. (ed.) Technology, Activism + Social Justice in a Digital Age, pp. 9–21. Oxford University Press (2018)

29. McNutt, J.G., Goldkind, L.: E-activisim development and growth. In: Khosrow-Pour, M. (ed.) Encyclopedia of Information Science and Technology, 4th edn., pp. 3569–3578. IGI Global (2018)

30. Belluomini, E.: Using digital self-advocacy to empower social work populations. New Soc. Work. (2014). https://www.socialworker.com/feature-articles/technology-articles/using-dig ital-self-advocacy-to-empower-social-work-populatio/

31. Council on Social Work Education: Envisioning the future of social work: report of the CSWE futures task force (2018). https://www.cswe.org/getattachment/About-CSWE/2020-Strate gic-Plan/Futures-Task-Force/CSWE-FTF-Four-Futures-for-Social-Work-FINAL-2.pdf.aspx

An Actionable Training: From Competency Model to Observable Behaviors While Empowering Communication

Elena Ciani[✉] and Andrea Laus

Lifelike SA, Chiasso, Switzerland
elena.ciani@skillgym.com

Abstract. There is a need for organizations to bridge the gap between competencies models and actionable behaviors. We argue that a structured training method, with measurable patterns, is needed to supply powerful behavioral tools that reflect competency models and skills directly into everyday working life. The method is an artificial intelligence or AI-driven interactive digital role-play game that allows participants to practice critical, emotionally charged conversations at work. Conversations take place in different scenarios with different characters based on personality psychology and neuroscientific findings. In this statement, we will focus on the architecture of the method, explaining why it is needed and how competency models and skills are put into action through communication, overcoming traditional learning tools.

Keywords: Learning Technology · Soft Skills · Communication

1 Introduction

As the Harvard Business Review [22] has noted, one of the biggest complaints about executive education is that the skills and capabilities developed don't find practical applications on the job. Indeed, traditional leadership development approaches no longer fully match the needs of organizations or individuals. According to the research, there are three reasons for this. First, organizations don't always benefit from leadership development. An organization may spend money to train people, and those people don't stay long enough for the organization to benefit from it. Second, providers aren't developing the soft skills that organizations really need. And finally, it's often difficult to apply to the real world what has been learned in classes. This gap is leading to the emergence of new approaches [22]. Studies by the Carnegie Institute of Technology shed more light on how the ability to "deal with people" has a strong impact on financial success. This data show that 85% of one's financial success arises from personal skills and the ability to lead people [6]. The ability to deal with people and to communicate efficiently is an essential skill in the workplace and the business environment, according to the worldwide-accepted Goleman's theory.

By facing the current realities and challenges of organizations, the method enables participants to practice a wide range of crucial skills strictly connected to the real world

D. Guralnick et al. (Eds.): TLIC 2022, LNNS 581, pp. 104–113, 2023.
https://doi.org/10.1007/978-3-031-21569-8_10

by simulating conversations in a working environment. This method aims to enhance performance by training people in critical conversations. Effectively practicing skills and behaviors enables individuals to contribute to the improvement of their organizations. Even though organizations hire people already trained, learning is a lifelong process in which individuals can actively change their behavior. We believe that fostering the ability to leverage fruitful and empowering conversation throughout the entire organization should be the new goal of any talent development professional since multidirectional and cross-functional leadership conversations are increasingly needed. It is important to relate the entire competency model to some actionable form of measurement and development to make sure that people in the organization can recognize it as a meaningful way to orientate behaviors along the way.

2 Problems

The architecture of the present method has been driven by two urgent interrelated needs:

- The first is the necessity for the training to be focused on measurable patterns. The method is a training program; therefore, to deliver effective and efficient training, there is a need for the different elements to be properly designed and correctly measured. Based on this, we argue that no training can be truly effective without a structured system that allows for identifying behaviors and measuring their evolution over time. Starting from a broader competency model framework, the method delivers a more detailed structure of measurable behavioral outcomes.
- The second is the necessity to bridge the gap between theories about competencies and skills, and specific daily behaviors to associate with these skills. Despite the huge body of literature about behavioral competencies, only generically actionable behaviors have been defined [23]. No standards of behavioral approaches are currently available to help in differentiating a superior performance from an average one [23]. The literature review doesn't provide conclusive and exhaustive answers to bridge this gap.

We argue that a methodology able to provide these kinds of practical insights has the potential to enable participants to practice skills directly in real-life, everyday situations—not just as a theoretical framework to aim toward.

3 Competence and Competency Models: Confusing Concepts?

In reviewing the literature about this topic, two points immediately pop out: A significant amount of literature and a wide range of meanings are attributed to these terms. Even though "competence" and "competency" concepts dominated the management strategy literature of the 1990s, the actual meanings are still a bit confusing. Despite the generally accepted distinction in the meaning of competence and competency, the terms can sometimes be interchangeably used, and they can have different meanings depending on who is using them [29]. The meaning of the concept of "competence" has been strongly debated, and it is almost impossible to identify a consistent theory or an agreed definition capable of accommodating and reconciling all the different uses of the term [17]. The terms "competence" and "competency" arise from different streams of thought, the generally accepted distinction might be "competence" defined as a description of work tasks or job outputs, while "competency" more often refers to a description of behavior [29]. In Boyatzis's book, The Competent Manager: A Model for Effective Performance [4], the concept of competency was a central theme, and it has been defined as "an underlying characteristic of someone which results in effective and/or superior performance in a job." The word is also strictly related to behavioral outcomes as a sequence of actions taking place in a system. Spencer and Spencer [27] provided an analysis of 650 jobs using the McClelland/McBer job competence assessment (JCA) methodology, pointing out that competency is an individual characteristic related to effective and/or superior job performance. Since an epistemological and historical debate about this topic is not the main purpose of the present statement, we will focus on the key concepts of our training method. The key concept of the present purpose is that competencies are characteristics related to skills and behavior in the workplace environment. A competency model is a behavioral job description that must be defined by each occupational function and each job [10]. Competency models have many advantages, including that they help organizations take a unified and coordinated approach. According to the SIES Journal of Management [29], there are various advantages to using a competency model. It helps to enhance the recruiting process, employee development, and performance management, as well as identify training needs. It is very useful in the unification of corporate culture across business units, establishing connectivity through integration of HR processes, and establishing clear expectations for success [29]. It is agreed upon worldwide that utilizing competency models helps with the strategic management of talent and leads to a profitable and successful organization [29].

3.1 Which Competencies Model? A Multidimensional Approach

As suggested by Le Deist and Winterton [17], there is a need to face the challenge of developing a consistent typology of competence in a context where, even within countries, there is a certain amount of diversity in the approaches. After exploring the three dominant approaches across the world that began relatively independently (USA, UK, France and Germany), Le Deist [17] proposed a comprehensive, holistic typology of competence model including: Functional, cognitive, social, and meta-competencies. Today, this concept of multidimensional frameworks of a competence model is widely accepted [17]. After a deep review of literature in an effort to clear up the concept, they suggest that many conceptions of competency now include knowledge and skills alongside attitudes, behaviors, work habits, abilities, and personal characteristics [11, 17]. Indeed, each of the dominant approaches has its strengths [17].

4   The Architecture

4.1 Behavioral Competencies Model to Meet EVERyone's Needs

To meet needs on a global scale, the method has been based on the holistic framework considering knowledge, skills, and behaviors as dimensions of the same outcome. The approach allows operation on each different level. It has been made by taking into consideration both well-known theories and more than 20 years of expertise in developing effective strategies for learning and development in companies. The result is a model that, over the years, has proven to match most of the competency models adopted or developed by the organizations in everyday reality. Designed to deliver an actionable approach for efficient and measurable training, much effort was dedicated to establishing a very clear interpretation and definition of each element of its structure. For instance, different areas of empowerment have been selected. In each of these areas, changes in behaviors can efficiently affect mutual interactions among individuals, which enhances the organization's results. Therefore, a key founding pillar of the present model's structure answers the question, "What areas of influence can be covered?" The areas of empowerment have been classified according to the role in the organization, and, most often, they overlap the traditional roles' competencies as designed in many corporate competency models. Built upon this initial four-area structure, we propose a complete model of eight key capabilities that blend perfectly within the four areas of engagement that inspire the design of our simulation scenarios (Fig. 1). For each area, the two capabilities balance the fulfillment of the need to improve oneself (Get) and support the improvement of others (Give).

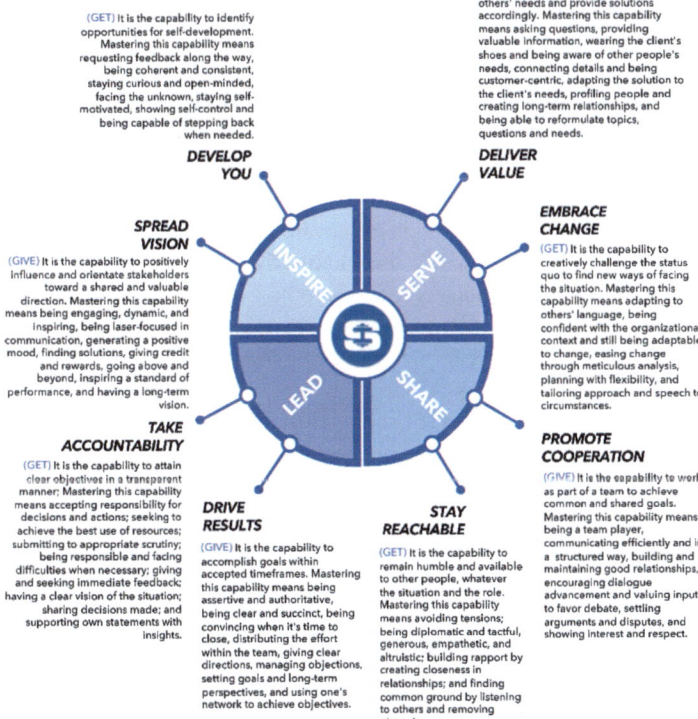

Fig. 1. Competencies model: Areas of empowerment and capabilities.

Given this framework, an inner layer has been developed to define and structure the way those competencies are detailed in more day-to-day actionable and specific skills and relevant behaviors.

4.2 Behavioral Model of Competence

Behavioral models of competence are powerful because they are part of the actual context in which behaviors are manifested [23]. Reviewing management literature, it is easy to infer the increasing interest in people skills and associated dimensions of behaviors underpinning proficient performance. Different studies and research have been conducted to provide insights about specific skills and associated behaviors for managers to adopt [9, 23] as suggested by Blackburn [2], Huemann [15], Dainty et al. [7], and Moore et al. [23]. Numerous methods were available to assess these competencies through behavioral event interviews [4, 20, 27], simulations, and assessment centers. There is general agreement about some fundamental skills required for an individual and (consequently) a company to function. McGregor [21] linked some methodologies for changing organizations to basic concepts of behavioral sciences, pointing out how an individual can grow and realize the basic goals for himself or herself, and at the same time, for the organization. Blake and Mouton [3] said that human needs are indispensable for mature and healthy relationships, so when managerial concerns for people are integrated into the team, the best long-term production is achieved and sustained. Likert and Hayes [18] also applied behavioral research to operating organizations, finding that an effective manager needs to show concern for people, involving trust and sympathy and people's emotions, for example, in solving problems. We find the work of Honey [14] very relevant for this, who, back in 1988, suggested that interpersonal skills—especially face-to-face skills—are particularly important for managers on different occasions, both formal and less formal. The author offered a way for people to make behaviors (verbal and non-verbal) a conscious process, being aware of their impact on others instead of simply applying them because that "comes naturally." Peters and Waterman [24] suggested that an effective people manager needs to be able to communicate well, inspire, lead, show empathy, and respect employees in a people-centered working environment. In more recent years [8], it appears that the behaviors that underpin competencies are being increasingly recognized as the driving forces that influence the effective management of people. Kets de Vries [16] suggests that in order to have well-functioning, healthy individuals in "vibrant companies," managers need to show a genuine interest in human nature and build an understanding of what is psychologically important to others. He refers to this as "authentizotic behavior."

Fisher [9], after a combination of literature review, face-to-face interviews, and focus group meetings, completed a search to objectively consider the skills and behaviors of an effective people project manager. The suggestions considered by his paper are not limited to application in any specific industries, such as for-profit and non-profit organizations and construction, or types of projects such as infrastructure and software development. Six different skills and related behaviors have been found. Skills on their own, including their applications, do not make an effective people project manager. Behaviors drive outcomes. For each skill, specific behaviors need to be applied to make these skills truly effective. It is the application of these behaviors that is preponderant and moves the needle for effective people project managers [9]. Essentially, the skills are related in order to understand the behaviors of others, be able to lead, support, influence, and inspire with visions and charisma, properly manage conflicts and be aware of cultural differences. Specific, appropriate behaviors are also explained, and the author also suggests that acquaintance and skills without associated behaviors are perhaps not as effective as previously thought [9].

4.3 Integration of Competencies, Skills, and Behaviors

Behavioral competencies help to define a path for behaviors following the latest findings. After the competency model has been defined, there is the need to integrate it into measurable, observable behavior. We argue that the model is totally consistent with the latest findings in the field (Fig. 2). For each capability, specific skills are then associated. For example, for the capability "Deliver Value" one of the underlying skills might be "Ask questions." The comprehensive model includes 78 skills. All of the skills are related to the essential skills required for employers and employees to function, based on literature research as previously described. Each skill is then de-scribed as a grade of observable behaviors made of four different levels of efficiency: Talented, Skilled, Overused, and Unskilled. Each skill and its underlying behavior can be practiced with different scenarios; however, of course, each scenario provides training only for a specific set of different skills. Practicing more scenarios together allows the user to practice a comprehensive set of skills to develop the competencies/capabilities for which the specific training was needed.

Fig. 2. The method: Explaining how different grades of observable behaviors directly represent measurable expressions of selected skills.

4.4 Skill-Behavior Association

A skill can be demonstrated through observable behaviors. According to the expertise level of the individual, different levels of observable behaviors can be attached to the same skill (Fig. 3).

The type and quality of behaviors adopted by individuals depend on the chosen strategy of communication and, of course, by their expertise in the specific skill. The method is based on a four-level scale to associate observable behaviors to each skill as a representation of different stages of proficiency:

Fig. 3. How skills can be expressed through behaviors.

- **Unskilled**. When a user might not be aware of their lack of proficiency. As an example: For the skill "ask questions," this level is associated with "being aggressive."
- **Overused**. Typically, this is applicable for users who start understanding the direction the conversation should take without the ability to handle it properly. It has been noticed that at this stage, the learner tends to over-behave. As an example: For the skill "ask questions," this level might be associated with "being confusing."
- **Skilled**. For example: For the skill "ask questions," this level might be associated with "barely open to listening."
- **Talented**. For example: For the skill "ask questions," this level is associated with "being genuinely interested in understanding."

Interestingly, this approach can be somewhat related to similar studies conducted in "The Stages of Competence" model by Thomas Gordon, designed to assess the learning process of the new skill. This model was introduced by Noel Burch in the 1970s. As in the Situational Leadership model, learners in the Stages of Competence model fall into one of four stages: Unconscious unskilled, conscious unskilled, conscious skilled, or unconscious skilled.

Conceptually, the Stages of Competence and the Situational Leadership models are parallel. The Stages of Competence model implies that all learners proceed in a sequential, somewhat predictable order through the four stages. At the unconscious unskilled stage, the individuals don't understand or they don't know how to do something. They don't necessarily recognize the deficit. Conscious unskilled is when the learners don't understand or they don't know how to do something, but at this point they are aware of the deficit. Conscious Skilled is when the individual understands or knows how to do something; however, demonstrating the knowledge or skill requires a certain effort. At the level of Unconscious Skilled, the individuals have had so much practice with that specific skill that it requires little thought, and it can be performed concurrently with other tasks.

4.5 Why Critical Conversations?

These are challenging times for organizations. Today, more than ever before, companies must remain competitive and survive the complex, unstable business environment.

Organizations need to develop strong systems able to endure the constant reformulations of strategic choices and deal with frequent mutations. Organizations need to invest in people, and employers and employees need to be flexible and able to adapt quickly. Essentially, people need to be emotionally intelligent [24]. Boyatzis and Goleman suggested applying a model of emotional intelligence in organizations. A model of emotional intelligence is based on the competencies that allow employees to intelligently use their emotions to manage themselves and effectively work with others [5]. This concept of emotional intelligence is particularly interesting because it offers a vision of how personality can be linked and related to a theory of job performance [12]. Goleman gave us a definition of "emotional competence" as a "learned capability based on emotional intelligence that results in outstanding performance at work." The integration of the work of Goleman [12] and Boyatzis [5] offers the opportunity to have a descriptive definition of emotional intelligence, in which its components (self-awareness, self-management, social awareness, and social skills) are properly applied to be effective in specific working situations. They argue that the implications of this integrated theory can be relevant to emphasize the importance of a "cluster of performance" (cf. Emotional Competence Inventory; ECI) in predicting performance [5]. It's been many decades since McClelland found that leaders with strength in six or more emotional intelligence competencies were far more effective than peers who scored lower. Those executives who scored low in emotional intelligence were rarely rated as remarkable in their annual performance reviews [13, 19]. The most common and observable way to express a skill during the daily business activity is through behaviors embedded in the dialogues (Fig. 4).

Fig. 4. How behaviors can be turned into dialogues.

Our method of improving people's capabilities through consistent practice in life-like dialogue-driven scenarios is strictly related to the emotional intelligence theory. As you might have already noticed in the previous chapters, particular attention has been dedicated to ensuring the skills are consistent with the emotional competencies. Based on all the findings, we believe that emotional intelligence, and its effects, is the key point for training competencies. It is not only a matter of self-improvement in communication skills, but communication is also directly related to the way competencies are expressed and how they can be empowered (Fig. 5).

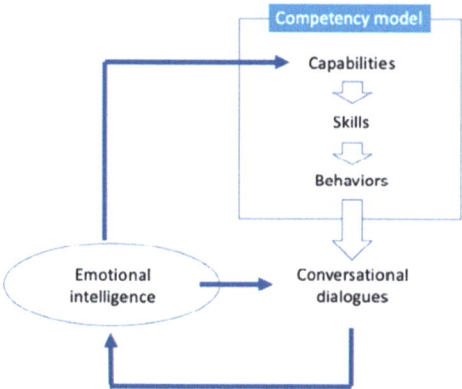

Fig. 5. Explaining the training method. Conversations are designed to express the observable behaviors of the competency model. The user manages the conversation using their emotional intelligence. Practicing conversations feeds and improves the underlying competencies of the user through the development of emotional intelligence.

5   Conclusion

In summary, we propose an actionable method that allows employers and employees to develop behavioral competencies, bridging the gap between theories and behaviors in an organizational environment. It provides measurable training to improve skills by practicing critical conversations. It overcomes traditional learning tools, allowing participants to experience various real-life situations and face different kinds of people in role-play, simulating the everyday work environment.

References

1. Adams, L. Learning a new skill is easier said than done. gordon training international. https://www.gordontraining.com/free-workplace-articles/learning-a-new-skill-is-easier-said-than-done/ (2011)
2. Blackburn, S. Understanding project managers at work (DBA Thesis), Henley Management College/Brunel University (2001)
3. Blake, R.R., Mouton, J.S., Bidwell, A.C.: Managerial grid. advanced management—office executive **1**(9), 12–15 (1962)
4. Boyatzis, R.E.: The competent manager: a model for effective performance. J. Wiley, New York (1982)
5. Boyatzis, R.E., Goleman, D., Rhee, K.: Clustering competence in emotional intelligence: Insights from the Emotional Competence Inventory (ECI). Handb. Emot. Intell. **99**(6), 343–362 (2000)

6. Carnegie, D.: How to Win Friends and Influence People. Simon and Schuster, New York (1937)
7. Dainty, A., Cheng, M.I., Moore, D.: A comparison of the behavioral competencies of client-focused and production-focused project managers in the construction sector. Proj. Manag. J. **36**(2), 39–48 (2005)
8. Fisher, E.J.P. Development of a new competence and behaviour model for skills in working with people for project managers, Ph.D. Thesis, Open University, Milton Keynes, Unit-ed Kingdom (2006)
9. Fisher, E.: What practitioners consider to be the skills and behaviours of an effective people project manager. Int. J. Project Manage. **29**(8), 994–1002 (2011)
10. Fogg, C.D.: Implementing your strategic plan: How to turn intent into effective action for sustainable change. Am. Manag. Assoc., New York (1994)
11. Gangani, N.T., McLean, G.N., Braden, R.A. Competency-based human resource development strategy, In: Academy of Human Resource Development Annual Conference, Austin, TX, 4–7 March, in: Proceedings, **2**, pp. 1111–1118 (2004)
12. Goleman, D.: Emotional intelligence: why it can matter more than iq. Bantam Books, New York (1995)
13. Goleman, D. Leadership that gets results (Harvard business review classics). Harvard Business Press (2017)
14. Honey, P. Face to face skills. Gower Publishing (1988)
15. Huemann, M. Individuelle Projektmanagement-Kompetenzen in Pro-jektorientierten Unternehmen, Univ. Econ. Manag.. Ph.D. Thesis, Vienna (2002)
16. Kets De Vries, M. Creating authentizotic organisations: well-functioning individuals in vibrant companies. Human Relations **54**(1), 101–111 (2001)
17. Le Deist, F.D., Winterton, J.: What is competence? Hum. Resour. Dev. Int. **8**(1), 27–46 (2005)
18. Likert, R., Hayes, S.P., Some applications of behavioral research. UNESCO (1957)
19. McClelland, D.C.: Testing for competence rather than for "Intelligence." Am. Psychol. **28**(1), 1–14 (1973)
20. McClelland, D.C.: Identifying competencies with behavioral-event interviews. Psychol. Sci. **9**(5), 331–339 (1998)
21. McGregor, D.: The professional manager. McGraw-Hill Int. Stud. Ed. (1967)
22. Moldoveanu, M., Narayandas, D.: The future of leadership development. Harv. Bus. Rev. **97**(2), 40–48 (2019)
23. Moore, D.R., Cheng, M.I., Dainty, A.R.J. What makes a superior management perform-er: the identification of key behaviours in superior construction managers, construction information quarterly, CIQ Paper, **5**(2), 6–9. 155 (2003)
24. Morone, M., Giorgi, G., & Pérez, J. F. Emotional and organizational competency for success at work: a review. Quality—Access to Success, 17.152 (2016)
25. Peters, T.J., Waterman, R.H.: In search of excellence: lessons from Ameri-ca's Best-Run companies. Harper & Row, New York (1982)
26. Peters, T.J., Waterman, R.H.: In search of excellence: lessons from America's Best-Run companies. Harper & Row, New York (1982)
27. Peel, J.L., Nolan, R.J.: You Can't Start a Central Line? Supervising Residents at Different Stages of the Learning Cycle. J. Grad. Med. Educ. **7**(4), 536–538 (2015)
28. Spencer, L.M., Spencer, S.G.: Competence at work: Models for superior performance. Wiley, New York (1993)
29. Tubbs, S.L., Schulz, E.: Exploring a taxonomy of global leadership competencies and me-ta-competencies. J. Am. Acad. Bus. **8**(2), 29–34 (2006)
30. Vazirani, N.: Review paper: Competencies and competency model–A brief overview of its development and application. SIES J. Manag. **7**(1), 121–131 (2010)
31. Woodruffe, C.: Competent by any other name. Pers. Manag. **23**(9), 30–33 (1991)

Literary History in Digital Teaching and Learning: The KoLidi-Project—Collaborative and Interactive Approaches for German Studies

Jens Ciecior[1] , Tanja A. Kunz[1] , Stephanie Wollmann[2] , Karima Lanius[3] , and Matthias Buschmeier[1(✉)]

[1] Bielefeld University, Universitätsstraße 25, 33615 Bielefeld, Germany
matthias.buschmeier@uni-bielefeld.de
[2] University of Wuppertal, Gaußstraße 20, 42119 Wuppertal, Germany
[3] Paderborn University, Warburger Straße 100, 33098 Paderborn, Germany

Abstract. KoLidi is a pioneering project in its attempts to use digital and collaborative tools to teach literary history at the university level. In this paper, we illustrate the framework conditions of our project, outline its scientific background, and present our methodological considerations. Literary studies are, at their core, based on mutual exchanges concerning texts, contexts, and discourses. Transposing this process into a digital environment poses new challenges regarding how to keep this conversation running. How can close (analogue) individual readings and digital collaborative learning be productively combined? In a team of three German universities, we started with this basic question from the outset in order to develop collaborative and interactive courses on German-language literary history from the Middle Ages up to the 20th century. KoLidi aims to enable students to delve into historical sources by offering them the opportunity to identify historical patterns and structures on their own. All media content that is created serves two major functions: to impart declarative knowledge to learners and to encourage students to integrate such knowledge into a larger context of problem-oriented knowledge acquisition through joint dialogue and teamwork beyond our institutions. To enable further exchange and development, all teaching materials are designed as open educational resources (OER).

Keywords: Literary history · Digital learning · Digital teaching · Collaborative learning

1 Aim of This Paper

We built our argumentation around four overarching questions to present and describe our KoLidi project: What? Why? How? And how could we make it sustainable? After explaining our project goals (what?), we illustrate its framework conditions, outline its scientific background (why?), and present our methodological considerations. To provide some illustrative examples, we will then describe some of our course elements (how?) before concluding with a preview of our evaluation efforts (sustainability?).

D. Guralnick et al. (Eds.): TLIC 2022, LNNS 581, pp. 114–124, 2023.
https://doi.org/10.1007/978-3-031-21569-8_11

2 About the Project

The KoLidi-acronym spells out "**Ko**llaborative **Li**teraturgeschichte **di**gital und **i**nteraktiv" (i.e., Literary History, collaborative, digital, and interactive). In our pioneering project, a team of scholars from the universities of Bielefeld, Paderborn, and Wuppertal are developing digital, collaborative, and interactive courses on (German) literary history from the Middle Ages to the 20th century. German-language literary history is an integral part of German literary studies curricula. As such, it is also part of the educational training for the teaching profession in higher education. Literary history as a genre and knowledge-organization technique provides a framework for students who are confronted with and often otherwise get lost in large textual landscapes; students often struggle to master broad masses of texts and materials in just a few semesters, to acquire sound contextual knowledge, and to apply this knowledge in their academic writing. KoLidi addresses these very common challenges in new ways. Using the power of digital means to facilitate student pathways through the historical material.

However, we do so in a way quite different from how literary history is imparted in German high schools, colleges, or even universities, where literary historical knowledge is mostly taught via top-down processes. Our students will be prepared to work in the field of literary history through their self-directed learning with the literary historical material. Far from offering learners a mere oversight of literary history topics and forcing them into a passive, consumptive learning attitude, our students actively plunge into various ways of reading (close/distant, critical/affirmative, etc.) to structure the field by themselves but with the support of our well-considered and strategically prepared setting. We guide them on their way through the uncharted territory, so to speak. One of our central goals is that students emerge from that journey with a self-drawn map of German literary history.

With our digital course, students encounter an innovative approach to literary history. We provide a manageable but demanding pre-selection of sources, enriched with guided reading tasks and supplemental materials. To our colleagues, we offer these courses as an opportunity to (re)use, discuss, and develop materials on a professional level.

Our target group are bachelor's degree students in their first and second year of study. Our curriculum earmarks methods and practices as learning goals for developing an exemplary foundational—and especially orientational—knowledge of German literary history. Students are expected to acquire basic techniques and concepts for literary historical research: e.g., historicity, concepts of epochs and epochal thresholds, literary evolution, and the tensions between canonical and non-canonical texts. Our courses offer students new ways toward these learning goals by using digital tools and techniques. For instance, we include easy-to-use tools to let students visualize literary historical processes. Moreover, our digital learning environment is also designed to enable first steps into research-oriented learning and hands-on experiences in collaborative digital workflows. Various tools help study groups to coordinate their work and to communicate and write together. Each course also includes an interactive presentation, whereby students can learn about the pros and cons of these tools, such as with respect to data privacy, for example.

Our course materials will be available internationally and discussible for professionals on the state-funded platform "ORCA.nrw" by the Digital University of North Rhine Westphalia (DH.NRW). In accordance with the Open Educational Resources (OER) approach, the courses will be open-access, reusable, and modifiable [1]. Our courses are implemented on the learning management system (LMS) "Moodle". Moreover, all courses are also implementable on other network-accessible LMS.

3 Background

It was before the COVID-19 pandemic that German Universities have been accelerating their digital transformation. In the past decade, the federal government of Germany has fiercely supported the "digital change" within research and academic teaching [2]. Since then, such a digital agenda has also been discussed in the Conference of Ministers of Education and Cultural Affairs. The conference is a standing consultative body that sets general guidelines for Germany's educational sector. There, in 2016, the importance of digital teaching for education and the everyday working life of students was set out in a strategy draft [3]. Furthermore, a paper from 2019 urges the promotion of interdisciplinary exchange in order to develop suitable concepts for the integration of digital elements in the curricula of teacher training, by extending digital teaching-learning formats [4]. With these strategic guidelines of the conference, the executive political board finally has begun following insights into the necessity of digital learning—although they have been discussed in higher education didactics for a couple of years, as the German Rectors' Conference stated [5].

Due to German federalism in educational matters, almost each state has founded digital platforms for the exchange of digital learning materials. Given this political framework, today, each university has its own digital teaching and learning agenda. Digital teaching and learning is pushed forward by centers for higher education didactics, and we see disciplinary initiatives accumulating. Indeed, digital learning has grown into a specialized field of research, both nationally and internationally [6, 7]. Not to drown the reader in too much of this abstract, highly specific knowledge at this point, we shall move on to taking a closer look at the discussions within our own discipline, German Studies.

To some extent the COVID-19 pandemic has become a catalyst for discussions on digital teaching methods in German Studies. In 2014, Schneider and Schöch [8] acknowledged that researchers in German Studies have expanded digital research methods and exchange opportunities in recent years. However, even if there are some initiatives to develop digital learning materials [9, 10], rarely does one hear about translocal interdisciplinary networking attempts, such as the Bochum event "DL_in_G", where, among others, German Studies scholars share experiences on their practical digital teaching attempts [11].

Another attempt to accelerate the discourse that had been delayed, if not outright rejected, within the discipline for a long time [12] is the "Digital Teaching in German Studies" portal "digitale-lehre-germanistik.de". In 2020, colleagues from Germany, England, Belgium, and the Netherlands came together to start an interprofessional exchange and launched this platform. They formed a consensus paper [12] with the aim of jointly reflecting the experiences made during the first digital "COVID-19 semester" and making them productive for further developments in German Studies and higher education didactics. Although the efforts made by the organizers are remarkable, the few resources listed on the website mainly refer to linguistic text analysis [12]. To summarize, the so-called Corona semesters revealed that digital learning formats, collaborations, and literary competencies for working in digital environments have widely been neglected in teaching the discipline.

So far, the teaching of literary history especially has not been addressed by any of the aforementioned developments. Moreover, it seems as if no professionally developed, interactive, distant-learning courses exist at the university level—apart from courses offered by the Open University Hagen, which are only accessible to its enrolled students. The KoLidi-project addresses this lack of an open-access digital learning environment for German-language literary history and embraces the discourse about it.

4   KoLidi in Context of Digital Teaching and Learning

According to Sarker et al.'s [6] extensive systematic literature review, there are multiple technology-based learning approaches that can help educators and learners in the teaching-learning process. They arrange teaching projects under established phrases like "Problem solving based learning," "blended learning," or "flipped learning." Their findings on "collaborative learning" and "distant learning approaches" correlate with our view on digital teaching and learning as an important component of leveraging digital technology to establish new ways of teaching humanities [6]. The research taken up in their paper showed that "Collaborative learning" can have a good impact on "academic performance" [13], and that "distance learning" can be successful if the instructional model is working [14]. This raises the question of what such collaborative and distant learning approaches might look like for German Studies.

With respect to teaching literature in a digital environment, one central challenge for the KoLidi project is the question of how to combine the long-established practice of intensive (analogue) individual readings and new ways of digital collaborative learning productively. Küchler describes this challenge briefly:

> Being able to summarize texts, to paraphrase knowledge, to give a reasoned value judgement about chosen phenomena and make decisions about [a text's] usefulness in a given context is maybe the most basic, yet also the most crucial work of academia. Learners, teachers, and researchers need to make informed decisions about the choice of knowledge and the exact methods of its application, about its relevance and importance, and, more importantly, about how to structure, categorize and evaluate available knowledge. This requirement demonstrates how automatizing summaries, categorization and retrieval may only have a rather limited reach in the humanities. Innovative knowledge and new ideas strongly depend on the adaptation, recombination, and transfer of knowledge to new usages, new fields or new contexts of application. This process of meaning-making seems to follow a disciplinary methodology, and yet an erratic, creative, and unpredictable pattern [15].

On a practical level, this means that digital courses should encourage users to dialogue about what they have read and learned. Literary studies as a discipline does not have broad parameterizable answer catalogues according to which a computer system can evaluate the answers. At the core of our discipline is the ability to find one's own intellectual voice and make it sound. To us, dialogue means the exchange of individual insights, and students need to be introduced to this way of thinking, speaking, and writing. This is one of the reasons why the classroom, the seminar, is still preferred over a digital meeting room or any LMS by so many of our colleagues. We do subscribe to the importance of these more classic methods. However, we also think that smart instructional design can take the disciplinary habits of older methods into account. Therefore, courses must work with open question-and-answer structures. Any result of (automatized) tasks, for instance, is an opportunity to get back to exchange and to foster discussion among our student users. Furthermore, we also believe that even a lively, open discussion among students can benefit from a teacher's voice. We think that the idea of fully computerized teaching will remain a tech fantasy distant to our field, at least within the realm of the current possibilities.

Scientific knowledge is in continuous flow, needless to say. Thus, it is necessary that teaching-learning materials be modifiable. The OER approach guarantees the meeting of this requirement, while also allowing instructors to adjust the materials to their own approaches and needs [16]. We would like to stress that point, since there also is a long tradition in our field of equating teaching with a kind of controlled authorship. For many of us, it is still a very awkward thought not only to share teaching material but to encourage colleagues to change and improve it. Yet this is also a very easy economic calculation; to produce digital learning materials is expensive and time consuming. Why, then, should we not share our limited resources for the best? KoLidi allows everyone to use, expand, and modify our materials—which, theoretically and technically speaking, should be easy to do.

Digital teaching can also be a way to make our classrooms more inclusive. We prepare our media so that it meets the needs of impaired students. KoLidi is barrier-free accessible. What is more, our course materials can be used modularly; this means an instructor might use only parts of it in a given class or course. Given the very heterogenous circumstances in our institutions, this seems to be the most appropriate approach for making our digital teaching environment most inviting. Furthermore—and this is no small consideration—it might also be a benefit for some of us not be saddled with the tangles of copyright laws [17].

5   Methods and Procedures

Our lecture courses rely on a modified model of the "constructive alignment" approach after Biggs and Tang [18, 19]. In synopsis, we follow the three overarching questions: (1) What are the **learning outcomes** or learning goals expected in the course? The term "learning outcomes" describes what students know and what they are able to do at the end of a learning unit. (2) What kind of **teaching and learning** methods, as well as learning activities, are used to achieve the learning objectives? (3) How can we best assess the achievements of learning objectives and outcomes by **examination**?

5.1 Learning Outcomes

Our learners need to discuss literary texts based on a specific methodological apparatus. They are becoming familiar with these methods and ways of thinking while working on a specific task in the field of literary history. However, it is important to say here that there is no such thing as a singular, monolithic "literary history." We rather speak of "histories of literature" in plural mode. Any approach to literary-historical processes is shaped by the perspective of its observer. Our learning outcomes do speak to this very basic but still complicated-to-learn insight. We would like to see students able to (1) read, analyze, and systematize a large amount of text; (2) index the material using discipline-specific terminology; (3) formulate hypotheses about aesthetic and topological changes in literary history; (4) work according to disciplinary norms, by conducting meaningful bibliographical research, as well as reviewing and applying research literature; (5) write in discipline-specific and-appropriate style; (6) communicate critically about literature in their study groups; (7) know how to use digital tools in literary studies; and finally (8) reflect on their own learning processes.

5.2 Teaching and Learning

Task Types. The teaching and learning structures of our digital courses are oriented towards basic hermeneutic processes of understanding and learning that are fundamental to scientific argumentation and research. To that end, students are guided by various task types, which are set at different levels.

First, we need to ensure the students' understanding of the given texts. Comprehensive-oriented questions guide students in their reading, helping them to establish an overall understanding of the literature. They draw their attention to key passages in the text that particularly illustrate the chosen thematic focus of the course. These questions, which tend to be small-scale, provide concrete clues as to how students should approach certain aspects of texts.

Secondly, we seek to deepen the students' basic knowledge through certain exercises. Exercises enable students to acquire, deepen, and practice basic knowledge about historical, aesthetic, and/or methodological aspects of the works they are reading. Comprehensive-oriented questions and knowledge exercises do not follow a strict chronology. Depending on the specific features of a certain text, students need to build a foundational understanding first before continuing with the reading. Or, the other way around, students need to comprehend the text before they can even proceed with any form of knowledge-based exercise. Therefore, exercise sequences and comprehensive-oriented questions are repeatedly integrated into interpretation and discussion tasks, as well as into small writing tasks (see below).

Third, discussion questions help students to exchange ideas in groups and to practice scientific language. Group discussions make it possible to gather different perspectives on an idea. Also, students are given the opportunity to socialize and network with each other [20].

Fourth, interpretive questions aim to guide students in distancing themselves from texts and analyzing them from the perspective of their acquired historical, aesthetic, and methodological knowledge. Interpretive questions can build on applied knowledge to foster synthetic competencies.

Fifth, small writing tasks serve to solidify the results from discussion and interpretation. They can take a wide variety of formats (timelines, other forms of visualizations, blog entries, etc.). In general, they point to the interconnectedness of certain aspects. Additionally, the students gain practice with writing at an academic level. These assignments—sixth—prepare the students for larger writing tasks, such as essays or papers in which they can prove that they are now able to present an original idea with respect to the material they have worked on in the course.

Media. Having a variety of media types available helps different learner types to process the large amounts of texts and tasks they encounter [21]. We therefore combine analogue and individual readings with digital elements and collaborative tasks. To keep the media inputs focused, the digital elements are inspired by the idea of "microlearning" [22]. With clear instructional design, various media are used to enable students to resolve complex issues with small learning units

Primary Sources. The historical texts that we chose for our courses belong predominantly, but not exclusively, to the German literary history canon. Although our discipline fiercely debates the problems of non-diversified curricula, most of our students will have to teach the canon after their studies as teachers. As a matter of fact, we often see that new students of the field enter our programs with only a small pre-existing familiarity with any canonical texts. We do think, however, that in order to challenge and enrich the canon, students should know what the canon actually is [23]. Familiarity with canonical texts and the ability to situate them within a history of aesthetic forms and societal power structures, after all, are the preconditions for questioning the long-established conventions of literary historiography. To ensure such a critical perspective then, all courses are accompanied by canon-critical voices and readings that challenge dominant historical narratives [24].

Videos. We are using video input to introduce professional debates about the courses' subject matter. In addition, short lectures cover historical and scientific contexts. Although reading research literature is part of the courses' exercises, research literature itself can only factor into the course design in a very limited, selective form, since students already need to conquer a large amount of historical reading besides. Video inputs are designed to close that gap, to some extent.

Audio. Our short audio recordings serve very different functions. They range from subject-related matters to methodological accesses on the field. The audio files within a single course vary. For example, in a course on the transformations of the dramatic form in the 19th Century, these range from the explanation of metatheatrical structures in Friedrich Schiller's *Maria Stuart*, to the description of other versions of the Herrmann "myth" in drama, beyond Heinrich von Kleist's *Herrmannsschlacht*, up to the distinction of different modes of representation in literary and non-literary texts [25].

Images. Images are mainly used to illustrate the contemporary reception of the reading assignments. We also present images of manuscripts and first-print editions. In addition to the instructional aspect, images brighten up "flat" textual surfaces—a response to how reading habits in a digital environment differ from those on the printed page.

Interactive formats. While the forums and other digital media formats can also be described as "interactive," this section specifically refers to our H5P-Formats. H5P is an open-source and free-to-use tool that is meant to enrich presentations, interactive videos, quizzes, or images with interactive elements [26]. We mostly use H5P to present additional declarative knowledge. For example, students may comprehend philological editing processes of a handwritten manuscript through the tool titled "Image Juxtaposition." With the help of this interactive content, students can overlay two editions to compare them closely.

5.3 Examination

Any course includes different kinds of assessments of a student's efforts, as we mentioned above. In addition to these smaller tasks, we also incorporate exercises that point to a more general review of the learning outcomes, as is common in our disciplinary field. Writing assignments combine "declarative" and "functional knowledge" on different structural levels [18]. Not only can the students thereby recapitulate knowledge learned (declarative), but they should also be able to reflect this knowledge and relate it to other texts and models that they have encountered in of the whole course (functional).

Therefore, at the end of the course, students are encouraged to work on two out of three larger writing assignments. We expect students to draw on multiple readings and research questions addressed in the course. Thus, they may show that they managed to identify changes and tendencies in literary history and that they can outline them in a disciplinary-appropriate way. We also supply some exemplary approaches, giving initial feedback to students before their instructor provides deeper commentary.

6 Outlook

KoLidi responds to the question about the sustainability of the project on two different levels: by student "process evaluation" [27] and by feedback from disciplinary colleagues. Students' evaluations and feedback about the user experience form the basis for further optimizations and changes to our courses. These evaluations, in other words, function as quality control. Three aspects of course design are of vital importance to us: (1) usability of the digital content, (2) rating of the social interactions, and (3) assessment of the general conditions under which students complete the course. Our first evaluation will take place in the summer semester of 2022. Its focus lies on evaluating existing instructional designs, (interactive) media types, and group activities. It will provide development perspectives for existing and future courses. Even though it would be interesting to know what the courses' impacts are on any possible increase in competencies, this particular outcome cannot be measured at the time of the first evaluation; competencies and knowledge will be evaluated, instead, in the final exams. Further research in this direction will have to be carried out in a different framework.

Feedback from colleagues in the discipline is important. Our network of project-involved scientists provides us with assessments of the materials that we create in smaller feedback loops. Furthermore, and prospectively, scientific feedback may be given through classic channels, such as conferences or workshops. We hope that discourse spaces such as the ORCA.nrw platform or the digitale-lehre-germanistik.de site will be used to discuss our materials and courses. Our vision of teaching and learning in the digital age aims to "democratizing" these processes [28]. Such an objective requires that a community of practice evolves, shares, and develops learning materials.

7 Conclusion

KoLidi argues that individual reading and digital collaborative learning can be productively combined in our field of literary history studies. We believe that digital courses indeed help our students to conquer central tasks in our curricula with respect to literary history and methodological reflection. The COVID-19 pandemic has opened a broader space for dialogues on digital teaching-learning concepts in our field. We aim to make an innovative contribution to the integration of digital content into university teaching—professionally developed and ready to deploy. To our users, we offer learning paths inspired by the "constructive alignment" model. Still, we are teachers in the field of humanities. Mutual exchange is at the core of our discipline, and our audio-visual and interactive learning materials address this crucial point. Students effectively work in study groups towards the end of the course exam, guided by their individual instructor, who adjusts the material we have provided to the needs of the class being taught.

To secure the sustainability of the project and to adjust the instructional design, there will be continuous process evaluations. These aim at the assessment of individual media elements, assignments, and overall course quality for learners and teachers. We seek any feedback from colleagues and would be delighted to collaborate with teams from other disciplines, in the humanities and beyond, to keep pace with further developments in digital teaching. We wish to be part of a community of practice in digital teaching and learning that has yet to be fully established within our discipline. We are confident that KoLidi is indeed an attractive contribution to facilitating greater dialogue for such a community.

Acknowledgments. We are thankful to the Digital University of North Rhine Westfalia for funding the KoLidi-Project in the context of the development of the ORCA.NRW platform. We are also grateful to committed students and colleagues beyond the project team, who are willing to contribute with their feedback and expertise to put KoLidi forward. For more information about the extended team, see: https://literaturgeschichte-kolidi.de/das-team/.

References

1. KoLidi-Website: Über das Projekt. https://literaturgeschichte-kolidi.de/das-projekt/. Accessed 3 Nov 2021
2. The Standing Conference of the Ministers of Education and Cultural Affairs of the Länder in the Federal Republic of Germany: Bildungsstandards im Fach Deutsch für die Allgemeine Hochschulreife (2012). https://www.kmk.org/fileadmin/Dateien/veroeffentlichu ngen_beschluesse/2012/2012_10_18-Bildungsstandards-Deutsch-Abi.pdf. Accessed 9 Nov 2021

3. The Standing Conference of the Ministers of Education and Cultural Affairs of the Länder in the Federal Republic of Germany: Strategy "Education in the Digital World (2016). https://www.kmk.org/fileadmin/Dateien/pdf/PresseUndAktuelles/2018/Digitalstrategie_2017_mit_Weiterbildung.pdf. Accessed 3 Nov 2021

4. The Standing Conference of the Ministers of Education and Cultural Affairs of the Länder in the Federal Republic of Germany: Recommendations for digitisation in higher education teaching (2019). https://www.kmk.org/fileadmin/Dateien/pdf/PresseUndAktuelles/2019/BS_190314_Empfehlungen_Digitalisierung_Hochschullehre.pdf. Accessed 3 Nov 2021

5. Kreulich, K., Wortmann, D.: Kompetenzen für die digitale Gesellschaft und Arbeitswelt. In: Digitale Lehrformen für ein studierendenzentriertes und kompetenzorientiertes Studium. Tagungsband: eine Tagung des Projekts nexus in Zusammenarbeit mit dem Center für Digitale Systeme (CeDiS) der Freien Universität Berlin, 16. und 17. Juni 2016. nexus Tagungsband, 1st edn., pp. 27–34. CeDiS; HRK Hochschulrektorenkonferenz Projekt nexus - Übergänge gestalten Studienerfolge verbessern; Waxmann Verlag, Berlin, Bonn, Münster (2018)

6. Sarker, M.N.I., Wu, M., Cao, Q., Alam, G.M., Li, D.: Leveraging digital technology for better learning and education: a systematic literature review. IJIET (2019). https://doi.org/10.18178/ijiet.2019.9.7.1246

7. Zawacki-Richter, O., Latchem, C.: Exploring four decades of research in computers & education. Comput. Educ. 136–152 (2018)

8. Schneider, L., Schöch, C.: Literaturwissenschaft im digitalen Medienwandel: Einleitung. PhiN - Philologie im Netz - Beiheft 1–17 (2014)

9. Kock, K., Lahn, S., Meister, J.C.: Treffpunkt AGORA. Literaturlehre und -forschung mit der integrierten e-Learning und e-Science-Plattform der Hamburger Geisteswissenschaften. Zeitschrift für Germanistik 91–103 (2011)

10. Boatin, J., Sina, K., Trilcke, P., Volmari, C.: eLearning by doing. Göttinger Projekt einer elektronischen Publikations- und Vernetzungsplattform für Studierende. Zeitschrift für Germanistik 104–113 (2011)

11. Kleinwort, M.: Corona als Brennglas. Denkanstöße zur literaturwissenschaftlichen Lehre im Jahr 2020. undercurrents – Forum für linke Literaturwissenschaft 1–13 (2020)

12. Forschungsverbund Marbach Weimar Wolfenbüttel: Das brauchen wir: 8 Anforderungen an die zukünftige Lehre in der Germanistik (2020). https://www.digitale-lehre-germanistik.de/konsenspapier. Accessed 2 Nov 2021

13. Aloraini, S.: The impact of using multimedia on students' academic achievement in the College of Education at King Saud University. J. King Saud Univ. Lang. Transl. (2012). https://doi.org/10.1016/j.jksult.2012.05.002

14. McCutcheon, L.R.M., Alzghari, S.K., Lee, Y.R., Long, W.G., Marquez, R.: Interprofessional education and distance education: a review and appraisal of the current literature. Curr. Pharm. Teach. Learn. (2017). https://doi.org/10.1016/j.cptl.2017.03.011

15. Küchler, U.: Digital learning and the humanities. 178–189 Pages/PraxisForschungLehrer*innenBildung. Zeitschrift für Schul- und Professionsentwicklung., vol. 2, No 4 (2020): Standards—Margins—New Horizons. Teaching Language and Literature in the 21st Century (2020). https://doi.org/10.4119/pflb-3504

16. OERInfo: Abschlusspublikation der Informationsstelle Open Educational Resources (OERinfo) (2020). https://open-educational-resources.de/wp-content/uploads/OER_Abschlusspraesentation_201102_final.pdf. Accessed 7 Nov 2021

17. Patzer, Y., Sell, J., Pinkwart, N.: Anforderungen und ein Rahmenkonzept für inklusive E-Learning Software. In: Lucke, U., Schwill, A., Zender, R. (eds.) DeLFI 2016 - die 14. E-Learning Fachtagung Informatik der Gesellschaft für Informatik e.V. 11.-14. September 2016 Potsdam, Deutschland. GI-Editio: [...], Proceedings, vol. 262, pp. 257–268. Gesellschaft für Informatik e.V, Bonn (2016)

18. Biggs, J.B., Tang, C.S.: Teaching for quality learning at university. What the student does, 4th edn. SRHE and Open University Press imprint. McGraw-Hill Society for Research into Higher Education & Open University Press, Maidenhead, England, New York, NY (2011)
19. Leibniz-Institut für Wissensmedien: Constructive Alignment (2020). https://www.e-teaching. org/didaktik/konzeption/constructive-alignment. Accessed 19 Jan 2022
20. Geramanis, O., Hutmacher, S., Walser, L. (eds.): Kooperation in der digitalen Arbeitswelt. Verlässliche Führung in Zeiten virtueller Kommunikation. uniscope. Publikationen der SGO Stiftung. Springer Gabler, Wiesbaden, Heidelberg (2021)
21. Meyerhoff, J.: Fachwissen lebendig vermitteln. Das Methodenhandbuch für Trainer und Dozenten, 4th edn. Springer Fachmedien Wiesbaden, Wiesbaden (2016)
22. Tipton, S.: Microlearning As a Framework. In: Brusino, J. (ed.) ATD's 2020 Trends in Learning Technology, pp. 7–16. ATD Press, Alexandria, VA (2020)
23. Cooper, G.: What good is the Canon for a diversified and decolonized curriculum? Die Unterrichtspraxis/Teaching German (2020). https://doi.org/10.1111/tger.12138
24. Cornelißen, C.: Zentrum für Zeithistorische Forschung Potsdam: Erinnerungskulturen, Potsdam (2012)
25. KoLidi: KoLidi-Podcast: Margreth Egidi zur Unterscheidung literarischer und nicht-literarische Texte (2021). https://youtu.be/fGn4jAjf_Uk. Accessed 27 Jan 2022
26. Sinnayah, P., Salcedo, A., Rekhari, S.: Reimagining physiology education with interactive content developed in H5P. Adv. Physiol. Educ. (2021). https://doi.org/10.1152/advan.00021. 2020
27. Gollwitzer, M., Jäger, R.S.: Evaluation kompakt. Mit Arbeitsmaterial zum Download, 2nd edn. Kompakt. Beltz, Weinheim, Basel (2014)
28. Blayone, T.J.B., vanOostveen, R., Barber, W., DiGiuseppe, M., Childs, E.: Democratizing digital learning: theorizing the fully online learning community model. Int. J. Educ. Technol. High. Educ. 14(1), 1–16 (2017). https://doi.org/10.1186/s41239-017-0051-4

Learning and Performance Science for Digital Transformation

Gary J. Dickelman[✉]

EPSScentral LLC, Boynton Beach, FL 33437, USA
gdickelman@epsscentral.net

Abstract. Select any industry and you will find scores of articles detailing how the SARS-CoV-2 pandemic fostered an explosion of online learning over the past two years. Within social distancing guidelines, organizations were able to meet many challenges of the learning and performance space. At the same time, training budgets have shrunk to offset pandemic headwinds as organizations rally to survive. Focus has shifted instead toward digital transformation of big data, to unify, manage, and visualize the flow of structured business data and, thus, improve overall efficiency and outcomes. While impressive innovations have emerged to focus learning and performance (L&P) on what is most pertinent, L&P content science lags woefully behind data science. Scalability of L&P content remains largely elusive. This paper outlines a vision for an L&P science to foster a cogent digital transformation on par with modern data science.

Keywords: Learning · Performance · Training · Digital transformation · Data science · Knowledge ecosystem · Performance support · Analytics · Compliance · Privacy · Scalability

1 Caveats

The L&P science vision addressed here is not about the many successful L&P lifecycles that reflect decades of progress. Nor is it about learning management systems (LMSs), content controllers, authoring tools, or data management around learner/performer competency. L&P management has surely benefited from advances in big data science. What lags behind is an intelligent *content* pipeline analogous to intelligent data pipelines. Sentences, paragraphs, media representations, and other fragments that flow into developing polished L&P solutions have lifecycles of their own, which are barely codified, normalized, or automated. The fundamental difference between data science and the envisioned L&P science is that the latter deals broadly with semantic structures and associated challenges around machine understanding and human-computer interaction. Data consists of short, digitized fact strings that fall into (mostly) unambiguous categories. Consider the contrast with semantic content. Some organizations have begun to architect semantic layers above the many disparate repositories, which provides "a business view of complex knowledge, information, and data and their assorted relationships in a way that can be visually understood" [5].

D. Guralnick et al. (Eds.): TLIC 2022, LNNS 581, pp. 125–135, 2023.
https://doi.org/10.1007/978-3-031-21569-8_12

Generally, L&P has not been a significant benefactor of such architecture, even when it exists in an enterprise. Common applications are around visualizing data relationships and applied to compliance, marketing, sales, governance, machine learning (ML), and more. Current applications of semantic architecture do not encompass all of "knowledge, information, and data," but they are focused primarily on data, as defined above, sometimes including "documents."

2   Data Science Today

An entire vocabulary has emerged around data science, including *digital transformation, artificial intelligence (AI), ML, scalability, low-code/no code solutions, predictive analytics, shift-left, contextualization*, and so many more. One expert recently published a list of 101 terms that data analysts should know [4].

The aim of data science is to continuously extract gems of wisdom from big data to improve business processes and outcomes. It requires intelligent data pipelines and specialized technologies to "…unify, manage, and visualize the flow of structured business data … with the goal of improving the overall efficiency of a business [5]." As computing devices have become relatively inexpensive and powerful, businesses have captured and stored huge volumes of data, quickly overwhelming their capacity and know-how to make sense of it all. Consequently, data science has evolved at a breathtaking rate. From a product and service perspective, it is today an enormously crowded marketplace. The typical business shakedown that leaves only a few of the best solutions is in its early stage. Channeling Geoffrey Moore, data science solutions have not yet crossed "the chasm" [25].

Data science is now a coveted discipline, taught across colleges and universities. It is the subject of many Gartner, Forrester, Nielsen, IDC, (and other expert) insights. As a result, there has been an explosion of data science products and services, all quoting the industry experts. The most common of the many buzzwords include "digital transformation," "democratization," "low-code/no code solutions," "contextualization," "continuous compliance," "shift left," and "edge computing." Search the Web on "shift compliance left," for example, and you'll get hundreds of products riding Gartner's and Forrester's coattails. There's nothing more satisfying to a technology vendor than being in Gartner's crosshairs, or better, occupying the upper right quadrant of a magic square [24].

Digital transformation is all about extracting *wisdom* from data as it moves through the pipeline:

- **Data** consists of small strings of digitized facts.
- **Information** is data with context.
- **Knowledge** is actionable information.
- **Insight** is generality abstracted from knowledge, where patterns of success emerge from individual solutions.
- **Wisdom** is insight rolled up into sound operating principles for continuous improvement.

Big data's automation technologies interact with different parts of the pipeline to ultimately arrive at wisdom:

Data -> Information -> Knowledge -> Insight -> Wisdom.

Some offer holistic, "full stack" solutions. Organizations capture almost everything, and solid business decisions are made using such tools without humans ever having to look at typical data elements. One of wisdom's most significant attributes is business's ability to predict and, thus, operate at much higher efficiency, continuously so.

3   Toward an L&P Science

The vision is to evolve an L&P science that extracts wisdom from *content* by borrowing from data science lessons and protocols, at least in part. That is different from simply applying data science to L&P *data*, as some professionals have argued is sufficient. But if we examine data science attributes and protocol detail, gaps quickly emerge as we compare and contrast data pipelines with L&P *content* pipelines.

Consider the following questions, where the context is a typical handoff between a product manager and L&P manager, communicating what is important in a new product or upgrade for L&P consideration:

- When the L&P solution is complete, can you trace its components back to the original communication from the product manager?
- Can you identify the communication's content all the way through development and evaluation cycles, forward and backward?
- Is the content codified such that you can determine what in the L&P taxonomy it supports?
- Can you connect the content to learner/performance evaluation for updating, clarifying, improving, etc.?
- Can you determine how the content supports and informs rigorous, quantitative performance evaluation?
- Is there an automated, continuous improvement cycle for the content pipeline?
- How much of the L&P development cycle associated with the content is automated? For example, is the L&P solution automatically updated if the product manager updates the communication?

To be sure, some elements of an intelligent L&P content pipeline already exist, but there is no apparent agreement on a comprehensive framework. In many respects, the SARS-CoV-2 pandemic has driven L&P practitioners more toward affectivism (i.e., focus on the affective domain) than on an intelligent L&P content pipeline. The narrative can be redirected by considering 10 critical data science attributes as they may apply to an L&P science, and where gaps and opportunities emerge. They are:

1. Codifying Content and Knowledge
2. Content Quality Management
3. Content Object Reuse
4. Semantic Encoding

5. Continuous, automated content monitoring and evaluation
6. Compliance, privacy, and trust
7. Prediction
8. Contextualization
9. Democratization
10. Cross-Discipline Cooperation

The remainder of this paper examines each attribute and summarizes how they manifest in data science, how we might apply the associated practices to an L&P science, where are the gaps, and how might they be filled.

3.1 Codifying Content and Knowledge

If you could be a fly on the L&P wall of any enterprise, you would see content flowing in from email, scattered among many individual workstations, in shared file servers, in content and document management systems like Sharepoint and Google Docs, in knowledge management systems like Oracle Knowledge, ServiceNow, Zendesk, and KNOWMAX, and from relevant blogs. To be sure, there is CMIS—Content Management Interoperability Services—that "provides a common data model covering typed files and folders with generic properties that can be set or read. There is a set of services for adding and retrieving documents ('objects') [6]." CMIS provides a means to search multiple, disparate content sources. By all measure, its utility has been around data and documents, but not the range of what comprises the L&P ecosystem.

A core issue is that organizations do not typically make crisp delineations between content and knowledge. Delineation is mostly nonexistent, obscured, or misunderstood. Repositories that comprise the L&P ecosystem—learning modules, knowledge articles, documents, application help, virtual assistants, job aids, checklists, blogs—are mostly separate entities of distinct domains, scattered across disconnected instances of SharePoint, Teams, Google Docs, within applications, and more. These digital entities are, unfortunately, used most often as simple filing cabinets, with not much more than keyword search. Discipline around versioning documents, for example, is problematic. Consider the confusion in how SharePoint and Teams are misapplied as knowledge management systems vs. the content management systems they were designed to be.

Douglas Weidner, founder of the Knowledge Management Institute [7], defines knowledge as "information in action." It means that information is not knowledge unless it solves relevant problems and answers questions of the day. Knowledge has present value and benefits; content may not. Knowledge is accessible, actionable, and current, which is what L&P is all about. By this distinction, it is clear that content often languishes in its repositories as it becomes less relevant, incomplete, and inaccurate as business and market conditions change. A well-designed Sharepoint or Teams system might classify content accordingly and through its lifecycle. But how many such instances are used that way? Like an intelligent data pipeline, an intelligent content pipeline would not have content languishing in obscure repositories.

Strategies for knowledge design and management differ from those of content. Taxonomy and ontology serve both, especially if there is to be enterprise search optimization.

At a minimum, L&P science requires clear distinction between content and knowledge framed within an ontological structure.

3.2 Content Object Reuse

Aside from SCORM and xAPI fostering great system interoperability for L&P objects, the vision of taxonomy-based construction and reuse articulated circa 1997 have not been realized. For example, relationships between raw content and how it ultimately manifests through the pipeline to published learning objects are lost in development cycles. Interoperability protocols surely exist and afford great utility for taming the content beast, but interoperability absent L&P taxonomy and ontology does not get to the root cause of unacceptably long development cycles, poor quality management, or the inability to assess actual behavior and performance.

The movement in big data science toward contextualization and visual representation (discussed in detail later) has not yet found its way into L&P practice. Education science conducts item analyses and calculates discrimination indices to determine efficacy and validity of test items. In the workplace, tasks are analogous to test items. Is there a similar protocol for distinguishing poor performance from ill-framed tasks? Adding more learning/training to improve performance on such tasks yields little to no return on investment. It is perhaps time for a Rasch model applied to tasks [1, 8].

3.3 Content Quality Management

There does not exist L&P content hygiene similar to that of data quality management. While some useful protocols exist in knowledge management systems—synonyms, stop words, concepts, keywords, proximity, relevancy, and the like—what about the rest of the L&P ecosystem? Publish exactly the same content to Sharepoint and Oracle Knowledge and see how quickly you can raise it through the same search strategy in each system. Not many organizations actually tune their SharePoint or KM systems for search optimization, or at least not at the level to support L&P science. It requires a more comprehensive service, like CMIS, across the entire ecosystem.

Data quality metrics and tools have evolved to address *accuracy, completeness, consistency, integrity, currency* (is the data up-to-date?), and *relevance* to the business problems at hand. In many respects, data cleansing and profiling are analogous to environmental ecology: If you do not clean out the old growth and attend to the new, hundreds of thousands of acres might burn uncontrollably. "Wildfires" downstream of poor data ecology manifest in lost competitive advantage, decreased market share, misdirected business leads, loss of venture funding, and leaving business opportunities on the table, to name just a few.

Key benefits of data quality to digital transformation include competitive advantage, razor-sharp sales and marketing focus, rapid system rollouts, and efficiently integrating disparate and emerging technologies. Most importantly, data quality enables organizations to draw a straight line from the investment in data management to its impact on the bottom line. L&P ecosystem quality management would reap the same benefits, plus increase business performance at lower support costs. It is not a stretch to say that wildfires in the L&P content pipeline too often burn out of control. An ecology for L&P

science analogous to that of data science is elusive. In the big data space, tools featuring human-centered graphical representations are available for continuous data quality monitoring and improvement. Why not for the L&P ecosystem?

Data science includes metrics like the number of errors per unit amount of data, or the rate at which parsing fails per unit amount of data. Such measures are correlated with those of business outcome quality, making the end-to-end cycle from data acquisition to business decision adaptive—which is, after all, the goal of ML. Has L&P established metrics around content quality? There are perhaps protocols with respect to formatting content for a particular learning strategy, but what about content stripped of style? Semantic encoding and markup have found their way into content management, and that is perhaps a starting point for quantifying content quality. If L&P is serious about object reuse and automation, then the basic attributes of data quality management are critical to content quality management.

3.4 Semantic Encoding

Quality data elements fall mostly into unambiguous categories. As discussed, L&P content is semantic. Protocols of the Semantic Web are applicable to the entire L&P ecosystem, especially those that inform Ontology and Trust, the top tiers of Berners-Lee's Semantic Web layer cake [2]. Where is L&P with all of this, like ML, natural language, and other applications of AI? Can our machines find, read, and reason about chunks of content? There is some success around semantic encoding of specific classes of documents, but we are certainly not there across the entire L&P ecosystem.

According to the American Psychological Association, Semantic encoding is "...cognitive encoding of new information that focuses on its meaningful aspects as opposed to its perceptual characteristics. This will usually involve some form of elaboration" [9]. Elaboration in big data science involves contextualization, which means surrounding it with related information to make it more meaningful and useful. According to Gemini, a solution provider for data contextualization and visualization tools [10], the benefits of contextualization include improved insights, reduced dependence on deep data science experience, simplicity, cost-effectiveness, lower cost, and manageability. Big data tool vendors seem to agree on these benefits, especially since being listed by Gartner [11], Forrester [12] and all the rest, as trending. More on contextualization follows.

3.5 Continuous Content Monitoring and Evaluation

Data science includes interoperable frameworks (e.g., for compliance) that enable continuous monitoring, prediction, and improvement. Gartner's "shift left" means evaluating data at the earliest possible moment, and continuously, through an intelligent pipeline [13]. Evaluating data late in the game by extracting, analyzing, and reporting in spreadsheets and presentation decks is not only inefficient, but the results are often obsolete within moments of publication. L&P processes still extract data—mostly from learner evaluations—and generate point-in-time reports in spreadsheets and presentation decks.

Do organizations really employ rigorous, quantitative performance evaluation? Data science is now able to predict customer behavior and buying habits. Where is the "shift-left" and "predictive analytics" for the L&P ecosystem?

Continuously monitoring, evaluating, predicting, and improving L&P content requires frameworks analogous to those employed for data compliance. For example, the Department of Defense (DoD) recently published cATO (continuous authority to operate) guidelines for its IT contractors per the Risk Management Framework (RMF) [14]. The goal of organizations in this category is to automate continuous compliance so that authority to operate is never revoked. For example, RegScale [15]—spawned from the highly successful C2 Labs' full stack compliance solutions [16]—provides a scalable, holistic compliance management solution of this nature.

The analogy is to continuously monitor the L&P ecosystem for meaningful semantic encoding, which could be baked into a conformance or compliance framework. That has been done successfully for L&P solution interoperability (e.g., SCORM manifest and conformance testing). Why not for a deeper dive into the L&P ecosystem? That may be a stretch for a universal framework, but it is certainly achievable at the edge (close to the content source and reflecting organization specifics or verticals) [17].

3.6 Privacy, Compliance, and Trust

What is L&P ecosystem compliance vis-à-vis the EU's General Data Protection Regulation (GDPR) [18] and the litany of similar privacy regulations that have emerged worldwide? L&P's aim is often to personalize, where the system in which you work observes you, calls out performance gaps and then delivers just-for-you remediation, in real time. It serves to continuously improve business performance through human performance. But are we able to embrace privacy compliance such that it does not defeat L&P individualization? "Anonymization"—a key requirement of data privacy regulation—is the antithesis of personalization. How do we know when we are throwing the personalization baby out with the privacy wash water? What are the critical compliance issues across the L&P ecosystem?

As an example, consider software simulations, which have been a staple for learning to use applications. Tools such as Epiplex [19] and Adobe Captivate [20] have automated simulation development around transaction-based systems in many industries. The automated capture process initially grabs screen images that may include customer names, demographics, and account numbers in some cases. Depending on how sims are made available to learners, content may need to be anonymized for compliance. The problem has been partially addressed by automating redactions in the capture process, but that's just a simple example in one class of regulations.

How could, or would, organizations address continuous, automated compliance across the L&P ecosystem? The techniques in big data science include generalizing and suppressing, anonymizing, and permuting (i.e., de-linking relationships between data attributes without modifying them), and more. The primary goal is to remove the risk associated with legal consequences of non-compliance, to which the L&P ecosystem is subject.

3.7 Predictive L&P

Learners and performers are the customers of the L&P ecosystem. Why, then, is there not predictive L&P analytics like we have, for example, in predictive data analytics, such as predictive advertising? Which are your best candidates for filling critical competency gaps for maximum business improvement? Big data science combines first and third party data to target best marketing and sales candidates [21]. The same techniques could be applied to L&P.

To illustrate how first and third party data is applied, consider the familiar example of visiting your doctor. You were probably given a logon to a personal health record (PHR) or electronic medical record (EMR) portal. There you have space to enter personal details and health history. Results of lab tests and doctor visits are added automatically. Behind the scenes, predictive analytics look at groups similar to you, combined with what you entered, to create healthcare recommendations that are not one-size-fits-all. That's predictive analytics making your doctor so much smarter and focused about your healthcare recommendations, with the best options for you to improve your health. In this example, first-party data are the details you entered into the portal, plus lab and exam results. Third-party data are all the health details about everyone like you: Demographics, healthcare trends, maladies, morbidity, risks, impact, and mitigation. Combining first and third party data enriches context while narrowing the target to just-for-you.

Predictive analytics around L&P customers and content would foster lower L&P solution costs, increase performer productivity, and yield a much higher return on the L&P investment. Predictive analytics applied to advertising uncovers target marketing and sales personae, extending the value of customer relationship management (CRM). It predicts buyer behaviors and, thus, fosters more successful campaigns. Predicting behavior tells you who are your best target customers, and which you should step over. A corresponding predictive analytics for L&P would tell us more precisely what needs to be trained or supported, for whom, to realize the greatest return on investment.

3.8 Contextualization

These days analysis tools contextualize data to expose relationships and provide deeper insights, often with meaningful graphics for visualization. It is a critical step in the journey from data to wisdom. According to Gartner, "By 2023, graph technologies will facilitate rapid contextualization [of data] for decision making in 30% of organizations worldwide" [22]. What about the L&P ecosystem? To what extent are its objects ever contextualized? How is it possible to enable, measure, and improve performance if there are no explicit mappings between performance taxonomies and personae context? Without contextualization, so much insight is obscured, and prediction is severely limited. Organizations pay dearly for over generalizing, developing more content than is needed, and delivering to much broader audiences than necessary.

Competency is demonstrated know-how across the performance taxonomy for a job. Competency assessment requires further specifics around individual behaviors as he or she performs tasks within essential use cases or job scenarios. For computer-mediated tasks, L&P objects that support task completion can be mapped to the actions and behaviors of job incumbents. These are often called context maps, which enable knowledge support to be delivered at the time of need.

An approach that has been successfully applied (albeit infrequently) is to establish context maps for primary and secondary personae. The protocol assesses worker actions as he or she navigates the virtual job space and compares it to best-practice use cases. When errors are detected, support is delivered from the knowledge ecosystem. Contextual support of this nature can be effective for a majority of use cases that fall in the range of intermediate complexity. For this approach to be practical and successful, the entire knowledge ecosystem needs to be contextualized with respect to performance taxonomies, which binds knowledge objects to performer actions.

In the big data world, tools have emerged to facilitate rapid contextualization and provide visualization of relationships otherwise hidden. That is precisely the utility required of the L&P ecosystem.

3.9 Democratization

Democratization is about putting "no code/low code" tools in the hands of subject matter experts or generalists to perform tasks that would otherwise require technology savvy specialists. L&P has enjoyed decades of democratization in authoring and management tools, but not for an intelligent content pipeline. So many L&P tools are desktop utilities rather than enterprise solutions, hence content flow from development to publication breaks critical connections, leaving many valuable artifacts out of the realm of L&P management and analytics. Organizations serious about enterprise L&P content and knowledge management still require deep technical savvy to pull it off. There is little democratization of L&P content pipeline tools, let alone an intelligent content pipeline.

3.10 Cross-discipline Cooperation

In the course of calling out the list of data science attributes in the preceding paragraphs, a number of distinct disciplines were identified, including big data science, the semantic Web, privacy regulation and compliance, predictive analytics, ML, AI, learning and performance management, content authoring, performance evaluation, performance taxonomy, and more. Individually, and in some subsets, they point to elements of an L&P ecosystem science. But there are apparently no compelling incentives for cross-discipline cooperation to accelerate its development.

Who in the L&P space has not felt the sting of drastic budget cuts when a business imperative of higher perceived value gets the attention? "Training is the first thing to go" is a common mantra. Or worse, technical development trumps usability and customer experience, hence performance gaps associated with poor designs are expected to be filled with "training." During the pandemic, there were many positive innovations in e-learning to accommodate social distancing and working from home. On the other hand, there is much hastily cobbled together virtual learning that caused more pain than performance. By analogy, at some point organizations realized the business imperative of cognitive science. User experience and customer experience, for example, have in many organizations become mainstays. Why not L&P, as well?

In the 1990s, Howard Rheingold wrote forward-thinking books about "mind-expanding technologies." He was talking about how our intelligence can be amplified with new technologies and how powerful cross-discipline cooperation could be in fostering such advances. Consider his thesis on The Virtual Museum of the Frog, from his book *Virtual reality* [23].

A properly curated display would not include just frogs. It would focus on a distinct species, placed in its natural habitat, with all its flora and fauna. You would get a complete picture of where it lives, how it lives, what it eats, its predators, reproduction cycles, and so forth. Unfortunately, repositories of frog media, frog biology, frog ecology, habitat weather, food chain, and so many more, were stove-piped. For there to be a meaningful virtual museum of the frog, cooperation among the repository stakeholders and their wares is paramount, argued Howard Rheingold.

The virtual museum of the frog is visual. You see which plants and insects it eats, which foliage it uses for camouflage, how it swims and hides in water, etc. Such high fidelity visuals featuring those relationships would not be possible without synthesizing the various disciplines and data sources. Computing power, argued Rheingold, was not (and is not) the limiting factor. Rather, cross-discipline cooperation is imperative. Create incentives for collaboration, said Rheingold, and you might just get everything you need for the virtual museum. The same goes for L&P science.

4   Stay Tuned…

Is L&P science moving in a positive direction? Somewhat, albeit way behind the data science curve. To be sure, there is far more complexity to the L&P ecosystem than to data, and L&P content is more voluminous. It is recognized that data science encompasses some content types, like documents, but it far from addressing the L&P ecosystem. Perhaps with improved cross-discipline cooperation the mostly disparate practices discussed here will converge to produce a cogent L&P science. Imagine stakeholders of data science, the semantic Web, privacy regulation and compliance, predictive analytics, predictive advertising, ML, AI, learning and performance management, content authoring, performance evaluation, performance taxonomy, and more, convening an L&P science summit to establish standards, technology frameworks, and incentives for cross-discipline cooperation. That would be a great starting point, and a subject for further investigation.

Acknowledgments. I am deeply indebted to Diane (DiBernadino) Parker, Charlie Sullivan, and Amile Samarakoon, for paving the way toward an L&P ecosystem during my tenure at Travelport (https://www.travelport.com/). Without the lead staff members for each component of Travelport's L&P ecosystem—especially R. Mark Moore, Lori Guillory, Jack Denion, Ines Hedenus, Sandra Yanez, and Laurenne McAteer—this vision of an L&P science would not have been possible. I would also like to thank TJ Bramblett, founder of Van Bram Marketing (vanbram.co), for the opportunity to gain insights around big data science.

References

1. Bond, T., Fox, C.: Applying the Rasch Model, 3rd edn. Routledge, New York (2015)

2. Passin, T.: Explorer's Guide to the Semantic Web, 1st edn. Manning Publications Company, Greenwich, CT (2004)
3. Enterprise Knowledge Homepage. https://enterprise-knowledge.com/what-is-a-semantic-arc hitecture-and-how-do-i-build-one/. Accessed 11 March 2022
4. Whizlabs Homepage. https://www.whizlabs.com/blog/big-data-terms/. Accessed 17 Feb 2022
5. Gemini Homepage. https://www.geminidata.com/why-your-business-needs-intelligent-data-pipelines/. Accessed 24 Feb 2022
6. Wikipedia Content Management Interoperability Homepage. https://en.wikipedia.org/wiki/Content_Management_Interoperability_Services. Accessed 01 March 2022
7. KMI Homepage. https://www.kminstitute.org/. Accessed 21 Feb 2022
8. Wikipedia.: Rasch model homepage. https://en.wikipedia.org/wiki/Rasch_model. Accessed 15 Feb 2022
9. APA Dictionary of Psychology Homepage. https://dictionary.apa.org/semantic-encoding. Accessed 10 March 2022
10. Gemini Home Page, https://www.geminidata.com/, last accessed 2022/03/05
11. Gartner Contextualization Resources Homepage https://www.gartner.com/en/search?key words=Contextualization, last accessed 2022/03/14
12. Forrester Contextualization Resources. https://www.forrester.com/allSearch?query=Contex tualization&s=relevance&dateRange=365. Accessed 19 Feb 2022
13. Gartner Homepage. https://www.gartner.com/en/documents/3953675/take-a-shift-left-app roach-to-testing-to-accelerate-and. Accessed 17 Feb 2022
14. Secretary of Defense memo, Continuous authorization to operate (cATO). https://media.def ense.gov/2022/Feb/03/2002932852/-1/-1/0/CONTINUOUS-AUTHORIZATION-TO-OPE RATE.PDF. Accessed 20 Feb 2022
15. RegScale Homepage. https://regscale.com/. Accessed 20 Feb 2022
16. C2 Labs Homepage. https://www.c2labs.com/. Accessed March 2022
17. Gartner Homepage. https://www.gartner.com/smarterwithgartner/gartner-predicts-the-fut ure-of-cloud-and-edge-infrastructure. Accessed 02 March 2022
18. Intersoft Consulting Homepage. https://gdpr-info.eu/. Accessed 09 March 2022
19. Epiance Homepage. https://www.epiplex500.com/. Accessed 14 March 2022
20. Adobe Homepage. https://www.adobe.com/products/captivate.html. Accessed 27 Feb 2022
21. Treasure Data Blog. https://blog.treasuredata.com/blog/2021/07/28/the-difference-between-first-party-second-party-and-third-party-data/. Accessed 08 Feb 2022
22. Gartner Homepage Articles. https://www.gartner.com/smarterwithgartner/gartner-top-10-tre nds-in-data-and-analytics-for-2020. Accessed 15 Feb 2022
23. Rheingold, H.: Virtual Reality, 1st edn. Simon and Schuster, New York (1991)
24. Gartner Homepage. https://www.gartner.com/en/research/methodologies/magic-quadrants-research. Accessed 14 March 2022
25. Moore, G.: Crossing the Chasm, 3rd edn. HarperBusiness, New York (1991, 1999, 2002, 2014)

A Conceptual Model for Meeting the Needs of Adult Learners in Distance Education

Anne Fensie[(⊠)] [iD]

University of Maine, Orono, ME 04469, USA
`anne.fensie@maine.edu`

Abstract. More than 40% of undergraduate students are 24 years of age or older [1], and over half of these students are enrolled in distance education [2]. Yet, adults do not fare as well as traditional-aged college students who are four times as likely to graduate [3]. Understanding the needs of the adult learner in distance education is important for improving their experience and outcomes. By combining cognitive, social, and emotional factors, sensitive to the impact of context, we can develop programming that meets the needs of the whole learner. Drawing from the science of learning, I will outline the components of my conceptual model for meeting the needs of adult learners in distance education. There is a diverse body of evidence-based instructional practices to support each of the factors in this model, but additional research may show significant interaction effects that may be especially beneficial for adult learners in distance education.

Keywords: Adult learners · Distance education · Learning sciences

1 Introduction

Instruction for adult learners in distance education needs to be more efficient and effective so that adult learners can meet course objectives without spending additional time that they generally do not have. Lee [4] reported that students often cite falling behind in their course work as a reason for dropping out. In their study of adult learner behaviors in distance education, Yin and Lim [5] found that more than half of the students reported spending an average of at least 12 h on completing an assignment, with nearly a third of adults spending more than 21 h on each assignment. This is in addition to up to five hours a week of self-study that more than half of the adult learners spent on each class. The majority of students in this survey held full-time jobs in addition to their enrollment in higher education. For students who maintain several concurrent course enrollments to meet financial aid requirements, this can amount to a second full-time job. I question whether workload in courses should be determined arbitrarily according to traditional notions of seat time or if assignments should be designed to maximize the limited time that adults have available to learn.

Brinthaupt and Eady [6] found that while most faculty held positive attitudes toward their adult learners, they did not adapt their instruction to meet their needs. However, those with the most favorable attitudes, the most experience working with adult learners,

© The Author(s), under exclusive license to Springer Nature Switzerland AG 2023
D. Guralnick et al. (Eds.): TLIC 2022, LNNS 581, pp. 136–149, 2023.
https://doi.org/10.1007/978-3-031-21569-8_13

and the most interest in learning about adult learners made more accommodations for these students. Panacci [7] found that faculty did not adapt their instruction for adult learners, despite the emphasis from their adult learners about their unique needs. Faculty approaches to instruction can have an impact on the success of adult learners [8, 9], but measuring this success can be difficult. Fong and Jarrat [10] identified three challenges to measuring the academic success of nontraditional learners: "lack of consensus on key definitions and metrics; insufficient coordination among industry stakeholders; inadequate resources at the institutional level" (p. 4). Adult learners in higher education, particularly distance education, are often not captured in reporting which limits the ability of researchers to analyze this population effectively [11]. In this paper, I will describe salient characteristics of adult learners, argue for the importance of the study of this population, and outline a model of factors from the learning sciences that should be considered in meeting the needs of these learners in distance education.

2   Characteristics of Adult Learners in Higher Education

In addition to the financial, personal, and family challenges that adults juggle, adding student identity into the mix poses additional challenges [12–14]. Rather than focusing their identity on being a student, as is the case for many traditional-aged college students, adults often incorporate "student" as one of their many identities into their already complex lives [13, 15]. Often, their engagement in higher education follows a life crisis, adding to the emotional burden of developing a new student identity [15]. Kasworm [15] argued, "When time is perceived as limited, emotion-related goals assume primacy," continuing to say that this "captures the paradoxical place of the adult student as an emotional self and a knowledge-striving self" (p. 30). She also noted the important role faculty play, as adults find learning successes and the validation from faculty about their worthiness and belonging particularly powerful.

There are several changes in cognition throughout the lifespan that are relevant to learning, including processing, brain structure, and memory. While many developmental changes in the brain are linear, some follow different trajectories [16, 17]. Generally, crystallized intelligence, or depth of knowledge and wisdom, is maintained or increased throughout the lifespan, and fluid intelligence, or reasoning and inference that does not rely on background knowledge, begins to decline after adolescence [16, 18, 19]. A synthesis of this research by Park and Gutchess [20] outlined these trajectories across the lifespan with more detail.

It should be noted that there is wide variability in cognitive functioning of adults, and some of this is culturally mediated. For example, when Chua et al. [21] studied visual patterns of Chinese and American participants as they examined images with focal content and complex backgrounds, they discovered that Americans tended to spend more time looking at the focal content, finding it more quickly, while the Chinese tended to spend more time observing the background of the image with more movement in their gaze. These kinds of cultural patterns of behavior can have lasting effects on the brain, worldview, and knowledge [20–22]. While there is some commonality in brain functioning patterns throughout the lifespan, it is important to understand learner variability to optimize instruction [23, 24].

The ability to quickly manipulate and integrate information and efficiently switch tasks peaks in adolescence and slowly declines throughout adulthood [16, 22]. More specifically, working memory performance begins to suffer in middle age, but adults recruit different and additional neural networks to complete tasks than children and adolescents, suggesting that their brains are adapting to the situation, but that this may require more effort [22]. Some of these changes may be due to changes in brain structure. For example, myelination of neurons, which increases efficiency of signaling, continues throughout childhood in several areas of the brain and is not completed in the frontal cortex, where higher order thinking occurs, until an adult is in their 30s [16]. At that point, white matter in that area begins to shrink. Craik and Bialystok [16] explained the importance of this, saying, "Efficient cognitive functioning depends on the degree of myelination and integrity of white matter, on the density and richness of synaptic connections, and on the specificity of synaptic pruning caused by fruitful interactions with the external environment" (p. 132). In addition to structural changes in the brain, lifespan changes can also be seen in the neurotransmitters that facilitate brain functioning. For example, dopamine, which is necessary for executive functions, has been shown to decline with age [25].

Lifespan development includes a gradual decline in episodic memory (memory of our experiences) in the third decade of life, increasing more rapidly after age 60 [22]. Fandakova et al. [26] suggested that this is due in part to a diminishing ability to bind specific knowledge with experiences, and Craik and Bialystok [16] noted the challenges associated with age in accessing stored memories. Fuzzy-trace theory posits that we simultaneously encode verbatim (exact details) and gist (essential meaning) memories from the same stimulus, and that children are more likely to use verbatim memory in decision making while adults show a preference for gist memory [27]. The shift to gist-based memory also assists adults in seeing the "big picture" and realizing implications when presented with new information [22]. Coupled with a bias toward pattern completion, adults have a strength in integrating across experiences and knowledge [22]. While some cognitive changes put adults at a disadvantage in higher education, others could be utilized to assist with learning if they are considered in instructional design.

3 Marginalization of Adult Learners in Higher Education

Many initiatives, policies, and reports in higher education focus primarily or solely on traditional aged students, such as a recent report sponsored by the Pell Institute for the Study of Opportunity in Higher Education, *Indicators of Higher Education Equity in the United States: 2020 Historical Trend Report* [1]. This report only looks at students who complete their degree by age 24, completely omitting adult learners.

> "When those who have the power to name and to socially construct reality choose not to see you or hear you ... When someone with the authority of a teacher, say, describes the world and you are not in it, there is a moment of psychic disequilibrium, as if you looked in a mirror and saw nothing. It takes some strength of soul — and not just individual strength, but collective understanding — to resist this void, this non-being, into which you are thrust, and to stand up, demanding to be seen and heard" [28, p. 2].

Some scholars have labeled adult learners in higher education a marginalized population, often noting their absence from the discourse. Tinto [29] noted, "In some respects, the experience of adult students is not unlike that of minority students. They too can feel marginal to the mainstream of institutional life" (p. 76). Terenzini and Pascarella [30] observed the absence of adult learners in the literature, calling this a "substantial" bias in the research (p. 152), noting this gap again in their 2005 research on the impact of college on undergraduates [31].

In a systematic review of the literature, Donaldson and Townsend [32] categorized the discourse on adult undergraduates with four labels: (1) invisible (the traditional student experience is presented as universal), (2) acknowledged but devalued (portrayed as "deficient, problematic, different, or other"; p. 37), (3) accepted (treated as a separate homogenous population), and (4) embraced (intragroup differences are acknowledged and their value in higher education is described). The authors looked at all articles in seven leading journals in higher education from 1990–2003 and found that only 1% of them were about adult undergraduates, as indicated by their titles. Similarly, in their content analysis of the literature on diversity, Sims and Barnett [33] found "a gross omission in the literature concerning adult students" (p. 9). Their searches identified only two journal articles that addressed adult learners in terms of diversity, and five books that provided recommendations for working with adults; however, none of these works addressed the intersectionality of adult identities. "Minority, female, gay, military, and disabled college student experiences are widely discussed in diversity sources, yet those same students are disregarded in the literature based on their age, educational background, family status, or life experiences. All of these dimensions of diversity are not all mutually exclusive and should be discussed together" (p. 9). Moreover, adult students of color are more likely to have negative past experiences with school and complete degrees at a much lower rate than White adults [34]. Understanding the beliefs faculty have about adult learners will be important as this will affect how they teach and whether they value the assets their adult learners bring to the classroom [35]. Faculty pedagogical preparation can have an especially significant impact on Black students and Pell-eligible students, closing the achievement gap in some cases [36].

4   Conceptual Model for Meeting the Needs of Adult Learners in Distance Education

There is a range of cognitive, social, emotional, and contextual constructs that are likely to impact adult learners in distance education (see Fig. 1). Several of these overlap multiple domains, but they all interact with the adult learner context which shapes their impact [22, 37, 38]. In this section, I will describe how prior knowledge, cognitive load, working memory/cognitive processing, attention, executive function, self-regulated learning, motivation and self-efficacy, social identity, teacher presence, and life roles are important considerations for the design of instruction for adult learners in distance education.

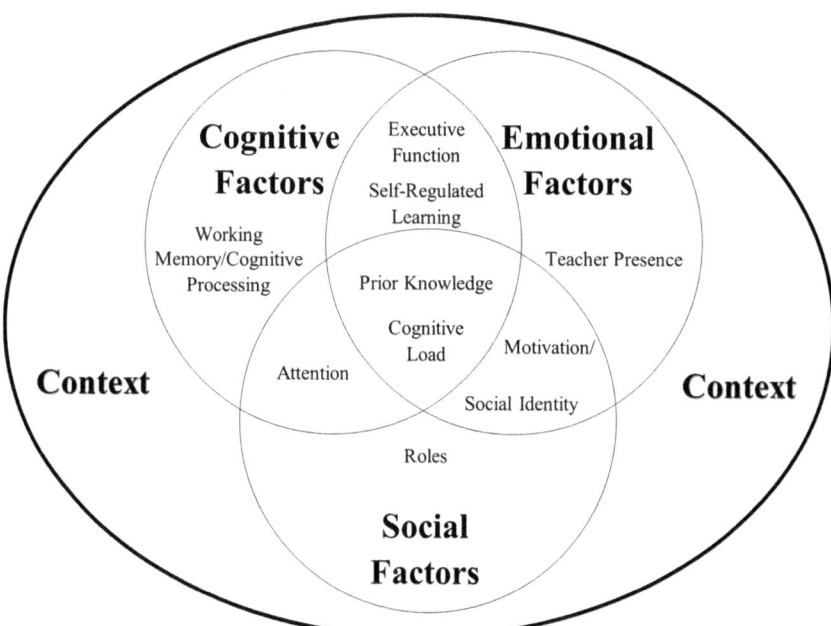

Fig. 1. Essential factors from the science of learning for meeting the needs of adult learners in distance education.

4.1 Cognitive Factors

Overlapping all domains are prior knowledge and cognitive load. Ambrose and Lovett [39] defined prior knowledge to include "content, skills, and beliefs, because all are 'knowledge' in that they result from past experiences and impact subsequent learning and performance" (p. 7). Knowledge is a function of information and experiences stored in long-term memory which can be later retrieved and used [40]. It includes memories of our experiences (episodic memory), the knowledge we have distilled from these experiences (semantic knowledge), and our skills (procedural knowledge) [22, 41].

We form memories through a process of encoding (the creation of memory traces through connecting new information, experiences, and meaning making with mental schema), storage (consolidation and maintenance of memory), and retrieval (using cues to recall and use the knowledge, reconstructing and strengthening the memory trace) [22, 40]. These knowledge stores are maintained and can be increased throughout adulthood [42, 43], although there is some evidence that identifying the source of the information becomes more difficult with age [18]. As adults learn and grow through their formal education and life experiences, they accumulate a network of mental schema that can be an asset to them as students if tapped during learning activities [22]. However, it can also be a deficit as incorrect knowledge is harder to unlearn, biases us away from new information, and we often rely on these misconceptions to solve new problems [22, 44, 45]. The prior knowledge and life experiences of adults effect not just their content and abilities but is a component of their social and emotional interactions in learning, as well as a product of their background and social context [22]. Universal Design for Learning can be an effective pedagogical approach for addressing this learner variability [46].

Cognitive load is the other central factor to be considered in meeting the needs of adult learners in distance education. Cognitive load theory posits that working memory and mental processes have a limited capacity but are a combination of somewhat independent subprocesses interacting with mental schema from long-term memory [41]. These processes include intrinsic load (due to the complexity of the information), extraneous load (distractors or the processing required to access the information), and germane load (the remaining load available to devote to the intrinsic load). The load is variable for each learner because of their background knowledge [41], level of social and emotional skill and involvement [47, 48], and their background and situational context [49, 50], but there are instructional strategies that can be developed to mitigate extraneous cognitive load [51]. Cognitive load is an especially important consideration in distance education because of its interactions with cognitive and social presence [52], and for adults in particular as their working memory capacity and speed declines with age [18, 43].

There are several models of working memory, but the one most appropriate to distance learning is the cognitive theory of multimedia learning [53]. Mayer explains that words and images are presented in multimedia instruction which we perceive using our sensory memory. Our working memory helps us to (1) select the appropriate words and images to focus on, (2) organize the verbal and pictorial models, and (3) integrate these with prior knowledge from our long-term memory. Beginning in early adulthood, working memory and processing speed begins a gradual decline throughout adulthood, indicating that these three tasks are less efficient in adult learners [18, 43]. There is evidence associating this decline with a loss of volume in the prefrontal cortex [54, 55]. Mayer's theory of multimedia instruction can be used to mitigate the demands of mental processing while other research suggests that the ability to pace their own learning has positive benefits for adult learners [42, 56]. There are several other learning strategies that are particularly useful for limited working memory capacity, such as retrieval practice [57] or making non-threatened identities more salient for those that experience stereotype threat [58].

Selecting which information makes its way into working memory is the role of attention [59]. Attention is a finite resource susceptible to depletion with high demands [60]. Because adult learners in higher education often have multiple roles [13, 61], their attention is more likely to be divided or used to maintain these other roles. Switching between tasks has a high cost in time and memory which negatively impact learning, putting learners at a disadvantage in distance education [62]. Knowing in advance that there will be an interruption or switch allows the learner to clear their working memory and prepare to focus on something new. If the task is switched without preparation, like with an interruption, the impact on attention will be more negative in terms of cost. This supports other research that demonstrates the benefits of self-paced learning for adults as well as the benefits of chunking course content and learning activities into smaller segments [53, 56].

The control processes responsible for "planning, assembling, coordinating, sequencing, and monitoring other cognitive operations" are referred to as executive functioning, a metaphor of a business executive who does not specialize in a particular task but manages all tasks" [63, p. 566]. Craik and Bialystok [16] explained that executive functions "overcome the prepotent 'default mode' of automatic behavior and allow the person to attend selectively, to concentrate on a particular task, to make choices in line with current goals, and to facilitate new learning and adaptive responding" (p. 134). Executive functions are potential mediators of age-related cognitive declines, yet there is evidence that these skills can be improved with practice [64]. While many of these functions are naturally supported in the traditional face-to-face classroom, they are often less present in the design of asynchronous online learning. Addressing executive functions in distance learning is important as these skills underlie effective self-regulated learning [65], which are significant predictors of academic achievement [66]. Social-emotional aspects of self-regulated learning, such as self-efficacy, may be particularly relevant for adults in distance education [67]. Self-regulated learning is not improved simply by participation in online learning [68], but these skills can be taught and improve outcomes for adults in distance education [69].

4.2 Social Factors

Working adults in higher education are twice as likely to consider themselves "employees who study" than "students who work," identifying first as an employee [13]. Employees who study are older, more likely to be married and have children, attend school part-time, and work full-time. Their goals for participation in higher education are to improve their career prospects, but also for personal enrichment [23]. Adults in higher education balance multiple roles, such as work, caregiving for children and older relatives, community engagement, and volunteer work [61]. Role strain can be particularly challenging for working mothers in higher education who can experience conflicting demands, overload, and guilt [70]. It is important to consider the multiple roles of adult learners as these carry a time and attention burden that is not present for most traditional-aged college students [71]. These roles contribute to the amount of attention that can be devoted to learning activities as well as extraneous cognitive load. Instructors can consider streamlining course content to minimize the time and attention required of the adult learner, as well as using authentic contexts for course content that are personally relevant to the learner and align with their life roles.

Marginalized identities place additional burdens on adult learners. When a learner experiences stereotype threat, or the fear that they will meet negative expectations set by society based on their identity [72], this taxes executive functions, increases worry and anxiety about failure, which "robs the person of working memory processes" [22, p. 128]. The effect is even greater in situations where the learner is especially motivated [73]. Students with marginalized identities may attribute challenges or failures in higher education to a belief that they don't belong in higher education, not thinking that challenges are common to most college students [22, 74].

Learning is socially contextualized because it involves "the experiences, social relationships, and cognitive opportunities as subjectively perceived and emotionally experienced by the learner", whether the learner is working alone or with others [22, p. 29]. Online presence can be defined as "a student's sense of being in and belonging in a course and the ability to interact with other students and an instructor although physical contact is not available" [75, p. 22]. This has been studied in terms of cognitive presence (intellectual connection to course content), social presence (connections to other students), teaching presence (sense of connection to the instructor), and other forms of presence [76]. Social presence of any kind online can be defined as "the degree of feeling emotionally connected to another intellectual entity through computer mediated communication" [77, pp. 1738–1739]. Indicators of social presence can be affective (expressions of emotions or mood), interactive (acknowledgement of another), and cohesive (things that build or sustain group cohesion) [78]. Adult learners find teaching presence to be essential for their learning and seek deep interactions with content rather than surface learning, with peer interactions a bonus [79, 80]. Mayer [53] suggested that this connection helps to foster deeper processing during learning. The Community of Inquiry (CoI) model can be helpful in designing for cognitive, social, and teaching presence and addressing the emotional needs of adult learners [81, 82].

4.3 Emotional Factors

Adults are increasingly motivated to obtain emotional meaning from life [23, 81] and their brains are too efficient and overloaded to learn something that is not meaningful—if there is no emotional connection to the content, such as curiosity or motivation to learn, the information will not be remembered [83]. Motivation is an important consideration for distance learning as some research has found lower levels of motivation in online learners with high correlations to course performance [84]. A meta-analysis of the research on motivation in adult learners found that adults who chose online courses were intrinsically motivated and had high self-efficacy, while instructional design strategies that were associated with high motivation included building on learners' life experiences and allowing for personal control in when, where, and with whom to learn [85].

One particularly relevant component of motivational theories for adult learners is self-efficacy [23] and its role in situated expectancy-value [38] as growth and accomplishment are important factors in motivation for adults [22]. Self-efficacy has been defined as "the belief one has in their own capabilities to perform certain tasks" and has a significant impact on the academic performance of adult learners [8, pp. 113–114]. To be an adult means to be competent, although when adult learners enter higher education, "they compartmentalize their prior understandings of self-efficacy and competence in their adult life worlds, believing that their backgrounds have limited or no value in the academic world" [15, p. 31]. Many adults will try to protect their self-image by taking steps to avoid failure, rather than take risks to improve learning [88]. Educational environments that encourage risk-taking and provide psychological safety can allow for adults to engage more deeply in learning. Strategies for improving self-efficacy in learners include helping students to set appropriate goals, helping them to break them down into subgoals, and providing feedback on progress so students can attribute their success to their own efforts [22].

Eccles and Wigfield [38] contend that significant impacts on performance come from the individual's expectancies for success and subjective task values. Expectancies of success, or the learner's belief about how well they will do on a task, are influenced by their beliefs of personal efficacy, academic self-concept, and perceptions of task difficulty. Subjective task values are determined by intrinsic value (anticipated enjoyment or interest), attainment value (identity-based importance), utility value (means to an end), and cost (including the amount of effort needed, time away from other valued tasks, and emotional costs related to anxiety and failure) [38]. Faculty can use this framework to design learning tasks that their adult learners will value and complete competently.

Social identity is an important factor in the success of any learner, but there are particular components of identity and self-concept that are relevant for adults in distance education. Higher education poses challenges for learning, and adults who have a more fully developed self-concept may be sensitive to their changing abilities and are likely to protect this self-concept through avoidance and compensation [23]. There is evidence that error management training can be beneficial for adults where they are instructed on the benefits of committing errors during learning and are given the freedom to make these errors [86, 87]. Threats to social identity can produce a cognitive load [58, 88], but efforts to remind students of their other life roles and de-emphasize the salience of the threatened identity can mitigate this threat [89, 90].

5   Conclusion

Understanding the needs of the adult learner in distance education is important for improving their experience and outcomes. By combining cognitive, social, and emotional factors, accounting for unique contexts, we can develop programming that meets the needs of the whole learner. While the extant literature shows support for the individual factors in this model, further research should be conducted to test the significance of this model as a whole. Research in these areas has primarily been conducted on school-aged children or traditional college-aged students in classroom or lab environments, so ecological validity and focus on adult learners will be necessary. There is a diverse body of evidence-based instructional practices to support each of the factors in this model, but additional research may show significant interaction effects that may be especially beneficial for adult learners in distance education.

References

1. Cahalan, M.W., Perna, L.W., Addison, M., Murray, C., Patel, P.R., Jiang, N.: Indicators of higher education equity in the United States: 2020 historical trend report. The Pell Institute for the Study of Opportunity in Higher Education, Council for Opportunity in Education (COE), and Alliance for Higher Education and Democracy of the University of Pennsylvania (PennAHEAD) (2020)
2. Snyder, T.D., de Brey, C., Dillow, S.A.: Digest of education statistics 2018 (NCES 2020-009). U.S. Department of Education National Center for Education Statistics. https://nces.ed.gov/pubs2020/2020009.pdf (2019)
3. Miller, C.: College graduation statistics. EducationData. https://educationdata.org/number-of-college-graduates (2019)
4. Lee, K.: Rethinking the accessibility of online higher education: a historical review. Internet High. Educ. **33**, 15–23 (2017). https://doi.org/10.1016/j.iheduc.2017.01.001
5. Yin, H.Y., Lim, W.Y.R.: Educating adult learners: bridging learners' characteristics and the learning sciences. In: Sanger, C.S., Gleason, N.W. (eds.) Diversity and Inclusion in Global Higher Education, pp. 97–115. Palgrave MacMillan (2020)
6. Brinthaupt, T.M., Eady, E.: Faculty members' attitudes, perceptions, and behaviors toward their nontraditional students. J. Contin. High. Educ. **62**(3), 131–140 (2014). https://doi.org/10.1080/07377363.2014.956027

7. Panacci, A.G.: Adult students in mixed-age postsecondary classrooms: implications for instructional approaches. Coll. Q. **20**(2) (2017). http://collegequarterly.ca/2017-vol20-num02-spring/adult-students-in-mixed-age-postsecondary-classroom-implications-for-instructional-approaches.html

8. Bowser, J.M.: Anxiety, preconceived negative perceptions, and self-efficacy: impact on adult learners' performance in introductory accounting courses (2021–61325–255) [Dissertation]. ProQuest Information & Learning (2021)

9. Mayhew, M.J., Rockenbach, A.N., Bowman, N.A., Seifert, T.A.D., Wolniak, G.C., Pascarella, E.T., Terenzini, P.T.: How College Affects Students: 21st Century Evidence that Higher Education Works. Wiley (2016).

10. Fong, J., Jarrat, D.: Measuring post-traditional student success: institutions making progress, but challenges remain. UPCEA (2013)

11. Advisory Committee on Student Financial Aid Assistance: Pathways to success: integrating learning with life and work to increase national college completion [Policy Brief]. https://lincs.ed.gov/professional-development/resource-collections/profile-158 (2012)

12. Banks, K.L.: Identifying online graduate learners' perceived barriers to their academic success utilizing a Delphi study [Dissertation]. ProQuest Information & Learning (2018)

13. Berker, A., Horn, L., Carroll, C.D.: Work first, study second: Adult undergraduates who combine employment and postsecondary enrollment (NCES 2003-167; Postsecondary Education Descriptive Analysis Reports, p. 94). U.S. Department of Education National Center for Education Statistics (2003)

14. Ross-Gordon, J.M.: Research on adult learners: Supporting the needs of a student population that is no longer nontraditional. Peer Rev. **13**(1). https://www.aacu.org/publications-research/periodicals/research-adult-learners-supporting-needs-student-population-no (2011)

15. Kasworm, C.E.: Emotional challenges of adult learners in higher education. New Dir. Adult Contin. Educ. **2008**(120), 27–34 (2008). https://doi.org/10.1002/ace.313

16. Craik, F.I.M., Bialystok, E.: Cognition through the lifespan: mechanisms of change. Trends Cogn. Sci. **10**(3), 131–138 (2006). https://doi.org/10.1016/j.tics.2006.01.007

17. Fjell, A.M., et al.: Critical ages in the life-course of the adult brain: nonlinear subcortical aging. Neurobiol. Aging **34**(10), 2239–2247 (2013). https://doi.org/10.1016/j.neurobiolaging.2013.04.006

18. Murman, D.: The impact of age on cognition. Semin. Hear. **36**(03), 111–121 (2015). https://doi.org/10.1055/s-0035-1555115

19. Salthouse, T.A.: Selective review of cognitive aging. J. Int. Neuropsychol. Soc. JINS **16**(5), 754–760 (2010). https://doi.org/10.1017/S1355617710000706

20. Park, D., Gutchess, A.: The cognitive neuroscience of aging and culture. Curr. Dir. Psychol. Sci. **15**(3), 105–108 (2006). https://doi.org/10.1111/j.0963-7214.2006.00416.x

21. Chua, H.F., Boland, J.E., Nisbett, R.E.: Cultural variation in eye movements during scene perception. Proc. Natl. Acad. Sci. **102**(35), 12629–12633 (2005). https://doi.org/10.1073/pnas.0506162102

22. National Academies of Sciences, Engineering, and Medicine: How People Learn II: Learners, Contexts, and Cultures. National Academies Press (2018). https://www.nap.edu/catalog/24783

23. Kanfer, R., Ackerman, P.L.: Aging, adult development, and work motivation. Acad. Manag. Rev. **29**(3), 440–458 (2004). https://doi.org/10.2307/20159053

24. Tare, M., Cacicio, S., Shell, A.R.: The science of adult learning: understanding the whole learner. Digital Promise (2021)

25. Karrer, T.M., Josef, A.K., Mata, R., Morris, E.D., Samanez-Larkin, G.R.: Reduced dopamine receptors and transporters but not synthesis capacity in normal aging adults: a meta-analysis. Neurobiol. Aging **57**, 36–46 (2017). https://doi.org/10.1016/j.neurobiolaging.2017.05.006

26. Fandakova, Y., Lindenberger, U., Shing, Y.L.: Deficits in process-specific prefrontal and hippocampal activations contribute to adult age differences in episodic memory interference. Cereb. Cortex **24**(7), 1832–1844 (2014). https://doi.org/10.1093/cercor/bht034

27. Reyna, V.F.: A new intuitionism: meaning, memory, and development in Fuzzy-Trace Theory. Judgement Decis. Mak. **7**(3), 332–359 (2012). https://doi.org/10.4135/9781412971980.n155

28. Rich, 1986 in Sims, C. H., Barnett, D.: Devalued, misunderstood, and marginalized: why nontraditional students' experiences should be included in the diversity discourse. Online J. Work. Educ. Dev. **8**(1). https://opensiuc.lib.siu.edu/cgi/viewcontent.cgi?article=1175&context=ojwed (2015)

29. Tinto, V.: Leaving college: Rethinking the causes and cures of student attrition, 2nd edn. University of Chicago Press (2012)

30. Terenzini, P.T., Pascarella, E.T.: Studying college students in the 21st century: meeting new challenges. Rev. High. Educ. **21**(2), 151–165 (1998)

31. Pascarella, E.T., Terenzini, P.T.: How College Affects Students: A Third Decade of Research. Jossey-Bass, An Imprint of Wiley (2005)

32. Donaldson, J.F., Townsend, B.K.: Higher education journals' discourse about adult undergraduate students. J. High. Educ. **78**(1), 27–50 (2007)

33. Sims, C.H., Barnett, D.: Devalued, misunderstood, and marginalized: why nontraditional students' experiences should be included in the diversity discourse. Online J. Work. Educ. Dev. **8**(1) (2015). https://opensiuc.lib.siu.edu/cgi/viewcontent.cgi?article=1175&context=ojwed

34. Person, A.E., Bruch, J., Goble, L.: Why equity matters for adult college completion. In: Mathematica [Education Issue Brief]. https://files.eric.ed.gov/fulltext/ED607733.pdf (2019)

35. Sissel, P.A., Hansman, C.A., Kasworm, C.E.: The politics of neglect: adult learners in higher education. New Dir. Adult Contin. Educ. **2001**(91), 17–28 (2001). https://doi.org/10.1002/ace.27

36. Snow, M.: Course completion gap closed for Black students and gap in passing courses closed for Pell-eligible students taught by ACUE-credentialed faculty at Broward College (Research Brief No. 13). Association of College and University Educators. https://acue.org/wp-content/uploads/2020/06/ACUE-Research_Brief_13_final.pdf (2020)

37. Daniel, D.B., Chew, S.L.: The tribalism of teaching and learning. Teach. Psychol. **40**(4), 363–367 (2013). https://doi.org/10.1177/0098628313501034

38. Eccles, J.S., Wigfield, A.: From expectancy-value theory to situated expectancy-value theory: a developmental, social cognitive, and sociocultural perspective on motivation. Contemp. Educ. Psychol. **61**, 101859 (2020). https://doi.org/10.1016/j.cedpsych.2020.101859

39. Ambrose, S.A., Lovett, M.C.: Prior knowledge is more than content: skills and beliefs also impact learning. In: Benassi, V.A., Overson, C.E., Hakal, C.M. (eds.) Applying Science of Learning in Education: Infusing Psychological Science into the Curriculum, pp. 7–19. Society for the Teaching of Psychology (2014). http://teachpsych.org/ebooks/asle2014/index.php

40. McDermott, K.B., Roediger, H.L.: Memory (encoding, storage, retrieval). In: Butler, A. (eds.) General Psychology (Fall 2018), pp. 117–140. Valparaiso University. https://core.ac.uk/reader/303864230 (2018)

41. Sweller, J., van Merriënboer, J.J.G., Paas, F.: Cognitive architecture and instructional design: 20 years later. Educ. Psychol. Rev. **31**(2), 261–292 (2019). https://doi.org/10.1007/s10648-019-09465-5

42. Beier, M.E., Ackerman, P.L.: Age, ability, and the role of prior knowledge on the acquisition of new domain knowledge: promising results in a real-world learning environment. Psychol. Aging **20**(2), 341–355 (2005). https://doi.org/10.1037/0882-7974.20.2.341

43. Salthouse, T.A.: The processing-speed theory of adult age differences in cognition. Psychol. Rev. **103**(3), 403–428 (1996). https://doi.org/10.1037/0033-295X.103.3.403

44. Ambrose, S.A., Bridges, M.W., DiPietro, M., Lovett, M.C., Norman, M.K., Mayer, R.E.: How Learning Works: Seven Research-Based Principles for Smart Teaching. Wiley (2010)

45. Chen, A., Lee, W.O., Chen, Z., Sadik, S., Lim, W.Y., Bhowmick, S.: Taskforce on the future of adult learning research Singapore: consultative paper by the subgroup on 'towards the science of adult learning.' Singapore University of Social Sciences. https://www.ial.edu.sg/content/dam/projects/tms/ial/Research-publications/future-of-adult-learning-research/ANNEX%202%20-%20Towards%20the%20science%20of%20adult%20learning.pdf (2020)

46. Rogers-Shaw, C., Carr-Chellman, D.J., Choi, J.: Universal design for learning: guidelines for accessible online instruction. Adult Learn. **29**(1), 20–31 (2018). https://doi.org/10.1177/1045159517735530

47. Feldon, D.F., Callan, G., Juth, S., Jeong, S.: Cognitive load as motivational cost. Educ. Psychol. Rev. **31**(2), 319–337 (2019). https://doi.org/10.1007/s10648-019-09464-6

48. Plass, J.L., Kalyuga, S.: Four ways of considering emotion in cognitive load theory. Educ. Psychol. Rev. **31**(2), 339–359 (2019). https://doi.org/10.1007/s10648-019-09473-5

49. Choi, H.-H., van Merriënboer, J.J.G., Paas, F.: Effects of the physical environment on cognitive load and learning: towards a new model of cognitive load. Educ. Psychol. Rev. **26**(2), 225–244 (2014). https://doi.org/10.1007/s10648-014-9262-6

50. Sweller, J., Ayres, P., Kalyuga, S.: Cognitive Load Theory. Springer, New York (2011).https://doi.org/10.1007/978-1-4419-8126-4

51. Moreno, R., Mayer, R.E.: Nine ways to reduce cognitive load in multimedia learning. Educ. Psychol. **38**(1), 43–52 (2003). https://doi.org/10.1207/S15326985EP3801

52. Kozan, K.: The incremental predictive validity of teaching, cognitive and social presence on cognitive load. Internet High. Educ. **31**, 11–19 (2016). https://doi.org/10.1016/j.iheduc.2016.05.003

53. Mayer, R.E.: Multimedia instruction. In: Spector, J.M., Merrill, M.D., Elen, J., Bishop, M.J. (eds.) Handbook of Research on Educational Communications and Technology, pp. 385–399. Springer, New York (2014). https://doi.org/10.1007/978-1-4614-3185-5_31

54. Kirchhoff, B.A., Gordon, B.A., Head, D.: Prefrontal gray matter volume mediates age effects on memory strategies. Neuroimage **90**, 326–334 (2014). https://doi.org/10.1016/j.neuroimage.2013.12.052

55. Raz, N., et al.: Selective aging of the human cerebral cortex observed in vivo: differential vulnerability of the prefrontal gray matter. Cereb. Cortex **7**(3), 268–282 (1997). https://doi.org/10.1093/cercor/7.3.268

56. Callahan, J.S., Kiker, D.S., Cross, T.: Does method matter? A meta-analysis of the effects of training method on older learner training performance. J. Manag. **29**(5), 663–680 (2003). https://doi.org/10.1016/S0149-2063_03_00029-1

57. Agarwal, P.K., Finley, J.R., Rose, N.S., Roediger III, H.L.: Benefits from retrieval practice are greater for students with lower working memory capacity. Memory **25**(6), 764–771 (2017). https://doi.org/10.1080/09658211.2016.1220579

58. Beilock, S.L., Rydell, R.J., McConnell, A.R.: Stereotype threat and working memory: mechanisms, alleviation, and spillover. J. Exp. Psychol. Gen. **136**(2), 256–276 (2007). https://doi.org/10.1037/0096-3445.136.2.256

59. Gazzaley, A., Nobre, A.C.: Top-down modulation: bridging selective attention and working memory. Trends Cogn. Sci. **16**(2), 129–135 (2012). https://doi.org/10.1016/j.tics.2011.11.014

60. Kaplan, S., Berman, M.G.: Directed attention as a common resource for executive functioning and self-regulation. Perspect. Psychol. Sci. **5**(1), 43–57 (2010). https://doi.org/10.1177/1745691609356784

61. Fairchild, E.E.: Multiple roles of adult learners. New Dir. Stud. Serv. **2003**(102), 11–16 (2003). https://doi.org/10.1002/ss.84

62. Van der Stighel, S.: Dangers of divided attention. Am. Sci. **109**(1), 46–50, 52–53 (2021)

63. Salthouse, T.A., Atkinson, T.M., Berish, D.E.: Executive functioning as a potential mediator of age-related cognitive decline in normal adults. J. Exp. Psychol. Gen. **132**(4), 566–594 (2003). https://doi.org/10.1037/0096-3445.132.4.566

64. Diamond, A.: Executive functions. Annu. Rev. Psychol. **64**(1), 135–168 (2013). https://doi. org/10.1146/annurev-psych-113011-143750
65. Hofmann, W., Schmeichel, B.J., Baddeley, A.D.: Executive functions and self-regulation. Trends Cogn. Sci. **16**(3), 174–180 (2012). https://doi.org/10.1016/j.tics.2012.01.006
66. Richardson, M., Abraham, C., Bond, R.: Psychological correlates of university students' academic performance: a systematic review and meta-analysis. Psychol. Bull. **138**(2), 353–387 (2012). https://doi.org/10.1037/a0026838
67. de Fátima Goulão, M. (2014). The relationship between self-efficacy and academic achievement in adult learners. Athens J. Educ. **1**(3), 237–246. https://doi.org/10.30958/aje.1-3-4
68. Barnard-Brak, L., Paton, V.O., Lan, W.Y.: Self-regulation across time of first-generation online learners. ALT-J Res. Learn. Technol. **18**(1), 61–70 (2010). https://doi.org/10.1080/096 87761003657572
69. Lee, K., Choi, H., Cho, Y.H.: Becoming a competent self: a developmental process of adult distance learning. Internet High. Educ. **41**, 25–33 (2019). https://doi.org/10.1016/j.iheduc. 2018.12.001
70. Webber, L., Dismore, H.: Mothers and higher education: Balancing time, study and space. J. Furth. High. Educ. 1–15 (2020). https://doi.org/10.1080/0309877X.2020.1820458
71. Kahu, E.R., Stephens, C., Zepke, N., Leach, L.: Space and time to engage: mature-aged distance students learn to fit study into their lives. Int. J. Lifelong Educ. **33**(4), 523–540 (2014). https://doi.org/10.1080/02601370.2014.884177
72. Steele, C.M.: Whistling Vivaldi: How Stereotypes Affect Us and What We Can Do (Reprint edition). W. W. Norton & Company (2011)
73. Steele, C.M.: A threat in the air: how stereotypes shape intellectual identity and performance. Am. Psychol. **52**(6), 613–629 (1997). https://doi.org/10.1037/0003-066X.52.6.613
74. Steele, C.M., Aronson, J.: Stereotype threat and the intellectual test performance of African Americans. J. Pers. Soc. Psychol. **69**(5), 797–811 (1995). https://doi.org/10.1037/0022-3514. 69.5.797
75. Picciano, A.G.: Beyond student perceptions: issues of interaction, presence, and performance in an online course. Online Learn. **6**(1) (2019). https://doi.org/10.24059/olj.v6i1.1870
76. Garrison, D.R., Arbaugh, J.B.: Researching the community of inquiry framework: review, issues, and future directions. Internet High. Educ. **10**(3), 157–172 (2007). https://doi.org/10. 1016/j.iheduc.2007.04.001
77. Sung, E., Mayer, R.E.: Five facets of social presence in online distance education. Comput. Hum. Behav. **28**(5), 1738–1747 (2012). https://doi.org/10.1016/j.chb.2012.04.014
78. Rourke, L., Anderson, T., Garrison, D.R., Archer, W.: Assessing social presence in asynchronous text-based computer conferencing. J. Dist. Educ./Revue de l'ducation Distance **14**(2), 50–71 (1999)
79. Angelaki, C., Mavroidis, I.: Communication and social presence: the impact on adult learners' emotions in distance learning. Eur. J. Open Dist. E-Learn. **16**(1), 78–93 (2013)
80. Ke, F.: Examining online teaching, cognitive, and social presence for adult students. Comput. Educ. **55**(2), 808–820 (2010). https://doi.org/10.1016/j.compedu.2010.03.013
81. Carstensen, L.L., Fung, H.H., Charles, S.T.: Socioemotional selectivity theory and the regulation of emotion in the second half of life. Motiv. Emot. **27**(2), 103–123 (2003). https://doi. org/10.1023/A:1024569803230
82. Majeski, R.A., Stover, M., Valais, T.: The community of inquiry and emotional presence. Adult Learn. **29**(2), 53–61 (2018). https://doi.org/10.1177/1045159518758696
83. Immordino-Yang, M.H.: Emotions, Learning, and the Brain: Exploring the Educational Implications of Affective Neuroscience. W. W. Norton & Company (2015)
84. Stark, E.: Examining the role of motivation and learning strategies in student success in online versus face-to-face courses. Online Learn. **23**(3), 234–251 (2019)

85. Styer, A.J.: A grounded meta-analysis of adult learner motivation in online learning from the perspective of the learner [Dissertation, Capella University] (2007). http://search.proquest.com/docview/304723729/abstract/3C49AFD969F743F1PQ/1

86. Carter, M., Beier, M.E.: The effectiveness of error management training with working-aged adults. Pers. Psychol. **63**(3), 641–675 (2010). https://doi.org/10.1111/j.1744-6570.2010.01183.x

87. Wong, S.S.H., Lim, S.W.H.: Prevention–permission–promotion: a review of approaches to errors in learning. Educ. Psychol. **54**(1), 1–19 (2019). https://doi.org/10.1080/00461520.2018.1501693

88. Croizet, J.C., Després, G., Gauzins, M.E., Huguet, P., Leyens, J.P., Méot, A.: Stereotype threat undermines intellectual performance by triggering a disruptive mental load. Pers. Soc. Psychol. Bull. **30**(6), 721–731 (2004). https://doi.org/10.1177/0146167204263961

89. Aronson, J., Fried, C.B., Good, C.: Reducing the effects of stereotype threat on African American college students by shaping theories of intelligence. J. Exp. Soc. Psychol. **38**(2), 113–125 (2002). https://doi.org/10.1006/jesp.2001.1491

90. Gresky, D.M., Eyck, L.L.T., Lord, C.G., McIntyre, R.B.: Effects of salient multiple identities on women's performance under mathematics stereotype threat. Sex Roles **53**(9–10), 703–716 (2005). https://doi.org/10.1007/s11199-005-7735-2

Adapt to Learners: Practitioner Levels and Practice Support Methodologies

Alicia Haulbrook[✉]

Cumming, GA 30028, USA
AliciaHaulbrook@gmail.com

Abstract. You have a learner who does not meet the prerequisites. How do you adapt? You can list prerequisites and/or require them. Inevitably, there are learners who enter the classroom without the appropriate background knowledge for a successful learning experience. Technical training, where the class is relatively short and learners need to be able to return to their job and implement what they learned, adds another layer of difficulty. It is your job as a trainer to do whatever you can to ensure learners are prepared for success when they leave your classroom. In a class with a wide range of experiences, you must be able to pivot with your learners. Resulting from an immediate need to solve for various experience levels in a technical training classroom, and inspired by Ken Blanchard's Situational Leadership II®, a Learning and Development department created a theory for identifying and then adjusting to various learner levels in one training session. Further tested through classroom experiences and then adjusted, Practitioner Levels and Practice Support Methodologies is a proven approach to help ensure that learners are ready for work once they leave the classroom. To apply the theory, trainers first identify the practitioner level of each learner through simple questions and then adapt based on various coaching techniques ranging from detailed coaching to pushing more experienced learners to dig deeper independently.

Keywords: Practitioner levels · Adapt to learners · Practice support

1 Introduction

1.1 Background

The learning and development team (LDT) of a large financial technology company, further described as "FinTech," is responsible for training their clients how to use the company's software. The team is made up of 6 instructional designers, 30 trainers, and 5 managers. Their audience (learners) is financial institution (FI) employees who use the FinTech's systems.

The LDT is a revenue-generating learning and development team. Unlike most learning and development teams, LDT charges for continuing education and advanced training. Given the FinTech's revenue goals, offering free training is rare and is used to promote other fee-based training. Additionally, training is viewed as a service and not a requirement to use the software clients purchase. In other words, they train the FI's employees to use their systems while meeting high profit goals. Turning away clients (learners) is unheard of; if a learner comes to class, they are trained.

© The Author(s), under exclusive license to Springer Nature Switzerland AG 2023
D. Guralnick et al. (Eds.): TLIC 2022, LNNS 581, pp. 150–159, 2023.
https://doi.org/10.1007/978-3-031-21569-8_14

1.2 The Problem

LDT recommends prerequisite courses, but requiring them is not an option for several reasons. At one time, they did not have the means to track the prerequisites. Once they did, there were often exceptions to prerequisites based on prior system experience, e.g., if a learner had used the system for a few years, they did not have to take the prerequisite courses. Learners and their internal account managers would often escalate prerequisite requirements to upper management who would, in turn, ask LDT to waive them. Lastly, since LDT is a revenue-generating department; it is not in the best interest of the department to turn away learners, either.

Historically, the trainers of LDT employed a teaching style that focused on training system functionality with an emphasis on explaining every field and option. Courses were delivered with little or no regard to adult learning theory. Topics were presented through behavioral training methodology, which focused on easily forgettable facts and procedures that were prone to change on systems that are frequently updated. The team relied significantly on lecture-based training.

At the start of 2015, their emphasis changed from a focus on "training the system" to learner centric facilitation which focuses on the day-to-day job of the learners. Courses also needed to be trained using adult learning principles. Facilitation techniques transitioned to those that support cognitive learning theory which emphasizes problem solving, building on existing knowledge, assimilating new knowledge, and how and why a system works the way it does; this closely aligns with how adults learn.

LDT leveraged *Writing Training Materials That Work: How to Train Anyone to Do Anything* [1] as the groundwork for their new approach to training. The Cognitive Training Model Foshay et al. [1] explains that techniques fall into one of five categories: Select, Link, Organize, Assimilate, and Strengthen. For this paper's purpose, a brief description of three of the five phases is needed. Within the Select phase, trainers should answer, "What's in it for me?" (WIIFM) for the learner and provide "You can do it" (YCDI) statements; both are designed to help learners select which information they should attend to and encourage the learner through more difficult tasks. In the Link phase, a trainer must help learners recall existing knowledge of the topic and relate it to the new information (Recall and Relate). In the final phase, Strengthen, trainers use practices and give feedback to the learners followed by asking questions that give the learn an opportunity to apply the information to their job [1]. LDT uses this time for open ended questions to ensure learners know the system and how it works. In most circumstances, the above has proven effective through final assessments.

Knowing that most FIs use the system slightly differently, LDT cannot enforce prerequisites, and that the learners coming to class have various experiences, the need to adapt to various levels of learners is immediate. The same WIIFM for one learner is not helpful to another. The same YCDI for someone new to the subject was condescending to someone who had experience. Often recalling prior knowledge meant the difference between recalling the alphabet versus recalling how to diagram sentences.

2 Solution

The solution was easy: Just adapt to the individuals of the room. Executing the solution was not so easy. The trainers were not only learning new training techniques for their entire approach to training, but they were also now being asked to adjust those new techniques to individuals. LDT needed to provide guidance to their trainers on how to adapt.

Upon research, there was not much on adapting to learners based on much other than knowledge levels. Bloom's taxonomy addresses various levels in learning, but not necessarily how to adapt if there are various levels in the room, and most certainly not in a technical training environment where the learners are working professionals learning the tool to do their job. What they needed were techniques to move learners from Bloom's Knowledge level to the Evaluation level.

2.1 SLII® as a Solution

Around the same time, FinTech leveraged Ken Blanchard's SLII® model to leadership. ("SL" stands for "Situational Leadership" and the "II" is the Roman Numeral equivalent of two denoting this is the second iteration of the model.)

The SLII® approach encourages leaders to manage with agility, change leadership style based on individual situations, and empower their teams to succeed. SLII® seeks to diagnose development levels of employees based on tasks. According to the model, employees fall into one of four development categories based on the metrics of competence and commitment:

- D1—Enthusiastic Beginner—Low Competence, High Commitment
- D2—Disillusioned Learner—Low to Some Competence, Low Commitment
- D3—Capable, but Cautious Contributor—Moderate to High Competence, Variable Commitment
- D4—Self-reliant Achiever—High Competence, High Commitment [2]

Based on the employee development level, leaders change their Directive Behavior based on commitment and Supportive Behavior based on competence. The higher the commitment, the less support an employee needs. The higher the competence, the less direction an employee needs [2] (Fig. 1).

2.2 Adapting SLII® to Classroom Attendees

SLII® provided the foundation of what is now known as the Practitioner Level Theory (PL) and Practice Support Methodologies created by the author and other LDT staff.

Based on the author's experience, knowledge level and confidence seemed to be the best metric for diagnoses. "Confidence" came from several learners observed by the author. These learners served as the "go to" for FI's operations staff. They needed "to know everything because people are going to ask me." Many learners had a frantic approach to learning and needed to know "everything today." Based on the author's experience, the details for all diagnoses are as follows (Fig. 2).

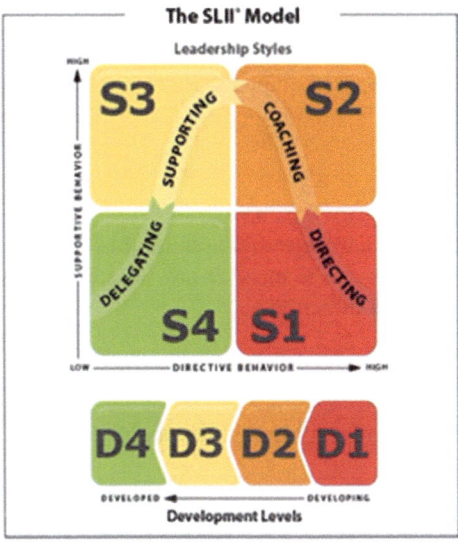

Fig. 1. Diagram of SLII® Model as published by The Ken Blanchard Companies [2].

	Practitioner 1 (P1)	Practitioner 2 (P2)	Practitioner 3 (P3)	Practitioner 4 (P4)
Knowledge Level	• New to the subject • Inexperienced in related subjects	• Some knowledge or prior experience • Unsure of how prior skills/knowledge relate	• Has experience with product or similar • Seeks higher skills and beyond basics	• Skilled in several techniques or extensive use of similar product • Experience with trouble-shooting problems
Confidence Level	• Low • Very tentative in exercises • Usually very quiet	• Med-high • Eager to try and apply new skills • Asks questions openly	• Low-med • Unsure of where to start or what to do next • Avoids questions so as to not appear ignorant	• High • Achieves desired results in labs • Tries new ways to achieve same result

Fig. 2. Original PLs based on knowledge and confidence level metrics.

Based on the PL, the trainer employed different Practice Support Methodologies based on the PL's goal.

- PL1—Lead

  - Goal: Transfer the necessary skills needed for desired job performance.
  - Coaching:

    Provide click by click instructions
    Answer questions directly
    Provide learner with the "Why"
    Explain thought process you went through to get there

- PL2—Directive Coaching

  - Goal: Leverage confidence to complete tasks to build knowledge.
  - Coaching:

    Ask: What would you do next?
    Ask: How did you get to where you are?
    Ask: In what situations at work would you use this?
    Ask for their thought process to uncover any missteps, ask well-framed questions to point them in the right direction

- PL3—Predict

  - Goal: Build confidence by confirming knowledge.
  - Coaching:

    Ask: What will happen if you....?
    Ask the learner questions about what they should do next and why
    Confirm accurate explanations and follow up with positive reinforcement
    For inaccurate explanations ask for their thought process to uncover any missteps, ask well-framed questions to point them in the right direction

- PL4—Self-Discovery

  - Goal: Leverage confidence and knowledge to upskill learner.
  - Coaching

    Give them a variable to work through

  - Add additional, advanced steps if available
  - Encourage them to try new techniques

  In theory, as the trainer coached the various practitioner levels, the learner would move from having basic knowledge of a subject to apply and analyze the knowledge to various situations.

3 Implementation

3.1 Train the Trainers

After implementing Foshay's et al. [1] Cognitive Training Model, other changes in technical training were becoming more popular; one example was story-based learning. LDT leadership determined LDT trainers needed a review of the Cognitive Training Model with adjustments such as Story-based learning and incorporating adjustments based on Practitioner Levels.

The variances mostly applied to WIIFMs, YCDIs, and Recall and Relate. LDT found that those who were P1s and P2s needed more detailed WIIFMs as they did not understand the purpose of some of the software or the techniques they were learning. P3s and P4s had existing knowledge of the software or a similar software therefore they knew why an FI would use such a system. YCDIs are meant to encourage learners. P1s and P4s were thought to be either enthusiastic about learning or had proved they could perform tasks based on prior experience, respectively. Telling an advanced learner something along the lines of "you can do this" seemed patronizing, and P1s needed significantly more detail, e.g., why the trainer believed they could perform the task. Was it because of the prior lesson, something that was learned in the past, etc.? Recall and relate continued to vary based on prior knowledge.

Lastly, trainers were instructed to either answer learner questions directly or with a reframed question based on PLs. In theory, a trainer could rephrase a question back to a P3 or P4 and, given their prior experience, this would trigger a different thought that would lead to the answer.

Training Clients Using Practitioner Levels. Trainers were asked to implement variances based on PLs with mostly positive results. However, as time went by, there were more and more reports of situations when the PL did not align with the confidence level. LDT received several testimonials similar to the examples below.

- "I had a P3 that was confident. They knew everyone relied on them and realized they couldn't learn everything, but they would 'figure it out' with more practice and situations as they arise."
- "The learner was angry because the 'system doesn't do what [we] thought it would.' I thought they were a P4 because they had 'used the system for years', but when I rephrased a question back to them, they got mad thinking the software was missing functionality. They were actually more like a P2."
- "I kept rephrasing questions back to the class. During a practice activity, I asked a P2 who was struggling if they had a question. They said, 'I don't want to ask it. You'll just repeat it back to me.'"
- "I rephrased questions back to a P3. After doing that a few times, they asked if I knew the software."

LDT quickly realized changes were needed to the PL theory. Confidence seemed to vary by personality and high or low confidence was found with every knowledge level. Trainer expertise was also called into question when they didn't answer questions directly.

4 Adjusting Practitioner Levels and Further Testing

4.1 Review Existing PLs

LDT surveyed several trainers asking them what they found to be true about the PL theory and what was not. Knowledge level held true, but confidence did not seem to be the correct, second metric. Some trainers felt that a "growth versus fixed mindset" would work. Others thought SLII® had it right with competence and commitment; for

LDT's purposes, competence equated to knowledge level, so commitment seemed to be the likely second metric. LDT took a step back and looked at andragogy according to Malcolm Knowles (Fig. 3).

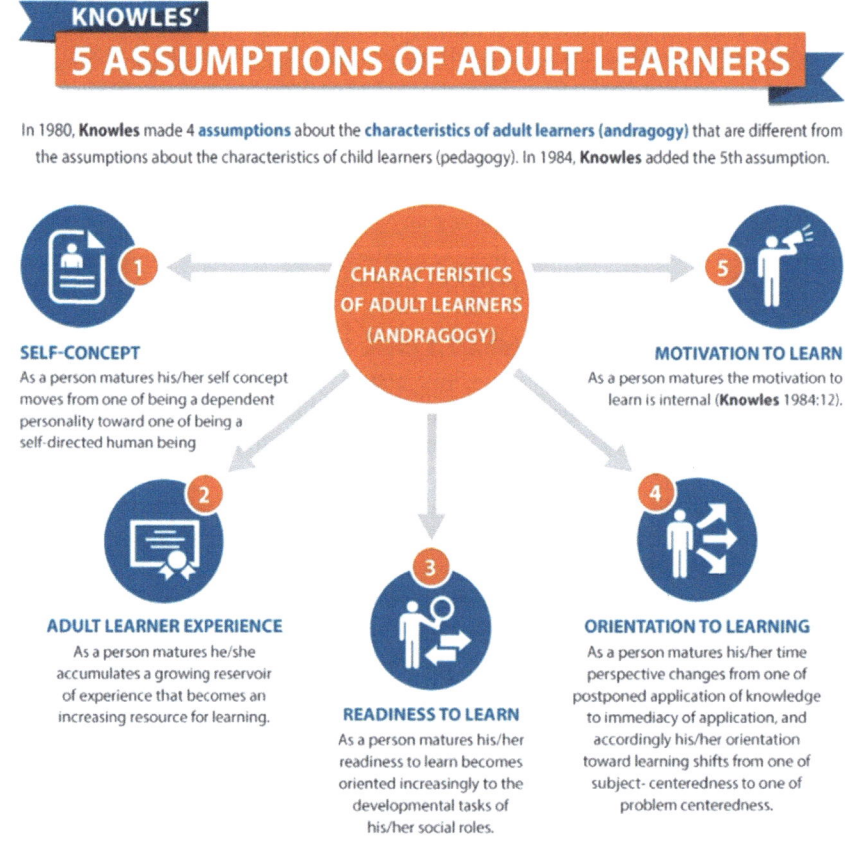

Fig. 3. Knowles' 5 assumptions of adult learners as presented by eLearning Industry [3].

All assumptions were considered. "Self-concept" seemed to support confidence. "Readiness to learn" seemed to line up with a growth versus fixed mindset; however, many people simply do not have a growth mindset. That brought LDT to "Motivation to Learn". While Knowles' theory states that an adult's motivation to learn is internal [3], there are times that adults are motivated by external forces. In the case of LDT learners, this aligns perfectly: their job requires them to learn a new system.

4.2 Practitioner Levels 2.0

The existing Practitioner Level Theory and Practice Support Methodologies now takes into consideration "Readiness to Learn." Each PL has supplemental support directions based on a low readiness to learn.

- P1s may require additional WIIFMs and digging deeper to make Recall and Relate more effective
- P2s may need a highly specified Recall and Relate based on personal experiences as well as additional real-world examples
- P3s may need additional YCDIs and to be reminded their job is not changing, just the tool
- P4s may need for the trainer to quickly establish expertise to build trust while the trainer uncovers negative experiences and then provides solutions (Fig. 4)

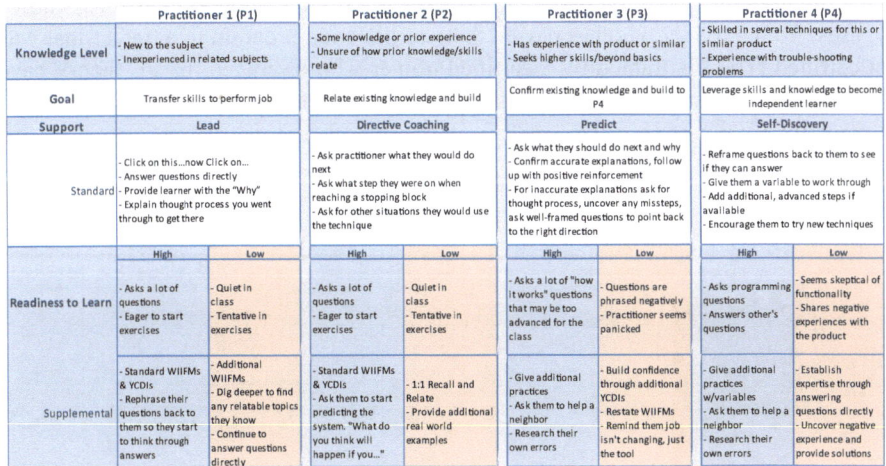

Fig. 4. Revised PLs with supplemental support for low readiness to learn.

4.3   Further Testing

The updated Practitioner Levels and Practice Support Methodologies need to be reviewed with LDT trainers and implemented. After the trainers have used the new supplemental support, LDT can review the findings and make further adjustments as needed.

Testing In a Controlled Environment. The ideal testing environment would be a controlled environment where trainers could take multiple groups made up of similar demographics where one group is taught with the supplemental support and the other is not. Given the nature of the FinTech's business and LDT's need to provide the best training possible so learners are successful on the job, LDT will continue to adjust to learners quickly, to the best of their knowledge and ability.

5   Application and Conclusions

The Practitioner Level Theory can be applied in most learning situations but is ideal for adult learning in technical training environments that include hands-on practice. Because of the focus and coaching given to individuals during practices, it is recommended class sizes do not exceed fifteen participants. To apply the PL theory, trainers need to monitor the individual learners and adapt accordingly. Larger class sizes hinder a trainer's ability to do so.

Other learner variances should be considered. If learners are at cognitive capacity, it is best to simply answer a question versus rephrase back to the learner. Trainers should establish and maintain content expertise which also means answering questions directly. As a learner moves through a course, their P level may change and require additional adaptations.

LDT knew the initial theory worked as it is the basis of their award-winning certification program. LDT compared the volume of customer support questions, known as cases, from a trainer who has passed their certification program to a trainer that was not certified. The FIs trained by a certified trainer who leveraged the PL theory have 46% fewer post-training support questions than those who are not certified and do not leverage the PL theory (Fig. 5).

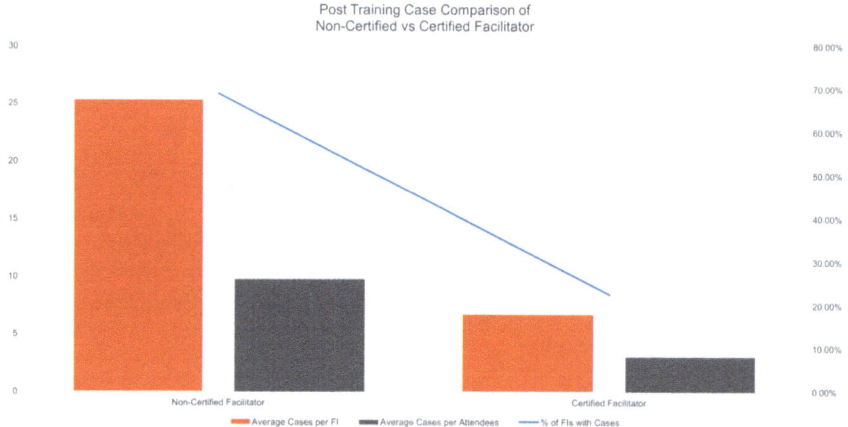

Fig. 5. Client questions, known as cases, from FIs who were trained by a certified trainer who leveraged the Practitioner Level Theory versus a non-certified trainer who does not leverage the Practitioner Level Theory.

Given the above evidence and the revised Practitioner Level Theory and Practice Support Methodologies, which is based on the andragogical principle of "Readiness to Learn," it is reasonable to believe the theory works for the betterment of the learner and learning retention.

Acknowledgments. Trainers of LDT tested the theory and advised on the suggested changes. The FinTech and trainers are not named due to privacy restrictions.

References

1. Foshay, W.R., Silber, K.H., Stelnicki, M.: Writing Training Materials That Work: How to Train Anyone to Do Anything. Pfeiffer (2003)
2. The Ken Blanchard Companies Solutions page. https://www.kenblanchard.com/Solutions/SLII. Last accessed 25 March 2022
3. eLearning Industry Instructional Design the Adult Learning Theory—Andragogy—Of Malcolm Knowles page. https://elearningindustry.com/the-adult-learning-theory-andragogy-of-malcolm-knowles. Last accessed 25 March 2022

Implementation of a Signature Pedagogy in an Online Course for Music Teachers

Svetlana Karkina[1]($\boxtimes$) and Elena Dyganova[2]

[1] University of Salamanca, 37008 Salamanca, Spain
svetlana_karkina@usal.es
[2] Kazan Federal University, 420008 Kazan, Russia
elena.dyganova@kpfu.ru

Abstract. This paper presents the implementation of the signature pedagogy at a university by the means of online learning used for music education as a teaching strategy. Based on the literature review, the gap in research was stated to cover learning strategies in music education, which would allow students to take a more active role in gaining knowledge in accordance with personal needs. Due to the shifting the educational process to the online way in the emergency situation COVID-19 pandemic, the relevance of online courses which will satisfy future music teachers' educational goals extremely increased. The comparative analysis of music education standards in USA, European countries and Russia let us determine the set of teachers' professional activities in general school music class. The main contribution was the design of a small private online course based on the signature pedagogy, which was delivered by the set of lectures, collaborative work in the professional field, and discussion of the learning experience. This course was implemented at Kazan Federal University. The experimental work used a questionnaire of students' self-perception of their professional readiness, along with structural equation modeling based on estimation of the results of assessment of students' musical skills and online activity. The obtained results proved the effectiveness of the signature pedagogy online course which provided learning activities according to the specific features of music teacher professional practice.

Keywords: Music · Education · Teacher · Online learning · Signature pedagogy · University · Learning strategy

1 Introduction

The rapid development of technical progress opens new perspectives in all educational fields, including Humanities and Arts. The implementation of technological tools promotes upgrading teacher strategies in the educational process. Currently, educators need new approaches for delivering professional knowledge and training students' skills for preparing future specialists. In this regard, the design of online courses based on a fundamental system of professional preparation suggests bridging the gap between global professional perspectives and teaching methods in educational practice.

The COVID-19 pandemic noticeably stimulated the teachers' attention to the online tools. In order to reach educational purposes, teachers were encouraged to select the best decision among the variety of different types of online courses. Above the most popular MOOC (Massive Open Online Course) [1] researchers characterized wide range of online courses, such as BOOC (Big Open Online Course), DOCC (Distributed Open Collaborative Course), LOOC (Little Open Online Course), MOOR (Massive Open Online Research) and SPOC (Small Private Online Course) [2]. The SPOC also was divided into several varieties, including COOC (Corporative open on-line courses), courses designed for companies; SOOC (Small open online courses), low-audience courses due to the extreme specialization of the subject; and NOOC (Nano open online courses), characterized by requiring less than 20 h of dedication [3].

Researchers pointed the preparing effective online course is a very expensive process [4]. It requires from the academics or teachers' specific skills and spending long time for creating the content and video records. What is more, in many cases, they need to be not only subject experts and educators, but instructional designers and Web developers also [4]. To reduce the cost of educational process and save human efforts for loading content in distance learning environments, Rumble [5] suggests shifting the point of view from teachers' activity in preparing study material, to students' active interaction by the means of online resources.

Fostering an active role of students in their professional development is provided by the signature pedagogy, invented by Shulman. This approach was characterized as a teaching style based on the fundamental dimensions of any profession such as thinking, performance, and acting with integrity [6]. Shulman's key idea based on the statement that professional knowledge is more than academic discipline and needed in special methods in order to teach how to think like a real professional including values and hopes of the profession. Structuring the educational process based on the signature approach allows for the systemization of professional activity and the implementation of them into the learning process by the means of online tools.

2 Background

2.1 Signature Pedagogy Within Subject Fields

The term "signature pedagogy" was coined by Shulman for the explanation of educational methods for professional preparation in the fields of medicine, law and clergy. Follow him, other researchers implied this approach in different subject fields including teacher education, psychology, history, journalism and arts. According to professional activity, researchers suggest specific methods for each subject field study. For initial education, they promote the method of activization of multidimensional thinking skills [7]. The improvement of school leadership was accomplished through watching the specific movies in the middle school based on the signature pedagogy for the development of the critical thinking skills [8]. For the postgraduate students, the signature pedagogy was implements through the research method [9].

The signature pedagogy was widely used in Arts and Humanities. For the development of students' writing skills, the method of critique was applied [10]. In the high school history class, the method of case study showed its efficiency [11], as well as in an orchestral composer workshop [12]. For the future journalists, preparing the integrated learning as a signature pedagogy was implemented [13]. The method of critique was the most popular in art classes, including teaching theater performing [14], graphic design [15] and modern choreography [16]. Despite that, we did not find an educational experience demonstrated to attempt to cover, by signature pedagogy, the full set of professional activities in the learning process.

2.2 Design of an Online Course by Using SPOC

Among a wide range of online course types, the educational trend to design a resource for a small number of participants to satisfy their specific learning needs appeared. This trend was supported by the creation of SPOC, invented by Fox [17] at the University of Berkeley. A new course presented the adaptation of MOOCs tending to personalize learning [18]. This was the next stage in education technologies in comparison with MOOCs, which spread out the general knowledge for large groups. Close interaction of tutor with students in SPOC facilitates intensive student engagement through intimate relationship between all the participants. In addition, SPOC has kept on providing access to information resources, including electronic libraries, provides knowledge through lectures, evaluates learning outcomes and creates an educational community for the exchange of learning experiences.

Among the variety of MOOCs providing music education—including history and theory of music, playing instruments or vocal singing, creating compositions or improvisations through the platform of Coursera—no one of them provides personalization in this field. At the same, time the research work of Pike in teaching by Skype for playing the instruments demonstrated positive results and advantages of the online interaction in comparison with traditional face-to-face practice [19]. The gap between the high development of modern computer tools and online courses for future music teachers covering the complexity of their professional activity has been established based on a literature review.

2.3 Standardization of Music Teacher Professional Activity

General school education includes music as a mandatory class in the curriculum in all civilized countries. Despite some elements of the content of this class differs depended on national traditions, the core of music activities follows the constant standard of music education. So, the physical activity as a priority in national culture of Spain defines the main trend of music education in this country, based on the close linkage between music and dance [21].

The fundamental set of children's activities in general school music class demonstrates the USA National Standard for Music Education [22], which includes:

- Singing, alone and with others, a varied repertoire of music.
- Performing on instruments, alone and with others, a varied repertoire of music.
- Improvising melodies, variations, and accompaniments.
- Composing and arranging music within specified guidelines.
- Reading and notating music.

- Listening to, analyzing, and describing music.
- Evaluating music and music performances.
- Understanding relationships between music, the other arts, and disciplines outside the arts.
- Understanding music in relation to history and culture.

The implementation of this set implies that the main goal of music education is the improvement of musical intelligence, which is defined by Gardner as ability to perform, compose and appreciate music and musical patterns [20].

The crucial elements of this system are covered by Spain Standard for General School in the section of Music [21]. This document regulates students' vocal and instrumental performance, perception and attentive listening, identifying and analyzing the sounds, recognition and understanding musical language, reading and writing musical notation, improvising movements, interpretation and musical assessment, expressive practice, and creation. The Educational Standard in Spain promotes study of relationships of music with other arts by the linkage between music and dance through the way rhythmic movement and dance integrate body expression with musical elements.

The standard set of music lesson activities in Russian schools is very close to the USA list. It includes vocal singing and playing musical instruments alone and with others. Development of music listening skills, the study of history and theory of music and composer activity in order to foster the creative skills present the pivotal parts of the educational process. The study of relationships between music and other arts delivered by music-mediated and musical poly-artistic activities are required, too [23].

Based on the Shulmans' definition and the comparison of general school music education standards was stated, that the surface structure of music teacher signature pedagogy have to provide the complex set of schoolchildren activities including vocal and instrumental performing, composing and improvising, reading, notating, listening, analyzing, describing and evaluating music and music performances, as well as understanding music in relation to other arts, disciplines outside the arts, to history and culture. The deep structure of signature pedagogy tends to intensify students in professional activity through their engagement study process developing and implementation new methods. The implicit structure ensures the students awareness in values and dispositions of real professionals, who implement the standard point in school practice.

Based on studied background reflected the relevant issues in the fields of professional training, online learning and standardization in music education, the objectives were stated of the investigation:

1. Which online tools can replace face-to-face learning in time of emergency in music teacher professional preparation?
2. Is the SPOC based on signature pedagogy affects future music teachers' learning outcomes and engagement?

2.4 Design of SPOC Based on Signature Pedagogy

To bridge the gap between academic knowledge and professional practice, the small private online course on MOODLE for future music teachers based on signature pedagogy was designed. This course applied of three fundamental dimensions of any profession: surface structure (all forms of interaction between educational process participants), deep structure (a set of assumptions about how to impart a subject knowledge), and implicit structure (values and dispositions of real professional activity) [24].

Following Shulman's model, the surface structure was implemented by a full set of schoolchildren music lesson activities according to the standard of music education. For training vocal and instrumental performing skills, students recorded themselves and sent their records to a teacher for feedback. In the frameworks of performance practice, students prepared ensemble records by the means of telematics technology. Students had been training their creative skills in the preparing art compositions of combining poetry with music. The skills of listening to music, understanding it and evaluating it were trained through the method of essays written by students. Workshops, forum and chat for online student-teacher interaction were used, also. By the means of online tools, teacher delivered individual and group consulting, provided personal advice for improvement skills and managed the process.

The implementation of the deep structure was provided by using specific methods for stimulation of music teacher pedagogical skills. For this purpose, the project work was organized. Students designed some didactic improvements for general school educational processes. They put themselves in the position of a music teacher; they had been creating new methods and means to foster school children music skills. Besides, the artistic projects were prepared. Student recorded solo and ensemble instrumental and vocal compositions by using online tools. These tools allow creating the video with a split-screen effect. By using such a tool, students created the video to follow the teacher's instruction separately and afterwards, they united their musical records in one video with a split-screen effect. In this work, students trained their creative and performing skills as well as the pedagogical skills of organizers, who decided collectively how to improve their music records. All this work had taken place in online way only.

According to Shulman's definition of implicit structure, the specific forms of learning activity were included into the course. These forms provided the professional development through awareness in music teacher values, attitudes and dispositions. For this purpose, interaction with schoolteachers who have real practice was organized. This interaction included watching video of school music lessons, visiting specify sites, interviewing schoolteachers, reading information about them, and organizing events for school children. The discussions of critical issues, which are strongly relevant in modern general school music education, had taken a pivotal role. The arguing of actual problems and discussion of contrary arguments to promote effective solutions were organized in online workshops and off-line forums. Through all these forms of activity, student engagement in the core of music teacher professional experience was increased through fostering of their awareness in moral aspects of the real school practice.

The SPOC based on signature pedagogy allowed us to gather in one course systematically the music teacher professional competence. The fundamental set of general school children's activities in music class and three levels of Shulman's signature pedagogy for promoting the professional knowledge, skills and ability to work with integrity was implemented by the means of MOODLE tools in online environment.

3 Materials and Methods

3.1 Study Design and Settings

Due to the emergency case COVID-19 pandemic, the education process shifted into online methods. The exploratory study design was used to uncover the effectiveness of the SPOC based on signature pedagogy in the replacing the music teacher professional training in remote way. This study was conducted at the Kazan Federal University. The study population comprised 56 undergraduate students who studied music teacher curricula during the academic year 2020–2021. Data were collected from one academic semester during September 1, 2020, to January 31, 2021. All these students were enrolled to the SPOC for training special music teacher professional skills. The effectiveness of the course was studied by questionnaires through an online tool.

Ethical clearance to carry out the study was attained from the Kazan Federal University policy regarding the processing of personal data. Students were informed about the purpose of the study and were given the option not to complete the questionnaire. Those who did agree to participate signed electronic consent forms. The survey was anonymized to protect the identities of the students [25].

3.2 Instruments

The data were collected by the quantitative study using the method of questionnaire. For this purpose, two questionnaires were designed for evaluation students' self-perception of professional readiness and for assessment their learning outcomes.

The first questionnaire consisted of six trends in music activity, with 20 items and an overall item (overall satisfaction). This questionnaire encompassed six trends namely, "Solo Performance (SP)" (04 items), "Ensemble Performance (EP)" (04 items), "Listening and evaluating (LE)" (03 items), "Artistic creation (AC)" (04 items), "Reading, analyzing and describing (RAD)" (02 items), and "Understanding music in context (UMC)" (03 items). The students' agreement level towards the items of each factor was recorded through a five-point Likert scale. All participants filled out the informed consent before completing the survey, and those were assured with privacy and anonymity before gathering the data from them.

The second questionnaire allowed to estimate relations between students' musical intelligence and their online activity results. Two lists of criteria were used:

1. Artistic idea, vocal skills, technical music instrumental skills, creativity, music literacy, ability to understand the meaning of music and the interpretation, artistic skills, digital music skills, and knowledge of music history and culture.
2. Performance, composing, music lesson design, discussion, essay, music-poetry compositions, project work.

Each criterion was assessed by teachers by using 10-level scale. The highest level of a criterion students received 10 marks, and the lowest received 0 marks. The first group of criteria demonstrated the impact on the musical intelligence. The criteria of the second group demonstrated the effectiveness of implementation of the signature pedagogy by using SPOC.

3.3 Statistical Analysis

In this study, the trends (variables) of the first questionnaire were observed with the Cronbach's alpha (α) indicating the adapted questionnaire is a reliable tool to measure the self-perception of readiness to music teacher professional practice among students. To prove the obtained data, statistical methods of median and standard deviation was used for establishing the diversity between students responds.

The structural equation modeling (SEM) was executed through the software named "Ωnyx" for creating and estimating structural equation model to reveal how well the factors (variables) of signature pedagogy in music education (SPME) influence the students' musical intelligence, which impacted by music activity.

4   Results

Out of 56 participants, all of them responded to the survey, demonstrating a response rate of 100%. Among the respondents, males accounted for 21.4% (n = 12), and females accounted for 78.5% (n = 44).

4.1 Reliability of the Questionnaire

In this study, the trends (variables) of the questionnaire were observed with the Cronbach's alpha (α) value as follows: SP (0.92), EP (0.97), LE (0.99), AC (0.97), RAD (0.95), and UMC (0.97). Furthermore, the overall Cronbach's alpha (α) value of all questionnaire items was observed as 0.99. Since the $\alpha > 0.90$, the internal consistency of the questionnaire could be rated as "excellent", indicating the adapted questionnaire is a reliable tool to measure the self-perception of readiness to music teacher professional practice among students.

4.2 Self-perception of Readiness to Music Teacher Professional Practice

The questionnaire included six scales: "Solo Performance (SP)", "Ensemble Performance (EP)", "Listening and evaluating (LE)", "Artistic creation (AC)", "Reading, analyzing and describing (RAD)", and "Understanding music in context (UMC)" and an overall satisfaction item. Questions were followed by five-point Likert scale questions (using strongly agree; agree; neither agree nor disagree; disagree; strongly disagree). For each answer, students received a mark from one to five points according to the Likert scale. The total result of each student could be from one to five. The results of the questionnaire were presented in the Table 1.

The obtained results of the median of the students' self-perception were not less than 3.8, while the total result was 4.8 (close to max 5). The results of standard deviation for each scale did not exceed 1.4, while the total result was 1.2983 (close to 1). This means there were no statistically significant differences among all the students responds.

Table 1. Students self-perception of professional readiness

Trend of professional practice	Average result (median) (max 5)	Standard deviation
Solo performance	4.7	1.3685
Ensemble performance	4.8	0.9875
Listening and evaluating	4.1	1.3256
Artistic creation	4.9	0.8984
Reading, analyzing and describing	4.9	0.5664
Understanding music in context	3.8	1.2983
Overall satisfaction	4.8	0.5683
Total result	4.8	1.2983

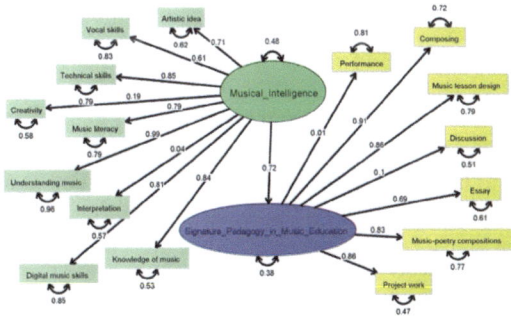

Fig. 1. A structural equation model of signature pedagogy implementation in online music education process.

4.3 Structural Equation Modelling for Signature Pedagogy in Online Music Education

A model obtained from SEM analysis is shown in Fig. 1. While assessing the proposed model, each factor of SPME showed a significant positive relationship with overall item, varying from 0.01 to 0.91. Besides, the assessing factors of musical intelligence (MI) as a main goal of music education showed positive relationship with overall factors, varying from 0.04 to 0.99. These findings indicate that the proposed model possesses the influence of students' online activity to their level of musical intelligence during study SPOC based on signature pedagogy.

5 Discussion

5.1 Signature Pedagogy in Music Education

Through last decades, the concept of signature pedagogy became very popular in the field of arts. Generally, they feature by using one or two methods for implementation of this approach in the professional education.

Therefore, most authors understand signature pedagogy as a critique. Research offers this method for different arts, including theater performance [12], graphic design [13], modern dance [14] or music performance [26]. They offer as a signature pedagogy arts integration for art educator preparation [27] and case study for development of creativity in an orchestral composer workshop [12] were used directly also.

Despite this fact, it seems appropriate to study any profession in all its complexity. As a part of educational experience, students participate in some sort of professional practice which helps them to learn the role of real professional. To reach the purpose of individual transition from the student into the professional role researchers offer to identify "the pivotal components within the education process" [28]. The complex of these components featured the structure of university professional practice used as the signature pedagogy will provide more intensive professional engagement of students.

Through the study of music education standards from different countries and continents, the core of general school music activity was identified. Different countries' music education standards emphasize close correlation between all forms of music activities. In order to reach children's cultural awareness and sufficient level of creative skills, they need systematical practice in vocal singing and instrument performance, as well as experience in composing and improvising, reading notation, listening to music, analyzing and describing it, and evaluating and understanding in the frameworks of history and culture. That is why the set of these activities was defined as the signature pedagogy in music education. Taking that into account, we can assume the similar approach to professional training in other subject fields based on their complexity will provide the effective teaching methods.

5.2 Implementation of Signature Pedagogy by the Small Private Online Course

Teaching performing classes vocal and instrumental remotely during the pandemic was the unique part among most of the educational practice. Only several examples were known in the resent years of teaching music online, particularly the piano by Skype [19]. Changing the teaching format required teachers to correct the approach for giving instructions. While in remote way, a teacher became unable to demonstrate to a student the artistic details directly, they ought to define objective students' mistakes and provide strict and effective solutions. Students also got more freedom in improving their performing style, including lack of physical contact with a teacher; they felt more comfortable and inspired by music. This fact supports the results obtained in Skype piano classes by Pike [19] and points the way to didactic enhancing of the teaching methods.

During the performing online practice, students prepared records with their solo and ensemble performances. Surprisingly, this work inspired them for collecting the records for social activity. Doing multiple attempts for improving the performances, students tended to create the best one to load as many records as possible to their own profile in social networks. The social activity became more than demonstrating learning achievements. Besides, it was appreciated by students in their desire to promote their own personal image, who put "likes" as positive marks and verbal comments with emotions and assessments.

Students' preferences were supported by teachers. The idea was to engage students in social life through the celebration of great national events or anniversaries of famous artists and musicians. Several topics of musical compositions according to the plan were offered, such as famous Tatar poet Gabdulla Tukay anniversary and memorable event of World War II. For the noticeable national memory, students recorded solo and in ensembles music art pieces. The best compositions were collected and loaded in the official pages on YouTube and Instagram, and they were assessed by the audience. The success of this work was noticed by a big number of views of the records, positive marks, excited comments and good wishes to performers.

What is more, university teachers recognized the critical social effect of this work. It allowed then to catch the attention of a global audience to noteworthy national events for refreshing the memory and sharing the historical knowledge among a young generation. Such work was done by engaging students in real social activity of music teacher, who spreads out the meaningful information about society life through the aesthetic images and awareness. Doubtlessly the impact of this work by using online resources increased while the remote teaching in pandemic enhanced the comprehension of this fact.

6   Conclusions

The research work results let us conclude that the online tools can replace face-to-face learning in a case of emergency, like COVID-19 pandemic. The design of a SPOC based on the signature pedagogy implemented by features of music teacher activity allowed students to improve their professional skills. What is more, the process promoted the student-teacher interaction and opportunity to offer for every student task in the field of his/her personal environment significantly increased. In this point, the perspective of design flexible, adaptive and meaningful educational resource to accommodate students learning, work, and life goals was noticed [29].

The experimental work obtained results demonstrating a high enough level of student self-perception of their readiness for future music teacher professional activity. Also, the positive correlation between students' results after studying signature pedagogy in an online course and the dynamic of their musical intelligence proved the influence. Based on these results, the effect of SPOC on learning outcomes and students' engagement was stated.

This research has some limitations, such as the number of students who were engaged in the experiment, while the same curricula studying by more students at the University in total. Therefore, only senior students participated and filled the questionnaire about their self perception of readiness to future music teacher professional activity and learning outcomes assessment.

Despite the limitations, some benefits of implementation of SPOC based on signature pedagogy were noted, such as unlimited access to learning content, close personal interaction between a teacher and students, and comfort levels leading to self-improvement in students' practice.

Acknowledgments. This research work is being carried out within the University of Salamanca PhD Programme on Education in the Knowledge Society scope (http://knowledgesociety.usal.es) [30].

References

1. Veletsianos, G., Shepherdson, P.: A systematic analysis and synthesis of the empirical MOOC literature published in 2013–2015. Int. Rev. Res. Open Distrib. Learn. **17**(2), 198–221 (2016)
2. Naert, F.: MOOCs, SPOCs, DOCCs and other bugs. European cooperation on MOOCs EADTU, pp. 64–74 (2015)
3. Olmos, S., Mena, J., Torrecilla, E., Iglesias, A.: Improving graduate students' learning through the use of Moodle. Educ. Res. Rev. **10**(5), 604–614 (2015)
4. Salmon, G.: E-moderating: The Key to Teaching and Learning Online, 2nd edn. Taylor & Francis, London (2013)
5. Rumble, G.: Flexing costs and reflecting on methods. In: Burge, E., Gibson, C., Gibson, T.: (eds.) Flexibility in Higher Education: Promises, Ambiguities, Challenges, pp. 264–301. Athabasca University Press, Athabasca, Alberta (2010)
6. Shulman, L.S.: Signature Pedagogies in the Professions. Daedalus (2005)
7. Tan, C.: A signature pedagogy for initial teacher education in Singapore. New Educ. **15**, 226–245 (2019)
8. Tan, C., Koh, K.: Signature pedagogies for educators using films: an example from Singapore. Teach. Educ. Q. **53**, 100–186 (2018)
9. Buss, R.R.: Using action research as a signature pedagogy to develop EdD students' inquiry as practice abilities. J. Transform. Prof. Pract. **3**, 23–31 (2018)
10. Heinert, J.L.: Peer critique as a signature pedagogy in writing studies. Arts Humanit. High. Educ. **16**, 293–304 (2017)
11. Walker, S., Gustavo Carrera, G.: Developing a signature pedagogy for the high school U.S. history survey: a case study. Hist. Teacher **51**(1), 65–88 (2017)
12. Love, K.G., Barrett, M.S.: Signature pedagogies for musical practice: a case study of creativity development in an orchestral composers? Workshop. Psychol. Music **47**, 551–567 (2019)
13. Woolley, B.: WIL-power: towards a signature pedagogy in journalism. Asia Pac. Media Educ. **28**, 237–249 (2018)
14. Kornetsky, L.: Signature pedagogy in theatre arts. Arts Humanit. High. Educ. **16**(3), 241–251 (2017)
15. Motley, P.: Critique and process: signature pedagogies in the graphic design classroom. Arts Human. High. Educ. **16**(3), 229–240 (2017)
16. Kearns, L.W.: Dance critique as signature pedagogy. Arts Human. High. Educ. **16**, 266–276 (2017)
17. Fox, A.: From MOOCs to SPOCs. Commun. ACM **56**(12), 38–40 (2013)
18. Lou, J.-Y., Zheng, P., Jiang, C.: The enlightenment of SPOC on teaching reform of higher education in China-based on the perspective of mastery learning theory. Sci. J. Educ. **4**(2), 95 (2016)
19. Pike, P.D.: Improving music teaching and learning through online service: a case study of a synchronous online teaching internship. Int. J. Music. Educ. **35**(1), 107–117 (2017)
20. Gardner, H.: The arts and human development: a psychological study of the artistic process (1973)
21. Real Decreto 1006/1991, de 14 de Junio, por el que se establecen las ensenanzas minimas correspondientes a la Educacion Primaria, https://www.boe.es/boe/dias/1991/06/26/pdfs/C00 003-00033.pdf. Last accessed 06 September 2021
22. National Standards for Music Education. https://old.philorch.org/sites/default/files/13-14-Nat-Arts-Achievement-Standards.pdf. Last accessed 10 September 2021
23. Abdullin, E., Nicolayeva, E.: Methods of Music Education, 1st edn. Music, Moscow (2006)
24. Shulman, L.S.: The Wisdom of Practice: Essays on Teaching, Learning, and Learning to Teach. Jossey-Bass, San Francisco (2004)

25. British Educational Research Association [BERA]. Ethical Guidelines for Educational Research, 4th edn, London. https://www.bera.ac.uk/researchers-resources/publications/eth icalguidelines-for-educational-research-2018. Last accessed 20 December 2021
26. Hastings, D.M.: With grace under pressure: how critique as signature pedagogy fosters effective music performance. Arts Human. High. Educ. **16**, 252–265 (2017)
27. Reck, B.L., Wald, K.M.: Toward a signature pedagogy for arts integration in educator preparation. Pedagogies: An Int. J. **13**, 106–118 (2–18)
28. Woeste, L.A., Barham, B.J.: The signature pedagogy of clinical laboratory science education: the professional practice experience. Labmedicine **37**, 591–592 (2006)
29. Dabbagh, N., Attwell, G., Castañeda, L.: Personal learning environments track introduction. In: Ninth International Conference on Technological Ecosystems for Enhancing Multiculturality (TEEM'21), Barcelona, Spain. ACM, New York (2021)
30. García-, F.J.: Formación en la sociedad del conocimiento, un programa de doctorado con una perspectiva interdisciplinar. Educ. Knowl. Soc. **15**(1), 4–9 (2014)

Neural Correlates of Creative Drawing: Relationship Between EEG Output and a Domain-Specific Creativity Scale

Sang Seong Kim[1] (ID), Sunhwa Hwang[2] (ID), and Eunmi Kim[3](✉) (ID)

[1] School of Transdisciplinary Studies, KAIST, 291, Daehak-ro, Yuseong-gu, Daejon, Republic of Korea

[2] Department of Physics, KAIST, 291, Daehak-ro, Yuseong-gu, Daejon, Republic of Korea

[3] Center for Contemplative Science, KAIST, 291, Daehak-ro, Yuseong-gu, Daejen, Republic of Korea

eunmikim@kaist.ac.kr

Abstract. Creativity is defined as the ability to develop novel and effective ideas, artifacts, or solutions. Neural correlates of the Kaufman domains of creativity scale (K-DOCS) have been studied to better understand the neurophysiological representations of specific mental processes. The K-DOCS is a self-report questionnaire that measures five domains of creativity: Everyday, Scholarly, Performance, Science, and Arts. In this study, twenty-two international undergraduate students at the Korea Advanced Institute of Science and Technology were first assessed based on the K-DOCS. Subsequently, the participants underwent electroencephalography recordings while solving picture completion problems from the Torrance Tests of Creative Thinking (TTCT). The relative alpha power was calculated for eight channels, and the correlations with the domains of the K-DOCS were analyzed using Spearman's rank correlation test. The only correlations regarded as significant were negative correlations between both the relative alpha power and the relative slow alpha power (in the temporal, occipital, and parietal sites) and the measures associated with the K-DOCS Performance domain. In particular, the relationship between alpha synchronization in parietal and occipital sites and the suppression of distracting information flows from the visual system, the devotion of cognitive resources to memory search and retrieval, and the outcome of interviews about participants' experiences while executing a specific task were analyzed. In this regard, the high score in the Performance domain might indicate a tendency to rely on visual inputs rather than memory when searching for ideas during creative tasks. This study contributes to widening the scope of quantitative assessments of creativity.

Keywords: Creative drawing · EEG alpha · K-DOCS · Domain-specific creativity scale · Neural correlates of creativity

D. Guralnick et al. (Eds.): TLIC 2022, LNNS 581, pp. 172–180, 2023.
https://doi.org/10.1007/978-3-031-21569-8_16

1 Introduction

Creativity is a powerful attribute of the human mind that has contributed to the blossoming of civilization. In view of the emergence of unprecedented disasters, such as global warming and pandemics, human creativity will continue to be an essential asset in the future. There have been numerous attempts to analyze creativity in various contexts such as cognitive science, education, and social psychology [1–3]. Neuroscience is also an arena of pioneering research on creativity.

Creativity is generally defined as the ability to generate novel and valuable ideas and actions [4]. Based on this definition, neuroscientific research on creativity is divided into three main topics [5]: (i) divergent thinking, (ii) artistic creativity, and (iii) insight. The first topic presupposes that creativity originates from generating multiple ideas and widespread thinking [6]. Research on artistic creativity focuses on the correlation between neurophysiological changes and artistic performance, such as playing musical instruments or painting. Research on insight focuses on the Eureka experience, which refers to the moment when people (metaphorically) say "aha" and gain new enlightenment as a part of the complicated creative process.

Several neuroscientists have used electroencephalography (EEG) to study creativity [7–9]. These studies mostly focused on brain alpha power expressed during creative tasks. Stronger alpha synchronization (especially in posterior sites) during various creative tasks are due to the nature of the task demanding divergent thinking (rather than convergent thinking) [7, 8], higher difficulty (rather than simple association) [8, 9], higher originality of the output [10], and higher creativity of the individual participating in the task [11], even though the contrasting results showing alpha desynchronization during creative ideation [12, 13]. However, in some cases, the results of the studies were inconsistent or contradicted each other [13, 14]. Although the interpretations of these findings were not consistent, it is clear that creativity is a complex process and does not depend on simple brain activation patterns [4, 5, 15].

In view of the limitations of existing studies, new approaches are required to advance creativity research. The Kaufman domains of creativity scale (K-DOCS) offers promise for triggering a new research paradigm in this respect. The K-DOCS is a self-reporting, domain-specific measure assessing how a person perceives him- or herself to be creative in activities that fall in one of five circumscribed domains: Self/Everyday, Scholarly, Performance, Mechanical/Science, and Arts [16]. The Self/Everyday domain includes intrapersonal and interpersonal creative behaviors in everyday life. The Scholarly domain reflects engagement in deep analysis and pursuits involving gaining knowledge. The Performance domain captures creativity in activities such as music, writing, and acting. The Mechanical/Science domain includes science, engineering, and math-related creative behavior. Finally, the Artistic domain includes art-related activities. Based on the K-DOCS domains, it is possible to subdivide the complex process of creativity and attempt to undertake precise research.

In this study, creativity research was conducted using two new strategies to overcome the limitations of previous studies. First, we analyzed the correlation between the K-DOCS and EEG signals generated while participants performed a drawing task. Second, we assembled a sample group of healthy STEM college students. This dual approach expanded the diversity of the study by forming a sample group of new features that differed from those of previous studies.

2   Methods

2.1   Participants

Twenty-two male international undergraduate students at the Korea Advanced Institute of Science of Technology (KAIST) were recruited as part of a larger experiment where five of them would attend the Mindfulness-based stress reduction (MBSR) course for a semester, and the remaining seventeen would act as controls. The mean age of the students was 21.1 years, and the standard deviation (SD) was 1.7 years. All measurement protocols and measures to ensure the protection of personal information were certified as safe and adequate by the KAIST Institutional Review Board with approval number KH2021-050. The data from both groups measured before the start of the MBSR course were used for correlation analysis in the present study, ensuring that both groups had no prior experience with breathing meditation.

2.2   Protocol

All participants completed the K-DOCS questionnaires through an online survey form and were instructed to perform the following seven tasks, during which the EEG signals were recorded (see Fig. 1).

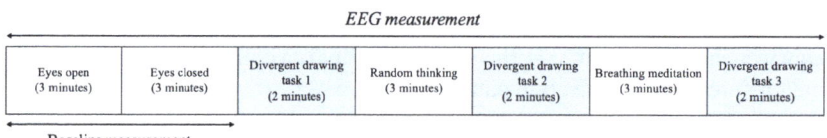

Fig. 1.   Protocol sequence diagram.

The subjects were instructed to keep their eyes open and close them for three minutes each for baseline measurement. Subsequently, they solved a single picture completion question selected from TTCT three times total, two minutes each. The task involved completing the initial picture with a short, partial curve into a recognizable landscape or object and giving a good title for the work. The subjects were asked to stay still, remain relaxed, and do random daydreaming with their eyes closed between the first and second drawing tasks. Between the second and third drawing tasks, they were instructed to practice breathing meditation over three steps of relaxing, finding the location in their body where the feeling of the breath was the strongest, and focusing their attention on that region.

2.3   Measurements

Electroencephalogram (EEG). EEG data were recorded using the BIOS-ST system (Biobrain, Daejeon, South Korea) with eight dry electrodes, a sampling rate of 1,000 Hz, a resolution of 24 bits, and a 50/60 Hz notch. The eight dry channels were attached to the Fp1, Fp2, T3, T4, O1, O2, Fz, and Pz regions of the brain, and ground channels were attached to each earlobe. Spectral analysis of EEG recordings obtained during divergent drawing tasks was performed, producing power spectra of various frequency bands for each channel (Fp1, Fp2, T3, T4, O1, O2, Fz, and Pz) (see Fig. 2). By dividing these respective power spectra by the power spectrum of the entire frequency band, the relative alpha power spectrum (8–13 Hz/4–50 Hz), relative slow alpha power spectrum (8–11 Hz/4–50 Hz), and relative fast alpha power spectrum (11–13 Hz/4–50 Hz) were computed. The mean values of the three power spectra from each divergent drawing session were used for correlation analysis.

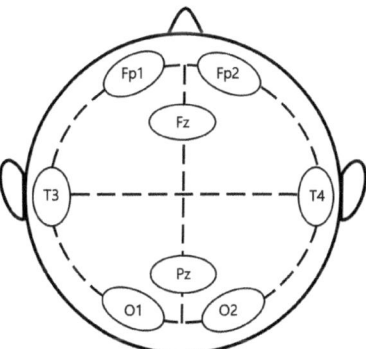

Fig. 2.   Eight-channel location map.

K-DOCS. The participants' creativity was measured using ninety-four developed and validated items of the K-DOCS [16]. The five domains of the K-DOCS were eluci-dated as follows: the Self/Everyday factor assesses self-perception of intrapersonal and interpersonal creative behaviors, the Scholarly factor assesses the gaining of knowledge; the Performance factor assesses self-perception of creativity in music, writing, and act-ing performance; the Mechanical/Science factor assesses self-perception of creativity in science, engineering, and math; and the Artistic factor assesses self-perception of creativity in art activities [16]. Responses are given on a 5-point Likert scale, where 1 represents very low levels of creativity and 5 represents very high levels of creativity, compared to people of similar age and life experience as the respondent [16].

Torrance Test of Creative Thinking (TTCT). The test is divided into two versions based on either language or figures. For the latter version, the test consists of three activ-ities: (i) completing a picture including the given line, (ii) including incompletely drawn figures, and (iii) adding lines to a pair of parallel straight lines. The levels of creativity are quantified by evaluating completed pictures using the following four criteria: flu-ency, flexibility, originality, and elaboration. Notwithstanding the fundamental question of whether divergent thinking and creativity constitute the same function of the brain, this test is one of the oldest used to measure creativity [5].

Statistics and Analysis. The correlations between the alpha power spectra and the average scores for each K-DOCS domain were analyzed. Because of the lack of normality of the data, Spearman's rank correlation test was used rather than the Pearson correlation test. The *SpearmanRankCorrelation* function in the Wolfram Mathematica software was used. The function performs the following calculation:

$$\rho = \frac{\{\frac{(n^3-n)}{6} - T_x - T_y - \sum r_i^2\}}{\sqrt{\frac{(n^3-n)}{(6-2T_x)}\frac{(n^3-n)}{(6-2T_y)}}} \tag{1}$$

where n is the number of participants, r_i is the rank difference between x and y of the i-th participant, and T is the correction term for the ties in each xlist and ylist [17].

3 Results

3.1 K-DOCS

All participants completed the survey. The descriptive statistics and approximate distributions of the data are presented in Table 1 and Fig. 1. Participants scored higher in the Self/Everyday, Scholarly, and Mechanical/Science domains than the Performance and Artistic domains (see Table 1 and Fig. 3).

Table 1. Descriptive statistics of K-DOCS scores.

	Self/everyday	Scholarly	Performance	Mechanical/science	Artistic
Mean	3.58	3.42	2.14	3.37	2.74
SD	0.74	0.64	0.66	0.84	0.69

3.2 Correlations between EEG and K-DOCS Values

Significance levels and Spearman's correlation coefficients were calculated (see Tables 2 and 3). Significant correlations were only detected between both the relative alpha power and relative slow alpha power (for the T4, Pz, and O2 channels) and the values associated with the Performance domain. Scatterplots of the Performance domain scores and relative alpha power spectra are displayed in Fig. 4.

4 Discussion

Participants in this study scored significantly lower in the Performance and Artistic domains than in the other three domains. Students of STEM-focused colleges are prone

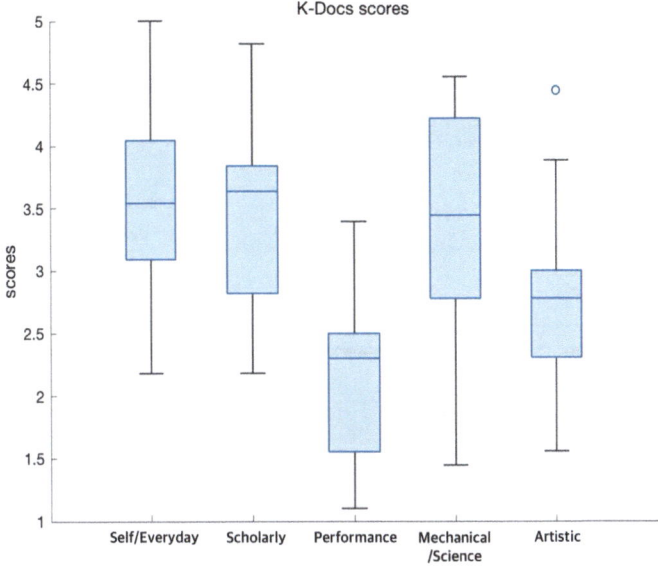

Fig. 3. Box-and-whisker diagram of K-DOCS survey.

Table 2. Significance levels and Spearman's correlations of the relative alpha power.

	Fp1	Fp2	T3	T4	Fz	Pz	O1	O2
Self/everyday	–	–	–	–	–	–	–	–
Scholarly	–	–	–	–	–	–	–	–
Performance	–	–	–	0.025* (−0.49)	–	0.018* (−0.51)	–	0.028* (−0.48)
Mechanical/science	–	–	–	–	–	–	–	–
Artistic	–	–	–	–	–	–	–	–

(Note) –: non-significant, *: $p < 0.05$ significance levels, (xx): Spearman's correlations.

to lack experiences associated with the Performance and Artistic domains, resulting in lower self-rated creativity in such activities.

The alpha power values corresponding to the T4, O2, and Pz channels were negatively correlated with the score of the K-DOCS Performance domain. Alpha synchronization during various creativity-related tasks, such as creative ideation, is often seen over the posterior parietal and occipital sites and has typically been interpreted as suppressing distracting information flow from the visual system. In addition, parietal alpha synchronization can reflect efficient memory processing, where cognitive resources are devoted to effective memory search and retrieval [4].

Many participants reported having tried to recall memories when filling up the test picture, such as the landscape of their hometown or scenes from fairy tales. Others first

Table 3. Significance levels and Spearman's correlations of the relative slow alpha power.

	Fp1	Fp2	T3	T4	Fz	Pz	O1	O2
Self/everyday	–	–	–	–	–	–	–	–
Scholarly	–	–	–	–	–	–	–	–
Performance	–	–	–	0.037* (−0.46)	–	0.027* (−0.48)	–	0.023* (−0.49)
Mechanical/science	–	–	–	–	–	–	–	–
Artistic	–	–	–	–	–	–	–	–

(Note) –: non-significant, *: $p < 0.05$ significance levels, (xx): Spearman's correlations.

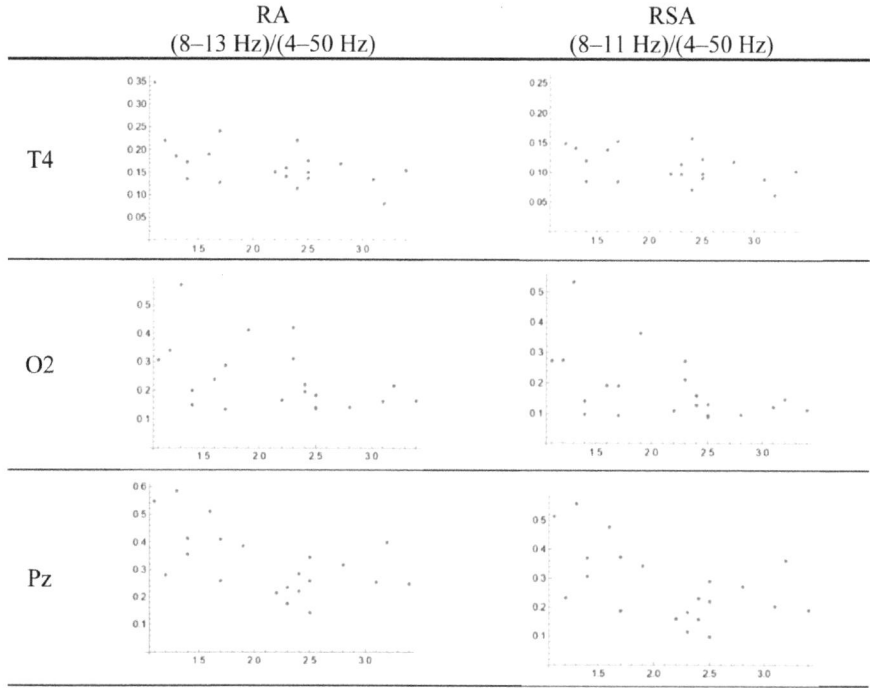

Fig. 4. Scatterplots of the correlations between K-DOCS values and EEG Indices.

elongated the initially provided curve and then repeated gazing at the picture until they generated the idea for the next expansion. Thus, weaker alpha power in O2 and Pz exhibited by individuals who scored high in the Performance domain might be due to a weaker tendency of relying on one's memory rather than on incoming visual information as a source of creative ideas.

However, such tendency during the divergent drawing tasks would only be related to the Performance domain. Questions are raised: Why is the negative correlation only observed with statistical significance for the Performance domain and not for the Artistic domain, even though the K-DOCS scores were similar for these two domains? Which characteristics of individuals with high levels of creativity in the Performance domain would generate such a result? A few observations engage with these questions.

First, the compact distribution of low scores in the Artistic domain could have made correlation analysis between various EEG parameters and the score of this particular domain less effective compared to the others. However, according to Kaufman's study on the correlation between five personality factors and five K-DOCS domains, the Performance and Artistic domains of the K-DOCS showed only a weak correlation [16]. Second, the study revealed that Performance was the domain with the strongest correlation with the extraversion personality factor, which could potentially be related to the tendency of extraverted persons to turn outwards rather than inwards for ideas.

Nevertheless, no significant increase in alpha power in any region was related to any of the K-DOCS domains. In contrast, some evidence supports the idea that self-rated originality is related to alpha synchronization [18]; thus, the relationship between alpha synchronization and self-rated creativity in each domain requires further investigation.

Acknowledgments. This study was supported by the KAIST Institute of Technology Value Creation, Industry Liaison Center (G-CORE Project) grant funded by the Ministry of Science and ICT (Project N11210200); and the KAIST POST-AI Research Institute (N11210157). Biobrain (Daejeon, South Korea) provided technological consulting on EEG signal processing.

References

1. Ward, S.M.S.T.B., Finke, R.A.: The creative cognition approach. MIT Press, Cambridge, MA (1995)
2. Sawyer, R.K.: Educating for innovation. Think. Ski. Creat. **1**, 41–48 (2006)
3. Mabile, T.M.: The social psychology of creativity: a componential conceptualization. J. Pers. Soc. Psychol. **45**, 357–376 (1983)
4. Fink, A., Benedek, M.: EEG alpha power and creative ideation. Neurosci. Biobehav. Rev. **44**, 111–123 (2014)
5. Dietrich, A., Kanso, R.: A review of EEG, ERP, and neuroimaging studies of creativity and insight. Psychol. Bull. **136**, 822–848 (2010)
6. Guilford, J.P.: The structure of intellect. Psychol. Bull. **53**, 267–293 (1956)
7. Mölle, M., Marshall, L., Wolf, B., Fehm, H.L., Born, J.: EEG complexity and performance measures of creative thinking. Psychophysiology **36**, 95–104 (1999)
8. Fink, A., Benedek, M., Grabner, R.H., Staudt, B., Neubauer, A.C.: Creativity meets neuroscience: experimental tasks for the neuroscientific study of creative thinking. Methods **42**, 68–76 (2007)
9. Jauk, E., Benedek, M., Neubauer, A.C.: Tackling creativity at its roots: evidence for different patterns of EEG alpha activity related to convergent and divergent modes of task processing. Int. J. Psychophysiol. **84**, 219–225 (2012)
10. Fink, A., Neubauer, A.C.: EEG alpha oscillations during the performance of verbal creativity tasks: differential effects of sex and verbal intelligence. Int. J. Psychophysiol. **62**, 46–53 (2006)
11. Jaušovec, N.: Differences in cognitive processes between gifted, intelligent, creative, and average individuals while solving complex problems: an EEG study. Intelligence **28**, 213–237 (2000)

12. Razoumnikova, O.M.: Functional organization of different brain areas during convergent and divergent thinking: an EEG investigation. Brain Res. Cogn. Brain Res. **10**, 11–18 (2000)
13. Rominger, C., Papousek, I., Perchtold, C.M., Weber, B., Weiss, E.M., Fink, A.: The creative brain in the figural domain: distinct patterns of EEG alpha power during idea generation and idea elaboration. Neuropsychologia **118**, 13–19 (2018)
14. Jia, W., Zeng, Y.: EEG signals respond differently to idea generation, idea evolution and evaluation in a loosely controlled creativity experiment. Sci. Rep. **11**, 2119 (2021)
15. Pidgeon, L.M., et al.: Functional neuroimaging of visual creativity: a systematic review and meta-analysis. Brain Behav. **6**, e00540 (2016)
16. Kaufman, J.C.: Counting the muses: development of the Kaufman Domains of Creativity Scale (K-DOCS). Psychol. Aesthet. Creat. Arts **6**, 298–308 (2012)
17. SpearmanRankCorrelation—Wolfram Language Documentation. https://reference.wolfram.com/language/MultivariateStatistics/ref/SpearmanRankCorrelation.html
18. Grabner, R.H., Fink, A., Neubauer, A.C.: Brain correlates of self-rated originality of ideas: evidence from event-related power and phase-locking changes in the EEG. Behav. Neurosci. **121**, 224–230 (2007)

Experiential Learning in Digital Contexts—A Case Study

Christoph Knoblauch[(✉)]

Ludwigsburg University of Education, Reuteallee 46 / D-71634, Ludwigsburg, Germany
christoph.knoblauch@ph-ludwigsburg.de

Abstract. Focusing on the intimate relation of experience and education, this paper discusses evaluation findings from a digital project-based course in the higher education sector. The course engages students in the implementation and reflection of projects in the field of education. The empirical findings, therefore, focus on students' experiences in digital project-based settings. This paper analyses the planning, execution, and critical reflection of projects in the context of experiential learning, considering the various digital settings of the course and the projects. It thereby discusses the structure, the digital methodology, and the outcomes of the course with a focus on the experience of the participating students. The study uses digital qualitative interviews for the evaluation research. By doing so, the study investigates and reflects the quality of students' experiences and their possible influences upon learning. It also discusses the question of how these experiences can be constructively implemented to improve future digital or blended-learning scenarios in the higher education sector.

Keywords: Digital project-based learning · Experiential learning · Experience · Higher education sector · Empirical qualitative evaluation

1 Experience, Learning, and Digital Environments

The process of experiential learning in digital environments has become increasingly important and has changed considerably in the face of new challenges. Experiential learning in the contexts of new information technologies has become an integral feature of societies and their educational systems all over the world. Against this backdrop experiential learning has been discussed as an important approach in various educational contexts with a special emphasis on the concepts of experience and reflection [1][1]. Within the last years, the ideas of experiential learning have been exposed to new opportunities and challenges by the development of digital learning environments [2]. The world of learners and teachers has become more flexible and interconnected and, at the same time, more isolated as a result of digital learning environments. *"...People*

---

[1] This publication "Experiential learning in digital contexts – a case study" was produced in the project "Schools of Education as Agents of Change. Coping with Diversity in the Digital Age," which is part of the Baden-Württemberg-STIPENDIUM for University Students – BWS plus, a programme of the Baden-Württemberg Stiftung.

© The Author(s), under exclusive license to Springer Nature Switzerland AG 2023
D. Guralnick et al. (Eds.): TLIC 2022, LNNS 581, pp. 181–191, 2023.
https://doi.org/10.1007/978-3-031-21569-8_17

who collaborate with us in our projects, inhabit and co-constitute environments in which digital technologies and media are inextricably entangled.... The people we meet in the course of our projects move through worlds that are at once on-line and off-line, and … are never separated from the digital or material elements of life" [3]. Accordingly, digital learning environments are understood as the various digital contexts in which learners are active, such as eLearning, interactive digital collaboration, and digital research [4]. While, for example, collaboration options in digital settings may be readily accessible, the qualities and characteristics of digital and in-person collaboration are different [5]. Digital learning environments influence experiences in learning, offering new opportunities but also challenging existing potentials. This is discussed by many academics whose interests lie within the scope of experiential and digital learning scenarios [6]. In this regard, it shows that experience has to be respected as a crucial factor in digital learning environments as it is fundamental to learning processes in general. Experiential learning can be regarded as one of the most original learning processes, deeply rooted in human nature [7]. Of course, not all experiences are equally educative; however, personal experience and education are often closely related [8]. Thus, the quality of experience, especially in digital environments, has to be discussed in depth. Identifying experiences in digital environments, which are agreeable and have a positive influence on further learning processes, seems to be a challenge for future research. Looking at the intimate relation of experience and education in non-digital environments, the question arises of how exactly digital contexts influence experience in learning. However, well-researched resources offering guidance on how to promote experiential learning in digital contexts and how to design experience-oriented digital courses for higher education remain rare.

2   Experiential Learning in Higher Education

This paper analyses and discusses students' learning experiences in project-based digital settings in the higher education sector. Therefore, it evaluates digital and experiential learning scenarios, focusing on possible relations between digital learning and experience. Against this backdrop, the study reports on the development, implementation, and evaluation of a course in the Master's program "Teacher Education"[2] at Ludwigsburg University of Education (LUE) in Germany. Participants were twenty-six students enrolled in teacher education programs at LUE in the summer term of 2021. Students' attitudes, practices, and preferences concerning learning experiences in digital settings play a major role in the design of the courses and its evaluation. The course discussed shows a mix of synchronous and asynchronous course sessions [9]. In addition to this, the participating students reserve several weeks within the semester for a project-based, autonomous development and execution of a small-scale project in the field of education. The focus of this paper is to discuss possible potentials for experiential learning in digital environments to present results on the possible impacts of digital learning scenarios on experience in learning. To facilitate an insight into the learning experiences of students in the described digital settings, the paper discusses qualitative studies focusing on the reflections of students towards personal learning experiences in digital contexts.

---

[2] https://www.ph-ludwigsburg.de/7684+M5054de7a952.html (accessed on 2021/4/8).

By doing so, the study assesses current practice and analyzes the course mentioned in detail.

The discussion of fundamental pedagogical characteristics [1] that influence learners' experiences combined with the perspectives of a digital environment leads to a theoretical framework for the development and analysis of the qualitative interviews [10]. (a) *Personalisation* in terms of customized activities within the course and the project can lead to experiences of ownership and control. (b) *Collaboration* with other students and partners can create experiences of membership and responsibility as students work together with common goals. (c) *Authenticity* through project-based scenarios can provide contextualized learning scenarios that can produce experiences of motivation and involvement [11]. (d) *Empathy* can include the student emotionally in the learning process and might lead to an experience of belonging [12]. (e) *Experiences in digital environments* can influence experiences in the learning. The research process is open for further characteristics which might be developed inductively.

3 Experiential Learning in Digital Contexts: Description of the Course Design

The course "Experiential and project-based learning in (religious) education" serves as the basis of this study and was conducted at LUE in the summer semester of 2021. The main foci of the course are (a) the discussion of project-based and experiential learning, (b) the autonomous implementation of a small-scale project in the field of education, and (c) the presentation and reflection of the projects. The character of the course aims for the interconnection of students with their future work area, their future colleagues, and their future students through a project in the field of education. The course was conducted digitally. Approximately 50% of the digital course sessions were conducted in a synchronous way using mostly videoconferences. The other 50% of the course was carried out in a digital asynchronous way, using learning management systems such as Moodle. The synchronous session mostly offered guided discussions on the asynchronous content and the planning and implementation of projects. The videoconferences were based on students' questions and reports, offering group discussions and individual tutoring in break-out rooms. The asynchronous sessions offered a blend of learning arrangements; readings, podcasts, audio presentations, interactive forums, videos, and chats were used. To these ends, (a) apt presentations with audio commentary were created; (b) a podcast with experts was offered, (c) common topics were deepened through a selection of pertinent literature; (d) videos of professors and students working in project-based settings were provided; and (e) digital forums were established to offer an ongoing interactive exchange of ideas [13]. The project-based focus of the course aims at helping students to gain experience in digital settings by actually implementing a project in the field of education. Students can gain up to three credit points (as defined by the European Credit Transfer and Accumulation System, or ECTS) for the course and can use their projects as a basis for their master's thesis.

4 Qualitative Design of the Study

The study focuses on the reflections and discussions of the participating students of the course, concerning their individual experiences in the described learning contexts. Therefore, it uses a creative qualitative design, which is conducted digitally. The complex research focus—students' experiences in learning in digital environments—asks for an innovative approach observing multiple perspectives through (a) synchronous dialogue-based interviews and (b) individual asynchronous feedback. Against this backdrop, the study looks for individual, subjective feedback, and group discussion to gain a more comprehensive view. Semistructured qualitative interviews were carried out (a) in a synchronous digital way, using video calls and (b) in an asynchronous digital way. The digital asynchronous way of interviewing students is a rather new technique in qualitative research, and it encourages respondents to reflect on their answers by allowing them to structure their ideas and responses beforehand [14].

The (a) synchronous interview situations offered the possibility to discuss questions in deep, to clarify ambiguities, and to develop a constructive dialogue between the interviewer and the participants. In these processes, the study follows established structures of qualitative research and analysis in manifold ways [15].

The (b) asynchronous way offered the possibility to record answers as audio files independently. The participants could take as much time for reflection as they individually considered appropriate. The completed audio files were then sent to the research team via a digital transfer system [14].

Using this combined method, the study looks for data, which offers a comprehensive view through individual, reflected, or subjective feedback and dialogue [16]. The semi-structured questionnaire shows different categories which were developed in a deductive procedure: (a) Personalization, (b) Collaboration, (c) Authenticity, (d) Empathy, and (e) Experiences in digital environments. These deductive categories were built based on the reflection of theoretical frameworks focusing on experiential learning. Each category includes several key questions, which are supported by specifying impulses or further questions. The questionnaire mainly focuses on the project-based and experience-oriented character of the course. The questions furthermore focus on students' attitudes towards and usage of the digital environment. Reflections on learning strategies and perceptions of change within learning are also part of the questions.

5 Findings of the Study

The qualitative study focuses mainly on learners' experiences in (a) project-based contexts and (b) digital settings. The analysis and discussion of the data show different combined categories, which are based on the deductive categories and new inductive impulses found within the data. The findings are structured and discussed according to these categories. Several answers and reflections show links to more than one category and are therefore discussed in various contexts. Experiences in digital environments are discussed in all categories and a special focus is given to these experiences in the category "Experiencing digital environments."

5.1 Experiencing Personalisation

Within the project-oriented digital settings, students had various possibilities to customize their learning activities, experience ownership over their learning arrangements, and control over their projects. The digital contexts of the course and the projects enabled students to plan, organize and implement activities efficiently. The students mostly used a mix of digital and face-to-face interaction[3] within their projects and appreciated the autonomous organization of this mix. Furthermore, many students report about the benefits of practical instruction and action-oriented collaboration within the course and their projects: *"Project-oriented work offers intensive learning experiences because the implementation of projects is not just theoretical but also practical."*[4] The students report that the experience of intensive learning experiences is mostly due to factors such as autonomous working, practical instruction, authenticity, interdisciplinarity, and collaboration.

In addition to this, many students see a great importance of project-based experiences for their future work fields. The reasons for this are mainly found in the interdisciplinary and social character of project-based learning: *"...it (project-based learning) is very much about social learning and interaction."* The fact that many of these interactions took place in digital environments did not seem to be problematic to most of the students.

The interviewed students appreciate learning contexts that can be personalized through individual project-oriented learning processes. The responsibility of controlling an individual project seems to increase an experience of ownership in the sense that students are running a project by themselves: *"...we were able to reach certain goals by implementing our individual project. Therefore, the learning processes are more effective and bigger—on top of that, the learning is sustainable and interdisciplinary."*

5.2 Experiencing Collaboration

Collaboration plays a major role in all projects as students work together with different partners in schools and communities. The various collaborations led to experiences of membership when a group of people worked together with a common goal for a certain time. Some students report that their projects involved a network of people who were able to develop and foster relationships: *"...I realized that the project had far-reaching impacts when parents of the participating children became interested and seniors wanted to join."* It shows that the students developed a broader understanding of collaboration and community while working on their projects: *"The whole social environment, whether it be friends, work or family...different areas came together in the projects: Such as school, hobbies, and family."* Other students report that some collaborations laid foundations for long-lasting relationships. Students experienced that they were responsible for the implementation of a project but also that their partners developed individual momentum, which outlasted the project.

---

[3] Face-to-face interaction in this article means analog, physical collaboration.

[4] Results, quotations, transcriptions, and analysis phases of the study are available through the author.

It is noticeable that most collaborations were established digitally. However, the students do not discuss this fact in detail. On the contrary, it seems almost conventional to the participating students to use exclusively digital ways for establishing collaborations.

5.3 Experiencing Authenticity

Authentic experiences in learning are often connected to the individual contexts of the learner and, thus, can lead to a high level of motivation and involvement [17]. The fact that students were able to develop projects individually motivated them to choose fields and topics connected to their lifeworld. Many students report that they experienced a strong involvement in their projects as they could incorporate personal interests in their projects: *"…because one had so many possibilities to connect topics and subjects and add a pedagogical value."* Additionally, the interviewed students experienced close relationships between different themes and subjects: *"I think the learning processes are intense as I experienced that learning is not one-sided but closely connected to many different fields."* The discussed connectedness is not limited to topics and subjects but also seems to play a role in collaboration. Some students report about succeeding projects that developed based on their work. This is interpreted as proof for authenticity as these new projects grew out of the personal interest of participants: *"At a later stage the project dealt with completely different topics. Children and seniors discussed cultural and pedagogical themes and the children developed new ideas about the concept of aging."* In this example, the initial project developed pen-friendships between children in kindergartens and seniors in retirement homes. Based on these projects the participants discussed topics that were authentic to them individually.

It shows that students appreciate a high level of authenticity, which can be created by connecting the learner's lifeworld with the actual learning processes. A high level of individual freedom in planning the projects seems to be necessary to create experiences of authenticity. Digital contexts are not discussed explicitly in this category. However, the interviews indicate that authentic experiences can also be created in digital contexts. Students report about according experiences during the digital implementation of their projects.

5.4 Experiencing Empathy

An emotional inclusion in the learning process can create experiences of belonging and, thus, lead to sustainable learning effects [18]. Against this backdrop, the development of empathy towards the project and its participants is crucial. In this context, empathy is understood as an act of deeper understanding and the development of a reflected awareness for the project's topic and the involved participants [19]. Many students discuss experiences of emotional involvement and belonging connected to their projects and the participants: *"…the learning experiences are or can be emotional. These emotions affect one's awareness and because of this, one perceives things differently."* Experiencing empathy in learning has the potential to foster a variety of competencies: *"It (project-oriented learning) offers intensive learning experiences, as it deals with more than specific subject matters. In the projects social learning, communication and encounters are taking place. This leads to a much broader learning."* In this context, the factor of "encounter" is mentioned by many students. The participating students did not differentiate between digital and face-to-face encounters in their projects. It seems that in the discussed projects both, digital and face-to-face encounters, led to the development of emotional experiences and empathy: *"Especially the (digital) collaboration between children and seniors helped to develop empathy for the other."* Furthermore, students observed changes in the perspectives of participants through digital encounters: *"The children gained a deeper understanding for the position of the other. Looking at Dewey's idea of 'experience', these might be sustainable positive experiences for them."*

Empathy seems to be an important factor for intensive learning experiences in this study. The interviewed students report many times about experiences of belonging to the project and the group of participants. Additionally, the development of empathy in project-based settings seems to change individual perspectives, when it is connected to encounter. The type of encounter does not seem to be important for the students, as digital and face-to-face encounters are discussed without any further differentiation.

5.5 Experiencing Digital Environments

The interviewed students discuss experiences in digital environments in all categories and in connection to different topics and processes. They used different strategies for the implementation of their projects in digital environments: Some students conducted their projects completely digitally, whereas others balanced online and face-to-face modes.

Students who chose completely digital ways mostly report about positive effects such as (a) the possibility to establish contact easily, (b) the opportunity to schedule meetings quickly, (c) the possibility to gather and share information online and simultaneously, (d) the chance to meet without using a car or public transport, and (e) the time-saving aspects of digital communication [13].

In contrast, students who chose face-to-face modes emphasize the necessity of face-to-face interaction as many participants cannot handle the required digital devices on their own. Additionally, some students report that they experienced an unobstructed flow of information and communication in face-to-face encounters and that the use of certain materials (e.g., pictures or items) may have a better effect.

Students explicitly report about their experiences with various digital tools: *"I advertised my project digitally. And I created a messenger group for the parents (of the participating children)."* Digital tools were often used for establishing contacts, organizing schedules, and research: *"...we were looking for pictures and stories online."* Additionally, students used video calls and messengers to prepare and discuss their projects: *"...to stay connected—especially in Covid times, all stayed in different places. Because of his we had to connect digitally to communicate constructively."* However, the use of digital tools in the students' projects also shows limitations: *"...to offer video calls instead of handwritten letters would have been too far from my initial idea of a pen friendship."* Analog options sometimes seem to be more suitable and constructive: *"There was one child who painted on the back of the letter and the paintings added even more value. I don't think this would have been possible with digital tools."*

All interviewed students appreciate and make use of digital environments, when planning and implementing their projects. Within the projects, digital tools seem to be used especially for establishing contacts, researching materials, and organizing schedules. Furthermore, the students balance digital and analog processes within their projects.

6 Discussion of the Findings

The qualitative approach of this study is looking for the reflection of individual experiences in digital and project-based courses in the higher education sector. The qualitative perspectives were implemented to gain deeper knowledge about the individual experiences of the participating students, to reflect on the quality of students' experiences and possible influences upon learning. It also discusses the question of how these experiences can be constructively implemented to improve future courses in digital environments. The methods employed largely focus on the same phenomena and the findings are discussed in this analysis, by triangulating the different qualitative approaches (synchronous, dialogue-based, asynchronous) [20].

Three main foci could be established during the analysis process and are discussed in the context of experiential learning: (I) experiences in project-based learning in digital environments; (II) the role of encounters in project-based digital contexts; (III) use of digital environments in project-based settings.

(I) Project-based learning in digital contexts offers a variety of learning experiences. Students can experience a high level of autonomy as they can pursue individual interests and control fundamental processes within their projects. Students can personalize learning experiences, as they can customize activities and processes in the course and the project [13]. Furthermore, digital project-based approaches can offer a high level of contextualization. Projects offer the inclusion of lifeworld experiences and interests and thus can lead to experiences of involvement and belonging. These experiences seem to motivate students and can lead to a strong connection towards projects and their participants. Students can develop empathy and are included in the learning processes emotionally, as they are in charge of their projects and collaborate with different participants.

Digital environments seem to foster these experiences as they enable students to customize their activities even more. Additionally, contact is established and maintained easily in digital environments. However, the potentials of digital environments in experiential learning vary; while the development and organization of projects seem to profit from digital environments, the establishment of relationships through encounter does not necessarily.

(II) This shows when experiences with collaboration and coherent encounters are discussed. Especially encounter seems to play a crucial role in project-based learning, as students hardly work on their own but are rather connected to several partners and participants [21]. Collaboration seems to create experiences of membership and offers potential for emotional involvement. This can lead to the development of strong relationships between the participants.

In this context, face-to-face contacts seem to play an important role, whereas experiences with encounters in digital environments seem to be less sustainable [22]. Digital tools seem to be more efficient, though, when it comes to logistical tasks, like organizing and structuring collaboration. It shows that digital environments are of great importance for collaboration and encounter. However, the benefits for and quality of encounters are different from face-to-face situations.

(III) There is a strong and partly natural use of digital environments in project-based settings. Establishing contact, advertising projects, organizing processes, and structuring activities mostly happen via digital tools. Within the digital environments, learners appreciate and use the manifold opportunities offered through digital tools. Students report various positive experiences with the flexible blending of various digital methods. Concerning this, digital environments seem to encourage independent learning as online content can be retrieved and reused self responsibly. Additionally collaboration can be established with chosen partners outside the classroom and online interaction can take place anytime [13]. Students use digital environments in an effective, flexible, and largely independent manner. Still, students tend to experience face-to-face encounters as more authentic and sustainable.

7 Outlook

Project-based learning is contextual, emphatic, individual, authentic, collaborative, and action-oriented [23]. Students show a high level of involvement when working in project-based contexts: *"…and in my opinion, you only learn things, when you actually do things. This is why I appreciated the projects so much."* Digital environments offer new potentials for project-based learning as many processes can be handled more efficiently with digital tools. Digital project-based learning can generate a high level of motivation in students and encourage them to do work autonomously and self-responsible: *"…we had the chance to work on our own, completely independent of time and place, which was good. And afterward, we could share in class our solutions and our thoughts (…)."* However, certain processes seem to have different characteristics, when implemented digitally: The quality of digital interaction for example seems to have a different quality than face-to-face interaction. This is especially important, as collaboration and interaction are crucial factors of project-based learning. The development of relationships and the experiences of membership and belonging seem to be more intense in face-to-face interactions.

Digital environments can make project-based learning more efficient, contextual, and personalized. Relationships can also be established digitally; however, they are probably not as intense and sustainable as face-to-face collaboration. Experiences in collaboration tend to be more powerful if the interaction is face-to-face. Project-based courses in digital environments should take these findings into account. Thus, project-based courses require a structure, which enables students (a) to find themes and develop projects individually through online content; (b) to share and discuss ideas online; (c) to establish contacts online; (d) to develop shared knowledge through collaboration online; (e) to meet online and face-to-face with partners and participants. Moreover, the entire learning environment should offer different digital and face-to-face learning methods. A combination of self-regulated learning, interaction, and diversified working options should be developed to meet the students' needs and provide digital and non-digital options. A flexible combination seems to be especially valuable as it offers efficient and suitable tools for different needs [24]. Given this, digital environments have to be discussed in terms of their functionality and efficiency for individual projects. Experiences in the fields of personalization, authenticity, collaboration, and empathy [8] can be helpful for this reflection and offer a structure for discussing project-based learning in digital environments.

References

1. Miettinen, R.: The concept of experiential learning and John Dewey's theory of reflective thought and action. In: International Journal of Lifelong Education (19:1), 54–72 (2000)
2. Yates, A., Starkey, L., Egerton, B., Flueggen, F.: High school students' experience of online learning during Covid-19: the influence of technology and pedagogy. Technol. Pedagog. Educ. **30**(1), 59–73 (2021)
3. Pink, S.: Foreword. In: Frömming, U., Köhn, S., Fox, S. Terry, M. Digital Environments, 9 (2017)
4. Iskakov, I., Kovalenko, B., Turovskaia, M., Getmanova, G.: Online blended learning in the digital environment. In: Journal of Physics: Conference Series 1691 (2020)
5. Brooks, E., Selander, S.: Designing for collaboration. Frameworks for learning. In: Brooks, E., Dau, S., Selander, S. Digital learning and collaborative practices. lessons from inclusive and empowering participation with emerging technologies, 1–4 (2022)
6. Bouilheres, F., Le, L.T.V.H., McDonald, S., Nkhoma, C., Jandug-Montera, L.: Defining student learning experience through blended learning. Educ. Inf. Technol. **25**(4), 3049–3069 (2020). https://doi.org/10.1007/s10639-020-10100-y
7. Boud, D.: Forward. In: Warner Weil, S., McGill, I. Making sense of experiential learning: Diversity in theory and practice. London: Open University Press (1989)
8. Dewey, J.: Experience and education. Touchstone, New York (1997)
9. Farros, J.M., Shawler, L.A., Gatzunis, K.S., Weiss, M.J.: The effect of synchronous discussion sessions in an asynchronous course. In: J. Behav. Educ., 1-13 (2020)
10. Hansen, R.E.: The role of experience in learning: giving meaning and authenticity to the learning process in schools. In: J. Technol. Educ. (11/2): 23–32 (2000)
11. Kearney, M., Schuck, S., Burden, K., Aubusson, P.: Viewing mobile learning from a pedagogical perspective. In: Res. Learn. Technol., **20**(1), 14406 (2012)
12. Holmberg, B.: The evolution, principles and practices of distance education. BIS-Verlag Carl von Ossietzky Universitat. Retrieved February 7, 2022 from https://uol.de/fileadmin/user_upload/c3l/master/mde/download/asfvolume11_eBook.pdf (2005)
13. Knoblauch, C.: Digital project-based learning in the higher education sector. In: Guralnick D., Auer M.E., Poce A. (eds) Innovations in learning and technology for the workplace and higher education. Lecture notes in networks and systems (349). Springer, 170-179 (2021)
14. Salmons, J.: Qualitative online interviews. Strategies, design, and skills, 2nd edn. SAGE, Los Angeles (2015)
15. Ehlers, U.: Qualitative onlinebefragungen. In: Lothar Mikos und Claudia Wegener (Hg.): Qualitative Medienforschung Ein Handbuch (2): 327–339 (2017)
16. Berg, B., Lune, H.: Qualitative research methods for the social sciences (9): 21–30 / 172–174 (2017)
17. Roach, K., Tilley, E., Mitchell, J.: How authentic does authentic learning have to be? In: Higher Education Pedagogies 3(1): 495–509 (2018)
18. Kasl, E., Yorks, L.: Do I Really Know You? Do You Really Know Me? Empathy Amid Diversity in Differing Learning Contexts. In: Adult Education Quarterly **66**(1): 3–20 (2016)
19. Md Hashim, A., Syed Aris, S.R., Chan, Y.F.: Promoting empathy using design thinking in project-based learning and as a classroom culture. In: AJUE **15**(3): 14–23 (2020)
20. Denzin, N.K.: The research act. A theoretical introduction to sociological methods (2). New York: McGraw-Hill (1978)

21. Marjan, L., Seyed Mohammad, G.: Benefits of collaborative learning. In: Procedia—Social and Behavioral Sciences (31): 486–490 (2012)
22. Liu, S. Student Interaction experiences in distance learning courses a phenomenological study. In: Online J. Distance Learn. Adm., **11**(1). Retrieved February 7, 2022 from https://www.learntechlib.org/p/158563/
23. Krajcik, J.S., Blumenfeld, P.C.: Project-based learning. In: The cambridge handbook of the learning sciences, 317–333 (2006)
24. McGuinness, C., Fulton, C.: Digital literacy in higher education. a case study of student engagement with E-Tutorials using blended learning. In: JITE:IIP 18: 1–28 (2019)

Experiences on Creating Personal Study Plans with Chatbots

Matti Koivisto[✉]

South-Eastern Finland University of Applied Sciences, 50100 Mikkeli, Finland
matti.koivisto@xamk.fi

Abstract. With the ever-increasing number of the college students, universities face a growing need for different kinds of counselling services. To fill this demand, the higher education institutions have tested and applied many sophisticated solutions including chatbots, which are Artificial Intelligence-based (AI-based) computer programs designed to simulate the discussion with the human users. This paper analyzes on the suitability of chatbots or virtual advisers in personal study plan creation and course selection. The empiric part of the study reports the observations of the experiment conducted in a Finnish university of applied sciences. In the study, post-graduate engineering students (n = 53) used a tailor-made chatbot while creating their personal study plans. During the task, the students were able to request suggestion for optional study modules, using different criteria, such as personal interest, learning needs, and course popularity. The findings of the study indicated that chatbots could, to some extent, improve student counselling, and the main advantages of the chatbots were scalability and unlimited service hours. However, students did not see that AI could at least now fully remove the need for human counselling. The main reported shortcomings of the chatbot were the minor significance of the individuality, lack of inspiring effect, and general attitude towards automated services.

Keywords: Digital assistant · Chatbot · Course selection

1 Introduction

Tutoring and counselling are essential support services universities provide to their students. Traditionally, the members of the staff have had an essential role in providing these services, but over the past few decades, the work has shifted more to dedicated specialists [1]. Human tutoring services have many benefits, but they are very labor intensive, and the tutors can be used only during their service hours. With the ever-increasing number of students, universities have sought to streamline their own operations with automation and artificial intelligence (AI). Earlier studies have indicated that universities have used the AI mainly in the following four application areas: adaptive systems and personalization, assessment and evaluation, profiling and prediction, and intelligent tutoring systems [2]. In this paper, we focus on the application of AI in tutoring and, in particular, the application of chatbot technology in personal curriculum planning and course selection.

© The Author(s), under exclusive license to Springer Nature Switzerland AG 2023
D. Guralnick et al. (Eds.): TLIC 2022, LNNS 581, pp. 192–200, 2023.
https://doi.org/10.1007/978-3-031-21569-8_18

The structure of the paper is as follows. In Sect. 2, we introduce the principles of the chatbots and some earlier studies on using chatbots in higher education. Since practically all students must select courses to determine their study program, our aim is that the virtual assistant is accepted and used on a large scale. Thus, in Sect. 3, we briefly cover the main theories on technology acceptance and the metrics created by scholars and practitioners to predict the success of new services. Our empirical study is presented in Sect. 4. In our experiment, we first develop a student counselling chatbot called Vivian and then analyze post-graduate students' opinions on its suitability to provide advice on course selection and curriculum planning. Finally, the work ends with the conclusions of Sect. 5.

2 Chatbots and Virtual Assistants

2.1 Introduction to Chatbots

A chatbot is an AI-based computer program designed to simulate the discussion with humans. Chatbots can be used for many purposes, but they can be classified to two main categories: question-answering assistants and social bots [3]. The question answering bots are task-oriented, and they provide answers and solve specific problems based on their information [4]. Many organizations use them in different kinds of customer service tasks, and according to some estimates, in 2020, already 25% of the buyers used chatbots to communicate with businesses [5]. Social chatbots are instead intelligent dialogue systems that can engage in empathetic conversations with humans [6], and these systems are supposed to provide personalized responses for the users in their areas of interests. Some popular examples of social chatbots are Woebot Health, a personal mental health ally [7], and XiaoIce, a Chinese social chatbot with more than 660 million users [8].

2.2 Chatbot Design and Application

Chatbots communicate with humans using natural language either with a text or voice interface [9]. Although many organizations have published different kinds of guides and rules regarding how to design a successful bot, there are still many open questions in chatbot design [10]. However, natural and successful chatbots must meet the following requirements [4]:

- They must have conversational capabilities beyond short answers.
- They must offer semantically correct information and use context specific vocabulary.
- They must give meaningful responses and answers.
- They must be trained on a domain-specific dataset.

These requirements highlight that chatbot designers must pay attention not only to functionality, but also to the social aspects [11] and to the form features of the chatbots [12]. These features together affect user experience and play an essential role in the user satisfaction and service acceptance.

Over the past few decades, universities have introduced chatbots in many areas, including education and student counselling. Earlier studies have identified both the benefits and limitations of chatbots in higher education institutions. Some scholars have found that chatbots increase students' learning motivation [13] and attention [14] and that they support peer communication [15] and collaborative learning [16]. However, some researchers have reported the limitations of the chatbots. For example, current implementations seldom provide enough personalization [17], and users seem to trust less to chatbots in more complicated tasks [18].

In this paper, we focus on the use of chatbots in personal curriculum planning and course selection. Course selection is a sequential decision-making process in which students determine their study program [19]. During the selection process, students typically need some form of guidance. Traditionally, this support has been provided either by faculty members or by designated counsellors. Although human counselling has many benefits, it has also some limitations. Earlier studies have identified for following challenges in traditional advising: recommendations can be inconsistent across counsellors [20], can be time-consuming [21], and may not be available outside office hours [22], and the advising sessions are limited in time, especially during the highest-demand periods [21].

3   Acceptance of Information Systems

3.1   Acceptance Models and Theories

Scholars and practitioners have investigated the factors affecting the user perceptions of information systems for many decades. The findings of these studies have indicated that the system adoption and use are affected by many factors and processes. Researchers have tried to explain the acceptance of new services and technologies with multiple models. The most cited models are Innovation Diffusion Theory (IDT) [23], Theory of Reasoned Action (TRA) [24], Theory of Planned Behavior (TPB) [25], Technology Acceptance Model (TAM) [26], and Unified Theory of Acceptance and Use of Technology (UTAUT) [27]. Although these theories have different acceptance determinants, the basic concept underlying them links individual reactions and intentions to the actual use of the system [28]. In organizational settings, the acceptance of the new services does not occur in isolation, but social influence plays also an important role. It is conventional wisdom that friends and peers have a large influence on us. This holds true, especially when an individual's underlying behavior is social or networked [29].

3.2   Metrics

Researchers have developed a range of feedback metrics and methods for predicting service acceptance. However, we concentrate only on three metrics aimed at understanding the relationship between the service and its customer. These metrics are Customer Satisfaction (CSAT), Customer Effort Score (CES), and Net Promoter Score (NPS), which are currently among the most popular techniques for measuring the strength of the relationship between the company's services and its customers [30].

CSAT uses a five-point Likert scale to measures the short-term user satisfaction. It targets a "here and now" reaction, and therefore, it has its limitations. Scholars have found out that it has limited capability to measure user's long-term relationship with the service or service loyalty [31]. CES, instead, measures how much or how little effort is needed from the user to use the service and it uses Likert scale ratings from 1 to 5 [31]. It is important to understand that CES analyzes the user experience from a very different perspective than CSAT. Instead of focusing on maximizing the satisfaction it pays attention to minimizing the effort [30]. Out third metric, NPS measures long-term loyalty and determines which customers are likely to promote our service to others; therefore, it is often used as a growth indicator. NPS has a scale from 1 to 10, and it groups the responders to the following four categories: "promoters" (9–10 rating—extremely likely to recommend), "passively satisfied" (7–8 rating), and "detractors" (0–6 rating—extremely unlikely to recommend) [32].

4 Empirical Part of the Study

4.1 Introducing Vivian

The aim of our study was to analyze the suitability of the course selection chatbot in higher education. To be able to do this, we first developed a simple study advice chatbot Vivian; Vivian offers her suggestions and advice on questions related to personal curriculum planning and course selection questions. Vivian was implemented using IBM's Watson Assistant cloud service. Watson Assistant uses AI and "has the ability to answer commonly asked questions, based on domain and industry specific information" [33].

The basic structure of our chatbot is shown in Fig. 1. Users interact with Vivian using their web browsers on their mobile phones or laptops. All communication between the user and Vivian is text based. When the chatbot receives a message from the user a logical structure called a dialog skill interprets the user input and directs the flow of the conversation. The dialog skill consists of three components, which are intents, entities, and dialogs (or nodes). As the name implies, the task of the intent is to recognize the intent of the user or, in other words, what the user wants to know. Vivian detects for example the following intents: #Structure (what is the structure of the curriculum), #Mandatory (what are the compulsory courses), #Optional (what are the alternatives for optional courses), #Recommendation (course recommendations using different criteria), #Thesis (details of master's thesis), and #Skills (how to learn a specific skill). When an intent is created, the designer provides the initial configuration, but because our chatbot is based on AI, it learns new ways to detect the user's intent.

Entities are used to catch essential information from the user input. They can be considered as keywords. Some examples of Vivian's entities are @Study_type (possible values: compulsory, optional, thesis) and @interest (which detects the specific interest of the student. Some example values: leadership, entrepreneurship, research methods, decision-making, digitalization etc.). Finally, dialogs are nodes in a dialog tree. They are placed between two automatically created nodes, Welcome and Anything else. Welcome node is the starting point of all conversations, and it contains some form of greeting. Anything else is the last resort, and it is used if the bot does not understand the user

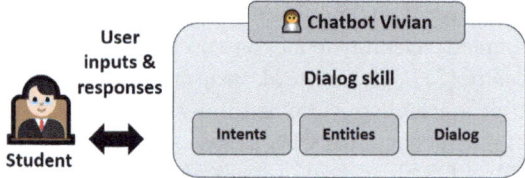

Fig. 1. The basic structure of the chatbot.

input. If Vivian fails to interpret the user's intent, it will first ask the user to rephrase the question. Because Vivian tries to mimic a human assistant, it does not always use the same phrases but varies its messages. If Vivian is not able to understand the user after the second time, it recommends that the user sends an email to his or her tutor.

The relationship between the dialog skill and the messages sent to the user are demonstrated in Fig. 2. At the beginning of the conversation, Vivian first introduces herself with Welcome node and then asks the name of the user with Get name dialog.

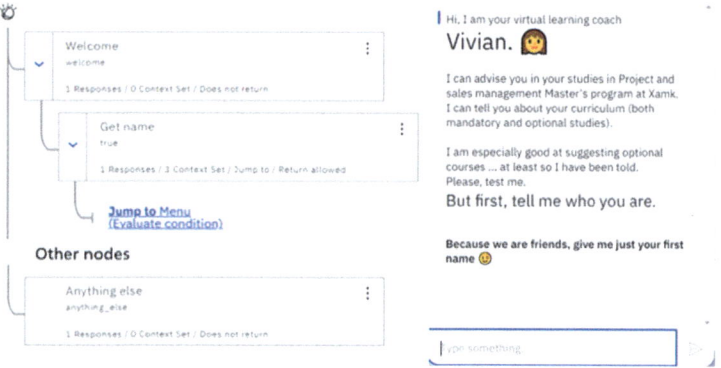

Fig. 2. Welcome and get name dialogs or nodes in the dialog skill and their outcome.

In typical conversations, students first asked Vivian to show the structure of the curriculum, and then they quickly looked at some of the mandatory courses. However, Vivian's main task was to advise students to choose optional courses; therefore, students spent most of their time using this feature. Students were able to ask Vivian for advice based on their own interest, either using some pre-defined categories or with open questions. Vivian analyzed the request, and if she was able to find a suitable course, she responded with her suggestions. If Vivian did not find a matching course, she apologized and offered three options for moving on. These alternatives were sending a new question, reviewing the courses offered by other universities of applied sciences, or sending an email to the tutor.

4.2 Study Design

In the empirical part of our study, we carried out a small test at a Finnish university of applied sciences to analyze the suitability of the chatbot to our students' needs. In our experiment, post-graduate engineering students used the chatbot when they planned their studies and selected optional studies from a relatively large pool of alternatives. We introduced our chatbot to our students during a normal online learning session. After that, they were asked to use the service if they wanted and then fill the feedback questionnaire. In total, 57 students took part on the online session, and 53 of them used Vivian and filled the feedback form. Our questionnaire had three parts. In the first part, we asked about the user experience and used the earlier introduced metrics CSAT, CES, and NPS. The second part concentrated on chatbot specific issues and the questions were based on the requirements of the natural and successful dialogue discussed in Sect. 2.2. In the third part, students were asked to evaluate the strengths and weaknesses of the chatbots in the course selection task.

4.3 Results

The results of the part one of our feedback questionnaire are shown in Table 1. Because all three metrics used different scales, we present the results both with original values and normalized to the scale 0 to 1. Our data revealed that the students found Vivian quite easy to use (CES = 0.80), but they were not as satisfied with it (CSAT = 0.68) or were not as likely to recommend it to other users (NPS = 0.63) and the difference is statistically significant (p = 0.05). Because we do not typically use the mean when reporting the finding of the NPS survey, we also calculated the NPS score. The NPS score of our chatbot was remarkably low –0.38, or –38%, because there were only 3 promoters and 23 detractors among students.

Table 1. Findings of the general metrics.

Metric	Original scale	Mean (Original scale)	Mean (Scale 0 to 1)	SD (Scale 0 to 1)
CES	1–7	5.60	0.80	0.16
CSAT	1–5	3.40	0.68	0.12
NPS	1–10	6.26	0.63	0.19

The findings of the second part of the questionnaire are shown in Table 2. According to the results, students gave the lowest score to Vivian's conversational capabilities. The other dimensions had quite similar ratings (in the range from 0.73 to 0.76). The difference between conversational capabilities and other metrics was statistically significant (p = 0.05).

In the third part of the survey, students identified the strengths and weaknesses of the chatbot in course selection task with open-ended questions. Four most often mentioned Vivian's strengths and weaknesses are listed in Table 3. The frequency describes how many students from all participants mentioned this topic.

Table 2. Findings of the chatbot specific metrics.

Metric	Mean (Original scale)	Mean (Scale 0 to 1)	SD (Scale 0 to 1)
Conversational capabilities	3.11	0.62	0.16
Correct information and vocabulary	3.66	0.73	0.12
Meaningful answers	3.74	0.75	0.14
Case specific information	3.87	0.76	0.20

Table 3. Strengths and weaknesses of the chatbot.

Strengths	Frequency (%)	Weaknesses	Frequency (%)
Service always available	63	I just don't like chatbots	42
No queues - can serve all students at the same time	38	System does not encourage or inspirate me to study	37
Quite easy to use	15	Not a personal service	29
Has useful information	12	I don't see what other students have selected	21

5   Conclusions

The aim of our study was to analyze the suitability of the chatbots or virtual advisers in student counselling. Our data reveal that the students found Vivian quite easy to use but they were not completely satisfied with it, and the difference was statistically significant ($p = 0.05$). The negative NPS score (–0.38) also clearly points out that Vivian is unlikely to be a viral hit among our students. That is a minor disappointment because the chatbot can remarkably reduce the need of human tutoring only if its widely accepted.

What could be the reasons behind our moderate success? We are not able provide full answer but only provide some possible alternatives. Firstly, the qualitative part of the study indicated that chatbots could improve student counselling and the main advantages of the chatbots were unlimited service hours and scalability. The main shortcomings of our chatbot were the minor significance of the individuality and the lack of inspiring effect. It was also interesting to recognize that many respondents had general negative attitude towards chatbots and automated services. Secondly, there were some flaws in our implementation, especially in its interaction. When Vivian's chatbot-specific characteristics were analyzed, the conversational capabilities got the lowest score, and this finding was statistically significant ($p = 0.05$). This area requires clearly further development.

Thirdly, students base their course selection to multiple factors like learning value, lecturer, course difficulty, prerequisite knowledge, and comfortability [19]. It is important to recognize that some of this information is not official and well documented, but it is only shared in informal contexts. Also, the findings of the earlier study [34] have indicated that "the occasional lack of adequate answers does not necessarily produce a bad experience, as long as the chatbot offers an easy path for follow-up with human customer service representatives." Both the transfer of unofficial silent information and the smooth transition from chatbot to human counsellor are important development areas.

All in all, the feedback clearly pointed out that Vivian is still a work in progress. To make it more useful, we must address the issues identified in this study. In the next version, Vivian must be connected better to other information systems of the university, and it should offer more flexible ways to share peer information and recommendations. Even after these modifications, we do not believe that Vivian replaces the need for human tutoring. Study planning is a complex decision-making process where students need the insight and advice of experts in addition to facts also in the future.

References

1. Schulenberg, J.K., Lindhorst, M.J.: Advising is advising: toward defining the practice and scholarship of academic advising. NACADA J. **28**(1), 43–53 (2008)
2. Zawacki-Richter, O., Marín, V.I., Bond, M., Gouverneur, F.: Systematic review of research on artificial intelligence applications in higher education—where are the educators? Int. J. Educ. Technol. High. Educ. **16**(1), 1–27 (2019). https://doi.org/10.1186/s41239-019-0171-0
3. Gao, J., Galley, M., Li, L.: Neural approaches to conversational AI. Found. Trends® Inf. Retrieval **13**(2–3), 127–298 (2019)
4. Nuruzzaman, M., Hussain O.: IntelliBot: a dialogue-based chatbot for the insurance industry. Knowl.-Based Syst. **196**(May) (2020)
5. Drift: 2020 state of conversational marketing (2020)
6. Zhou, L., Gao, J., Li, D., Shum, H.-Y.: The design and implementation of XiaoIce, an empathetic social chatbot. Comput. Linguist. **46**(1), 53–93 (2020)
7. Prochaska, J.J., Vogel, E.A., Chieng, A., Kendra, M., Baiocchi, M., Pajarito, S., Robinson, A.: A therapeutic relational agent for reducing problematic substance use (Woebot): development and usability study. J. Med. Internet Res. **23**(3) (2021)
8. Xu, S.: Microsoft chatbot spinoff Xiaoice reaches $1 billion valuation. Bloomberg, July 14 (2021)
9. Dale, R.: The return of the chatbots. Nat. Lang. Eng. **22**(5), 811–817 (2016)
10. Meyer von Wolff, R., Hobert, S., Schumann, M.: How may I help you?—state of the art and open research questions for chatbots at the digital workplace. In: Proceedings of the Hawaii International Conference on System Sciences (2019)
11. Feine, J., Morana, S., Maedche, A.: Designing a chatbot social cue configuration system. In: Proceedings of the International Conference on Information Systems (2019)
12. Rietz, T., Benke, I., Maedche, A.: The impact of anthropomorphic and functional chatbot design features in enterprise collaboration systems on user acceptance. In: Proceedings of the 14th International Conference on Wirtschaftsinformatik, (2019)
13. Hwanga, G.-J., Chang, C.-Y.: A review of opportunities and challenges of chatbots in education. Interact. Learn. Environ, July (2021)
14. Song, D., Young Oh, E., Rice, M.: Interacting with a conversational agent system for educational purposes in online courses. In: Proceedings of the 10th International Conference on Human System Interactions (2017)
15. Kowalski, S., Hoffman, R., Jain, R., Mumtaz, M.: Using conversational agents to help teach information security risk analysis. In: Proceedings of the First International Conference on Social Eco-Informatics (2011)

16. Bii, P.: Chatbot technology: a possible means of unlocking student potential to learn how to learn. Educ. Res. **4**(2), 218–222 (2013)
17. Stone, P., Brooks, R. Brynjolfsson, E., Calo, R., Etzioni, O., Hager, G., Hirschberg, J., Kalyanakrishnan, S., Kamar, E., Kraus, S., Leyton-Brown, K., Parkes, D., Press, W., Saxenian, A., Shah, J., Tambe, M., Teller. A.: Artificial Intelligence and Life in 2030. One Hundred Year Study on Artificial Intelligence (Report of the 2015–2016 Study Panel). Stanford University, Stanford, CA (2016)
18. Nadarzynski, T., Miles, O., Cowie, A., Ridge, D.: Acceptability of artificial intelligence (AI)-led chatbot services in healthcare: a mixed-methods study. Digital Health, Aug (2019)
19. Babad, E., Tayeb, A.: Experimental analysis of students' course selection. Br. J. Educ. Psychol. **73**, 373–393 (2003)
20. Castells, J., Mohammad, P.-D., Galárraga, L., Méndez, G., Ortiz-Rojas, M.: A student-oriented tool to support course selection in academic counseling sessions. In: Proceedings of the 15th European Conference on Technology Enhanced Learning (2020)
21. Assiri, A., AL-Malaise, A., Brdesee, H.: From traditional to intelligent academic advising: a systematic literature review of e-academic advising. Int. J. Adv. Comput. Sci. Appl. **11**(4) (2020)
22. Méndez, M.G., Arguello, G.: Best practices of virtual advising: the application of an online advising portal. FDLA J. **5**(1) (2020)
23. Rogers, E.: Diffusion of Innovations, 4th edn. The Free Press, New York, USA (1995)
24. Fizbein, M., Ajzen, I.: Belief, Attitude, Intention, and Behavior: An Introduction to Theory and Research. Addison-Wesley, Reading, USA (1975)
25. Ajzen, I.: The theory of planned behavior. Organ. Behav. Hum. Decis. Process. **50**(2), 179–211 (1991)
26. Davis, F.: Perceived usefulness, ease of use, and user acceptance of information technology. MIS Q. **13**(3), 319–340 (1989)
27. Venkatesh, V., Morris, M., Davis, G., Davis, F.: User acceptance of information technology: toward a unified view. MIS Q. **27**(3), 425–478 (2003)
28. Koivisto, M.: Mobile information system adoption and use: beliefs and attitudes in mobile context. Helsinki University of Technology (2009)
29. Kim, Y.J.: Social influence process in the acceptance of a virtual community service. Inf. Syst. Front., July (2006)
30. Bleuel, W.: CSAT or CES: does it matter? Graziadio Bus. Rev. **22**(1) (2019)
31. Dixon, M., Freeman, K., Toman, N.: Stop trying to delight your customers. Harvard Business Review, July–August (2010)
32. Reichheld, F.: The one number you need to grow. Harvard Business Review, December (2003)
33. IBM, Watson Assistant: The shortcut to great customer service. White paper. IBM (2016)
34. Følstad, A., Skjuve, M.: Chatbots for customer service: user experience and motivation. In: Proceedings of the 1st International Conference on Conversational User Interfaces (2019)

The *Notebook* to Reflect on the Meaning of Life: An Educational Proposal for the Guidance of Young Migrants

Concetta La Rocca(✉) and Massimo Margottini

Roma Tre University, Roma, Italy
`concetta.larocca@uniroma3.it`

Abstract. This work presents the use of the *Notebook to Reflect on the Meaning of Life* in the CPIA 2 e 3 (Provincial Centers for Adult Education) in Rome (Italy). The experience is part of the FARO (FAre Reti e Orientare - Make Networks and Guidance) Project, co-financed by the EU and the Italian Ministry of the Interior. The aim of the project is the construction of an integrated network to respond to the training and work needs of Third Countries citizens living in some Italian territories. Through the application of educational tools and methodologies, a connection was made between knowledge, skills, and metacognitive competencies. The *Notebook to Reflect on the Meaning of Life* fits into this framework and places the theme of reflection as a key development in young migrants of the awareness of being protagonists of their own educational and professional path. The *Notebook* is compiled on word files and is published online in a Google presentation. It consists of six pages, each of which contains specific exercises to reflect on the values and on the meaning of life. Fifty young migrant students who attended CPIA courses were involved; the analysis of quantitative data showed that most of them were interested in the activities they joined under the guidance of the teachers. In addition, the summary of the work carried out by Bah'a young migrant girl from Congo who migrated to Italy one year ago-is reported; the link to Bah's work is also available.

Keywords: Reflection · Life values · Meaning of life

1 A Brief Description of the FARO Project

The FARO Project (FAre Reti e Orientare - Make networks and Guidance) was co-financed by the EU and the Italian Ministry of the Interior (Asylum, Migration, and Integration Fund 2014–2020).[1] In addition to the Department of Education at Roma Tre

---

The paper is the result of the collaboration of the two authors. Massimo Margottini is the au-thor of paragraph 1; Concetta La Rocca is the author of paragraphs 2, 3, 4.

[1] The FARO project (08/07/2019–31/03/2022) was coordinated by Massimo Margottini (Department of Education at Roma Tre University, Italy) and follows the path traced by the previous CREI Project (CREI: CreareREti per gli Immigrati - Creating Networks for Immigrants) (10/01/2017–31.03.2018), also co-financed by the EU and the Italian Ministry of the Interior, and always coordinated by Massimo Margottini.

© The Author(s), under exclusive license to Springer Nature Switzerland AG 2023
D. Guralnick et al. (Eds.): TLIC 2022, LNNS 581, pp. 201–212, 2023.
https://doi.org/10.1007/978-3-031-21569-8_19

University, many training agencies from the Italian Regions of Lazio (Central Italy), Lombardy (Northern Italy), and Puglia (Southern Italy) joined the partnership. The project, which ended on 31/03/2022, had the objective of experimenting the construction of an integrated network aimed at responding to the training and work needs of citizens of Third Countries in some Italian territories. The massive migratory flows led the host countries to seek an answer to the need for integration of newcomers through the design of new intervention methods and guidance actions. People with a migrant background are known to pose a challenge to the guidance system, whose task is to develop personalized tools and paths to encourage integration and social inclusion. On the contrary, the logic of the emergency often pushes to look for an immediate solution, but it is of little value in terms of personal improvement: the individual can satisfy basic needs (e.g., survival, search for means of sustenance), but they are not given the chance to develop their true potential and the right to emancipate. The host society can offer several job opportunities, which can only be seized by those young migrants who have the learning skills and professional adaptability required to proactively address uncertainties and problems. For these reasons, many scholars [3, 4, 7, 8, 16–18, 28] argue that investments in training and employment are a key strategy for the integration of newcomers, since, in addition to being a source of financial security, they are resources that facilitate integration into other domains of life as well. To better understand the situation and find appropriate solutions, research and projects will have to study actions that are not limited to the professional training of young migrants, but also aimed at the development of strategic skills that are strongly correlated to personal success and career. The enhancement of an adequate and conscious process of self-construction and professional identity is essential for personal fulfillment, to contrast the risks of discomfort or discrimination and favor integration. This process encourages psychological well-being, the motivation to develop one's skills and the acknowledgement of social and working challenges. To achieve the development of an "adaptive competence" [22], it is essential to promote constructive, self-regulated and collaborative learning. In recent decades, the new guidance paradigm of the Life Design [23–25], founded on Career Construction [23] and on Life Construction [9] proposed a new intervention model aimed at supporting the individual to build their own individuality in the present and to project in an imagined future. A new approach to guidance is currently being conceived, which is meant to support those persons who consciously plan their life and profession consistent with their interests and way of being. In this context, training and professional guidance are a useful strategy to encourage development and capacity building in young migrants and to enhance formative and professional needs consistent with their migration project and full social inclusion.

1.1 The Structure of the Project: Training of Operators and Application of Tools

Through the application of different educational tools and methodologies, a connection was made between knowledge, skills, and metacognitive and meta-reflective skills [5, 10]. Such a connection is the starting point for the development of a person who can act, reflect, change, and constantly adapt their actions in order to achieve personal goals and fulfill life and career expectations [9], in a lifelong, life-wide, and lifespan [1] perspective. The project involved the use of multiple methodologies and tools, developed by Roma Tre research group, and submitted in training modules to the operators of local agencies for the instruction and training of young migrants. There were seven modules, all characterized by two closely related aspects: a theoretical section for study and reflection, and a laboratory section, for the practical use of methodologies and tools.

The training modules are listed below:

- Module 1 - Survey of skills, qualifications, and professional experiences
- Module 2 - Personal history: narration about oneself, personal experience, and migratory project
- Module 3 - Analysis and development of strategic skills to direct oneself in learning and work
- Module 4 - Strengthening citizenship and intercultural competences
- Module 5 - Detection of professional interests
- Module 6 - Elaboration of a training and professional project
- Module 7 - Building a network to support training and work

Also, an ePortfolio was to be created, which was to collect all the results obtained after administering the research tools. However, due to lack of time and of adequate technological equipment, this activity has been postponed and will be carried out in the future.

In each module, participants are asked to use a personal computer to fill in the provided tools. The *Notebook* to Reflect on the Meaning of Life is the tool presented in Module 2. It focuses on the narration of personal history and of the migratory project. The *Notebook* will be discussed in the next paragraph.

2   The *Notebook* to Reflect on the Meaning of Life: A Practical Use

In the theoretical framework, a main principle has been set out: the measures to guide young immigrants cannot be reduced to the exclusive goal of identifying their professional skills to facilitate integration into the labor market; rather, they should aim to bring out desires and projects, expectations, and interests. Of course, we wanted to try not to entrap young people into an image that, although real, can sometimes fuel prejudice and stereotypes on migrants. It was decided that the participants would be invited to reflect on themes concerning their overall vision about the meaning of life and about the meaning attributed to their own life. The work of investigation, reflection, and revision of the various sections of the *Notebook to Reflect on the Meaning of Life* was shared between university researchers and the teachers/educators who work within the training structures involved in the project.

According to the *Notebook* conceptual framework, reflection is the starting element from which to develop those metacognitive skills that can enable youth and young adults to become aware of their own educational path and professional growth [20, 21]. Dewey [6] strongly emphasized the importance of reflection in education. As he pointed out, it is not experience but, rather, reflection on experience that significantly changes personal beliefs, thus producing a cognitive modification, such a learning. Mezirow [19], Bernaud [2] and Mancinelli [14], who are the authors explicitly referred to in the *Notebook* theoretical framework, underline the importance of reflection, as a key concept, to drive an individual to critically analyze their own beliefs and reformulate the justification process that sustains them. The authors also highlight how appropriate it is to encourage the development of soft skills that allow to acquire that flexibility of thought and action that is necessary to face specific and complex contexts of life.

The *Notebook* is made of six pages. Pages 2–3–4–5 were built based on the tools developed by Jean-Luc Bernaud [2]; page 1 is based on the concepts stated by Maria Rosaria Mancinelli [14] and on the Zimbardo and Boyd's ZTPI [27]. Page 6 contains instructions on how to share personal reflections in a group.

The teachers supported the young migrant students in filling the pages and used a grid to observe the way in which they approached the work. At each meeting, the teacher emphasized that the work is not done, and that the person could add any other elements over time.

Each *Notebook* was compiled on a word file and assembled using a Google drive presentation.

As part of the FARO project, the *Notebook* was tested in CPIA2 and CPIA3 (CPIA: Provincial Center for Adult Education) in Rome. The administration period lasted six months and involved mainly four teachers from CPIA2 and three teachers from CPIA3. A total of 50 migrant students from different parts of the world participated. The *Notebook* was given to all the young migrants attending CPIA classes with a level of Italian between basic and intermediate.

For each page of the *Notebook*, the exercises are briefly described below; also, the quantitative results of the administration are reported.[2]

2.1 Page 1 - The Imagined Future

Methodology: the teacher uses tools and materials to allow the person to build their image of the future to show themselves to the teacher and to whoever else they want (Table 1).

Table 1. Results of *Notebook* - Page 1

The inserted image is:	Respondents
– a collage	18
– a handmade representation photo (drawing, painting, etc.)	2
– an image downloaded from the internet	29
– a photo taken by the author of the *Notebook*	1
There is no image	4

This page has been remarkably successful: only four students did not fill it out. Interestingly, most of the students used images downloaded from the internet to give a picture of their future.

---

[2] For a more analytical description of the exercises contained in the pages of the *Notebook*, see La Rocca (2018, 2019), La Rocca and Margottini (2018).

2.2   Page 2 - Finding the Meaning

Methodology: the teacher leads the person in doing two exercises: 1. Praise a character or a person considered exemplary; 2. Describe an artwork of various kinds (painting, film, image, video, or music etc.) and the reflections arising from its use. The artwork is chosen by the teacher and presented to the person (Table 2).

Table 2. Results of *Notebook* - Page 2

First exercise: *praise a character or a person considered exemplary, the author has chosen:*	Respondents
– a famous person	21
– a family person	18
– a friend	3
The exercise was not carried out	5
The data has not been entered	3
Second exercise: *regarding artworks shown by the teacher, the author has:*	Respondents
– made a description of the work	42
– expressed the emotions s/he felt	41
– indicated the message he drew from the work	29
The exercise was not carried out	7

Page 2 was also successful. In the first exercise, most of the young migrants chose a famous person as an example to follow, even if a large part referred to a family member.

In the second exercise, almost all described the artwork and expressed their feelings. Only 29 people spoke about the message of the work: a very good result, given the difficulty of the task and the weak Italian spoken by most participants.

2.3   Page 3 - The Analysis of the Values

Methodology: the teacher leads the person in carrying out three exercises: (1) detect and analyze the values that the person identifies as their own; 2_reflect on the place that the identified values occupy in their existence and (2bis) in relation to their time; and (3) place the values in the perspective of professional and life development (Table 3).

The third page was successfully completed. From the values listed, the young people sorted out those they most felt their own. Then they reflected on how these values could be present in their lives.

2.4   Page 4 - The Role of Values in ONE'S Life

Methodology: the teacher leads the person in carrying out two exercises: (1) imagine two possible life paths consistent with their values and (2) pay attention to the things that the person would like to do but is unable to do, identifying the causes (Table 4).

Table 3. Results of *Notebook* - Page 3

3 Exercises:	Respondents
(1) S/he has entered in the table the values s/he feels her/his own and has indicated their weight (± important)	50
(2) S/he reflected on the place of values in her/his life	40
(2bis) S/he has entered the values in relation to time (past, present and future)	46
(3) S/he has reflected on which elements could make her/his values come true in the future	38

Table 4. Results of *Notebook* - Page 4

First exercise: *imagine two different and possible life paths (two future autobiographies) consistent with one's own values*	Respondents
– S/he developed two life paths	25
– S/he developed only one life path	15
The exercise was not carried out	10
The data has not been entered	1
Second exercise: *list everything that you have not done in your life and that you would have liked to do, identifying obstacles, and looking for ways to overcome them*	Respondents
– S/he filled in the table by inserting more than three entries	16
– S/he filled in the table by inserting less than three entries	24
The exercise was not carried out	11

The first exercise on page 4 was fully completed by the majority; only 11 of them did not. The same happened with the second exercise: 11 people did not do it and the majority only partially completed it. As you progress through the pages of the *Notebook*, the exercises become more complex and therefore the level of the language probably makes it difficult to perform, even with the help of the teacher.

2.5   Page 5 - Developing the Art of Living

Methodology: the teacher supports the person in two exercises. In the first, a form consisting of three columns is presented: c1- statements regarding behavior linked to existence are inserted; c2- descriptions of the previous statements are inserted; and c3- using a scale, the person must weigh each statement/description. In the second exercise, a form consisting of two columns is presented: c1- the previous statements are reported; c2- the person describes how they could perform the behaviors described in the statements (adapted from [2], pp. 184–186) (Table 5).

Table 5. Results of *Notebook* - Page 5

First exercise: *in the third column, enter the level of importance for each statement*	Respondents
– S/he filled in the table entirely	37
– Partially filled in table (from 4 to 7)	5
– Completed the table minimally (from 1 to 3)	0
The exercise was not carried out	8
Second exercise: *imagine how to carry out, every day, a concrete action that has relevance to each of the statements listed in the previous exercise and how to implement it for 15 days*	Respondents
– S/he filled in the table entirely	23
– Partially filled in table (from 4 to 7)	6
– Completed the table minimally (from 1 to 3)	4
The exercise was not carried out	17

Though complex, the exercises on page 5 were fully done by the majority. However, 17 people did not carry out the second exercise. It was probably difficult for the young migrants to imagine how to put into practice the behaviors described in the statements. This can be understood because it is difficult for them to plan a vision of their future life.

2.6 Page 6 - The Sharing

Methodology: the teacher creates small groups and invites the persons to share their completed *Notebook* (Table 6).

Table 6. Results of *Notebook* - Page 6

The subject:	Respondents
– Actively participated by describing in depth her/his own work and commenting on that of others	1
– Participated in a partial way, briefly presenting her/his own work, and expressing few comments on the work of the others	7
– Did not participate in the group relationship	8
The exercise was not carried out	34

The activity foreseen on the sixth page was fully carried out by one person only. Teachers said that young migrants had difficulty in sharing the pages of their *Notebooks* with others. Compiling the pages led them to make a profound reflection on their values and on the meaning of their lives: perhaps they did not want to share such intimate reflections in a group openly. One of the girls involved in a group is Bah, the protagonist of the case study that will be presented in paragraph 3.

2.7 The Observation Grid Filled by the Teacher

While each student filled in the *Notebook*, the teacher observed and took note of their behavior in a grid. The results are reported below (Table 7).

Table 7. Observation grid and outcomes

In carrying out the activities for the compilation of the *Notebook*, the author showed	Not at all (%)	Adequate (%)	Highly (%)
Curiosity about the work	7.9	13.2	78.9
Attention to deliveries	5.4	5.4	89.2
Understanding of the task	5.4	24.3	70.3
Involvement in the execution of the task	5.4	16.2	78.4
Commitment in carrying out the task	5.4	8.1	86.5
Compliance with deliveries	5.4	8.1	86.5

The data obtained using the observation grid show a strong participation of the students in the activities for the compilation of the *Notebook*. In all the descriptors, most behaviors are marked with "Highly." However, 24.3% of the students did not fully understand the task; this data can be related to their familiarity with Italian and, perhaps, to the fact that young people are not used to reflect on their values and on the place, they occupy in their lives.

3   A Case Study: The Young Woman *Bah* Talks About Herself in the *Notebook* to Reflect on the Meaning of Life

3.1 Brief Methodological Note

In reference to the case study procedures indicated by Yin [26], the researcher analyzed Bah's *Notebook* with the teachers who supported the girl in producing her work. Then the researcher and the teachers examined the *Notebook* to find the links between the research objective and the result obtained from the field investigation. Finally, the researcher wrote a report to tell the case of Bah using the information taken from the *Notebook* and highlighting the links with the project goals.

3.2 BAH'S Notebook to Reflect on the Meaning of Life

Bah's *Notebook* cover is the image of a tree whose trunk and leaves are colored hands. The image of the tree also constitutes the final page: this is, however, the image of the "tree of life" with its colored circles bathed in colored rain goggles.

On the first page of her *Notebook*, Bah inserted photos to describe her imagined future: a woman walking barefoot on a cobbled street, a young smiling man beside a young smiling woman holding a child in her arms; photos of great strength and beautiful African nature; and an image of the growth stages of a seedling. Bah is a young woman from Guinea who has lived in Italy for a year and a half. In her country, she was subjected to violent cultural practices that made her feel the pain of not being free to make her own choices because she is a woman. Bah explained that on this page. She wanted to describe the love for the beauty of her country and, at the same time, the desire to flee from her country, feeling just like the woman in the image: barefoot on a land of stones. But pain and difficulties did not stop her because she feels like the seedling that slowly, over time, will grow, become stronger, and will bear its fruits: the desire to fulfill herself in love and motherhood, overcoming violence and deprivation of freedom.

On the second page, Bah introduces Ahmed Sékou Touré, the first president of Guinea who took his country out of French colonial rule. Bah believes that Ahmed is an exemplary person because he fought for the freedom of Guinea and proved that nothing is impossible and that we must fight for rights and freedom. For the exercise that involves commenting on a piece of art, Bah chose Pablo Picasso's painting "Guernica." The painting depicts the extermination perpetrated by the Nazi air force on the defenseless population of the little Spanish town. Bah comments: "when I read the description of the painting it was like a mirror to me: everything represents me." The woman described the piece of art in a detailed and poetic way and stated that this painting, like President Ahmed, taught her that one must not surrender to oppression and that one must always fight for one's ideas and freedom.

On page 3 of the *Notebook*, more than 30 values are listed which are an expression of our culture. Bah selected 6: love, courage, happiness, faith, dignity, and freedom. She attributed the highest level to each and entered them all for each period of her life indicated in the appropriate table. Hence, to Bah those six values are fundamental today, were fundamental 10 years ago and will be fundamental for the years to come. Bah inserted images representing her values: a heart of red petals *for love*; a person who jumps from rock to rock over an abyss *for courage*; a woman who smiles with the wind in her hair *for happiness*; the hands of an adult and a child enclosing an ear of corn against the backdrop of a sunset *for faith*; a poor child who returns a wallet *for honesty*; peasants who work and harvest potatoes *for dignity*; two fisted hands that break a rope holding them tied *for freedom*.

This last image is the greatest: Bah needed to voice out that freedom is her greatest value and her greatest passion. In fact, she also inserted the beautiful and famous poem "Liberté" by Paul Éluard. This poem was published in 1942, during the Nazi occupation of Paris; thousands of copies were dropped by allied aircraft into Nazi-occupied France. Bah ended this page of her *Notebook* with a reflection: "I have lost a bit of freedom, of dignity, because I ran away from my country. Now I have no documents, so I cannot have freedom and dignity. I never lack courage."

On page 4 of the *Notebook*, Bah imagined two possible events in her life that are consistent with her values. In the first, Bah imagined being able to receive a residence permit and work in a hotel. In the second, she imagined being able to marry and become a receptionist in a hotel. The two paths are presented as successive, from an evolutionary perspective. Bah also identified the things she would have wanted to do, and she has never been able to do: she never learned to swim out of fear and now she wants to overcome fear; she never learned to drive because she had a driver in her country and she wasn't allowed. Now, she wants to learn how to drive.

On page 5, Bah read the form in which 11 statements were inserted to help reflecting on the meaning she could give to her life, to her work and to the development of her spirituality-interiority. She highly agreed with all the statements, except for the last one, to which she gave a medium weight. This statement said: "To forget about yourself and consecrate yourself to others or to a cause," or "To be able to grant time, means and ideas to a person, to a group, to a cause." At present, Bah needs to focus on herself to build her new life, therefore dedicating herself to others is not among her priorities; nonetheless, in the following exercise she will say that volunteering will be her concrete action to turn the statement into practice.

In the last exercise, Bah imagined what to do in her daily life to perform the actions described in the statements. She said: "I want to: learn more; be less worried about the future; do the things I do well; study Italian better; have excellent relationships with everyone; do good deeds; not to complain because I have health and courage; go to the beach and look at the sea and at the people; be attentive to what I eat; continue to have my own spaces of freedom; go with Caritas volunteers to help people who live on the street."

As for the activities on page 6, the teachers reported that Bah was very active in the teamwork and wanted to share her *Notebook* with the other students.

Bah's *Notebook* ended with a poem she wrote in French and, as we have said, with the image of the "tree of life". Title of her poem: Hope.[3]

4   Short Closing Note

The path undertaken in the project has shown how important it is for young migrants to be able to express themselves and communicate even at a profound level. The teachers themselves reported that the young people showed interest and involvement and actively collaborated in filling the proposed tools.

In particular, this article clearly says that the young migrants gladly participated in the compilation of the *Notebook*. All the *Notebook*s achieved the aims of the research: allow young migrants to reflect on their future, on their values, on the meaning of their life. Many *Notebook*s turned out to be interesting and very well done. We chose to present Bah's *Notebook* because she filled it with great enthusiasm. She presented it herself at the Final FARO Project Conference. Bah's speech caused participation and emotion in the audience because she showed her great strength in wanting to win her freedom in a new life project to be pursued with optimism and courage. Her imagined future is now in her hands.

---

[3] We thank Ada Maurizio, the CPIA 3 school head, and the teachers who supported Bah: Antonella Bracalenti, Pamela Di Lodovico, Paola Russo. And we thank Bah. Bah's Notebook is written in Italian and can be viewed at the following link: https://express.adobe.com/page/10r73yENf RHP0/. Each page of the Notebook has been named "quadro".

References

1. Baltes, P.B.: Theoretical propositions of life-span developmental psychology: on the dynamics between growth and decline. Dev. Psychol. **23**(5), 611–626 (1987). https://doi.org/10.1037/0012-1649.23.5.611
2. Bernaud, J.L.: Psicologia dell'accompagnamento. Il senso della vita e del lavoro nell'orientamento professionale. Erikson, Trento (2015)
3. Catarci, M.: L'integrazione dei rifugiati. Formazione e inclusione nelle rappresentazioni degli operatori sociali. Franco Angeli, Milano (2011)
4. Colic-Peisker, V., Walker, I.: Human capital, acculturation, and social identity: Bosnian refugees in Australia. J. Community Appl. Soc. Psychol. **13**(5), 337–360 (2003). https://doi.org/10.1002/casp.743
5. Cesare, C.: Metacognizione e apprendimento. Il Mulino, Bologna (1995)
6. Dewey, J.: Experience and Education. Collier Boooks, NY. Tr.it. Esperienza e educazione (1949). Firenze: La Nuova Italia (1938)
7. Fiorucci, M.: La formazione interculturale degli insegnanti e degli educatori/Intercultural training of teachers and educators. FORMAZIONE & INSEGNAMENTO, XIII **1**, 55–69 (2015)
8. Fiorucci, M., Margottini, M.: (a cura di) Creare reti per immigrati. Milano: Franco Angeli (2020)
9. Guichard, J.: Career guidance, education, and dialogues for a fair and sustainable human development. Inaugural conference of the UNESCO chair of Lifelong guidance and counselling, Nov 2013, Wroclaw, Poland. ffhal-03240556f (2013)
10. Hacker, D.J., Dunlosky, J., Graesser, A.C. (eds.).: Handbook of Metacognition in Education. Routledge/Taylor & Francis Group (2009)
11. La Rocca, C.: Il Quaderno per riflettere sul Senso della Vita. Una pagina di ePortfolio. Ricerche Pedagogiche. Anno LII, n. 208–209, dicembre 2018, pp. 107–127 (2018)
12. La Rocca, C., Margottini, M.: The notebook to reflect on the meaning of life: a page inside the ePortfolio. Int. J. Human. Soc. Sci. Educ. (IJHSSE) **5**(6), 1–8. https://doi.org/10.20431/2349-0381.0506005 www.arcjournals.org (2018)
13. La Rocca, C.: Orientamento e valori: il quaderno sul senso della vita. Nuova Secondaria. Anno XXXVII, n. 2, ottobre 2019, pp. 45–47. ISSN 1828-4582 (2019)
14. Mancinelli, M.R.: Tecniche d'immaginazione per l'orientamento e la formazione. Franco Angeli, Milano (2008)
15. Margottini, M.: Orientare sé stessi nella vita e nel lavoro. In: Fiorucci, M., Margottini, M. (eds.) Creare reti per gli immigrati, pp. 65–75. Milano, Franco Angeli (2020)
16. Margottini, M., Rossi, F.: Un modello di orientamento per i giovani immigrati. In: Volpicella A.M., Crescenza, G. (eds). Educazione permanente e società interculturale, pp. 141–164. Edizioni Conoscenza, Roma (2019a)
17. Margottini, M., Rossi, F.: Un modello di orientamento formativo per giovani immigrati. Giornale italiano della Ricerca Educativa **22**, 179–198 (2019b)
18. Michael, M., Richard, W.: Social Determinants of Health. Oxford Scholarship Online (2005)
19. Mezirow, J.: Apprendimento e trasformazione. Il significato dell'esperienza e il valore della riflessione nell'apprendimento degli adulti. Raffaello Cortina Editore, Milano (2003)
20. Pellerey, M.: Le competenze individuali e il portfolio. Etas scuola, Milano (2004)
21. Pellerey, M.: Apprendimento e trasferimento di competenze professionali. In Orientare l'orientamento. Roma: Isfol, 305–323 (2007)
22. Michele, P. (Ed.).: Strumenti e metodologie di orientamento formativo e professionale nel quadro dei processi di apprendimento permanente. Roma: CNOS-FAP (2018)

23. Savickas, M.L.: The theory and practice of career construction. In: Brown, S.D., Lent, R.W. (eds.) Career Development and Counseling: Putting Theory and Research to Work, pp. 42–70. John Wiley, Hoboken, NJ (2005)

24. Savickas, M.L.: Career Counseling. American Psychological Association, Washington DC (2011)

25. Savickas, M.L.: Career Counseling. Guida teorica e metodologica per il XXI secolo. Erikson, Trento (2014)

26. Yin, R.K.: Lo studio di caso nella ricerca scientifica. Progetto e metodi. Armando Editore, Roma (2005)

27. Zimbardo, P.G., Boyd, J.N.: Putting time in perspective: a valid, reliable individual-differences metric. J. Pers. Soc. Psychol. **77**, 1271–1288 (1999)

28. Zizioli, E.: Una stanza tutta per noi. Letture collettive al Femminile. I PROBLEMI DELLA PEDAGOGIA, LXIV **2**, 331–349 (2018)

Effects of Game Elements on Performances in Digital Learning Games

Inbal Leuchter[(✉)] and Gila Kurtz[(✉)]

Holon Institute of Technology, Holon, Israel
inballeuchter@gmail.com, gilaku@hit.ac.il

Abstract. Studies have shown that digital game-based learning (DGBL) can stimulate learners and increase motivation. However, to accomplish these goals, we must understand the role and impact of the game elements. This study aimed to examine the effects of four-game elements on players' performance: instructions and assistance, narrative, competition, and challenge. The data are based on game performances of 3,281 users during the period 2015–2020. Users played as part of their visit to 'Musa,' a multidisciplinary museum of local cultural materials in Tel-Aviv, Israel, either in 'Family' game mode or in 'Multi-player' mode. Results show that players performed better on 'Multi-player' games. In addition, players' performances improved when narrative depth was significant and the play area was smaller. Separating the data into two groups led to additional results: players in 'Family' mode performed better when the game instructions included a video, while in 'Multi-player' mode, participants performed better when a human guide was available, to some extent. The results of the study and its implications can assist educators and game designers in planning more accurate and effective learning games.

Keywords: Game-based learning (DGBL) · Instructional game elements · Game performances · BIG DATA

1 Introduction and Background

Since the 1980s through the early 2000s, researchers have been interested in the new possibilities of enrichment video games can provide. They pointed out that this environment has the ability to contain enormous amounts of content, give immediate feedback, and motivate the participants intrinsically within a fun, safe, and risk-free environment [1–3]. Other scholars have stressed that video games could function as an excellent 'incubator' for developing life skills, such as identifying patterns, recognizing and solving problems, and making quick decisions [4]. Today, many studies explore the advantages of video games as an effective tool in various fields, including sport, medicine, math, and mental health [5–7]. Education has been one of the most popular areas for integrating digital games, offering both great promise and a fascinating research field for many educators and instructional designers [8].

Many studies have shown that digital games can be an effective and powerful tool for improving and strengthening learning [9–17]. To be precise, research has shown that integrating digital games into educational systems can often improve the learning process and usually improve students' motivation, making learning more appealing. Moreover, studies suggest that games promote a stronger positive emotion toward learning than traditional learning methods [18]. For these reasons, digital educational games have become popular in learning environments such as schools, military training facilities, and workplaces [19–21]. This study examines a set of nine educational digital games designed and operated within a museum environment to understand their effect on participants. In the following section, we review the research literature regarding the impact of specific game elements on players.

1.1 Motivation and Performance in Digital Learning Games

In contrast to classic video games, learning games are designed to achieve an educational goal beyond simple enjoyment [22, 23]. The topic of motivation in digital games has been extensively discussed and researched over the years due to its critical significance for retaining players in the game, producing better performances, and increasing players' participation [3, 24–26]. Flow, immersion, presence, and engagement filled the world of game research, as well as questions regarding how to keep players engaged for as long as possible [27, 30]. Game developers realized motivation (defined as the desire to participate and the intensity to succeed [31]) is a key factor for encouraging players to play more [32], and in the educational world, it is a key factor to deeper learning processes [33, 34].

In is important to note that despite the popularity of digital learning games, there have been studies that question whether or not they are as effective in improving learning outcomes as attributed [17, 35], yet there is hardly a question about their role in enhancing motivation. Studies consistently and unequivocally demonstrate that learning games increase learners' motivation and engagement in the study material and in-class [26, 35–37].

1.2 Game Elements and Their Impact on Participants in Learning Games

Game environments contain a wide range of elements, such as rules, feedback, goals, competitions, challenges, narratives, progress, collaboration, rewards, and stages [1, 38]. Werbach and Hunter [39] compared designing a new house (which requires knowledge of different dimensions and the professional skills to combine everything into a successful work) to the complexity of designing and executing a game. Many researchers have realized that in order to be able to plan a successful game, one must understand more deeply its mechanics. In order to do so, some researchers have chosen to isolate one or more individual game elements and conducted experiments to reveal their effect on gameplay [21, 36–38, 40–43]. They aimed to uncover which of the elements have a significant impact on the player and to understand the mechanism for this: which are key to effective learning, motivation, and enjoyment, and which are weak or not necessary at all. With this knowledge, they believed that educators and game designers could plan accurate game experiences, tailored appropriately to the learning context and the desired outcome. In this study, we focus on examining four common elements: Instructions and assistance (instruction method and in-game assistance), narrative integration (station order and narrative depth level), competition, and challenge.

Instruction's method: Research shows that the type of instruction given before reading a text can influence the cognitive process that takes place during reading and learning. Likewise, it is possible that instructions on participants' performances are important in learning digital games. Our study includes two types of pre-game instructions: one with a video clip and the other without it. We will compare the number of stations completed by participants with and without video, in order to examine the effect of this parameter.

Assistance: In numerous studies, teachers and instructors have been shown to play a crucial role in determining the success of a learning game [44, 45]. It is a challenging task requiring the instructor to have a broad set of skills including technical abilities, playful literacy, specialization in the material being taught, and a solid pedagogical foundation [46]. A high priority was also placed on the availability and quality of assistance provided to students during the game. The results of a study of about 250 high school and high school students, in which assistance was offered to players (individual and team) during the game, demonstrated conclusively that help influenced their performance and motivation [47]. Instructor assistance has been shown to be also crucial in university [48]. Clark and his colleagues [35] found that the more humane and massive the assistance provided in a game, the more positive and substantial the impact on players. As part of this study, we will examine whether instructor assistance is a factor influencing participants' performance. To this end, we shall examine two levels of accompaniment offered to players in the 'Group' mode—'close' and 'distant'. In the 'close' mode, the guide was physically present throughout the game activity, whereas in the 'distant' mode, the participants were made aware that the guide was available to them, but he was not in their immediate vicinity.

Narrative. According to researchers, a narrative element in a learning game produces a positive attitude and engagement towards the learning material, enhances authenticity and enjoyment, and leads to better participation [37, 40, 43, 49, 52].

The narrative is examined in this study using two parameters—order of stations and narrative depth.

Order of stations. In our study, all learning games will be classified as 'free' or 'linear' based upon the participants' ability to choose their path in the game. In 'free order' games, participants did not have to follow a predetermined station order and were free to decide their game path. In 'linear order' games, however, the order of stations was predetermined by us (the game designers), and participants could not alter it. 'Free' games typically featured a weak narrative of the game, since it was impossible to predict where the participants will go or adapt the narrative accordingly, while 'linear' games typically featured a strong, clear and well-constructed narrative.

Narrative depth. In this study, we separated the games into three categories based on their narrative depth—'weak,' 'medium,' and 'strong' Games without any narrative elements or with a slight narrative layer were classified as 'weak' narrative depth. We regarded narrative depth as 'medium' for games that were motivated by a story and reflected it to some extent in their graphics. Lastly, narrative depth was classified as 'strong' for games that had an evolving story throughout the game, which was also reflected in their graphics.

Competition. Competition is one of the most common game elements and usually involves points, tags, and scoreboards [19]. Competition can be between individuals or between groups, face-to-face or over a computer, in person, against oneself, against time, against luck, etc. [38, 42]. Although competition effectiveness in improving learning remains uncertain [21, 39], it is still considered one of the most popular game elements. According to many studies, competition improves motivation and learning performance, creates challenges, a sense of ability and meaning, and even encourages greater effort, teamwork, active participation, and enjoyment of games [21, 26, 36, 37, 41, 51]. In our study, we compare two different game options that were available in 'Group' mode: competitive and non-competitive.

Challenge. The challenge element is also a popular one in games [1, 3, 26, 38]. Studies show that challenges improve participants' learning, motivation, sense of flow, and performance [27, 52], as well as their level of enjoyment and satisfaction [34, 40]. Providing there is a balance between the abilities of the players and the game difficulty, it is expected that the players will experience levels of satisfaction and enjoyment [27, 53]. However, if the level of challenge is not appropriate for the participant (i.e., too easy or too difficult), motivation and interest are likely to decrease [2]. This study examines the challenge element using the parameter of 'size game area'. In general, the longer the museum game was, the greater the physical and cognitive effort the players had to exert. For this reason, we categorized our nine educational games into three sizes—'small,' 'medium,' and 'large.' A 'small' game area included a visit to only one exhibition. A 'medium'-sized play area included visits to two to four exhibitions. A 'large' game area included visiting five or more exhibitions.

2   Research Objective and Research Questions

The purpose of this study is to examine the effect of four game elements on motivation and performance in nine museum digital learning games. We asked the following research questions:

1. Are there differences in the number of stations completed by participants with 'Family' mode and participants with 'Group' mode?
2. Is there a correlation between the number of stations completed by the participants and the following variables: Instruction method, Level of narrative depth, Size of the play area?
3. Is there a correlation between the number of stations completed by the participants of the 'Family' mode and the following variables: Instruction method, Order of the stations, Level of narrative depth, and Size of the play area?
4. Is there a correlation between the number of stations completed by the participants of the 'Multi-player' mode and the following variables: Level of assistance, Competition, Level of narrative depth, Size of the play area?

3   Research Field and Data Collection

The nine museum digital learning games, that were included in this study (See Table 1), were created, and played between 2015 and 2020 in 'MUSA - Eretz-Israel Museum, Tel-Aviv' in Israel. The museum, which is one of the largest museums in Israel, showcases local cultures that lived in the area during different periods, next to contemporary Israeli art and craft (http://www.eretzmuseum.org.il).

We collected all data for this study from the 'Wandering' game platform, which automatically recorded each player's information and score. The study includes the performance of 3,281 users in nine educational games offered by the museum to its visitors. Due to the extent and diversity of the data and their source, they can be considered BIG DATA [54]. As mentioned before, participants for the games were instructed not to play alone, but to play together on one device, in cooperation with other players. Observations have revealed that this guideline was met by the vast majority of participants so that any registered 'user' can be considered as representative of a 4- to 3-player squad (3.5 on average). By multiplying these numbers, we arrive at an estimate of 11,515 players represented in the data. The data can be broken down into two categories: a group of players in 'Family' mode, comprising 1,694 users (about 52% of respondents) and a group of players in 'Group' mode, comprising 1,587 users (about 48% of respondents).

4   Findings

4.1   Research Question 1

We conducted an independent Samples t test to determine whether participants who played in the 'Family' mode completed more stations than those who played in the 'Group' mode. Statistical analysis revealed a significant difference ($p < 0.05$, t (2669.354) $= -9.075$). **Participants who played in the 'Group' mode** (average $=$ 5.47, standard deviation $= 3.97$) **completed more stations than those who played in the 'Family' mode** (average $= 4.40$, standard deviation $= 2.54$).

4.2   Research Question 2

We conducted an independent Samples t test to determine whether the number of stations completed by participants whose opening instructions included a video differed from those whose instructions did not. Statistical analysis revealed a significant difference (p < 0.05, t(2838.631) $= -3.910**$). **Participants whose game instructions included a video completed a greater number of stations** (average $= 5.16$, standard deviation $= 3.89$) **than those whose game instructions did not include a video** (mean $= 4.69$, standard deviation $= 2.73$).

Pearson correlations were also conducted to determine whether there is a correlation between the number of stations completed by the participants and the level of narrative depth and the size of the game area. However, no significant correlation was found in either case. As the order of the stations could not be tested, no analyses were performed on this study question.

4.3 Research Question 3

An independent Samples *t* test was performed on the 'Family' mode players to examine the connection between the number of stations completed to the instruction's method and the order of the stations in the game. The result was a slight but significant difference ($p < 0.05$, t (1218.403) $= -2.694**$). In other words, **those who played games with video instructions and linear station order completed slightly more stations** (average = 4.63, standard deviation = 2.04) **than those who played games without video entry instructions and free station order** (average = 4.31, standard deviation. = 2.71).

Additionally, Pearson correlation analyses were conducted in order to see if there was a connection between the number of stations completed and the level of narrative depth, and the size of the playing area. Calculation showed that the two elements contributed to the number of completed stations in a weak but significant way. **The number of stations completed was negatively related to the size of the game area** ($p < 0.01$, r $= -0.200**$) **while the number of stations completed was positively related to the narrative depth level** ($p < 0.05$, r = 0.058*). No analysis was conducted for the level of assistance and competition since these factors had a single value in 'Family' mode games.

4.4 Research Question 4

An independent Sample *t* test was conducted to examine the impact of guide assistance received by the participants in the 'Multi-player' mode on their completion rate. Analysis revealed that the group receiving remote accompaniment differed significantly from the group receiving close accompaniment ($p < 0.05$, t(1320.612) = 18.137**). **Those who had a remote instructor accompany their game completed almost twice as many stops** (mean = 6.30, standard deviation = 4.12) **as those who were accompanied by a close guide** (mean = 3.26, standard deviation = 2.40). Additionally, an independent Sample *t* test revealed a significant difference between the station completed by those who participated in competitions versus those who did not. Statistical analysis revealed a significant difference between the two groups ($p < 0.05$, t (1320.612) = 18.137**). The results demonstrate that **those who followed a competitive format completed more stations than those who followed a non-competitive forma**t (average = 6.30, standard deviation = 4.12). Pearson correlations were carried out to examine whether there was a correlation between the number of stations completed by players in 'Group' mode and the level of narrative depth and size of the gameplay area. We found that both had a weak but significant impact on the number of completed stations. There was a weak and significant negative relationship between the number of stations completed and the size of the game area ($p < 0.01$, r = $-0.219**$), while the narrative depth level revealed a very weak and significant positive relationship ($p < 0.05$, r = 0.079**). The opening instructions and the order of the stations in the game could not be tested for this research question.

5   Discussion and Conclusions

For several decades, research has shown that learning games have tremendous educational potential [1, 26]. Recently, researchers have begun to isolate specific game

Table 1. The nine museum digital games names and parameters included in this study

Participants age	Game area size	Competition	Narrative depth	Stations order	Level of assistance	Instruction method	Game Mode available	Game duration	No. of stations	Game Name
Children, adolescents, adults	Large	Family mode—Yes Multi-player—No	Weak	Free	Distant	With video clip	Family and Multi-player	Family mode – 40 min Multi-player mode – 50 min	104	'Around Musa'
Adolescents	Medium	No	Weak	Free	Close	With video clip	Multi-player	20 min	24 (4 for every squad)	'Need to product'
Adolescents	Large	Yes	Strong	Free	Distant	Without video clip	Multi-player	50 min	12	'Passcode'
Adolescents	Small	No	Weak	Free	Close	Without video clip	Multi-player	15 min	8	'The Map'
Families (diversed)	Large	No	Weak	Free	No assistance	Without video clip	Family	50 min	24	'Israeli Picnic'
Adolescents	Small	No	Medium	Free	Close	Without video clip	Multi-player	15 min	7	'Jacqueline Kahanov'
Families (diversed)	Large	No	Strong	Linear	No assistance	With video clip	Family	75 min	10	'The Archaeological Quest'
Families (diversed)	Small	No	Weak	Free	No assistance	Without video clip	Family	35 min	16	'The coins challenge'
Children	Medium	Yes	Medium	Free	Distant	Without video clip	Multi-player	50 min	16	'Greek Experience'

elements and study their impact on players' motivation and learning [37, 38, 41–43, 51]. This study is a direct result of such trends. Our objective was to examine the participants' performances and the game components in order to better understand the elements of the game and their effect on motivation. We explored these issues using Big Data that included the completion rates of thousands of participants, collected over 5 years, from nine learning games offered to museum visitors. According to our results, some of these gameplay elements had a large and significant impact, while some only had an only minor impact.

5.1 Game Mode

According to our analysis, players playing in the 'Multi-player' mode completed a higher number of stations than the 'Family' mode players, and in the multivariate analysis, the variable predicted (albeit weakly) the performance of the participants. These findings fit our observations in the museum, but unfortunately, we were unable to determine, based on the data available, the cause behind it. This issue can be examined in future research.

5.2 Instructions and Assistance

We found that the addition of a video clip to the opening instructions improved participants' performance and that this variable was found as a predictor in the 'Family' mode in the multivariate analysis. These results are in line with the findings of Liao et al. [55] who found that the use of video instruction improved participants' achievement in a digital learning game. Note, however, that the effect of opening instructions did not significantly contribute to the analysis of the factors predicting participants' performance using multivariate analysis. In addition, and contrary to the results of Chen and Law [47], our analysis showed that people who received remote assistance completed a significantly higher (almost double) number of stations and that this variable was the strongest predictor of the results for 'Group' mode participants. Based on our observations of game players, we suspect that distance supervision led to a sense of freedom, higher motivation, and better performance.

5.3 Competition

Competition was found to not be a predictor of the players' performance. Our findings are interesting in the light of Van Eck and Dempsey's [56] research, which showed a connection between assistance and competition. Sadly, our data prevented us from comparing our results to those of Van Eck and Dempsey [56], but our findings regarding the effectiveness of remote assistance (compared to close assistance) suggest a similar direction.

5.4 Narrative

Our finding showed that narrative was not a predictor for 'Family' mode. Another variable we looked at was the depth of the narrative. The multivariate calculation, however, found that the narrative depth parameter predicted participants' performance. When we examined this parameter separately in the 'Family' and 'Group' mode, we found that stronger narrative depths had a slight positive effect on participants' performances. We emphasize, however, that the effect was small.

5.5 Size of the Gaming Area

Our analysis revealed that there was no significant impact of the variable on the number of stations completed by the participants, nor on the prediction of performance. However, when we analyzed the effect of the variable on the 'Family' and 'Group' modes, we found that the size of the area had a weak (but significant) effect on the stations completed by participants (which indicated that a larger game area, and therefore a greater challenge level by our definition, would result in fewer completed stations). Furthermore, the multivariate test showed that the size of the game area plays a key role in predicting the performance of the participants in the game, especially in a 'Family' mode. We hypothesize (based on observations and conversations with participants) that larger area sizes were hard for participants (walking times, difficulty navigating, and longer playing time) and consequently led to a slower pace and even abandonment of the game. There is widespread agreement that a challenge must be suitable to a participant's abilities and not take too long to complete [2, 57]. In that sense, our large area games were probably too difficult and lead to weaker game performance.

In summary, this study's findings align with previous studies mentioned above. We want to emphasize that we chose to study players' activity in educational games based solely on actual game tracks, a relatively innovative methodology. This methodology allowed us to conduct causal (rather than correlative) research, as is often the case in perception studies. These data, collected from an automated system that documented players' performance, allowed us to investigate the impact of various instructional and gameplay elements in a way that attempts to clarify the impact of participants' perceptions on outcomes and to examine their performance objectively. Hopefully, this study will help game developers, and educators better understand the components of learning-learning experiences by increasing our understanding of learning games. This will undoubtedly result in better activities for active participation, enjoyment, learning, and a fascinating experience. Finally, our findings and observations suggest many nuances in player behavior that require further research. Also, we examined all nine games in this study as a whole. Further research should examine each of these games separately.

References

1. Prensky, M.: Digital Game-Based Learning. McGraw-Hill, New York (2001)
2. BECTa: Computer games in education project. becta.org.uk [Online]. https://cibermemo.files.wordpress.com/2015/12/edujoc2004.pdf. Accessed 13 Mar 2022
3. Malone, T.: What makes things fun to learn? A study of intrinsically motivating computer games. Cognitive and Instructional Sciences Series. Xerox, Palo Alto Research Center, California (1980)

4. VanDeventer, S.S., White, J.A.: Expert behavior in children's video game play. Simul. Gaming **33**(1), 28–48 (2002)
5. Kamboj, A.K., Krishna, S.G.: Pokémon GO: an innovative smartphone gaming application with health benefits. Primary Care Diabetes **11**(4), 397–399 (2017)
6. Page, Z.E., Barrington, S., Edwards, J., Barnett, L.M.: Do active video games benefit the motor skill development of non-typically developing children and adolescents: a systematic review. J. Sci. Med. Sport **20**(12), 1087–1100 (2017)
7. Birk, M., Wadley, G., Abeele, V., Mandryk, R., Torous, J.: Video games for mental health. Interactions **26**(4), 32–36 (2019)
8. De Sousa Borges, S., Durelli, V., Reis, H., Isotani, S.: A systematic mapping on gamification applied to education. In: Proceedings of the 29th Annual ACM Symposium on Applied Computing, pp. 216–222 (2014)
9. Quellmalz, E.S., Timms, M.J., Schneider, S.A.: Assessment of student learning in science simulations and games. Paper commissioned for the committee on the learning science: computer games, simulations, and education of the National Academics Board of Science Education. Academia (2009). https://www.academia.edu/15333866/Assessment_of_Student_Learning_in_Science_Simulations_and_Games. Accessed 13 Mar 2022
10. Chang, K.E., Wu, L.J., Weng, S.E., Sung, Y.T.: Embedding game-based problem-solving phase into problem-posing system for mathematics learning. Comput. Educ. **58**(2), 775–786 (2012)
11. Da Rocha Seixas, L., Gomes, A.S., de Melo Filho, I.J.: Effectiveness of gamification in the engagement of students. Comput. Hum. Behav. **58**, 48–63 (2016)
12. Connolly, T.M., Boyle, E.A., MacArthur, E., Hainey, T., Boyle, J.M.: A systematic literature review of empirical evidence on computer games and serious games. Comput. Educ. **59**(2), 661–686 (2012)
13. Su, C.H., Cheng, C.H.: A mobile gamification learning system for improving the learning motivation and achievements. J. Comput. Assist. Learn. **31**(3), 268–286 (2015)
14. Yildirim, I.: The effects of gamification-based teaching practices on student achievement and students' attitudes toward lessons. Internet High. Educ. **33**, 86–92 (2017)
15. Sitzmann, T.: A meta-analytic examination of the instructional effectiveness of computer-based simulation games. Pers. Psychol. **64**(2), 489–528 (2011)
16. Wouters, P., Van Nimwegen, C., Van Oostendorp, H., Van Der Spek, E.D.: A meta-analysis of the cognitive and motivational effects of serious games. J. Educ. Psychol. **105**(2), 249–265 (2013)
17. Tokac, U., Novak, E., Thompson, C.G.: Effects of game-based learning on students' mathematics achievement: a meta-analysis. J. Comput. Assist. Learn. **35**(3), 407–420 (2019)
18. Lamb, R.L., Annetta, L., Firestone, J., Etopio, E.: A meta-analysis with examination of moderators of student cognition, affect, and learning outcomes while using serious educational games, serious games, and simulations. Comput. Hum. Behav. **80**, 158–167 (2018)
19. Dicheva, D., Dichev, C., Agre, G., Angelova, G.: Gamification in education: a systematic mapping study. Educ. Technol. Soc. **18**(3), 75–88 (2015)
20. Byun, J., Joung, E.: Digital game-based learning for K–12 mathematics education: a meta-analysis. Sch. Sci. Math. **118**(3–4), 113–126 (2018)
21. Acquah, E.O., Katz, H.T.: Digital game-based L2 learning outcomes for primary through high-school students: a systematic literature review. Comput. Educ. **143**, 103667 (2020)
22. Iacovides, I., Cox, A.: Moving beyond fun: evaluating serious experience in digital games. In: Proceedings of the 33rd Annual ACM Conference on Human Factors in Computing Systems, pp. 2245–2254 (2015)
23. Davidson, D.: Beyond Fun: Serious Games and Media. ETC Press, Pittsburgh, PA (2008)
24. Van Roy, R., Zaman, B.: Need-supporting gamification in education: an assessment of motivational effects over time. Comput. Educ. **127**, 283–297 (2018)

25. Hanus, M.D., Fox, J.: Assessing the effects of gamification in the classroom: a longitudinal study on intrinsic motivation, social comparison, satisfaction, effort, and academic performance. Comput. Educ. **80**(Part C), 152–161 (2015)
26. Barata, G., Gama, S., Jorge, J., Goncalves, D.: Engaging engineering students with gamification. In: 2013 5th International Conference on Games and Virtual Worlds for Serious Applications (VS-GAMES), pp. 1–8 (2013)
27. Sweetser, P., Wyeth, P.: GameFlow: a model for evaluating player enjoyment in games. Comput. Entertain. (CIE) **3**(3), 1–24 (2005)
28. McMahan, A.: Immersion, engagement, and presence: a method for analyzing 3-D video games. In: Wolf, M., Perron, B. (eds.) The Video Game Theory Reader, pp. 67–86. Routledge, New York (2003)
29. Schoenau-Fog, H.: The player engagement process—an exploration of continuation desire in digital games. In: Proceedings of think design play—5th international DiGRA conference, pp. 14–17 (2011)
30. Cheng, M., Lin, Y., She, H., Kuo, P.: Is immersion of any value? Whether, and to what extent, game immersion experience during serious gaming affects science learning. Br. J. Educ. Technol. **48**(2), 246–263 (2017)
31. Garris, R., Ahlers, R., Driskell, J.E.: Games, motivation, and learning: a research and practice model. Simul. Gaming **33**(4), 441–467 (2002)
32. Gee, J.P.: What Video Games Have to Teach us About Learning and Literacy (Revised and Updated Edition). Palgrave Macmillan, New York, N.Y (2014)
33. Hamari, J., Koivisto, J.: Why do people use gamification services? Int. J. Inf. Manag. **35**(4), 419–431 (2015)
34. Hung, C.Y., Sun, J.C.Y., Yu, P.T.: The benefits of a challenge: student motivation and flow experience in tablet-PC-game-based learning. Interact. Learn. Environ. Learn. Technol. Learn. Environ. **23**(2), 172–190 (2015)
35. Clark, D.B., Tanner-Smith, E.E., Killingsworth, S.S.: Digital games, design, and learning: a systematic review and meta-analysis. Rev. Educ. Res. **86**(1), 79–122 (2016)
36. Hew, K.F., Huang, B., Chu, K.W.S., Chiu, D.K.: Engaging Asian students through game mechanics: findings from two experiment studies. Comput. Educ. **92**, 221–236 (2016)
37. Sailer, M., Hense, J.U., Mayr, S.K., Mandl, H.: How gamification motivates: an experimental study of the effects of specific game design elements on psychological need satisfaction. Comput. Hum. Behav. **69**, 371–380 (2017)
38. Vandercruysse, S., Vandewaetere, M., Cornillie, F., Clarebout, G.: Competition and students' perceptions in a game-based language learning environment. Educ. Tech. Res. Dev. **61**(6), 927–950 (2013). https://doi.org/10.1007/s11423-013-9314-5
39. Werbach, H., Hunter, D.: For the Win: How Game Thinking Can Revolutionize Your Business. Wharton Digital Press, Philadelphia (2012)
40. Aldemir, T., Celik, B., Kaplan, G.: A qualitative investigation of student perceptions of game elements in a gamified course. Comput. Hum. Behav. **78**, 235–254 (2017)
41. Brom, C., Buchtová, M., Šisler, V., Děchtěrenko, F., Palme, R., Glenk, L.M.: Flow, social interaction anxiety and salivary cortisol responses in serious games: a quasi-experimental study. Comput. Educ. **79**(Part C), 69–100 (2014)
42. Chen, C.H., Shih, C.C., Law, V.: The effects of competition in digital gamebased learning (DGBL): a meta-analysis. Educ. Technol. Res. Dev. **68**(4), 1855–1873 (2020)
43. Jemmali, C., Bunian, S., Mambretti, A., El-Nasr, M.S.: Educational game design: an empirical study of the effects of narrative. In: Proceedings of the 13th International Conference on the Foundations of Digital Games, pp. 1–10 (2018)
44. Hébert, C., Jenson, J.: Digital game-based pedagogies: developing teaching strategies for game-based learning. J. Interact. Technol. Pedagog. (15) (2019) [Online]. https://jitp.

commons.gc.cuny.edu/digital-game-based-pedagogies-developing-teaching-strategies-for-game-based-learning. Accessed 14 Mar 2022

45. Huizenga, J.C.: Digital game-based learning in secondary education. Doctoral dissertation, University of Amsterdam, Netherlands (2017)

46. Marklund, B., Taylor, A.S.: Teachers' many roles in game-based learning projects. In: European Conference on Games Based Learning, pp. 359–367 (2015)

47. Chen, C.H., Law, V.: Scaffolding individual and collaborative game-based learning in learning performance and intrinsic motivation. Comput. Hum. Behav. **55**, 1201–1212 (2016)

48. Hernández, A.B., Gorjup, M.T., Cascón, R.: The role of the instructor in business games: a comparison of face-to-face and online instruction. Int. J. Train. Dev. **14**(3), 169–179 (2010)

49. Garneli, V., Giannakos, M., Chorianopoulos, K.: Serious games as a malleable learning medium: the effects of narrative, gameplay, and making on students' performance and attitudes. Br. J. Educ. Technol. **48**(3), 842–859 (2017)

50. Malegiannaki, I.A., Daradoumis, T., Retalis, S.: Teaching cultural heritage through a narrative-based game. J. Comput. Cult. Herit. (JOCCH) **13**(4), 1–28 (2020)

51. Cagiltay, N.E., Ozcelik, E., Ozcelik, N.S.: The effect of competition on learning in games. Comput. Educ. **87**, 35–41 (2015)

52. Cornillie, F., Clarebout, G., Desmet, P.: Between learning and playing? Exploring learners' perceptions of corrective feedback in an immersive game for English pragmatics. ReCALL **24**(3), 257–278 (2012)

53. Csikszentmihalyi, M.: Flow: The Psychology of Optimal Experience. Harper Perennial, New York (1991)

54. Hershkovitz, A., Alexandron, G.: Understanding the potential and challenges of big data in schools and education. Tendencias pedagógicas **35**, 7–17 (2020)

55. Liao, C.W., Chen, C.H., Shih, S.J.: The interactivity of video and collaboration for learning achievement, intrinsic motivation, cognitive load, and behavior patterns in a digital game-based learning environment. Comput. Educ. **133**, 43–55 (2019)

56. Van Eck, R., Dempsey, J.V.: The effect of competition and contextualized advisement on the transfer of mathematics skills in a computer-based instructional simulation game. Educ. Technol. Res. Dev. **50**(3), 23–41 (2002)

57. Denisova, A., Cairns, P.: Adaptation in digital games: the effect of challenge adjustment on player performance and experience. In: Proceedings of the 2015 Annual Symposium on Computer-Human Interaction in Play, pp. 97–101 (2015)

Creating a Learning Environment for the Fifth Industrial Revolution

Crystal Loose[1(✉)], Michael Ryan[2], and Rose Jagielo-Manion[1]

[1] West Chester University, West Chester, PA 19383, USA
{ccloose,rjagielo-manion}@wcupa.edu
[2] Deleware State University, Dover, DE 19901, USA
mryan@desu.edu

Abstract. It has been argued that we have moved into the age of personalization. One can see this while ordering drinks at a local Starbucks where options are limitless. This personalization has been called the Fifth Industrial Revolution, a time noted for a deep, multi-level cooperation between people and machines. With emphasis on innovation, purpose, and inclusivity, this revolution calls for changes in the classroom setting to focus on relationships between human needs and lived experiences. So how do we prepare our students for this reality? Methods of instruction that create an engaging and collaborative learning community need to be considered when designing classroom experiences. The five facets of personalized learning will be examined through the lens of student research and application. Examples of products and the process in which they were created will be shared including Padlet, Mentimeter, and Google Jamboard. The learning instruments introduced in this paper will demonstrate ways to encourage individual and collaborative reflection in stimulating learning environments. Situated Learning Theory will be used to explain the necessity of learning in context for pre-service teachers.

Keywords: Fifth industrial revolution · Personalized learning · Technology tools

1 Introduction

Challenges in education confronted during the pandemic have caused teachers and researchers to study the way we have been instructing students both present day and prior to the COVID-19 pandemic. Although the pandemic inflicted setbacks in academic achievements, lessons learned about classroom environment will challenge current perspectives and future discussions about the way students learn. Connections made to recent realizations in the literature regarding personalized learning experiences, both the benefits and challenges, have cautioned educators and instructors in higher education to consider classroom experiences necessary to prepare our students for the Fifth Industrial Revolution (FIR) [25].

With investment in improving educator preparation during pre-service training, the authors examined teacher candidate training and the immediate impact on their comfort levels upon entering the workforce. Moreover, recent evidence suggests that the context

© The Author(s), under exclusive license to Springer Nature Switzerland AG 2023
D. Guralnick et al. (Eds.): TLIC 2022, LNNS 581, pp. 225–234, 2023.
https://doi.org/10.1007/978-3-031-21569-8_21

of teacher preparation is associated with important consequences for their later employment outcomes [5]. It has been determined that authentic learning environments can influence graduate success.

In recent years, there has been a growth in personalized learning contexts to match industry models such as Starbucks and Spotify. Personalized education has shown positive impacts on student education [10]. With this in mind, we can strengthen student academic achievement, through personalized in-depth training and reflective experiences [8].

Furthermore, the lessons learned because of the pandemic should inform educators' readiness to cultivate learning experiences for students who are going to be living in the FIR and beyond. The FIR represents a synergy between technology and people, developing what is being called cobots where people are interacting with and informing the ways that technology completes specific tasks [19]. This requires a workforce who are truly lifelong learners. Industry 5.0 calls for people who can leverage technological tools as they engage with data and work to solve problems to meet the specific needs or demands of society [15]. Students today need learning opportunities that help them to think holistically, make critical assessments, and decisions based on a full understanding of different situations. While these themes are certainly noted in the literature pertaining to educational opportunities, many environments are still designed to deliver learning better suited for the first industrial revolution rather than stay aligned to current needs [27].

Responsibility for preparing students to enter the workforce armed with the tools and skills for career success rests on university education programs. This paper will address background related to the FIR and personalized learning and discuss how programs can use this knowledge to successfully prepare educators for classrooms of today and tomorrow. With emphasis on innovation and inclusivity, strategies for collaboration and engagement in the classroom setting will be examined. Examples of specific technological tools that can be utilized to encourage all voices and varying perspectives to be heard, as well as ways to support both individual reflection and collective understanding of ideas will be presented.

1.1 Fifth Industrial Revolution

Rapid technological advances as well as continuing changes in industry shape the way we live. These shifts impact our lives in all ways and influence how we live, work, and behave [23, 25]. This creates a demand for a more dynamic workforce prepared to meet the challenges of the day rather than individuals working on isolated tasks using only typical methods. The learning process required to be responsive to these needs are shaped by the many technological breakthroughs, as well as the shifts in tasks required to meet the needs and demands of today's society [23]. Historically, schools and other learning institutions have been slow to shift and respond, often being accused of not adequately preparing individuals for the demands of the life and job market they will face.

Over the course of the last century, we have moved from an industry focused on standardization and mechanical production to the influence of technology becoming an integrated and essential part of our lives. These shifts have changed the nature of work and created a demand for learners who could actively solve problems using creative thinking

[23, 25]. The advent of the FIR takes the concept of the digitalisation of our lives to one where all technology is used strategically to personalize products and services to meet the varied needs of individuals [17]. Industry 5.0 increases the collaboration between humans and technology establishing a new role for each and identifying the need for different types of skills. In this framework, machines assume the role of repetitive and monotonous tasks and humans dedicate their time to recognize needs, apply creative reasoning and guide the application of technology to customize services and products [17, 25]. "This demand of human touch will be raising in the future much more because consumers seek to express their individuality through the products they buy" [17].

These shifts require educators to think critically about the ways they create learning experiences that provide students with opportunities to develop creative thinking and problem-solving skills. The structure of our current educational system is still set up to meet the needs of an industry focused on mass production, arguably meeting goals established by the first and second industrial revolutions [2]. Unfortunately, it is well documented that the culture of school institutions supports status quo thinking, making changes in practices challenging [4]. Additionally, the interplay of society, politics and student needs makes changes in the educational system complicated. While there may be a similar goal for education, each group often pushes for varied and conflicting reforms, all demanding immediate results [4]. Often there has not been sufficient time devoted to building understanding, creating capacity, and ensuring that there are adequate resources to enact change. As a result, most educational reforms over the past fifty years have failed to gain any type of traction (Fullan, 2016) [4].

Clearly, educational institutions have been slow to change. Most educators view change through a skeptical lens that includes a lack of sufficient supports and resources available to adequately implement new ideas and practices [4]. Consider that despite the advancement in technology during the past twenty years there has been a reluctance in acquiring, accepting, and using technology to facilitate teaching and learning [13, 16]. We contend that this reticence to change has allowed educational institutions to lag behind the demands of the workforce. The COVID-19 pandemic, however, created a critical incident for education that forced everyone to rethink their practices and make immediate changes [13, 29]. While there was much challenge during this time, there was also a need that led to greater interest in learning more about online and distance learning techniques [13, 29].

As we think about making the necessary changes to support learner needs for the FIR and beyond, we must identify practices that training facilitators can and should use to develop learners' skills in processing information, engaging in productive struggle, and thinking creatively. This gives us an opportunity to reimagine what and how we teach [29]. To best prepare learners to thrive in this new environment, we must engage students and teachers in critical dialogue about how and why we use technology, focusing on how it helps to support or meet a need [14, 23, 25]. This will help to bring meaning to the use and application of technology. Curricular changes should support learning experiences that are situated within a narrative that provides a context for thinking, problem solving and learning [2, 23]. The learning experiences need to be tailored to the specifics of individuals, context, problem or group, fostering engagement, a focus on personalization, and guiding students to develop the competencies needed for their future [14, 29].

Creating learning experiences that are future ready means that educational institutions need to find better ways to process changes to meet the needs of a globally interconnected world. We might consider starting by reframing our collective view of environments to be more of a learning organization that fosters lifelong inquiry and encourages both educators and learners to continually expand their thinking [2, 14]. That type of change needs to happen on a scale that impacts all facets of society and works to change mindsets and beliefs about the running and function of educational institutions [2]. However, we can start by making small changes one classroom and educator at a time. In the following sections, we offer suggestions we feel educators can use to help enhance their practices to help prepare future ready learners who can be successful in this and future industrial revolutions.

1.2 Personalized Learning in the Classroom

The world is changing, and students will no longer succeed by regurgitating facts. In the age where the answer to almost anything can be found on a handheld device, we need to change the way we present information to our students in higher education as we prepare them to enter the work force. Educators need to reimagine their curriculum post-pandemic to address not only social justice and empathy, but for inclusion of opportunities that allow for personal voice, choice, and active engagement. Classrooms of late have embraced reforms that move teaching and learning toward more personalized learning (PL) approaches [18]. PL reconsiders the traditional classroom setting with increased personalization based on student achievement, interest, and learning styles.

Findings from recent research confirm that optimal learning conditions include environments that give students choice, ownership, and voice through authentic learning experience [24]. Through personalized, flexible learning environments, educators meet each student within their own zone of proximal development. This zone, studied by Vygotsky, places student learning in an environment where scaffolded instruction supports higher level thinking [26]. This environment will be discussed through a lens that is applicable to higher education or workplace training settings. Although our research pertains to teacher preparation programs, the suggestions we make can be used to improve any classroom, curricular, or workplace training modification. It is in the university settings where change can be made in the way we are preparing our students to meet the needs of work force goals in the FIR. Through immersion in PL techniques within higher education courses, students not only learn the meaning but experience methods that they can utilize in real-life circumstances. PL environments reconsider the use of assessments and data, instructional rigor, student agency, project-based learning, and technology to support student learning paths while cultivating creative learning experiences [24].

For the intent of this paper, the authors have considered 5 facets of PL including classroom culture, assessments, instructional rigor, equity in education, and student agency. Each of these areas will be described with emphasis on implementation in the classroom or professional learning setting, with additional emphasis on contextual design. First, creating a classroom culture that inspires learning is the goal of education, as without a safe environment learning will not transpire. A student-driven model of education should empower learners to pursue their goals, investigate problems, design solutions, and develop curiosities that foster creative thinking and dialogue. All essential ingredients of a setting conducive to the FIR. The design of such a space should support peer accountability and self-assessment. Fostering an environment of collaboration means that not only is the group benefiting from each other, but there is awareness of how to support one another when there are trials. This type of ecosystem is not possible without the establishment of trust, where it is understood that it is okay to make and learn from mistakes. Promoting responsibility for learning means that there is a clear understanding of how to evaluate peers and self, and how to identify and locate resources to fill identified learning gaps. Hence the need to consider assessments, the second facet of PL.

Assessments should be meaningful and connected to content. There are several methods to assess students, including summative evaluations, necessary to document student foundations for learning and growth. However, it is necessary to consider what to do with this information. Using it to foster future learning opportunities will support the growth found in a classroom that strives to personalize experiences for students. The most successful classrooms utilize deliberate systems and structures to ensure that each student is afforded opportunities for higher levels of learning [22]. If students are to truly personalize their learning process, consideration of project-based learning experiences where students choose a topic and project idea will encourage innovative, choice-oriented ways to display learned content. When designing this form of assessment, instructors should consider a list of projects that would satisfy requirements for students to display their understanding of particular content areas. Through this process, students not only display their understanding but also become part of the learning process. As students develop projects pertaining to their area of study, they are empowered to take control of their own learning, while self-monitoring along the way. According to Hattie [6], students that contribute to their own learning process pave the way toward a highly effective learning environment. Project-based learning allows students to engage in work that connects to the world beyond their school and supports their interests. It supports student learning in the FIR as they engage with tools and technologies that foster creativity and problem solving. This type of learning lends itself to increased instructional rigor, the third facet of PL.

Instructional rigor found within classrooms may consider varied learning experiences, differentiated learning objectives, and personalized pathways. Rigor puts responsibility of learning on students and not the teacher. The teacher facilitates initial learning while students investigate and discover on their own. Instructional rigor can include the use of technology while fostering an environment that embellishes on meaningful student choice. This may stem from use of inquiry questions, media, or technology to explore within independent or collaborative groups that work to create a final product. This could center around project-based learning projects that result in authentic sharing of work. To support student learning, the instructor may have prepared co-planning tools, increased amounts of mini-lessons, conferencing techniques, modeling opportunities, and numerous accounts of feedback. Within the classroom setting, rigor is promoted as students apply what they have learned to new situations and varied learning experiences. Authenticity is at the forefront of learning through opportunities to connect to real-life experiences. For example, while completing a science lab, students may connect to digital simulations that offer enhanced conceptual understanding of science lessons and critical thinking skills. Furthermore, community connections are essential and should be part of the learning process that foster experience outside the classroom. To support PL and student engagement, differentiation is used to create opportunities for all types of students. Students are encouraged to follow their own interests and choose a learning pathway that is of particular interest to them. Instructional rigor is not possible without the consideration of PL experiences that create equitable opportunities in education.

Equitable PL experiences should be student centered, culturally responsive, with awareness of students' identities and needs. In an article by Bree Picower [20], six key frameworks are suggested for social justice curriculum design including self-love and knowledge, respect for others, issues of social injustice, examples of social movement and change, awareness raising, and social change. She suggests that preservice teachers are often exposed to social justice education theories but feel overwhelmed when considering how to use these theories when lesson planning [20]. Contemporary classrooms should consider cultural heritage projects, examination of history and the influence on present day situations, current events, and community connections. Through experiences in the field of education, we have found that instructional rigor is influenced using technology. However, not all students have equitable access to technology or adequate training in how to use technology in the classroom setting. PL strategies are not possible without student agency. Students, although empowered by instructors, need to take responsibility for their own learning.

Student Agency can be defined as the ability to take ownership and responsibility for learning. In a personalized space, this is developed through the establishment of relationships. Teachers form bonds with students through supportive environments that acknowledge and respect the individual and collective identity/identities in the community. Feedback within this community is not only between teacher and student but peer to peer to facilitate self-directed learning. Based on student goals, learning experiences are developed and carried out to satisfy personalized pathways. Students are encouraged to advocate for personal interests, as well as things that support the greater community and world.

Each of these areas will be further addressed as we consider technological tools to enhance classroom environments while considering the FIR and ways to increase intentional use of technology.

1.3   Adult Learning Principles

In addition to considering research related to the FIR [17] and PL [10], it was important for the authors to ascertain the challenges, needs, and goals of student learning with emphasis on adult learning principles. Recent needs and trends pointed to changing instruction to adapt to the new and most likely permanent place of virtual instruction, consideration of culturally diverse materials, the importance of transferring instruction to real world experience, and on teaching in inclusive environments.

Adult learning principles were considered while prompting changes in instruction and classroom environments. Adult learning theory, or andragogy [7] highlights the importance of students being in control of their own learning. Adult learners are self-directed and self-dependent, they assume a problem-centered approach to learning and, consequently, are typically internally motivated [7]. In fact, active adult learners who take initiative in learning tend to learn more and learn better than passive learners [11]. Research also supports self-directed learning as a vehicle for critical awareness [3, 11].

Andragogy fuses nicely with all aspects of PL as learners are asked to participate in self-directed learning to explore and expand upon topics of interest to them. As we strive to make changes to instruction and prepare students at the university level for future environments that are reflective of the FIR, it is necessary to consider facets of adult learning. Situated Learning Theory (SLT) gives meaning to learning in context and views learning as a recursive process that occurs through participation in a collaborative environment where adults act in and with context and tools [9]. SLT informs learning that is not only context dependent but rooted in the situation in which a person participates with support and guidance [12]. The authors provided authentic learning experiences to support learners as they tried new pedagogical tools. With emphasis on SLT, students were instructed on how to incorporate new instructional strategies by using the platforms in a classroom directed by their professors.

Intentional Use of Technology to Engage and Prepare. When considering the FIR [17], PL [10], andragogy [7], SLT [9], and what this means for the classroom content, the authors confronted the question of how can instructors leverage technology as a tool to not only engage, but to also foster innovation, purpose, collaboration, and critical awareness [28]. To maximize student learning, instructors must make intentional decisions related to learning outcomes and develop "constructive" and "interactive" learning activities involving the use of technology for engagement, problem-solving, and/or collaboration [21]. Moreover, in contemplating which specific technological tools might be most effective, factors such as anonymity of participant responses, opportunities for individual reflection and/or collaborative thinking, sharing of varying perspectives and/or experiences, and whether a process- or product-driven use of technology is best must also be addressed. Responses to these questions allow the instructor to create opportunities that maximize engagement and learning.

Many websites or applications, like Mentimeter, promote learner engagement through the use of a variety of response types. Participants can respond to open-ended questions, multiple choice questions, polls, and can even submit their own questions. Because participant responses are displayed anonymously in real-time, such technology tools encourage participants to be actively engaged in individual reflection and sharing of perspectives and experiences. At the same time, these tools can also foster collaborative understanding, often through intentional planning and facilitation by the instructor. Furthermore, Mentimeter provides instructors with the ability to export and analyze participants' responses after the learning experience to guide future instruction. Similarly, applications like Google Jamboard offer learners the opportunity to anonymously participate in individual reflection and collaborative interaction by adding responses using sticky notes, text boxes, images, or through drawing. The instructor serves as a facilitator of learning as appropriate using these tools.

One example of the use of Mentimeter to personalize learning for students was with an implicit bias workshop where they were asked to share their individual thoughts and reactions to displayed images while also analyzing and reflecting on the whole group's responses. This multi-layered approach asked learners to think about their own perspectives and experiences as well as consider their impact on others. This activity encouraged critical, individual, and collaborative reflection and discussion about how it is essential to understand the process for building and participating in culturally responsive practices.

Other technology tools offer flexibility for both process- and product-driven instructional purposes. For instance, Padlet can be used for personalized, self-directed aims while also inviting collaborative thinking. Participants add their thoughts and reactions to a topic or prompt on a virtual wall using words, images, links, or videos, and others are invited to respond to or comment on these posts. The Padlet wall can then serve as a future resource for participants. Using technology in this way supports choice, collaboration, and contextual learning to foster engagement and encourage deep, thoughtful reflection.

After reading about and discussing equitable teaching and learning, for example, the authors asked students to post their own personal goal(s) related to their own careers. Then, participants added resources to support each other's goals, including websites, literature, and instructional activities, to a Padlet wall. This use of Padlet included active student engagement through personal reflection and collaboration on how to best support diverse student populations. Additionally, Padlet was used to create a more personalized learning purpose. Students were instructed to choose a learning activity from a variety of suggestions to delve more deeply into a course topic related to diversity, equity, and inclusion. Suggestions included listening to a podcast, watching a documentary, interviewing a teacher, or attending a webinar or presentation on campus. After engaging in the activity, first year students were instructed to complete a "what, so what, now what" post on Padlet reflecting on what they learned from their activity and how they will apply this in their own lives. This activity cultivated personalized learning through a meaningful and authentic learning opportunity related to their future career or educational journey [24].

Through these intentional and focused uses of technology, instructors were able to blend principles related to PL, andragogy and the FIR while also addressing career preparation. As research suggests, these learning experiences were designed to meet the specific needs of students in personal, engaging, and collaborative ways to expand the competencies required for their future career [14, 29].

2   Conclusion

The FIR embraces the ubiquitous use of technology in our lives with the understanding that technology use must be guided by the depth and creativity of human thinking. Industry 5.0 celebrates the collaboration between humans and technology establishing a new role for each and presents us with a new set of competencies needed to be successful in the workforce. These include recognizing where machines can and should be used to complete repetitive tasks while humans dedicate their time to recognizing needs, creative thinking and identifying ways to customize services and products [17, 25]). This suggests a true need to make changes in curriculum and teaching practices to ensure that students are future ready [2, 16, 29]. Given the lessons learned about the possibilities for change from the COVID-19 pandemic we recognize that change and learning can happen in education [13, 29]. In this paper, we provided some suggestions for practices that can be used to foster the competencies that university students will need as they enter the workforce.

Acknowledgments. The authors wish to thank West Chester University for opportunities to investigate personalized learning strategies through a Provost Research Grant and research application opportunities in seminars and first year-experience courses.

References

1. DeMink-Carthew, J., Netcoh, S., Farber, K.: Exploring the potential for students to develop self-awareness through personalized learning. J. Educ. Res. **113**(3), 165–176 (2020). https://doi.org/10.1080/00220671.2020.1764467
2. Dintersmith, T.: What School Could Be. Princeton University Press (2018)
3. Freire, P.: Pedagogy of the Oppressed. Continuum International Publishing Group Inc., New York, MY (1970)
4. Fullan, M.: The New Meaning of Educational Change. Routledge (2016)
5. Goldhaber, D., Cowan, J., Theobald, R.: Evaluating prospective teachers: testing the predictive validity of the EdTPA. J. Teach. Educ. **68**(4), 377–393 (2017)
6. Hattie, J.: Visible Learning for Teachers. Routledge, New York, NY (2012)
7. Knowles, M.S.: Adult education: new dimensions. Educ. Leadersh. **33**, 85–88 (1975)
8. Kraft, M., Papay, J.: Can professional environments in schools promote teacher development? Explaining heterogeneity in returns to teaching experience. Educ. Eval. Policy Anal. **36**, 476–500 (2014)
9. Lave, J., Wenger, E.: Situated Learning: Legitimate Peripheral Participation. Cambridge University Press, New York (1991)
10. Lee, D.: How to personalize learning in K-12 schools: five essential design features. Educ. Technol. **54**(3), 12–17 (2014)
11. Loeng, S.: Self-directed learning: a core concept in adult education. Educ. Res. Int. 1–12 (2020). https://doi.org/10.1155/2020/3816132
12. Loose, C.: Practice-based professional development in education. IGI Global, Hersey, PA (2020)
13. Loose, C.C., Ryan, M.G.: Cultivating teachers when the school doors are shut: two teacher-educators reflect on supervision, instruction, change and opportunity during the covid-19 pandemic. In: Frontiers in Education, vol. 5, p. 231. Frontiers (2020, November)
14. Mogas, J., Palau, R., Fuentes, M., Cebrián, G.: Smart schools on the way: how school principals from Catalonia approach the future of education within the fourth industrial revolution. Learn. Environ. Res. 1–19 (2021)
15. Nahavandi, S.: Industry 5.0—a human-centric solution. Sustainability **11**, 4371 (2019). https://doi.org/10.3390/su11164371
16. Oke, A., Fernandes, F.A.P.: Innovations in teaching and learning: Exploring the perceptions of the education sector on the 4th industrial revolution (4IR). J. Open Innov. Technol. Market Complex. **6**(2), 31 (2020)
17. Paschek, D., Mocan, A., Draghici, A.: Industry 5.0—the expected impact of next industrial revolution. In: Thriving on Future Education, Industry, Business, and Society, Proceedings of the MakeLearn and TIIM International Conference, Piran, Slovenia, pp. 15–17 (2019, May)

18. Patrick, S., Worthen, M., Truong, N.: Rethinking state accountability to support personalized, competency-based learning in k-12 education. Issue Brief. iNACOL. Vienna, VA. Website: https://www.inacol.org. Last accessed 21 June 2017

19. Piacentini, R., Vega, M., Mujumdar, A.: Beyond industrial revolution 4.0: how industrial revolution 5.0 is related to drying technology. Dry. Technol. **39**, 4437–4438 (2021). https://doi.org/10.1080/07373937.2021.1875185

20. Picower, B.: Using their words: Six elements of social justice curriculum design in the elementary classroom. Int. J. Multicultural Educ. **14**(1), 1–17 (2012)

21. Sailer, M., Schultz-Pernice, F., Fischer, F.: Contextual facilitators for learning activities involving technology in higher education: the Cb-model. Comput. Hum. Behav. **121**, 1–13 (2021). https://doi.org/10.1016/j.chb.2021.106794

22. Stuart, T., Heckmann, S., Mattos, M., Buffum, A.: Personalized Learning in a PLC at Work. Solution Tree Press, Bloomington, IN (2018)

23. Sudibjo, N., Idawati, L., Harsanti, H.R.: Characteristics of learning in the era of industry 4.0 and society 5.0. Adv. Soc. Sci. Educ. Hum. Res. **372**(1), 276–278 (2019)

24. Thibodeauz, T., Harapnuik, D., Cumnmings, C.: Student perceptions of the influence of choice, ownership, and voice in learning and the learning environment. Int. J. Teach. Learn. Higher Educ. **31**(1), 50–62 (2019)

25. Voskoglou, M.G.: Thoughts for the future education in the era of the fourth industrial revolution. Am. J. Educ. Res. **8**(4), 214–220 (2020)

26. Warfold, M.K.: The zone of proximal development. Teach. Teach. Educ. **27**, 252–258 (2011)

27. World Economic Forum Annual Meeting, January 23–26 Davos-Klosters, Switzerland. https://www.weforum.org/events/world-economic-forum-annual-meeting-2018. Last accessed 7 Mar 2018/03/077

28. Young, S., Nichols, H.M.: A reflexive evaluation of technology-enhanced learning. Res. Learn. Technol. **25**, 1–13 (2017)

29. Zhao, Y., Watterston, J.: The changes we need: education post COVID-19. J. Educ. Change **22**(1), 3–12 (2021). https://doi.org/10.1007/s10833-021-09417-3

The 2CG® Poetry Machine—a Hybrid Approach to Human Capability Cultivation with Disruptive Artistic Impulses

Christina Merl[✉]

TalkShop/2CG®, 1190 Vienna, Austria
cm@christinamerl.com

Abstract. Human capability cultivation has been neglected in educational and professional contexts in the previous decades. The focus has clearly been on tasks and efficiency, on planning, scaling, and then executing according to plan. Professionals, students and pupils have become used to getting clear directions, simplified instructions and exact predictions and managers have continually looked for measurable output. Unrealistic goals imposed on students and workforce have increasingly led to frustration, drop-out and even burnout. Digitalization has added to this development. Cutting out all redundancy from our lives has prevented us from staying open to surprise, exploring unkown territory and embracing difference and diversity, whereby diversity does not only refer to color, race and sexual orientation but to different opinions and perspectives. As we live in a complex world and face increasingly complex challenges, however, we need to train these human capabilities, also referred to as 21st century skills. The current paper aims to introduce an educational approach that fosters human capability cultivation. Practice examples shall demonstrate how the 2CG® Poetry Machine, a disruptive educational model that combines artistic impulses with collaborative and social learning processes in communities of practice, enables learners across disciplines, cultures and hierarchies to practice their human skills, leave their mental models and tap into their full creative potential. The multi-method approach has been applied in in-person, online, live virtual and hybrid settings over the past 15 years. Evaluation has mostly been based on qualitative data collected through observation, customer surveys, peer feedback and expert opinion and analyzed by means of a customized value creation framework.

Keywords: 2CG® Poetry Machine · Human Capability Cultivation · 21st Century Skills · Community of Practice · Peer Exchange · Transdisciplinary Learning · Hyper-curriculum

1 Bringing Future Skills into Focus

1.1 Skills Needed to Meet Complex Challenges

We live in a complex world where dynamic variables interact in non-linear, non-predictable ways. Giving direction and effecting change in such complex environments

D. Guralnick et al. (Eds.): TLIC 2022, LNNS 581, pp. 235–247, 2023.
https://doi.org/10.1007/978-3-031-21569-8_22

requires a new approach to learning, problem-solving, collaborating, wellbeing and leading. What is more, we cannot draw on previous experiences to solve complex problems [1] such as climate change, the global pandemic, the threat of declining democracies and, last but not least, the impact of digitalization on how we will work and live in future. Navigating complexity [2] implies dealing with uncertainty, unpredictability and lack of control as well as not getting stuck in expectations that can no longer be met requires governments, organizations, educational institutions, societies and individuals to re-imagine leadership, to re-think educational concepts, to reframe uncertainty, and to focus on future skills development. Future skills, also referred to as 21st century skills, include learning skills (the 4 C's), literacy skills (IMT), life skills (FLIPS) [3] and 21st-century digital skills (see Table 1) [4], whereby digital literacy is seen as a mindset that enables learners to perform intuitively in digital environments, and to both easily and effectively access a wide range of knowledge embedded in such environments [4]. The Skills Framework as presented in Table 1 implies different skills models as there is no "one-fits-all" skills framework.

Table 1. 21st Century Skills Framework

Learning Skills or the 4 C's	
Critical thinking, creativity, communication, and collaboration	To acquire and build the 4 C's, learners need to train the mental processes that are required for connecting, collaborating, and complex problem solving, and which they will need to adapt and to improve upon modern work environments
Literacy Skills or IMT	
Information management, media literacy, and technology skills	Learners need to be able to discern facts, to validate information, and to transform data into effective stories. Emphasis needs to be put on determining trustworthy sources and to separate factual from misinformation that floods the Internet and print publications
Life Skills or FLIPS	
Flexibility, leadership, initiative, productivity, and social skills	These are needed to effectively build relationships, find purpose, adapt to constant change, and build inclusive, kind, sustainable and democratic societies
21st-Century Digital Skills	
Core 21st-century digital skills and contextual 21st-century digital skills	7 core skills (technical, information, management, communication, collaboration, creativity, critical thinking and problem solving) and 5 contextual skills (ethical awareness, cultural awareness, flexibility, self-direction and lifelong learning)

1.2   A Human-Centric Approach to Future Skills Development

The current paper aims to introduce an educational concept and approach aimed at supporting learners in developing 21st century skills and enabling them to leave their mental models and explore new pathways of thinking and doing. The 2CG® Poetry Machine is a human-centric multi-method approach to future skills development that delivers customized, emotionally engaging and collaborative learning experiences in in-person, live virtual, online, and hybrid settings [5] [6]. 2CG stands for content- and context-specific generic competency coaching and targets leaders, employees, apprentices, students and pupils across disciplines, cultures, generations and gender, who need to acquire 21st century skills to prepare themselves for dealing with complex challenges in future. The approach is grounded in learning science and in the fundamental pillars of communities of practice (CoP) as adapted from [7] (see Pillar 1– Pillar 4 in Table 2).

Table 2. CoP Pillars

CoP Pillar 1. Intrinsic motivation is crucial for learning	
Intrinsic Motivation	Learners are enabled to discover their passion and shared interest; they define their shared practice and find out what their purpose is. They need to define for themselves: What's the difference we want to make?
CoP Pillar 2. Learners engage in a shared practice	
Shared Practice	Learners continually reflect how their learning activities relate to their professional, real-life context and practice. Deep learning can happen when the learning is anchored in a shared practice, a shared interest, and a shared purpose
CoP Pillar 3. All learning is social	
Sense of Belonging	Learning is a social process, both in real life and in virtual settings. Social cohesion, trust and a sense of belonging are the foundation of fruitful learning and collaboration. For deep learning to take place, learners need to define how they want to work and communicate with each other
Pillar 4. Deep learning follows the principle of impulse—action—reflection	
Action Learning	Social constructivist learning processes are based on the action-learning principles of 'impulse—action—reflection' and incorporate the cognitive, emotional and psychomotoric dimensions of learning

1.3   Facilitated Learning Journey: Creating Value with Customized Content

The 2CG® Poetry Machine (see Fig. 1) is based on the assumption that surprising elements and disruptive impulses are an essential source of inspiration that can trigger new thinking in learners and thereby create added value for organizations and societies. It is linked to a professional practice and puts the needs of learners center stage. Elements of surprise as well as a variety of creative activities and somatic movement techniques

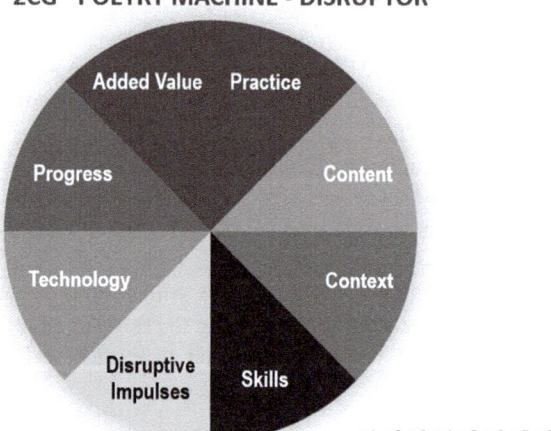

Fig. 1. The 2CG® Poetry Machine is linked to a professional practice and puts learners' needs center stage.

are aimed at enabling learners to tap into their full potential, which usually results in high learner engagement and positive learning outcomes.

Customized content and impulses from poetry, puppetry, literature, theater, film, painting and music are provided to learners in facilitated, tech-enabled in-person, virtual live, online or hybrid learning environments. These impulses are likely to trigger emotional reactions in learners, which can help them better understand and analyze complex issues [8], disrupt unhelpful thinking patterns, and leave their current mental models while developing future skills. A variation of fast and slow learning activities enrich the 2CG® learning journey (see Fig. 2).

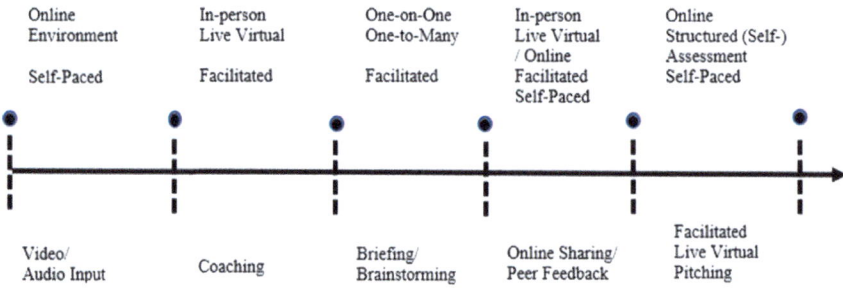

Fig. 2. Fast and slow learning activities provide 2CG® learning journey touchpoints.

It should be pointed out here that the 2CG® approach advocates for a mix of methods, and also for the freedom to create methods, rather than simply using pre-existing and fixed procedures [9] [6]. Starting points for 2CG® learning journeys can be half-day workshops, full-day workshops, deep dives, exploratory formats, retreats, semester courses, and ongoing community of practice meetings. Overall, the approach puts the needs of learners and learning organizations center stage and enables them to connect with their intuition, unlock their creative potential and embrace difference and diversity as a resource [10].

1.4 Value Creation Through Three-Dimensional Learning

The value creation cycles as defined in [11] and [12] (see Table 3) have provided a useful framework for analysing the impact of the three dimensions of 2CG® learning: 1. The practice layer (shared interest, shared undertaking, artefacts, outputs and outcomes); 2. The social collaborative dimension (peer learning, multiple learning loops, inputs); and 3. The reflective dimension of learning (evaluation and assessment) [13]. During their facilitated and tech-enabled learning journey, learners—and organizations—become aware of what skills they already have and what skills they need to further develop in order to stay relevant.

Table 3. Value Creation Framework

Cycle 1—Immediate Value	
Activities and Interactions	Networking, community activities and interactions have value in themselves
Cycle 2—Potential Value	
Knowledge Capital	Value of knowledge to be realized later—understanding and intention to apply relationships and resources; transformed ability to learn
Cycle 3—Applied Value	
Changes in Practice	Implementation of advice/insights; use of tools/innovation in practice
Cycle 4—Realized Value	
Performance Improvement	Reflection on effects on the achievement of what matters to stakeholders
Cycle 5—Reframing Value	
Redefining Success	Proposing new metrics for performance that reflect the new definition of success

1.5 Multiple Learning Loops Based on Disruptive Impulses

During their facilitated learning journey, 2CG® learners experience regular interactive learning sequences (see Fig. 3) that consider both, the learning process and the learning output, impact or outcome by following the action learning principles of 'impulse—action—reflection' [14]. These principles incorporate the cognitive, emotional and psychomotoric dimensions of learning, which can help balance social inequalities and different cognitive learning levels, as pointed out in [15]. Step by step, learners build awareness and skills, and they start to identify and analyze the problems they need to solve. Through multiple learning loops [16], they receive tailored feedback from their peers, facilitator and from experts (see Fig. 3). Like this, learners develop 21st century skills while being immersed in their practice and producing concrete outputs, such as prototypes, concepts, future scenarios, stories, manifestos, etc.

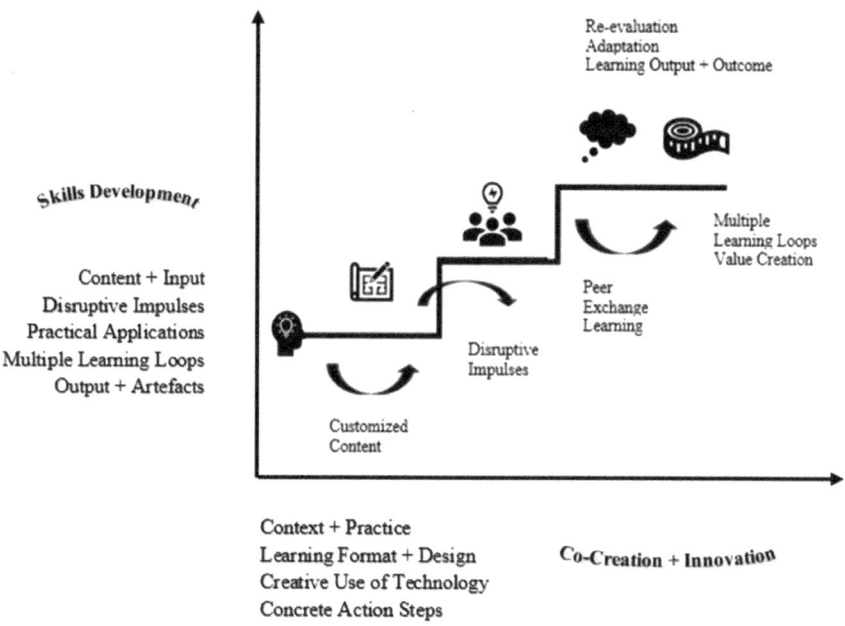

Fig. 3. Interactive learning sequences triggered by the 2CG® Poetry Machine.

The role of the 2CG® facilitator—coach and curator—is to provide guidance to learners and to support them on their exploratory and reflective learning journey. Technology is used creatively in that it supports collaborative and individual learning activities in in-person, live virtual, online, and hybrid settings.

2   2CG® Poetry Machine: Practice Examples

The 2CG® Poetry Machine is applied by decision-makers who know that their organizations and their people have to operate in a complex field of options to stay relevant. They make use of the Poetry Machine to enable their people to develop future skills and to explore new pathways of thinking and doing while sticking to the values of their organization, which can clearly provide a sense of stability and organizational coherence in a complex world.

Four practice examples are given in this paper to demonstrate how the 2CG® approach can support learners in building 21st century skills while exploring unknown territory. These practice examples were carried out in the past two years in different professional, educational and cultural settings and contain a variety of learning elements, such as in-person workshops, live virtual sessions, online learning modules and coaching sessions. Evaluation and value-tracking was mainly based on qualitative data gained through participant observation, learner feedback, participation and engagement and, most importantly, through change that could be effected in real life as a result of the learning experience. The main reporting methods were customer satisfaction surveys and written learner feedback; data gained from performance reviews; and personal stories about behavior change, implementation of new processes and tools in the professional context of learners. Quantitative data, such as completion rates of training, login rates, average time spent on the system, attendance, resource downloads and video views as provided by Moodle, social media and other technical solutions played a marginal role. They served as rough indicators of learning needs and learner priorities but were not of prior importance as they told little about the perceived value to participants and organizations (Table 4).

3   Evaluation and Qualitative Data Analysis

The 2CG® Poetry Machine practice examples given in this paper were implemented in the past two years in the framework of 1. a 24-day deep dive open to participants from different disciplines, cultures and hierarchies (age group: 30–63 yrs); 2. a combined technical apprenticeship and part-time bachelor study program for mechanical engineering (age group: 20–25 yrs) 3. a creativity workshop series in a technical college for programmers (age group: 17–18 yrs); and 4. a 3-day poetry and qigong retreat for leaders from different disciplines (age group: 50–65 yrs). They contained in-person and live virtual meetings, online modules and hybrid formats. Duration varied and consisted of half-day workshops, 3-day retreat, 24-day deep dive and a full semester course consisting of 8 half-day workshops. Participation was obligatory for bachelor students and college students; the other two formats were voluntary formats.

Evaluation was based on qualitative data analysis as laid out in [17] and on a value creation analysis through learning cycles as suggested in [11] and [12] (see Table 3). Data were collected by means of one-on-one interviews, anonymous learner feedback, learner surveys, focus groups, participant observation, traditional skills assessment practices, and expert opinions. More specifically, the data collected for evaluation and analysis of Practice Example 2 included video material, audio recordings, an electronic survey on subjective learner experience and satisfaction with the course content, impact, design and delivery as well as technological tools utilized. Learner assessment was based on traditional test and assessment methods via Moodle, oral presentations, audio and video recordings. Evaluation of Practice Example 3 was based on written learner feedback, a public workshop report, participant observation and expert opinion. Feedback questions ranged from personal satisfaction with the overall learning experience, skills acquired, course design, delivery, facilitation and technical equipment. Participants of the voluntary course formats—Practice Examples 1 and 4—gave personal feedback and participated in an anonymous online customer-survey that focused on the learning experience, impact, course content, design and delivery, facilitation, tools and technical equipment as well as the learning outcome.

Table 4. Practice Examples

Practice Example 1: Other Than Human: 24-Day Deep Dive	
Target Group Context + Practice	Mixed-gender, cross-generational group of learners across disciplines, cultures and hierarchies who wanted to connect with their creative power and find new inspiration through deep reflective practice, imagination and creativity exercises. (Age: 30–63)
Format + Assessment	Deep Dive format with a live virtual, in-person and self-paced online treasure hunt, object-based learning elements, artistic impulses and creativity techniques. Learner feedback, observation and expert opinion
Disruptive Impulse	Visual impulses as well as impulses from poetry and prose
Required Skills	Imagination, communication skills, empathy, critical thinking skills, collaborative team skills, digital literacy skills
Learning Output	Increased awareness of environmental issues; narrative of shared learning journey
Learning Impact	Participants improved their second language and reflective skills while trying out creative tools and reflective activities. They built awareness of environmental issues, discovered new facets of their personality and built confidence. To do so, they had to overcome their shyness and share their thoughts and ideas with people from different disciplines who they hadn't met before
Practice Example 2: Future of Gender Podcast, University of Applied Sciences	
Target Group Context + Practice	3rd semester mechanical engineering students of a part-time study programme that combines a technical apprenticeship with a bachelor's degree. This mixed-gender, cross-cultural group of learners needed to develop their soft skills with a focus on communication skills. (Age: 20–25)
Format + Assessment	8 half-day workshops (in-person, live virtual, online). The course was rated by means of learner feedback and a customer-survey. Student assessment was based on team assignment, individual assignment, and end-of-term test
Disruptive Impulse	Disruptive impulses included audio input, statistics, visuals, creativity techniques and reflective tasks
Required Skills	Language skills, communication skills, critical thinking skills, creativity, empathy, collaborative team skills, peer feedback skills, digital literacy, media literacy and reflective skills
Learning Output	Podcast episodes, video presentations, learning log
Learning Impact	Participants discovered new facets of their own personality, gained in self-confidence, built topic-specific awareness, and felt empowered

(*continued*)

Table 4. (*continued*)

Practice Example 3: Future Skills Workshop at Technical College	
Target Group Context + Practice	Pupils of 3rd form of a technical college/vocational school for computer science and programming. Learners were asked to come up with ten commandments for programming the future creatively and responsibly. (Age: 17–18)
Format + Assessment	Half-day workshop in hybrid format with on-site and live virtual facilitation. Learners had to submit a report and evaluated the workshop through an online survey and feedback
Disruptive Impulse	The narrative of Ian McEwan's novel "Man and Machine" served as framework for the workshop design. Learners had to complete reflective tasks and were exposed to works of art through a so-called 'knowledge huddle' aimed at triggering emotional reactions in learners
Required Skills	Critical thinking skills, empathy, creativity, imagination, collaborative team skills, connected thinking skills, communication skills, 21st-century digital skills
Learning Output	Workshop participants came up with an extensive catalogue of concrete action steps that need to be considered when creating a future that provides for democracy, equal opportunities and wellbeing in the era of AI
Learning Impact	Students realized that programming asks for social skills, such as empathy, critical thinking skills, collaborative team skills, creativity and communication skills. They connected with their creative power and intuition, which for many of them was an eye-opener
Practice Example 4: Poetry & Qigong Retreat	
Target Group Context + Practice	Western leaders across different disciplines who were ready to think and reflect on their decision-making and wellbeing strategy. Learners were expected to dive deep and embark on a journey to self. (Age: 50–65)
Format + Assessment	3-day on-site workshop format built on Asian body philosophies from Qigong and the metaphorical power of poetry. Participants were not assessed but evaluated the retreat by means of an online survey and individual feedback
Disruptive Impulse	Somatic approach combined with reflective activities, creativity techniques and impulses from poetry, placing value on individual experiences and perceptions and focussing on experimenting, observing and playing
Required Skills	Imagination, curiosity, empathy, engagement, focus, 3-dimensional thinking, reflective skills, willingness to take risks and embark on an exploratory journey

(*continued*)

Table 4. (*continued*)

Learning Output	Learners discovered how to activate their body and energy and tried out different reflective tools
Learning Impact	Through the movement and reflection experience, participants discovered new facets of self and practiced who they want to become. They started reframing their ideas, got new insights and continued building a growth and benefit mindset

3.1 Main Findings

Overall, participants across all contexts—universities of applied sciences, technical colleges, and participants from corporate and public sector organizations—report that they benefit from the individual and social collaborative learning experience as provided by the 2CG® Poetry Machine. The disruptive impulses in combination with the customized content, reflective activities and creative tasks help them embrace different perspectives, get new insights and come up with new ideas and solutions. Other important success factor of the interdisciplinary 2CG® approach are the facilitators, whose role is to give guidance and to act as coaches and curators, rather than traditional trainers or teachers. Most importantly, learners state that they enjoy both, the safe CoP space for sharing and exploring and the slightly irritating disruptive impulses that enable them to stay open to surprises, get curious, better understand complex topics, and tap into their creativity.

More concretely, participants of all 4 learning journeys (see Practice Examples 1–4) reported that they found immediate value in their community's activities and interactions. Also, all participants reported that getting inspiration from the arts and engaging in peer exchange learning helped them gain increased understanding and awareness of the topic at hand (see Table 3, Cycle 1). They observed a transformed ability to learn (see Table 3, Cycle 2). After a while, they were better able to identify and define the problems they needed to solve.

While all learners appreciated the opportunity to improve their 21st-century digital skills and other future skills, approximately 20% of respondents stated that they could not use the acquired skills in their immediate work context but saw the potential value of indirect performance improvement through new thinking, broader perspectives, different viewpoints and increased awareness (see Table 3, Cycle 2). About half of respondents stated that they saw applied value, such as the implementation of tools in their work practice (see Table 3, Cycle 3). Interestingly, the majority of participants found that course goals were not linked to strategic business goals, which they considered a missing link in performance improvement (see Table 3, Cycle 4). While almost none of them were in the position to redefine success for their organization, they reported that they could redefine their personal success indicators as a result of the skills development training (see Table 3, Cycle 5). Overall, learners stated that they had increased topical awareness, built confidence and were empowered as a result of their learning journey.

4 Discussion

Learner feedback and expert observation suggest that the interdisciplinary 2CG® Poetry Machine needs to be linked to a professional practice, a cause, societal or organizational development goals to be fully effective. Embedding 2CG® learning experiences in communication classes, knowledge management initiatives, awareness-building workshops or somatic movement sessions has shown fruitful learning outputs and outcomes. To effect change on a bigger scale, many such learning experiences are needed. However, it is definitely not sufficient to expose learners to art-works and hope that the mere exposure to artistic impulses will do the job. Learners need a lot of guidance and facilitation to grow on their learning journey.

One major personal insight gained through designing and facilitating these learning journeys has been that organizational issues are almost never about strategy itself but about communication—including misunderstandings, relationships, problem analysis, definition of the shared practice, purpose and learning goals. Individual learning and skills development should therefore increasingly be linked to organizational development. In fact, the gap between organizational development, individual learning goals and changed behavior seems to pose the biggest obstacle to overall learning progress. Future skills training can be most effective when leaders have a sense of direction and also the courage to explore new pathways, as well as the ability to understand the boundaries within which their organizations and people operate. To effect behavioral change and create value, leaders need to focus on linking organizational strategies and societal developments to individual learning goals and purpose. Human capability cultivation with the 2CG® Poetry Machine can serve them as a tool and support them in coming up to learners' increasing wish for transparency, trust, stability, and purpose in times where the focus clearly is on technology and AI.

Future effective hybrid combinations of in-person, live virtual, online, and blended learning formats need to be built on design elements that include customized content, elements of surprise, disruptive impulses, mixed-pace learning, professional facilitation, and an appropriate choice of media and communication channels. The interdisciplinary learning journeys presented in this paper show that learning environments that are based on the pillars of CoPs can provide a safe space for learning, sharing and exploring and support learners in embracing difference and diversity as a resource. As we can see, learners in general seek and appreciate opportunities to grow and learn. Organizations that want to stay relevant need to provide their people with many such possibilities for learning and upskilling. The examples given in this paper also demonstrate that organizations need to embed learning into fieldtrips, wellbeing retreats, water-cooler moments and so on. These learning experiences will help create a sense of belonging while at the same time allow for deep individual and collective learning. Learning needs to be perceived as a way of being—people need to be motivated to bring their whole self to the organization.

4.1 Concluding Remarks and Future Outlook

Learners as well as leaders need to understand that learning and growing is a way of well-being. Being well implies a sense of belonging, taking initiative, sharing responsibility and co-creating better solutions. Organizational cultures and societies that promote continuous learning and upskilling can be a driver of success. Learners who get possibilities to grow are energized—organizations and educational institutions that manage to tap into the full potential of their people are likely to stay relevant.

Acknowledgments. The author of this paper would like to express her thanks to the stakeholders in organizations, institutions of higher education, and schools who have supported the new way of teaching and learning. A special thanks goes to all the (working) learners for their ongoing commitment and invaluable contributions that have delivered important educational insights and many joyful moments of learning, sharing and innovating.

References

1. Zhang, L., Gläscher, J.: A brain network supporting social influences in human decision-making. Science Advances (2020); https://www.science.org/doi/https://doi.org/10.1126/sci adv.abb4159, last accessed: 2022/3/13
2. Battram, A.: Navigating complexity: the essential guide to complexity theory in business and management. Ind. Soc., London (1998)
3. World Economic Forum: The skills needed in the 21st century, new vision for education, (2015)
4. Van Laar, E., van Deursen, A., van Dijk, J., de Haan, J., The relation between 21st-century skills and digital skills: A systematic literature review, Elsevier Ltd. (2017)
5. Merl, C.: Fostering 21st century skills in engineering and business management students, ICL 2018: The challenges of the digital transformation in education, pp 145–156 (2019)
6. Merl, C.: Human Intelligence Cultivation with the 2CG® Poetry Machine. How to Boost Future Skills Development and New Idea Generation with Artistic Impulses in Lab 21. to be published in ijac (March 2022)
7. Wenger-Trayner, E., Wenger-Trayner, B.: Learning to make a difference: value creation in communities of practice, Cambridge University Press (2020)
8. Sowa, H., Gemeinsam vorstellen lernen. Theorie und didaktik der kooperativen vorstellungs-bildung, Glas, A., Heinen, U., Krautz, J., Miller, M., Sowa, H., Uhlig, B. Eds. München: kopaed, pp 65–69 (2015)
9. McNiff, S.: Philosophical and practical foundations of artistic inquiry: creating paradigms, methods, and presentations based in art. In handbook of arts-based research, edited by P. Leavy, 22-36. New York: Guilford (2017)
10. Poetry in business, Wie das Heute im Morgen entsteht, (2018). https://issuu.com/christina merl/docs/poetry_in_business_by_christina_mer, last accessed: 2022/3/13
11. Wenger, E., Trayner, B., de Laat, M.: Promoting and assessing value creation in communities and networks: a conceptual framework. Ruud de Moor Centrum (2011), https://www.asmhub. mn/uploads/files/11-04-wenger-trayner-delaat-value-creation.pdf, last accessed: 2022/3/13
12. Laurillard, D., & Kennedy, E.: The role of higher education in upscaling global professional development through open, online collaboration. In C. Callender, W. Locke and S. Marginson, Changing higher education for a changing world. London: Bloomsbury (2020)

13. Merl, C.: Lab 21—A Space for Learning, Sharing, Innovating. In: Guralnick D., Auer M.E., Poce A. (eds) Innovations in learning and technology for the workplace and higher education. TLIC 2021. Lecture notes in networks and systems, p. 349. Springer (2021)
14. Lewin, Kurt, Action research and minority problems, J. Soc. Issues, **2**(4) (1946)
15. Laloux, F.: Reinventing organizations, A guide to creating organizations, Nelson Parker (2014)
16. Shelley, A.: Knowledge Succession, e-book, Bus. Expert. Press. (2015)
17. Miles, M.B., Huberman, A.M., Saldana, J.: Qualitative data analysis: a methods sourcebook, SAGE Publication, 004 Edition; ISBN-10: 150635307X (2019)

Promoting Flourishing in Hard Times: Theoretical Reflections on Ethics of Care in Distance Learning

Luigina Mortari , Alessia Bevilacqua$^{(\boxtimes)}$, and Roberta Silva

University of Verona, 37129 Verona, Italy
alessia.bevilacqua@univr.it

Abstract. Although the scientific literature in the educational field has thoroughly investigated the ethical dimension of distance learning, few studies are currently focusing attention on the ethical issues that characterize emergency remote teaching and learning. And when this question is addressed, the ethical dimensions that emerged mainly concern the equity in access to educational technologies, and cheating behaviors, especially during the assessment phase. This document aims to present a theoretical reflection on the ethics of care in distance learning, with specific attention to emergency remote teaching and learning, where teachers must be attentive to both the technical and the human aspects by implementing a teaching approach inspired by the ethics of care for the construction of the individual. The dimensions that characterize this construct—attentiveness, responsibility, competence, and responsiveness—allow to understand how an ethically oriented learning environment promotes the flourishing of the learners and their well-being and not just the mere passage of information and instructions: their engagement, rather than simple participation; the birth of learning communities, rather than just aggregation; and a protected space where everyone cultivates their personal growth to feel good with the others, rather than a place of competition and self-affirmation. Since the ethics of care establishes that ethics is something that aims at a good life lived with and for the others in just institutions, it invites professionals to promote a caring posture not only oriented to the individual good, but also to the others' and institutional good. Specifically, in emergency remote teaching and learning, this means to be present anyway and stay in a relationship with the pupils and the families, not making them feel abandoned.

Keyword: Emergency remote teaching and learning · Ethics of teaching · Ethics of care

1 Taking Care: A Brief Theoretical Introduction

The paper aims to understand[1] how ethics-oriented practices of care can be implemented in emergency remote teaching and learning. For this purpose, starting from a recognition of the theoretical foundations of the concept of care, a reflection will be formulated

---

[1] The responsibility of single paragraphs should be attributed as follows: to Luigina Mortari paragraphs 1 and 4, to Alessia Bevilacqua paragraph 2, to Roberta Silva paragraph 3. The reference list is equally divided by the authors.

© The Author(s), under exclusive license to Springer Nature Switzerland AG 2023
D. Guralnick et al. (Eds.): TLIC 2022, LNNS 581, pp. 248–257, 2023.
https://doi.org/10.1007/978-3-031-21569-8_23

concerning good care practices in the educational field, as well as a review of the scientific literature that focuses attention on the ethical dimension in distance learning and emergency remote teaching and learning.

Although the theories of caring have ancient roots (in Plato's Apology, Socrates affirms that educating means to orient young people to care for themselves), only in the twentieth century has the thought of care developed, firstly with Heidegger [1] and later with such a number of female scholars (Sara Ruddick, Eva Kittay, Joan Tronto, Nel Noddings just to name the main ones) that caring was configured essentially as a feminine thought.

The term care is used daily. Consequently, over time, it has been attributed multiple meanings. Similarly, also on a scientific level, the reflections formulated on this topic are numerous and different from each other.

According to Heidegger [1], care is a primary dimension of human life. We can speak of the primacy of care since human being needs both to be the object of care and to take care for the others. It is this continuous exchange of care that makes life possible; for this reason, wisdom is, in its essence, to know how to take care—that means to take life to heart. The practice of care that takes life to heart deals with being there in all the situations of its occurrence, and in all situations, seeking the best: caring of oneself, of one's own becoming in its most proper potential; caring of the others, because to live is to live together; caring of nature, because we are part of the natural world; and caring of the world, namely, of material works, institutions and different contexts of life, because human artefacts structure the context where our humanity takes shape [2]. From a phenomenological point of view, caring, which presents itself as something acted out in the world with the others, is a practice. To be more precise, it is a practice that takes place in a relationship between a caregiver and a person who receives care. It is set in motion by the interest in the others, by the concern for his/her condition and his/her way of being, and it is oriented from the intention to provide well-being for the other. What has been stated so far highlights the caring's protensive nature. Taking care means being turned towards the other. Good care is a receptive and responsive practice that implies attention to the other, compassion to be understood as a feeling with the other and an available presence; it is a practice that urges to put the other in the conditions to experience a good quality of life [3]. However, it is necessary to distinguish between that ways of being that manifest themselves as caring and others that reveal worrying. The actions we take just to accomplish something without affective involvement and dedication can be defined as taking care of something. On the other hand, the actions that move from the desire to do something that makes the quality of the experience lived by oneself and the others good can be considered worrying [4].

1.1 Taking Care in Educational Settings, also in the Time of COVID

The contextualization of the concept of care in the wider educational landscape also has its ancient roots in the Socratic concept of education as epimeleia (the Greek word for care). According to Socrates, education means "*help*[ing] *students to cultivate the desire to care for themselves – that is, to accompany them through the process of building the cognitive skills and emotional attitudes they will need to be self-sufficient and enthusiastic on the path of their existence, so that their lifetime will be the actualization of a process of bestowing sense*" [5]. Mortari—assuming Nodding's theory [6], which identifies the key behavioural indicators of good care: receptiveness (that means to assume a passive posture trough which it is possible to listen to the others) and responsiveness (that means to answer to the others adequately)—explains that it is also necessary to understand what the relational postures are that allow the educator to be receptive and reactive [5].

- *being available both at a cognitive and emotive level*, because receptiveness and responsiveness requires a sensitive disposability, that consists of a readiness to bestow and spend our resources in relationships with others.
- *having empathy*, which means being able to feel the reality of the other means being capable of empath.
- *being attentive*, which implies maintaining a deliberately intense concentration on a phenomenon.
- *giving security*, welcoming and safeguard the other, giving him/her a scaffolding, makes the cared-for person perceive that the one caring is reliable.
- *being unobtrusive*, which means activating a kind of solicitude that should be continuous but discreet.
- *being capable of waiting*, to allow the other to realize his/her own way of being without forcing his/her in a direction.
- *cultivating positive and healthy sentiments*, because a necessary condition for good care is nourishing the caring relationship with healthy, vital sentiments such as hope, acceptance, trust, tenderness, confidence.
- *being reflective*, as caring for the other also means caring for the life of the mind, that is gives meaningful cognitive experiences to the cared-for.

Good educational care requires that an ethical instance characterizes the posture of the educator to make him/her devote mindful attention to the other by showing solicitude. This ethical substance, which arises from a concrete, embodied encounter with the other and not from abstract and universal principles, consists of:

- *being responsible for the other*, since he/she is aware of the other's dependence – that is, he/she is in need of the help of the carer in order to preserve and flourish in his/her life. Since ethicness does not arise from the application of abstract and universal principles generated by a theoretical reflection, but from a concrete and embodied encounter with the other, the face-to-face relationship which obliges one to be responsible [7]. Infinite responsibility is therefore defined in relation to the contextual and unique meeting with the other.
- *having respect*, because good care occurs when one knows how to safeguard the infinite value of the otherness and he/she also feels his/her vulnerability.
- *wishing for what is good for the other, being giving*, which can be expressed in different levels of involvement: looking after the other, worrying about the other, or giving attention and dedication [5].

The present time, marked first by the COVID-19 pandemic and now also by war conflicts, is a difficult time. It generates uncertainty about the rightness of actions and fear of not being able to return to life as it was lived daily. The above listed ethical principles and postures are always essential to be assumed and mostly in moments of crisis, since the ethics of care represents that way of being in the world that considers fragility, seeks to reduce vulnerability, cultivates the becoming of precisely being there while remaining faithful to the primary desire that moves the way, that is, the desire for good.

2 Distinguishing Ethics and Morality: A Compass to Guide the Analysis of Educational Practices

Given the widespread rise of ERTL in schools of every order and grade determined by the Covid19 health emergency and the great amount of scientific publications that followed, an issue on which it is worth to reflect is the ethical dimension of distance teaching and learning processes. Although the scientific literature in the educational field has thoroughly investigated the ethical dimension of distance learning [8, 9, 11], few studies are currently focusing attention on the ethical issues that characterize ERTL. The scarcity of publications on this subject is certainly partly determined by the fact that the phenomenon is recent or still ongoing. What strikes the attention is that, in publications that focus attention on the ERTL's ethical dimension, this phenomenon is predominantly read through the gaze of morality, understood as deontology and normativity, and not through the gaze of ethics as a tension towards good. Ricoeur [11] distinguishes ethics and morality: in his reflection, ethics has a teleological perspective, while morality has a deontological one. In other words, while ethics refers to an understanding of what is evaluated as good to do and what makes a good life, morality corresponds to what is right to do and therefore defines the rules and codes of behavior.

The proposed code of ethics for online learning during the crisis, which is the subject of analysis in the paper by Salhab et al. [12], for example, organizes its contents on the following themes: commitment to distance educational system, commitment to profession, teachers and students' responsibilities and respect and protection of digital dignity. The papers focusing on cheating behaviors during the assessment of learning processes are several. Hill, Mason, and Dunn [13], for example, examine academic integrity, ethics of student practices and illegal commercial activities. Parks-Leduc, Guay, and Mulligan [14] delve into the relationships among personal values, justifications, and academic cheating. Lee et al. [15] explain before-during-after strategies and technological tools adopted as deterrents to prevent cheating in remote assessments during the COVID-19 pandemic. The contribution of Pascault et al. [16] analyzes the terms and conditions of some selected online services used to deliver distance learning, focusing on copyright ownership, liability and content moderation.

Although it is also important to question the institutional and regulatory aspects of ERTL, in a moment of general emergency, in the educational field in particular, it is considered a priority to focus attention on ethics as "something that aims at a good life lived with and for the others in just institutions" [17]. This choice finds also legitimacy in the words of Ricoeur himself [11], who suggests replacing the term "aims" with the term "care," thus establishing ethics as a discourse on self-care, care for the others and care of institutions. Specifically, in ERTL, this means to be present anyway and stay in a relationship with the pupils and the families, not making them feel abandoned.

The most addressed issue within the broad framework of the ethical dimension is equity in terms of access to ERTL. The focus of the choral contribution by Czerniewicz et al. [18] goes on the role of equity and inequality concerning the access to distance teaching and learning in order to recognize the dangers and risk responses to prepare for the post-pandemic future. The authors focus attention on three different types of inequities that hinder the access to ERTL: (a) vital inequality, which refers to life chances, and indeed to survival rates; (b) resource inequalities, which manifest in this period more openly through material divisions, and which also affects the range of capital needed to negotiate and survive the crisis; and (c) existential inequality, which is described as the denial of equal recognition and respect. They resort to Tronto's thought [19] in order to explain how teaching and learning projects are relational and that none of these complex problems can be effectively solved by an autonomous component because each subject, each institution, lies within a complex network of relations. It is only by considering all the stakeholders that ethical and caring conditions can be created but such processes of deliberation and negotiation require time and trust, two resources that have been scarce in these times of crisis.

Moorhouse and Tiet [20] also focus attention on the relational dimension of learning to address the dimension of care in ERTL environments as their intent consists in improving online teaching practices by enacting a pedagogy of care. Referring to Noddings [21], they explain that if care is a relational and not an individual process, this means that care can and should be reciprocal. To this aim, it is necessary to establish an online presence on cognitive, social, and teaching levels.

Whited and Sisk's study [22] exploits the concepts of ethics of care and cosmopolitanism to support the students engaged in developing the skills required by the course of study and in meeting the additional demands that COVID-19 has brought with it. The project illustrated by the authors focuses on the intertwining of two similar theoretical frameworks. On the one hand, Tronto's concept of ethics of care is based on attentiveness, responsibility, competence, responsiveness, and plurality/solidarity [23, 24]. On the other hand, Emdin's cosmopolitanism philosophy [25] is used in addition to the ethics of care in order to ensure a complete, more comprehensive approach to the COVID-19 adaptations, as cosmopolitanism can be defined as "an approach to teaching that focuses on fostering socio-emotional connections in the classroom with the goal of building students' sense of responsibility to each other and to the learning environment." Therefore, their ethics of care practices rest on three strategies: (a) explicitly teaching empathy in the class; (b) promoting reflective practice to foster the socio-emotional connections emphasized in cosmopolitanism through the journey to understand and interpret feelings; and (c) involving graduate students in the clinic management by assigning roles and clarifying expectations to teach mutual care and socio-emotional connections.

Rabin's self-study [26] allows us to understand how to act on care ethics in online teaching based on modelling authentic [27] caring through story, practice and continuity, dialogue, and on addressing power and confirmation in assessment. The author highlights how authentic care is distinguished, within the broader framework of the ethics of care, in the attention paid to everyday subtleties; therefore, it requires a reactivity that the student perceives in satisfying his own needs and cultivating reciprocity and connection. It is considered useful, in particular, to underline how this attention is translated into the assessment practices. Demonstrating flexibility, welcoming revisions, transcending niceties to provide constructive feedback, reminding students of boundaries, and softening deadlines have led the students to experience a sense of confirmation, which is the premise of good intentions.

The articles presented in this short bibliographic review highlight how, at an international level, seeds of some educational practices oriented to the ethics of care can be identified. The reflections that the authors outline, however, remain focused on specific practices, which are also carried out almost exclusively in higher education.

3 Ethics-Oriented Care Practices in ERTL

The results of empirical research aimed at understanding teachers' experiences concerning the experience of ERTL [28] created by the Melete research group of the University of Verona (Italy) will be reinterpreted through the conceptual framework of the ethics of care in order to understand how ethics-oriented care practices can be implemented, transversally to the different school grades, in ERTL. This research stems indeed from the conviction that it is worth listening to the teachers' voices to bring lived reality to the evidence of political management. The goal is to contribute to elaborate a realistic balance sheet of these difficult months, in which ordinary school life has suddenly come to a halt and teaching has had to experiment with new forms of reorganization.

The element most frequently referred to when talking and writing about ERTL is the *care of relationships*. Many teachers worked hard to foster closeness even in the distance, to make the pupils and the families perceive that, even in that moment of emergency, they continued to be there as a reference presence. The care of relationships required significant time and commitment, unlike those implemented in learning environments characterized by physical presence [28]. As a matter of fact, the educational process takes place in the asymmetrical relationship between teacher and pupil. A good management of this asymmetry implies on the part of the adult a series of ways of being in the relationship that are essential to caring—i.e., receptivity and responsiveness, which have been previously described. These ways of being can also persist in a distance relationship only if the teacher can act in a posture based on attention, because paying attention is an essential premise for understanding the other, namely putting oneself in brackets to make room for the other. Relationships in the classroom are not just about the teacher-pupil dyad. For the children, it is important to experience the dynamics of the classroom; they need the exchange and the dialogue among peers. In distance learning, it is difficult to feel this sense of closeness. Therefore, in ERTL, it is necessary to continue to support the virtue of friendship, since being friends means, as in any other virtue, seeking the good, in this case, the good of a friend.

A second element to which attention must be paid is the *emotional relationality* that nourishes action, especially of younger pupils who need to feel welcomed by a real presence [28]. Thinking always lies within a feeling; that feeling, when it is positive, moves the things of the world. The teachers know that a learning context deprived of contact becomes also deprived of those exchanges of experiential intensity that forge both vital thoughts and ways of feeling that generate humanity. Therefore, they have to take care of the emotional life by promoting learning contexts, including virtual ones, in which educational subjects have the opportunity to practice in the investigation and self-understanding of their emotional experiences and in giving them a voice, as well as to practice in an understanding of the others' experience.

Concerning *instructional design*, specific difficulties emerged about the quality of teaching and learning strategies. The teachers note that distance learning makes curricular learning difficult and that, due to their specific characteristics, some disciplines are more challenging to teach. ERTL seems to penalize the educational aspect compared to the instructive one in all school grades. If the teachers could find effective strategies for the transmission of content, they find it more difficult to carry out the educational work that fosters the pupil's personal maturation and is based on the embodied relationship [28]. Even the online learning environment must take the form of laboratories of experience [29] in which the students can practice, taking care to sustain the students' development and their achievement of the educational goals previously indicated. Taking care of the life of the mind implies that cognitive activity is kept rooted in experience because, without active participation, the product of the activity of thinking loses its content and becomes inert. Therefore, even at a distance, it is necessary to reduce the weight given to the essentiality of the frontal lesson and enhance a school of experience, where pupils are invited not only to do things but also think reflexively about what they do.

Focusing on the *assessment design*, the ERTL has brought to light the need to rethink methods and strategies since traditional ones are no longer adequate for the teaching innovations that have been introduced. Indeed, to use the traditional way to evaluate pupils and students in this new situation lacks the reality principle. Furthermore, for some teachers, the difficulties of assessment in the time of an emergency become the starting point for a broader reflection on the need for strategies that are not merely summative but become truly formative [28]. It is, perhaps, in these terms that education is called to fully assume the meaning of the crisis and stay in it to accompany the flourishing of each person and of the world we inhabit. In this framework, the implementation of feedback literacy strategies can help the students not only to identify directions for growth but also to implement them, embracing the emotional significance that evaluation processes entail [30]. Researchers revealed a further unexpected factor concerning assessment: the intervention of the students' parents, which teachers often evaluated as more intense than in ordinary situations and, in some cases, incorrect. These behaviors of excessive support require a reflection concerning the alliance between school and family, without which the school educational project loses value. When there is no authentic alliance between school and family, ways of being ethically incorrect risk to prevail, invalidating thus educational actions. What seems to be missing is the virtue of measure. The practice of the ethics of care requires not only seeking the good but also doing it in the right measure, nothing less and nothing more. The right care requires finding the balance between excessive availability towards the other and the right absence when the other needs to experience himself. This means putting the other in the condition of gradually finding a way to take care of him/her self by himself, leaving him/her free.

The ERTL raises ethical issues concerning the *protection of people*: the teachers feel it puts their right to privacy at risk, and at the same time, they feel the importance and responsibility of protecting pupils' right to privacy [28]. As previously mentioned, this is not a question exclusively related to ethical and normative morality. This issue must be played on the level of strengthening school-family co-responsibility. In current times, marked by a void of ethics aggravated by a lack of attention to civic responsibilities, it is essential to design schools as training grounds for responsible citizenship, promoting learning contexts where to develop the ethics of responsibility. Teachers and students are equally asked to create communities permeated by the culture of hospitality and by rigorous conviviality. The value coordinates of the educational project must be respect, solidarity, the responsibility to choose and act consciously, the ability to take care of the world we inhabit. This also applies to civic responsibilities, which constitute an inalienable objective in terms of developing an ethics of public responsibility.

Finally, the *principal's role and the school's organization* were also perceived as strategic to promote a good ERTL. When this coordination is lacking, the teachers perceive it as a critical element [28]. However, competent leadership is required not only on an organizational level but also on a relational and ethical level. According to Ricoeur [11], the leader who acts on ethics of care pays attention to the bureaucratic-administrative aspects and, at the same time, is also committed to stimulate development, provoke thoughts, and evoke feelings. He/she tries to build sharing processes rather than imposing his/her own will, to motivate people and, above all, to dispense the culture of responsibility [31]. These guidelines are oriented to personal and professional growth, as an individual and as a group, of each teaching team member and of the whole school staff.

4 Conclusions

The argument exposed in this article starts from presenting the theoretical framework of the concept of care and how it can also be applied in the educational field, with specific attention to today's era characterized by a great fragility of the system and of the subjects who live in it. The subsequent clarification of the difference between ethics and morality allowed the authors to explain how ethics-oriented care practices can also be implemented in ERTL. The authors intentionally do not offer ready-to-use solutions, toolkits or good practices in ethics of care, but guidelines that the teachers can take as suggestions in order to rethink the practices they can implement in the specific learning contexts where they and their students live. Today more than ever, their essential heuristic action involves elaborating an educational project, realizing it and then examining it, describing as precisely as possible what happens, and interpreting the collected data. The objective consists in verifying if and how this action has made it possible to achieve the expected outcomes and then assessing whether these outcomes are 'good' to promote and sustain the students' flourishing.

References

1. Heidegger, M.: Essere e tempo. Longanesi, Milano (1976)
2. Mortari, L.: La politica della cura. Prendere a cuore la vita. Cortina, Milano (2021)
3. Mortari, L.: La cura nel tempo del "Covid." Nuova rassegna di studi psichiatrici **22**, 1–16 (2021)
4. Mortari, L.: The Philosophy of Care. Springer VS, Wiesbaden (2022)

256 L. Mortari et al.

5. Mortari, L.: For a pedagogy of care. Philos. Study **6**(8), 455–463 (2016)
6. Noddings, N.: Caring. University of California Press, Berkeley, CA (1984)
7. Lévinas, E.: Totality and Infinity. Duquesne University Press, Pittsburgh, PA (1969)
8. Almseidein, T., Mahasneh, O.: Awareness of ethical issues when using an e-learning system. Int. J. Adv. Comput. Sci. Appl. **11**(1), 128–131 (2020)
9. Blaga, P.: Ethical considerations in human resource training based on e-learning. Curentul Juridic **79**(4), 42–51 (2019)
10. Swartz, B.C., Gachago, D., Belford, C.: To care or not to care-reflections on the ethics of blended learning in times of disruption. South Afr. J. Higher Educ. **32**(6), 49–64 (2018)
11. Ricoeur, P.: Ethique et morale. Revue de l'Institut catholique de Paris **34**(avril-juin), 131–142 (1990)
12. Salhab, R., Hashaikeh, S., Najjar, E., Wahbeh, D., Affouneh, S.: A proposed ethics code for online learning during crisis. Int. J. Emerg. Technol. Learn. **16**(20), 238–254 (2021)
13. Hill, G., Mason, J., Dunn, A.: Contract cheating: an increasing challenge for global academic community arising from COVID-19. Res. Pract. Technol. Enhanc. Learn. **16**(1), 1–20 (2021). https://doi.org/10.1186/s41039-021-00166-8
14. Parks-Leduc, L., Guay, R.P., Mulligan, L.M.: The relationships between personal values, justifications, and academic cheating for business vs. non-business students. J. Acad. Ethics, 1–21 (2021)
15. Lee, J., et al.: Using technologies to prevent cheating in remote assessments during the COVID-19 pandemic. J. Dent. Educ. **85**, 1015–1017 (2021)
16. Pascault, L., et al.: Copyright and remote teaching in the time of COVID-19: a study of contractual terms and conditions of selected online services. Eur. Intellect. Prop. Rev. **42**(9), 548–555 (2020)
17. Ricoeur, P.: Oneself as Another. University of Chicago Press, Chicago (1992)
18. Czerniewicz, L., Agherdien, N., Badenhorst, J., Belluigi, D., Chambers, T., Chili, M., ... & Wissing, G.: A wake-up call: equity, inequality and Covid-19 emergency remote teaching and learning. Postdigital Sci. Educ. **2**(3), 946–967 (2020)
19. Tronto, J.: An ethic of care. In: Holstein, M., Mitzen, P. (eds.) Ethics in Community-Based Elder Care, pp. 60–68. Springer, New York (2001)
20. Moorhouse, B.L., Tiet, M.C.: Attempting to implement a pedagogy of care during the disruptions to teacher education caused by COVID-19: a collaborative self-study. Stud. Teach. Educ. **17**(2), 208–227 (2021)
21. Noddings, N.: Caring: a feminine approach to ethics and moral education. University of California Press (2003)
22. Whited, J., Sisk, A.: Adapting during a pandemic: using ethics of care and cosmopolitanism to train graduate students during the COVID-19 crisis. Online J. Interprofessional Health Promot. **3**(1) (2021)
23. Tronto, J.C.: An ethic of care. In: Cudd, A.E., Andreasen R.O. (eds.) Feminist theory: a philosophical anthology, pp. 251–263. Blackwell Publishing (2005)
24. Tronto, J.: Caring Democracy: Markets, Equality, and Justice. University Press, New York (2013)
25. Emdin, C.: For white folks who teach in the hood ... and the rest of Y'all too reality pedagogy and urban education. Beacon Pr (2017)
26. Rabin, C.: Care ethics in online teaching. Stud. Teach. Educ. **17**(1), 38–56 (2021)
27. Valenzuela, A.: Subtractive schooling: U.S. Mexican youth and the politics of caring. State University of New York Press (1999)
28. Mortari, L. (Eds): La scuola al tempo del Covid-19: i vissuti dei docenti. Cortina, Verona (2021)
29. Dewey, J.: Democrazia ed educazione. trad. it. La Nuova Italia, Firenze (1974)

30. Carless, D., Boud, D.: The development of student feedback literacy: enabling uptake of feedback. Assess. Eval. High. Educ. **43**(8), 1315–1325 (2018)
31. Mortari, L., Tomba, B.: The moral dilemmas of Italian principals. New Trends Issues Proc. Hum. Soc. Sci. **6**(7), 12–18 (2019)

A Project-Based Learning Experience Through a Double Interaction Between Virtuality and Reality

Luigina Mortari, Roberta Silva$^{(\boxtimes)}$ (iD), and Alessia Bevilacqua (iD)

University of Verona, Verona, Italy
`roberta.silva@univr.it`

Abstract. Collaborative learning is a crucial topic in the debate about higher education, and more recently, this issue has been addressed focusing on the opportunities connected to hybrid and blended learning. Although this issue has been addressed from several points of view, there are still few scholars who have analyzed it by comparing collaborative learning experiences conducted in presence with parallel online experiences. The contribution here proposed, starting from an experience conducted within a teaching course belonging to the Combined Bachelor and Master's Degree Course in Primary Education of the University of Verona (Italy); it analyzes a Project Based Learning experience that, following the new needs emerged during the pandemic, has been conducted in two parallel paths (one for in-presence students and one for online students). At the end of the course, a survey was proposed to the students with the aim to investigate the effectiveness of this "double experience" in order to optimize the path for the future.

Keywords: Collaborative learning · Project based learning · Hybrid learning

1 Collaborative Learning in the (Post) Pandemic Era[1]

Collaborative learning has been defined in different ways by different authors [1], but despite that, we can define collaborative learning as "an educational approach to teaching and learning that involves groups of learners working together to solve a problem, complete a task, or create a product" [2]. Moreover, a collaborative learning experience must have some specific characteristics to be defined as such: (a) it must imply a positive interdependence among the students (this means that the students are allowed to achieve their goals together or not at all); (b) at the same time, it must promote individual responsibility (each student must have a recognizable task for which he is responsible); (c) it must devote a significant amount of time to free the interaction among the students; (d) it must have a specific focus on the promotion of social skills; and finally, (e) the group must engage in a critical-reflective group processing activity focused on its

---

[1] The responsibility of single paragraphs should be attributed as follows: to Luigina Mortari paragraph 1, to Roberta Silva paragraphs 3, to Alessia Bevilacqua paragraph 2. The reference list is equally divided by the authors.

© The Author(s), under exclusive license to Springer Nature Switzerland AG 2023
D. Guralnick et al. (Eds.): TLIC 2022, LNNS 581, pp. 258–269, 2023.
https://doi.org/10.1007/978-3-031-21569-8_24

internal dynamics and on how the group has managed to achieve its goals, in order to process feedbacks on its effectiveness [3].

Starting from the end of the'90s, collaborative learning (which was previously mainly used in primary and secondary education) began to be more and more widespread in university contexts, as well. However, learning in the university environment has specificities that can impact the effectiveness of cooperative learning: for example, the students often don't know each other (as it occurs in primary and secondary learning); classes can also be very numerous, the evaluation of the activity has a wider impact than the final assessment, and so on. These aspects must be taken into consideration and problematized before designing a cooperative learning experience, in order to counteract the potentially problematic elements and maximize the effectiveness of the intervention [4].

The need to calibrate the collaborative learning intervention on the specific needs of the academic teaching has acquired even more intense nuances with the massive introduction of distance learning in higher education. In fact, if on the one hand, the digital dimension introduces new tools and offers new opportunities for comparison, on the other, it cannot be taken for granted that the teachers (and sometimes even the students) are adequately trained to exploit these resources, not only from a technical point of view, but also from a communicative and socio-relational point of view. Furthermore, the same platforms for distance learning often arise from the desire to "simulate" what happens in the classroom rather than exploit the possibilities offered by the technological means to "think outside the box." Overcoming this obstacle can only take place starting from an accurate didactic planning of collaborative learning, which uses digital tools as a method of experimentation for the realization of authentic tasks through a two-way dialogue that exploits the potential of digital platforms, without limiting yourself to them [5].

The pandemic that has hit the entire globe in the last two years has made the situation even more complex since, in some cases, it has required to "suddenly shift" teaching activities in remote mode or sometimes in hybrid mode [6, 7]. Hybrid teaching (which involves part of the students in presence and part in remote) represents a big challenge for the teacher, since it is not easy to manage a didactic action that "moves" on a double channel [8, 9]. Furthermore, carrying it out in emergency conditions and with the desire to create collaborative learning experiences makes the challenge even more demanding [10]. However, this challenge, although unexpected, can represent a valid opportunity for us to think about the conditions that allow greater effectiveness of this teaching strategy. Indeed, after the COVID-19 pandemic, "many universities and academic institutions have adopted hybrid or blended medium of instruction," and the "need for conducting studies to demonstrate the effectiveness of blended and hybrid instruction and how instructors can work on designing their classes making it a viable option during current times and as we prepare to teach in the post-vaccine and post-pandemic world" [10, p. 143] is strongly emerging. Starting from these considerations, we have here decided to investigate a collaborative learning experience (more precisely a Project Based Learning experience) conducted during the period of the health emergency in order to analyze, through a comparative approach, the feedbacks collected by "in-presence students" and "online students."

2 Project Based Learning: An Expression of Collaborative Learning

Before continuing, it is necessary to specify what Project Based Learning (PBL) means. Actually, Project learning has distant roots: at the beginning of the eighteenth century, in Europe, the final exams for engineering and architecture students were represented by the realization of projects that had to solve real and practical problems, and over the next century, this didactic modality also spread in high schools [11]. In 1914, with the Smith-Lever Act, the American government decided to promote this teaching method also in the United States, and the interest in it grew further after the publication of Kilpatrick's essay *The Project Method*, in 1918. Kilpatrick, student of John Dewey, highlights in his text how the realization of a project could promote the students' motivation by encouraging them to freely decide the aims they wanted to pursue [12]. Another scholar who lays at the basis of PBL is Vygotsky, and more precisely, his theories about the influence of social interaction on learning: according to this position, knowledge can be more easily understood, developed, and meaningfully "absorbed" when the students are engaged in significant interactions with peers and teachers. Indeed, according to Vygotsky, only in this case is it possible not only to achieve disciplinary goals, but at the same time, to gain the development of skills such as problem-solving, decision-making or critical thinking [13].

PBL refers to this framework and can be defined as a collaborative teaching strategy that provides the students with the opportunity to be engaged in a located and meaningful learning through the realization of an authentic task in an open-ended perspective [14]. The capability of PBL to relate theories to practice, through the concrete realization of a task strictly linked to a real dimension (real-world product) makes this tool particularly suitable for higher education, with particular reference to academic paths with a professional vocation [15].

What is the difference between a PBL and the simple realization of a project carried out by the students? Firstly, while the realization of a project is often mainly focused on the achievement of certain disciplinary learning objectives, PBL also implies the achievement of other learning goals, such as the development of decision-making skills, problem-solving skills, and socio-relational skills. For this reason, while in the first case, the teacher's instructions are very specific and detailed, in PBL, the teacher deliberately gives indications exclusively regarding the goal to be achieved, leaving the students free to find the way they deem most appropriate to reach it. Furthermore, the realization of a project is not necessarily linked to the realization of a real-world object, while in PBL, this is a fundamental element. Finally, regarding the evaluation phase, in the case of the realization of a project, the evaluation is often linked to the final product, while in PBL, it is essential that it is also based on the process [13, 16].

There are many models of PBL; however, all share a four-phase articulation. In the first phase, the students are involved in a collegial way in identifying the "guiding question" that will guide their work, starting from the need to carry out a project that can be spent in a real context [17]. In the second phase, the students, divided into small groups, autonomously search for the information they need to develop a project capable to answer their guiding question—a project that must be carried out in an articulated way and completed with all the materials that are necessary for its implementation.

Each student must have a specific task, both in the collecting of the materials and in the definition of the project, and must be responsible for it in front of the group [18]. In the third phase, each group presents its project to the community: within this phase, each student must take an active role and assume a specific responsibility. Every group receives, in this phase, feedback from its peers and from the teacher (or from other figures deemed appropriate to be involved in this phase) to guide the revision and possibly optimize its project. In the fourth phase, each group submits its project to the evaluation of the teacher and of the peers. The fourth phase involves the teacher's evaluation, the self-evaluation of the students, and the peer evaluation [19, 20].

These considerations make it clear that PBL is, to all intents and purposes, an expression of collaborative learning because (a) it implies a positive interdependence among the students; (b) it promotes individual responsibility; (c) it devotes a significant amount of time to free the interaction among the students; (d) it has a specific focus on the promotion of social skills; and finally, (e) it involves the students in a critical-reflective group processing activity.

3   A Learning Experience Between Virtuality and Reality

These premises were essential to contextualize the experience we intend to present here conducted within a course belonging to the Combined Bachelor and Master's Degree Course in Primary Education of the University of Verona (Italy). Within this course, a collaborative learning path inspired by Project-Based Learning had been proposed for several years: complying with the new needs that emerged during the pandemic, starting from 2020/2021, the path has been changed, and two parallel paths have been conducted: one devoted to the students who attended the course "in presence" and one devoted to the ones who attended the course online. The planning of the learning experience was guided by the need to modify the elements involved in the original path in order to keep them consistent with the specificities of the medium adopted, in compliance with the training objectives of the course. At the end of the course, a survey was proposed to the students with the aim to investigate the effectiveness of this "double experience" in order to optimize the path for the future.

3.1   The Course Description

The course "Learning Theories and Teaching Techniques" is one of the fundamental teaching courses within the Combined Bachelor and Master's Degree in Primary Teacher Education. It is scheduled during the first semester in the second year. The duration of the course is 60 h, divided into 20 lessons of three hours each. The contents of the course involve the main learning theories, approaches, and methods with a specific focus on Constructivism. The Teaching Techniques (TTs) will be briefly presented in this course, mainly explaining how a specific Teaching Technique is not a neutral tool but represents a budding consistent with specific Learning Theories. The detailed description of the main TTs connected to the theoretical framework predominantly espoused by this teaching course (Constructivism) will be the main content of PBL, which is developed in the second part of the course. In the first part of the course, each lesson is composed by two parts: during the first one, the teacher will present some concepts, while during the second part the students will be engaged in an activity that allows them to explore these concepts. The Learning Objectives of the course are represented in the figure below (Fig. 1).

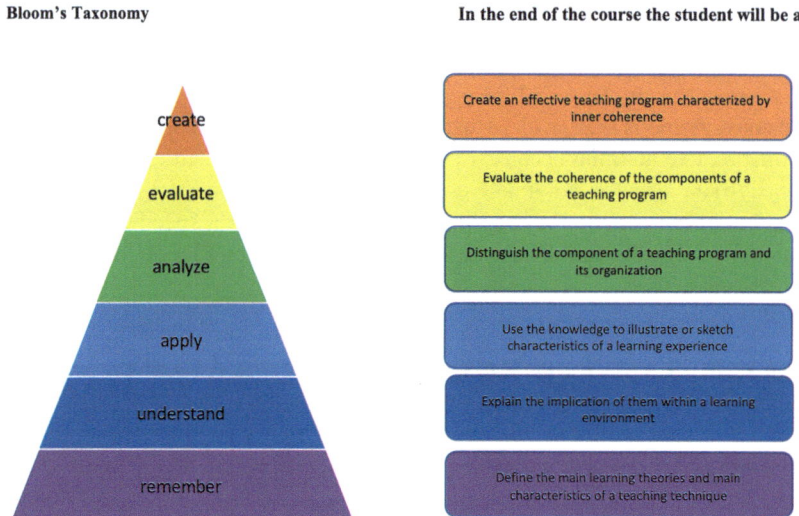

Bloom's Taxonomy

create

evaluate

analyze

apply

understand

remember

In the end of the course the student will be able to:

Create an effective teaching program characterized by inner coherence

Evaluate the coherence of the components of a teaching program

Distinguish the component of a teaching program and its organization

Use the knowledge to illustrate or sketch characteristics of a learning experience

Explain the implication of them within a learning environment

Define the main learning theories and main characteristics of a teaching technique

Fig. 1. The learning objectives of the course learning theories and teaching techniques

At the very beginning of the course, the students are divided into small groups: starting from 2020/2021, according to the new needs emerged during the pandemic, the students are divided into two macro-groups: one composed by the face-to-face groups and one composed by the remote groups. Therefore, the students are divided into 9 face-to-face groups and 9 remote groups, each made up of 4 or 5 students.

In the first lesson, the "guiding question" of each group is collegially defined. More specifically, each group is asked to deepen a specific Teaching Technique (different for each group) and design a teaching intervention, linked to a specific disciplinary object that is conveyed in primary school. This teaching intervention will be implemented during the second part of the course, involving all the classmates (since these students are all future primary school teachers, the accomplishment of this homework can be considered a "real-world task"). In order to carry out this task, during the first lessons, the students learn (a) what TTs are and what their characteristics are; (b) how to conduct a literature review focused on a learning topic; and (c) how to design a teaching activity. Furthermore, during the first part of the course, the groups are asked to work autonomously in order to design a teaching program, which is coherent with the disciplinary content and with the characteristics of the TT: within the group, each student assumes a specific task and is responsible for it (Fig. 2).

In the second part of the course, each face-to-face group is asked to realize the designed teaching activity during a lesson, involving the classmates in the activity that the group has planned: also in this case, each member of the group is asked upon to have

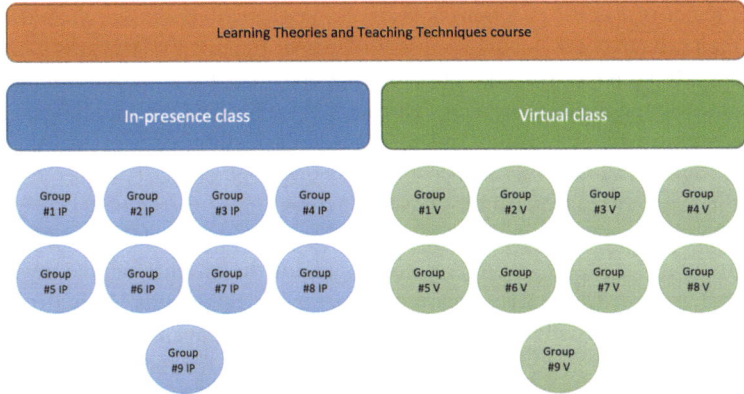

Fig. 2. Face-to-face and virtual groups

a specific role and a specific task. In the same occasion, each group receives feedbacks from its peers and from the teacher with the aim to collect insights useful to optimize its project. At the same time, the 9 remote groups are called to do the same work, but instead of carrying out the activity in presence, they carry out this task through the distance learning tools (specifically Moodle and Zoom). Then, the presentation of the projects, as well as the collection of the feedbacks from the individual groups, takes place within the macro-group to which they belong.

After that, at the end of the course, each group presents the final version of its project in a double form: anonymous and named (reporting the name of all the students of the group). While the teacher evaluates the named version of each project, the anonymous one is used for a peer-evaluation activity. Each student receives two reports produced by the other groups, and he or she is required to review them; these reviews are later shared. Also in this case, the groups are divided into two macro-groups (face-to-face groups and remote groups), and the peer review activity is conducted within the macro-group to which the student belongs. For each student, the assessment takes into account both the PBL activity and the peer evaluation activity.

Regarding the alignment between the activity and the learning objectives, (a) the use of PBL allows developing the students' improvement of self-directed learning skills as well as their problem-solving, analytical, and critical thinking skills; (b) the fact that the students must design and realize the work activity allows developing their teaching skills; (c) the discussion and the peer review activity develop the students' analytical and critical thinking skills. Moreover, the activity leads to the development of some important skills, such as working in a team, communication, and interpersonal skills.

3.2 Data Collection and Data Analysis

At the end of the course, the students are provided with a survey tool that aims to collect their experiences in order to identify useful elements to optimize the course starting from their direct experience. The tool chosen to collect the data is inspired by SWOT analysis. Born at the Stanford Research Institute in the 1960s, the SWOTS's purpose is to analyze the performance of specific programs or services in order to hypothesize changes that can improve their effectiveness. It pursues this aim by investigating the strengths of a specific experience (Strengths); its weak points (Weaknesses); its opportunities for improvement (Opportunities); and the risks it faces (Threats) [21]. This tool has rapidly gained great popularity within higher education, and many academic institutions use it to evaluate their Teaching Programs and identify possible areas of development [22], thanks to its ability to focus the gaze without mortifying the free expression of the involved subjects [23] and, therefore, allowing a strategic development perspective [22, 24]. Furthermore, this tool is particularly useful in optimizing educational pathways through the involvement of "practical" professions (physicians, teachers, etc.), thanks to its ability to highlight unexpected training needs [24]. In order to materialize its transformative potential, the insights deriving from the SWOT analysis must be used for the optimization of the Teaching Programs in a framework characterized by the flexibility and the ability to balance the needs of all actors [25].

According to the aim of this particular analysis, the data coming from the face-to-face groups and those coming from the remote groups were kept separated, in order to understand how the two different "ways" may have influenced the students' experience and how to modify both "sides" of the course according to the needs highlighted by the students themselves.

We have chosen a comparative approach because; although hybrid and blended learning have been addressed from several points of view, there are still few scholars who have analyzed collaborative learning experiences conducted in presence and parallel experiences conducted online through a comparative approach [6, 26].

Obviously, the data were collected after the end of the course, and the students were able to choose whether to join the survey or not, expressing their opinion in an absolutely anonymous way. The only profiling question they were asked was whether they had been part of a face-to-face group or of a remote group. Here, we analyze the data collected during the first year of experimentation of this hybrid modality (academic year 2020/2021). The survey was attended by 26 students belonging to the face-to-face groups and 24 students belonging to the remote groups. The collected answers were analyzed through the content analysis, which is suitable for a particularly flexible use [27]. It can be used to analyze the interviews deriving from SWOT analysis, producing coding organized in four categories (represented by the four areas of the tool); in each, the topics that emerge from the inductive examination of the material are inserted [28]. This organization was considered useful and therefore adopted in this case. The analysis led to the development of two coding systems: one for face-to-face groups and one for remote groups (Fig. 3).

The comparison between these coding allows for some interesting observations. Firstly, it can be noted that the identified strengths are very similar and, moreover, that they are essentially consistent with the aspects that the scientific literature identifies as the main strengths of collaborative learning in general and of PBL in particular (connection between theory and practice, support motivation, socio-relational dimension, etc.). However, it is significant that the in-presence students highlight how PBL gives them the possibility to create a more individualized relationship with the teacher, while this aspect is not underlined by the remote students, despite the presence of a forum of discussion

In-presence students' coding		**On line students' coding**	
Strengths	Connection between theory and practice	Strengths	Connection between theory and practice
	Involvement in stimulating activities that support motivation		Involvement in stimulating activities that support motivation
	Implementation of a challenging activity that allows the students to have an active role during the course		Implementation of a challenging activity that allows the students to have an active role during the course
	Promotion of peer collaboration		Promotion of peer collaboration
	More individualized relationship with the teacher	Weaknesses	Greater complexity linked to the implementation of activities in online mode (connection problems, loading materials, unexpected technical problems, etc.)
Weaknesses	Emotional handling of presentation within a large group (due to embarrassment, shyness, etc.)		
			Individual tasks not always done properly
	Poor participation of some peers during the feedback activities		Poor participation of some peers during the feedback activities
Opportunities	Creation of smaller groups in order to make the team working easier	Opportunities	Creation of groups with geographically close people (in order to make team-working easier)
	Possibility of "exceeding" the three hours of lessons for practical activities		Periodically organization of zoom meetings involving all the students
Threats	With a bigger number of the involved students, the management of the activities would be more complex	Threats	The lack of some technical skills (especially digital skills) could be a problem for some students

Fig. 3. Coding confrontation

and the possibility (underlined by the teacher) to organize zoom meetings (both for individuals and groups, as needed). With regard to this last aspect, it should be noted that not all students took advantage of this possibility: indeed, about half of the students did not ask for a direct comparison. Actually, some scholars have underlined how, for some students, it is difficult to be comfortable in building a relationship through the tools offered by distance learning contexts [29]. The data here confirm this assertion. Moreover, this element seems to resound in one of the labels of the "Opportunity" category of online students coding: in this label, indeed, the opportunity is indicated in order to schedule periodic meetings in advance, regardless of those voluntarily requested by the students. This request indicates the need, for some students, to have an external stimulus (or rather a "pre-existing pact") that acts as a catalyst to adhere to those moments of confrontation that are necessary for the construction of a profitable relationship. Coherently with these elements, in the redesign of the course, it was decided to foresee pre-established moments of meeting with the students of the remote groups.

Regarding the "Weaknesses" category, the comparison between the two coding reveals a single point of contact—namely, the low participation of some students in the feedback activities that followed the presentation of each group. Indeed, the scientific literature has been very concerned with the issue of peer feedback in PBL activities, and it is known that it represents a critical point [20, 30]. Therefore, one of the elements of attention in redesigning the course will concern the creation of a tool capable to support the expression of feedbacks on the work of classmates, both for face-to-face and remote activities: indeed, in both cases, an intervention will be conducted that, starting from a concrete situation, will show how the feedbacks from peers represent an important source of insights for the optimization of the final project. In addition, a gamification activity will be proposed to make this phase more engaging, declining it in a presence and in an online mode. It is interesting to note that the only other weakness highlighted by the face-to-face groups is in regard to the emotional handling of the presentation moments due to the difficulty perceived by the students in exposing themselves in front of a large audience. To solve this, during the redesigning of the course, activities will be included that make the students gradually familiarize with the moments of presentation. On the other hand, if we turn to online students, we realize that two other points of weakness emerge. The first is linked to difficulties related to the implementation of activities in online mode, and also in this case, it is an aspect that has been highlighted in the scientific literature and that is, for the most part, physiological [31]. The other aspect, on the other hand, is more particular and underlines, among the online groups, a difficulty in respecting individual tasks and assuming responsibility for them. This element is important since it focuses on a central aspect for collaborative learning in general and for PBL in particular: actually, the identification of this aspect effectively jeopardizes the effectiveness of PBL itself and requires strong corrective actions. For this reason, it has been considered necessary to introduce, in the redesign of the course, an intervention aimed at making individual actions more explicit and manageable within the group work.

Regarding the "Opportunities" category, for online groups we have already partially dealt with this theme with regard to the label "Periodically organization of zoom meetings involving all the students." However, it is interesting to note that both in-presence and online students have given suggestions about the groups' composition: the in-presence students asked for the formation of smaller groups (ideally, groups composed of 3 people), and the online students asked to organize the groups taking into consideration the geographical element. In particular, regarding the suggestion of students in presence ("Creation of smaller groups in order to make the team working easier"), this solicitation is linked to the Threat highlighted in the last category, that is the fear that the enlargement of the number of the involved students (i.e., with the entire group of the students of the course present during the lessons) could make the management of the activities more complex. Indeed, if the number of the students in the course was bigger, it would inevitably be necessary either to (a) increase the number of the students per group, contrasting the just highlighted suggestion or to (b) increase the number of groups, which would contrast with the second suggestion of the students in presence, or to give more time to practical activities in the classroom, even "exceeding" the hours formally dedicated to the lesson (which would be objectively difficult to organize from a logistical point of view as it would conflict with the scheduling of the other lessons). This analysis clearly highlights the complexity of the students' needs, which can sometimes even contrast with each other: hence, at the moment, it remains an open issue. However, beyond the specific requests, what emerges is the need to find strategies to facilitate group work and make the moments of discussion within the group more frequent and effective.

Finally, regarding the last element highlighted by online students, which belongs to the "Threats" category, it is pointed out that the lack of some technical skills (with particular reference to digital skills) could represent a problem for some students. In order to counter this risk, an initiative promoted by the University's Teaching and Learning Center, aimed at developing students' digital skills, was more publicized among the students.

3.3   In Conclusion

The analysis here conducted has put in evidence how the PBL conducted in-presence and the one conducted online share most of the positive elements highlighted by the students, and this is a sign that a PBL "in parallel" (in-presence and online) can represent a substantially effective choice. Despite this, the analysis has highlighted that there are some problems, in particular for the online PBL, which must be addressed in order to maximize its helpfulness, and from a design point of view, two problems in particular are the most relevant. The first element, which emerges both from the cross-analysis of the strengths and from an element highlighted in the "Opportunity" category of the coding relating to the online PBL, concerns the construction of meaningful relationships, both with the peers and the teacher. Indeed, it emerged that, despite having activated the channels recommended in the literature to encourage such relationships, they are still not at the optimal level in the online experience, and this is confirmed by some studies that have underlined how for some students it is not easy to build strong relationships through the tools offered by distance learning [29]. For this reason, accepting the requests of the students, it could be useful to give a greater structure to the meetings, by foreseeing pre-established moments of meeting with the students of the remote groups. The second element concerns the assumption of responsibility by all the members of the group: indeed, online students have underlined how, in some cases, it was difficult to manage a fair division of work and deal with cases of failure or poor assumption of responsibility about the assigned tasks by some members of the group. This is a significant critical issue because it focuses on a central aspect for collaborative learning in general and for PBL that can negatively impact on the effectiveness of PBL itself. It is, therefore, necessary for the teacher to plan in advance a work organization capable to support the students in the distribution of work and define in agreement with them solutions capable to trace the contribution of the individual within the collective work and increase the responsibility of the individual himself.

Acknowledgments. We thank all the students who participated in the course and who voluntarily chose to express their experiences at the end of the course.

References

1. Dillenbourg, P.: What do you mean by collaborative learning? In: Dillenbourg, P. (ed.) Collaborative Learning: Cognitive and Computational Approaches, pp. 1–19. Pergamon, Elsevier, Oxford (1999)
2. Laal, M., Laal, M.: Collaborative learning: what is it? Procedia Soc. Behav. Sci. **31**, 491–495 (2012)
3. Johnson, D.W., Johnson R.T., Smith K.A.: Cooperative Learning, pp. 88–100. Minneapolis, MN (2000)
4. Scager, K., Boonstra, J., Peeters, T., Vulperhorst, J., Wiegant, F.: Collaborative learning in higher education: evoking positive interdependence. CBE Life Sci. Educ. **15**(4), ar69, 1–9 (2016)
5. Reeves, T.C., Herrington, J., Oliver, R.: A development research agenda for online collaborative learning. Educ. Tech. Res. Dev. **52**(4), 53–65 (2004)
6. Syafril, S., Latifah, S., Engkizar, E., Damri, D., Asril, Z., Yaumas, N.E.: Hybrid learning on problem-solving abiities in physics learning: a literature review. J. Phys: Conf. Ser. **1796**(1), 012021 (2021)
7. Herrera-Pavo, M.Á.: Collaborative learning for virtual higher education. Learn. Cult. Soc. Interact. **28**, 100437 (2021)
8. Chen, B.H., Chiou, H.H.: Learning style, sense of community and learning effectiveness in hybrid learning environment. Interact. Learn. Environ. **22**(4), 485–496 (2014)
9. Hwang, A.: Online and hybrid learning. J. Manag. Educ. **42**(4), 557–563 (2018)
10. Singh, J., Steele, K., Singh, L.: Combining the best of online and face-to-face learning: hybrid and blended learning approach for COVID-19, post vaccine, & post-pandemic world. J. Educ. Technol. Syst. **50**(2), 140–171 (2021)
11. Knoll, M.: A marriage on the rocks: An unknown letter by William H. Kilpatrick about his project method. Eric-Online Doc. 511129 (2010)
12. Csikszentmihalyi, M., Csikszentmihaly, M.: Flow: The Psychology of Optimal Experience, vol. 1990. Harper & Row, New York (1990)
13. Kubiatko, M., Vaculová, I.: Project-based learning: characteristic and the experiences with application in the science subjects. Energy Educ. Sci. Technol. Part B Soc. Educ. Stud. **3**(1), 65–74 (2011)
14. Kokotsaki, D., Menzies, V., Wiggins, A.: Project-based learning: a review of the literature. Improv. Sch. **19**(3), 267–277 (2016)
15. Guo, P., Saab, N., Post, L.S., Admiraal, W.: A review of project-based learning in higher education: student outcomes and measures. Int. J. Educ. Res. **102**, 101586 (2020)
16. Blumenfeld, P.C., Soloway, E., Marx, R.W., Krajcik, J.S., Guzdial, M., Palincsar, A.: Motivating project-based learning: sustaining the doing, supporting the learning. Educ. Psychol. **26**(3–4), 369–398 (1991)
17. Krajcik, J.S., Blumenfeld, P.C.: Project-based learning. In: Sawyer, R.K. (ed.) The Cambridge Handbook of the Learning Sciences. Cambridge, New York (2006)
18. Karahoca, D., Karahoca, A., Uzunboylub, H.: Robotics teaching in primary school education by project based learning for supporting science and technology courses. Procedia Comput. Sci. **3**, 1425–1431 (2011)
19. Maher, D., Yoo, J.: Project-based learning in the primary school classroom. J. Educ. Res. **11**(1) (2017)

20. Lerchenfeldt, S., Mi, M., Eng, M.: The utilization of peer feedback during collaborative learning in undergraduate medical education: a systematic review. BMC Med. Educ. **19**(1), 1–10 (2019)
21. Hill, T., Westbrook, R.: SWOT analysis: it's time for a product recall. Long Range Plan. **30**(1), 46–52 (1997)
22. Dyson, R.G.: Strategic development and SWOT analysis at the University of Warwick. Eur. J. Oper. Res. **152**(3), 631–640 (2004)
23. Panagiotou, G.: Bringing SWOT into focus. Bus. Strateg. Rev. **14**, 8–10 (2003)
24. Gordon, J., et al.: Strategic planning in medical education: enhancing the learning environment for students in clinical settings. Med. Educ. **34**(10), 841–850 (2000)
25. Helms, M.M., Nixon, J.: Exploring SWOT analysis–where are we now? J. Strateg. Manag. **3**, 215–251 (2010)
26. Gutiérrez-Braojos, C., Montejo-Gamez, J., Marin-Jimenez, A., Campaña, J.: Hybrid learning environment: collaborative or competitive learning? Virtual Reality **23**(4), 411–423 (2018). https://doi.org/10.1007/s10055-018-0358-z
27. White, M.D., Marsh, E.E.: Content analysis: a flexible methodology. Libr. Trends **55**(1), 22–45 (2006)
28. Elo, S., Kyngäs, H.: The qualitative content analysis process. J. Adv. Nurs. **62**(1), 107–115 (2008)
29. Kutsenko, S.M., Malatsion, S.F., Saltanaeva, E.A.: Solving the problem of adaptation to online learning: the path to sustainable development of education. In: Proceedings of the International Scientific and Practical Conference on Sustainable Development of Regional Infrastructure (ISSDRI 2021), pp. 620–625 (2021)
30. Cook, A., Hammer, J., Elsayed-Ali, S., Dow, S.: How guiding questions facilitate feedback exchange in project-based learning. In: Proceedings of the 2019 CHI Conference on Human Factors in Computing Systems, pp. 1–12 (2019)
31. Cardoso, V., Bidarra, J.: Open and distance learning: does IT (Still) matter? Eur. J. Open Distance E-Learn. **10**(1), 1–6 (2007)

Models and Methods of Online Team Teaching

Gary Natriello[1]([✉]) [iD] and Hui Soo Chae[2] [iD]

[1] Teachers College Columbia University, New York, NY 10027, USA
gjn6@tc.columbia.edu
[2] New York University, New York, NY 10003, USA

Abstract. Although the majority of classroom instruction is conducted by a single teacher, practices such as co-teaching or team-teaching have evolved in face-to-face settings to expand the possibilities for delivering learning experiences. (Here "team teaching" refers to any of the various multi-person instructional arrangements for delivering learning experiences.) The online venue offers a new setting for team teaching that presents both challenges and opportunities for creative educators. On the one hand, the challenges include the need to coordinate within the constraints of digital environments while managing what is often a changing and unstable technical infrastructure. On the other hand, the needs connected to attending to complex multi-application and multi-platform configurations can make multi-instructor arrangements both essential and powerful. In this session we will consider the growing variety of models for team teaching online. Each model will be described and examined for its potential to contribute to high impact learning experiences in digital environments. This front-end learner perspective will be complemented by a discussion of the back-end technical means available to support multiple instructors, often in multiple locations and brought in by means of diverse applications.

Keywords: Team teaching · Online learning · Models

1 Introduction

In this paper, we consider a variety of models and methods for team teaching in networked and online environments with a goal of expanding the set of possibilities for facilitating learning. We aim to describe both the existing models that occur in online classes and additional models that may be borrowed from other sectors and venues in an effort to expand the possibilities. Our assumption is that, while no single model is the best option for all occasions, different learning goals and conditions will require different combinations of instructional staff to achieve the best outcomes.

1.1 The Networked and Online Environment

The unabated growth of learning in online and networked environments that we have experienced over the past several decades [7] has only been accelerated by the impact of the COVID-19 pandemic which forced many heretofore face-to-face educational institutions to move to online delivery [14]. The rapid transition to online delivery left little time for exploration and innovation, but it did highlight the need for such as we look to a future where online and networked environments will play a greater role in all types of education.

1.2 The Teacher at the Center

There are several dimensions of the shift to online delivery of education that motivate the need to consider models involving multiple teachers. Perhaps most important is the continuing premium placed on having live teachers to deliver the learning experience [14]. The delivery of learning experiences remains instructor dominated or oriented despite hopes that digital affordances might displace live instructors or at least diminish their role in formal educational processes [13].

The reinforcement of the teacher role may stem from the advent of video communications platforms such as Zoom, Meet, Teams, and Webex. These platforms, which came into widespread use as a result of the rapid shift to online learning due to the COVID-19 pandemic, place emphasis on synchronous delivery of educational experiences by a teacher. In many cases, these video communications platforms have displaced earlier asynchronous learning delivery platforms that emphasized the provision of curated resources, either text based or recorded video, often accompanied by student exercises in text formats. Such asynchronous formats promised to extend learning opportunities without a heavy teacher presence, but live video classes bring the teacher back to the center of things.

While the delivery of online learning through live video formats, offered a relatively straight forward translation of classroom-based, teacher-led activities, ambitious educators soon learned the limitations of point-and-shoot video production, and these limitations became more apparent as educators tried to introduce more innovative formats to keep students engaged and allow for more active participation. The demands of more complex delivery put pressure on educators to seek assistance from others and led, in some cases, to the formation of teaching teams. Such team configurations continue to evolve.

2 Models

The literature on team teaching or co-teaching encompasses a wide range of options for two or more teachers working together on a class session or a course [1, 5, 8]. Although there has been less discussion of team teaching online [4, 15], online delivery opens up an additional set of options for collaborating on instruction and presents some new approaches and models for consideration. In addition, the online environment suggests some additional sources for delivery models. In this section we consider three sources of such models: face-to-face learning settings, broadcast media, and production teams.

2.1 Models from Face-Face Learning Settings

Considering the range of practices that have grown up in face-to-face learning settings, Friend and Cook [3] use the term co-teaching to label situations where two teachers are working in a single classroom. They identify six different models for two-person teaming: (1) teaching/observing where one teacher provides instruction while another one watches student learning; (2) teaching/assisting where one teacher provides instruction to the entire class while a second teacher works with individual students who may need more help; (3) tag team teaching where both teachers provide instruction at the same time; (4) parallel teaching where two teachers provide instruction to two groups of students in the same class; (5) alternative teaching where one teacher provides instruction to the class while a second teaches a small group in need of additional help; and (6) station teaching where students rotate among teachers who each handle part of the curriculum.

Beyond these six types, there are several additional configurations common in k-12 educational institutions. For example, student teachers often begin working with a supervising teacher by joining them in their classroom. In such cases, the roles of the two will evolve over time with more responsibilities shared until the student teacher assumes the role of teacher alone. In many cases, k-12 classrooms may have a teacher aid to share some classroom duties. In other cases, special teachers may join a classroom to provide additional assistance to one or more students who need additional support. These arrangements may not always constitute team or co-teaching in the sense of two staff planning and coordinating their activities from the start.

The typology developed by Friend and Cook [3] is most directly drawn from experience in k-12 educational settings. Experience in higher education offers additional options [12]. These include working with one or more teaching assistants, composing teaching teams with colleagues from two or more specialties within a field, two or more disciplines, or from two or more departments or schools [11]. In addition, it is somewhat common to invite guest speakers into courses. Teaching teams may extend to include individuals in staff roles such as librarians [2].

2.2 Models from Broadcast Media

As online learning becomes more immediate with both video and audio elements, it starts to resemble the long familiar configuration of live broadcast media that we have experienced in many forms and formats over decades. Broadcast media has evolved to adopt a set of collaborative and interactive formats for on-screen talent that lead to engaging programming. These formats have become common for live broadcasts of both news and entertainment programs. They are worth considering as models because in addition to being engaging they are proven as sustainable models. As such, they may be quite useful for online live video class sessions.

The *host with a single guest model* is common in broadcast media. In this model the main player or host brings on a single guest for a conversation. The host leads the conversation to reveal something interesting about or in the experience or expertise of the guest. In an educational setting, this would allow the host teacher to engage the guest to inform the class on a topic. In comparison to the guest speaker model common in higher education, the model from broadcasting showcases the conversation between the two individuals led by the main instructor.

The *host with a guest panel model* expands the guests from one to several individuals who all participate in a conversation led by the host. The host directs the conversation. The members of the panel typically restrict their comments to the back-and-forth with the host. Avoiding conversations between panelists reduces the chances that members of the guest panel will speak over one another.

Co-anchors in the same location, generally at the same desk, are a familiar format for local news programs. In this model, the two individuals share responsibilities equally throughout a program. In some cases, they refrain from speaking directly to one another, but in other cases there is interaction between the co-anchors, often in the form of casual banter. Throughout the program, the co-anchors provide a continuous stream of information.

Co-anchors in different locations, sometimes in different studios and often in different cities, present another option. In this model, the two individuals again share responsibilities, but with this configuration, the responsibilities are sometimes divided based on relevance to location. With the anchors out of immediate face-to-face contact, the coordination challenges are somewhat greater and often rely on the audio-video connection to provide sufficient cues for smooth transitions.

Host with a Sidekick is another distinctive model in broadcast settings. This model includes the very specialized relationship and special role of the sidekick. With the term originally derived from the "kick" or the front or safest pocket from theft, the sidekick is a person's closest companion [16]. In much of literature and in broadcast settings, the sidekick can serve several unique purposes, including serving as an alternative to the hero or main character by doing things the main actor cannot or by being more accessible to the audience members.

Host with Band Leader is another model common in certain broadcasts. In this model, the host interacts with another individual who also directs the band or other group providing live music. Although this relationship and role is not as dominant as the sidekick, it functions in similar ways at times and offers the added element of live music as part of the interaction.

Of course, this list of models is not nearly complete over the course of broadcast programming, but these particular models do seem to have persisted over time. This may be because they are engaging for the audience. It may also be because these models can be readily enacted and maintained. As we approach the educational applications of such models, both impact and feasibility are important considerations.

2.3 Production Team Models

Focusing on broadcast media as experienced by viewers as a source of models for online team teaching necessarily causes us to focus on the front stage elements. This is clear in the prior discussion of models, where the emphasis is on talent in front of the camera or cameras. Of course, in many broadcast settings or even in reporting engagements, there are also individuals on the same team who are working behind the camera.

McKay [6] provides a general list of roles and responsibilities for television news shows. He includes the following roles behind the camera:

- Producer—producers oversee broadcasts and might write, edit and collaborate with reporters in the field
- Director—directors plan the broadcast, schedule content and maintain quality control
- Writer/Editor—writers and editors prepare scripts and produce content

- Camera Operator—camera operators set up and manage the camera or cameras in a broadcast
- Broadcast Technician—broadcast technicians ensure sound and video quality by monitoring broadcasts in real time
- Audio Engineer—audio engineers regulate volume levels and sound quality

In addition, there may be personnel handling the lights, props, makeup, media, and cue cards. These roles are assembled in different combinations with the same individuals often playing multiple roles on small production teams. Together, they provide a broad idea of what roles and responsibilities might be included in a team for our purposes.

3   Factors Affecting Model Selection

As we have shown, there are a number of models that might be available for those intent upon team teaching. Of course, all models are not equally useful or appropriate under all circumstances.

3.1   Initial Conditions or Constraints

It is almost always the case that there are some initial conditions or constraints that shape the possibility set for team teaching. In some cases, only one instructor is available or assigned; in other cases, there may be more staff but their status or capacity is defined. For example, an instructor and a teaching assistant (often a student not enrolled in the class) may be available. In such cases, the role of the teaching assistant is typically limited either by the nature of the appointment or by the expectations of the students in the class.

The availability of instructional staff is often constrained by resources since it is more costly to engage multiple staff for a class. Because instructors themselves often have little or no control over the initial staff configuration, they must adjust to the conditions presented. In some cases, this means accepting whatever is made available. In other cases, a lead instructor may be able to recruit some volunteers to augment the instructional team. Of course, depending on the conception of learning driving the course, students themselves may be enlisted to play various instructional roles. And we have all been in instructional settings, where a student has become the technical assistant for the less technically adept teacher.

3.2   Online Constraints

Because good communication, both in the planning phase and in the delivery phase, is essential for good collaboration and because such communication typically requires more effort at a distance, teaming for online instruction can present additional constraints. Some teaching teams will be co-located and deliver instruction to students at a distance, and in those cases, the collaboration will benefit from the physical proximity of the instructors. However, in other cases, the members of the instructional team themselves will be working at a distance from one another, and so will rely on online communication and collaboration technologies.

Despite the proliferation of online tools and platforms, the choices available to teaching teams for organizing their efforts are constrained by the design of the technical systems. It is important to recognize such constraints in advance when planning instructional activities. There are, for example, typically limits on the number of roles and rights associated with such roles that can be found in online video communication platforms. These limits may affect both the activities of instructors and the activities in which students may be asked to engage. In some cases, such limits make certain instructional strategies cumbersome while in other cases they become impossible to implement at all.

While the constraints facing instructors working online are real, it is worth noting that networked and online learning situations offer some additional possibilities for the instructional staff. When physical presence is not required because the class is online, the local pool of possible staff is no longer a constraint as individuals may be located anywhere. This sometimes allows individuals at a distance to participate as members of the instructional team.

3.3 Pedagogical Configurations

The range of pedagogical configurations intended or anticipated for a class or set of classes is often the primary consideration in making decisions about team teaching and the model adopted or created. Typically, the more elaborate the pedagogical arrangements, the greater the need for a team approach. Such arrangements may be important for responding to a diverse group of students with special learning needs (e.g., learning disabilities), for using labor intensive instructional strategies (e.g., personalization), or for engaging students in collaborative learning (e.g., groups) where there is a need for targeted coaching.

The online environment may interact with these and other pedagogical approaches to place additional pressure on instructional staff resources and options. In particular, online environments typically alter the visibility of instructional activities, making some aspects more visible and others less so, often in ways that reduce teaching options.

On the one hand, the greater visibility of all teacher-student interactions makes some interactions less feasible for one teacher to manage since all of the other students in the class have a front-row seat with mic'd audio and can listen in to what might more appropriately be a private conversation or at least one covered by the background hum of a face-to-face classroom. On the other hand, the more limited visibility of some online arrangements may make certain practices more cumbersome for a single instructor. For instance, group activities that can be managed by a single teacher in a face-to-face classroom where all groups can be monitored even as a teacher works directly with one group, are more difficult to handle for a teacher who must stop into each of a number of online small group breakout rooms in a single class session, leaving the others out of sight and hearing. In situations where one teacher does not have suitable visibility, a team might be a solution.

3.4 Emerging Technical Possibilities

Although, in many situations, instructors are limited in how they might think about teaming options by the technical platforms that they are using or have been assigned to use, there are new options emerging that might be exploited by those interested in leveraging them to create new combinations. As online technologies for communications and collaboration continue to evolve, there will be additional team teaching possibilities.

For example, with many online instructors using one of the rising video meeting platforms (e.g., Zoom, Teams, Webex), some are seeking to overcome some of the limitations by employing additional technologies that add flexibility that facilitates certain pedagogical strategies. The video meeting platforms offer certain capacities optimized for meetings. These include features such as breakout rooms, polling, and a chat window. Some of these are useful for teaching, but they typically support a single teacher model.

Several technologies have found their way into the practice of online teachers and some of these support team arrangements. For example, borrowing from the gaming community, some instructors have adopted Open Broadcaster System (OBS; https://obs project.com/), which allows them to compose a variety of scenes that can use multiple views within a single stream sent to a video meeting platform in the place of a webcam. In addition, the composed scenes can use multiple layers to create complex text and image overlays. It is also possible to transition from one scene to another in a single class session in the same video meeting system. For team teaching, these capacities allow multiple instructors based in the same location and captured on different cameras to be combined into a single scene or to appear in different scenes that can be transitioned throughout a class. The multiple scenes are not limited to camera views; in addition, it is possible to bring in views of one or more browser screens or other media. Like the camera views, these can be combined into a single view or rotated through a sequence of instructional activities. For example, the gaming community routinely features a video stream that shows a team playing a computer game, a narrator discussing the game play, and one or more branding overlays or multimedia elements punctuating the action. OBS, thus, provides a level of fine grained control over the composition and sequencing of activities on the screen that is not possible with the video meeting platform alone.

OBS alone is particularly useful for assembling and composing locally available resources, either individuals or media elements. As such it can be effective in creating team friendly online environments when the members of the team are in the same location. Other technical platforms can be used together with OBS to support distributed teams. NDI (Network Device Interface; https://www.ndi.tv) is a video-over-ip transport and protocol that allows remote video streams from any location on the internet to be brought into a production in a single location and then sent out as a single integrated program to a streaming service or a video meeting platform. In addition to bringing in local and distant cameras, NDI tools provide the functionality of a production studio with coverage anywhere there is connectivity. VDO Ninja (https://vdo.ninja/) offers another approach to bringing in remote video feeds into OBS using WebRTC, an open source project supporting real time communications to allow audio and video to work inside web pages.

The capability to bring remote video streams into a program like OBS provides the foundation for teachers to bring team members from anywhere with an internet connection into a live video session. Moreover, unlike video brought into a meeting platform, teaching team members from different locations along with resources can all appear in a single integrated scene.

4 Team Teaching in Action

The possibilities for team teaching online are many and growing as teachers gain more experience online. In our own case, we have had the opportunity to teach online over the past several years, and this has allowed us to try several configurations. Moreover, teaching together for multiple courses led us to refine our approach and to experiment with some new elements from time to time.

4.1 Initial Conditions

We did come with certain background and work within certain parameters that shaped our overall approach to teaming, and it is important to understand both. First, we had worked and taught together in face-to-face settings for many years before teaching together online. Second, the synchronous class sessions for the courses were offered on the Zoom video conferencing platform. Third, we would not be located in the same place for the duration of the class. This meant that we would be collaborating at a distance. Fourth, the classes typically included 20 to 25 students.

4.2 Zoom Options

The Zoom platform provides several features that we adopted for the class sessions. These included the roles of host and co-host, the ability to share a computer screen with the class, the ability to create breakout rooms for small groups, a polling feature, and a real time chat. We typically made use of these features throughout the courses.

4.3 Course Offerings

Our direct experience with team teaching is the result of co-teaching five courses over two years from fall of 2020 through spring of 2022:

- Networked and Online Learning
- Evaluation: Individuals, Groups, Institutions, and Society
- Data, Learning, and Society
- Learning in Small Groups
- Technology and Human Development

These offerings have been discussed in earlier work reporting on our efforts to create conditions for self-directed learning [9] and to employ project-based learning in an online course [10]. Here, we will discuss our approaches to team teaching.

Gary	Hui Soo
• Course designer	• Course technologist
• Content curator	• Session designer
• Session planner	• Session co-pilot
• Session leader	• Session resource sharer
• Reflection respondent	• Time keeper
• Project evaluator	• Participation monitor
• Session DJ	• Discussion board facilitator
	• Session photographer

Fig. 1. Role responsibilities.

4.4 Team Roles

One way to begin to describe how we approach team teaching online is to examine the roles we assumed in the classes (Fig. 1).

These responsibilities have evolved over the set of courses as the team arrangements have been refined. Early on, we had a teaching assistant who provided additional support to students outside of the class sessions, but we found it a bit too complex to work a third actor into the synchronous class sessions. This was a result of the rather informal hand-off practices we used during the class sessions. Because we had been teaching together for many years, these informal processes worked smoothly.

The role responsibilities varied a bit at times. The table shows the role responsibilities when Gary was the instructor of record. On those occasions when Hui Soo was the instructor of record, he assumed the session lead and feedback responsibilities and Gary moved to the session co-pilot role.

4.5 Teaming

Our team teaching approach continues to evolve, but there are certain patterns that seem to be consistent over time. We attribute this consistency to the fact that these patterns address and often solve a number of problems associated with teaching online.

A simple but important element is that, in addition to the planning that co-teaching instructors typically engage in to be ready for each class session, we always get online prior to the class start time and review the plans for the session. This is made possible by having a teaching venue that is available to use without interference and without the distractions of campus based life. Such pre-class review time and space is often not available due to tight class schedules in a particular classroom and to the fact that students often drift in before the start of the class. In the online venue, we stage the students in a waiting room until we are ready to begin, and we use the time before class to review our plans for the session and to do a final check on the technology platforms we will be using for the class. This kind of tech check is important to make certain that any technical problems are resolved before students join.

The monitoring process continues throughout the class session when it also includes paying attention to student activities on the platform, including student technology, chat messages, gestures, and expressions on video. All of these activities require attention, and it has proven very beneficial to have one instructor systematically engaged in these activities while the other moves through the planned session activities. Student responses in face-to-face classrooms are typically hard to miss but noting them in online platforms requires more active monitoring.

The team arrangement also provides an important option for the use of breakout rooms. When students are sent to breakout rooms for small group work instructors may be needed to provide assistance to a group. In this scenario with a multi-instructor team, one instructor can visit a breakout room on request while the second instructor can remain in the main class session in case members of other groups need to rejoin the main room for one reason or another. Of course, the second instructor can also monitor the groups and provide additional instructions to all groups during this time, as well.

In the most recent semester, we have adopted a regular session feature in which content is introduced by having one instructor interrogate the other about one or more topics of the week. This allows us to introduce content in the course of a conversation and allows for dialogue to clarify key concepts and address questions from the chat or comments from the discussion boards.

Having two instructors also has benefits for demonstrations where we are able to model the kinds of responses and interactions we hope to see students engage in during an activity. Students often need an active demonstration to illustrate how an activity is intended to work in a class, and such demonstrations seem more important in the online venue, since the activities are more likely to be ones they are encountering for the first time and since the online format sometimes makes it more difficult to communicate.

5 Conclusions

Teaming for online learning offers a number of advantages for those in a position to have a team teaching or co-teaching arrangement. There are all of the traditional advantages of team teaching such as the more expansive view of a field or multiplicity of viewpoints on key questions.

But beyond the advantages cited for team teaching in face-to-face settings, there are some additional advantages of team teaching online. We have already noted the possibility of having different members of the team assume different responsibilities. Dividing instructional responsibilities allows for a more diverse repertoire for teaching that seems to make synchronous class sessions more engaging for students. It also allows the teaching team to model more diverse and informal interactions around course content for students, which in turn, encourages them to engage in more such interactions themselves. This heightens student active learning.

Because online learning with synchronous sessions requires the equivalent of live broadcasting, a team teaching or co-teaching arrangement relieves a single teacher of some of the tasks associated with managing the session. This allows each member of the team to focus on some key dimensions of the session while monitoring the outcomes of the efforts of other team members, and at times providing corrective feedback on the fly.

Perhaps the most important advantage of team teaching for online learning is that it allows instructors to aspire to more ambitious class sessions involving multiple pedagogical and technical practices that are simply not possible for a solo teacher. Ambitious trajectories such as these promise to advance online learning in the years to come.

References

1. Anderson, R., Speck, B.: Oh what a difference a team makes: Why team teaching makes a difference. Teach. Teach. Educ. **14**(7), 671–686 (1998)
2. Bharuthram, S., Mohamed, S., Louw, G.: Extending boundaries: Team teaching to embed information literacy in a university module. South Afri. J. Inform. Stud. **37**(2), 1–17 (2019)
3. Friend, M., Cook, L.: Interactions: Collaboration Skills for School Professionals. Pearson, New York (2016)
4. Fuller, R., Bail, J.: Team teaching in the online graduate environment: collaborative instruction. Int. J. Inf. Commun. Technol. Educ. **7**(4), 72–83 (2011)
5. Kariuki, M., Jarvis, D.: Co-teaching in graduate education. In: Jarvis, D., Kariuki, M. (eds.) Co-Teaching in Higher Education: From Theory to Co-Practice, pp. 202–221. University of Toronto Press, Toronto (2017)
6. McKay, D.: Who works in a television newsroom? The Balance Careers (2019). Online at: https://www.thebalancecareers.com/tv-news-careers-525690
7. McPherson, M., Bacow, L.: Online higher education: beyond the hype cycle. J. Econ. Perspect. **29**(4), 135–153 (2015)
8. Minett-Smith, C., Davis, C.: Widening the discourse on team-teaching in higher education. Teach. Higher Educ. **25**(5), 579–594 (2020)
9. Natriello, G., Chae, H.: Designs for supporting self-directed learning online. Paper presented at the Innovate Learning Summit Online (2021)
10. Natriello, G., Chae, H.: Taking project-based learning online. Paper presented at the Learning Ideas Conference. New York (2001)
11. Nungsari, M., Dedrick, M., Patel, S.: Team teaching an interdisciplinary first-year seminar on magic, religion, and the origins of science: a 'piece-by-piece" approach. J. Scholarsh. Teach. Learn. **17**(1), 24–36 (2017)
12. Plank, K. (ed.): Team Teaching: Across the Disciplines, Across the Academy. Stylus Publishing, Sterling, VA (2011)
13. Roblyer, M., Castine, W., King, F.: Assessing the Impact of Computer-Based Instruction: A Review of Recent Research. The Haworth Press, London (1988)
14. Rodriguez, C.M., Cobo, C., Munoz-Najar, A., Ciarrusta, I.: Remote Learning During the Global School Lockdown: Multi-Country Lessons (English). Washington, D.C.: World Bank Group (2020). http://documents.worldbank.org/curated/en/668741627975171644/Remote-Learning-During-the-Global-School-Lockdown-Multi-Country-Lessons
15. Scribner-MacLean, M., Miller, H.: Strategies for success for online co-teaching. J. Online Learn. Teach. **7**(3), 419–427 (2011)
16. Sidekick: Wikipedia (2022). https://en.wikipedia.org/wiki/Sidekick

The Use of Comics as a Teaching and Learning Tool

Jenny Pange[(⊠)]

Laboratory of New Technologies and Distance Learning, School of Education, University of Ioannina, Ioannina, Greece
jpagge@uoi.gr

Abstract. Nowadays, there is great interest in using comics for teaching and learning. Comics are used by many educators for face-to-face or online teaching. They are important educational tools, and students develop visual literacy skills. Students enjoy creating comics because they freely express themselves, use realistic expressions, and make emotional stories. This study aims to discuss the use of information and communication technologies (ICT) to design comics for children in kindergarten for learning statistics. A group of undergraduate students, in the Department of Early Childhood Education at the University of Ioannina Greece, developed comics for teaching basic concepts of statistics to children in kindergarten. Students were free to create a comic to teach children the notions of mean, median, and mode. The most popular free online ICT tools used by undergraduate students for creating comics were: "Storyboard That," "Pixton," "MakeBeliefs-Comix," and "Scratch." Students used the microlearning approach to present their comics in the classroom. The main advantage of the application of comics in the teaching process is that they can present basic statistical concepts, with a very challenging and humorous approach using visual arts.

Keywords: Comics · Digital storytelling · ICT · Learning · Teachers · Students

1   Introduction

Learning in the 21st century requires teachers and students to be more creative, forward-thinking, flexible, and able to use innovative methods for teaching and learning. Most students nowadays are digital natives, and Information Communication Technologies (ICT) play an important role in their learning process. Additionally, students using technology strategically upgrade their communication skills, their collaboration skill, their creativity, and their critical thinking [1, 2]. These days, traditional teaching methods changed and include more ICT skills and online learning. Teachers empower students continuously with new ICT skills, for their greater involvement in the learning process. Amongst ICT tools which support critical thinking, collaboration, motivation, and creativity are digital storytelling and specially comics. Digital comics are digital tools for storytelling which combine graphical and textual elements. They are goal-oriented, eliminate stress for learning and make learning more concrete in a digitized environment [3].

© The Author(s), under exclusive license to Springer Nature Switzerland AG 2023
D. Guralnick et al. (Eds.): TLIC 2022, LNNS 581, pp. 281–289, 2023.
https://doi.org/10.1007/978-3-031-21569-8_26

Digital storytelling is a process using digital media platforms for fiction and non-fiction stories, involving interactivity [4, 5]. Digital storytelling is not a new concept [6], but due to the recent increase of the use of distance learning, it has nowadays started to appear in many school curricula [7, 8]. Additionally, storytelling appears either as a stand-alone approach or is included into other pedagogical approaches, such as project-based learning [9]. This process requires the learning material to be reduced and presented in a well-defined digital form [8, 9].

Digital comics are in a way digital books and they are used in digital storytelling. In education, they are considered as creative activities and useful ICT tools for both teachers and students. Many students nowadays are digital narratives [10, 11]. Students who like to develop comics gain knowledge, skills, and practices, like photomontage, text creation, and drawings. This way they present their thoughts or ideas as a digital story with pictures or drawings in separate panels [12, 13]. According to reference [14], comics have the "ability *to collapse time and space in one panel*" and develop critical thinking. Consequently, digital comics include many digital forms like text, dialogues, narration, and so on and have the structure of webcomic or mobile comic [15, 16]. Nowadays, there is a great interest in schools using comics for teaching either in face-to-face or online processes, as they support learning and students' need to collaborate [14]. A recent study on the experiences of creating comics by prospective teachers shows that engaging students in collaborative activities, through comics, supports building new knowledge and concepts, through critical thinking and role playing [10]. There are many subjects like mathematics, statistics, and new technologies where we can use comics. They are a challenging teaching process, since comics require the active engagement of students and offer control in the learning process [12]. In a recent study [12], comics are great tools for teaching math as they can convey the idea of the visual reference of x and y axes accurately. Additionally, students develop digital skills via comics and learn through entertainment [17].

Other recent studies refer to students' engagement in comics as a cognitive, emotional, and behavioral responsibility [13, 15]. In cognitive engagement, the promotion of learning and skills is obvious. Comics uphold the emotional engagement of students for their teacher, or their school and offer behavioral engagement which includes the student's participation in school-related activities or academic duties [18].

Nowadays, there are many free online tools for comic creation like, *Pixton, Make-BeliefComix, Storyboardthat,* and *Scratch* where students can make their own choice for use. Some online tools for comics need a lot of training and others do not. To evaluate students' performance on comics creation, we must follow an assessment which need to analyze specific learning goals [19, 20]. Amongst these assessment tools there is the "*Dimension Star*" model, and it is described [19, 21, 22, pp. 318] as a model which includes important learning factors like, the user contribution, the structure, the concreteness, the virtuality, the cognitive effort, the spatiality, the collaboration, the continuity, the control, the interactivity, the immersion, and the coherence (see Fig. 1).

Amongst digital comics creation tools, "Pixton" is widely used as a comic-making application with "subject-specific content packs" (https://www.pixton.com/). This tool can be used by students and teachers in many educational levels from elementary schools up to universities [21, 23–25]. According to these studies, "Pixton" empowers students'

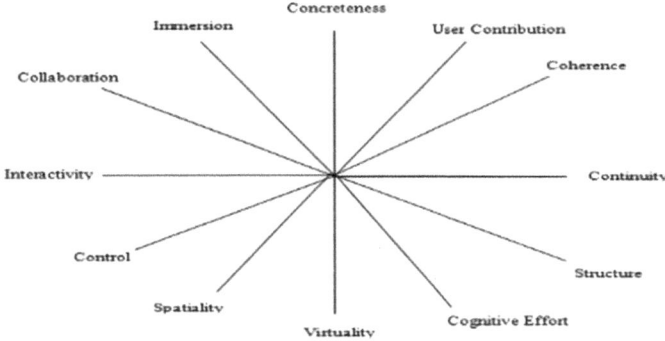

Fig. 1. The *"Dimension Star"* model [22].

skills development, reveals their creative and writing ability, and brings fun to the learning process.

Another online comic creation tool is "MakeBeliefsComix" (https://makebeliefsc omix.com/). This is an in-browser comic creator tool for teaching and learning, as well as a tool for developing personal skills. "MakeBeliefsComix" provides features that have the potential to be effective in supporting students to explore the learning subject in a stimulating and fun way [26].

Additionally, "Storyboardthat" (https://www.storyboardthat.com/) is a platform for creating digital stories. "StoryboardThat" allows people of different skill levels to create visual objects for teaching, learning, and communication. Researchers consider "StoryboardThat" as a platform for developing creative skills [27].

The well-known at school level "Scratch" (https://scratch.mit.edu), is an educational programming tool from the MIT. "Scratch" allows students to create their own digital stories, games, and animations developing at the same time computational thinking, problem-solving skills, creative learning, and cooperating skills [28, 29]. Additionally, "Scratch" develops use-of-media skills and human-computer interaction [28].

The aim of this study is to explore how undergraduate students in Educational Departments can use digital comics to create digital stories and prepare courses for children in kindergartens to learn basic statistical notions.

2   Materials and Methods

This case study took place during pandemic, and it was applied in a distance learning process. Undergraduate students had to create for young children online comics with concepts of mathematics-statistics. These comics should be short in panels, and students made them for children in kindergarten. The sample of this pilot study was self-selected and consisted of 10 undergraduate students (8 females and 2 males) aged 21 up to 24 years of age. Their main task was to create simple digital comics using statistical concepts, in a project-based learning process. All students received one-week training on ICT tools for comics creation. Students were free to add text in their comics. Either the preschool teacher or the students had to read it to children during the presentation of the comic in the kindergarten. Students were free to choose any version either of "Pixton," "MakeBeliefComix," "Storyboardthat," or "Scratch" to create their comic.

For this study, a mixed model design was applied using qualitative and quantitative research methods accordingly. To study the cognitive engagement as promotion of learning and skills, the emotional engagement as the feelings of students for their teacher, and the behavioral engagement of students as active, passive, or not engaged responsibility, qualitative research methods were applied.

All students for one month were actively participating in creating their own comic. Students joining this study had to collect information from online sources about the introduction of statistical notions to children using digital environments. At first, they collected published papers on this topic. Then they discussed how to use the detailed national curriculum for ICT skills and statistical ideas for preschool children. Afterwards, they cooperated with the teacher and any classmates they liked to develop the story for the comic. Ethical issues were discussed with the teacher and applied in the comics.

As rubrics for the final evaluation of the comics, we used characteristics from the "*Dimension star Model*" and we considered their presence or absence in the comic. Specially, we examined the user contribution (role), the logical structure of the comic, the concreteness of the statistical idea, the virtuality of the comic, the cognitive effort, the spatiality, the collaboration, the control of the comic, the interactivity, and the possible engagement of children.

3   Results and Discussion

All comics were designed by each student, individually. The length of all comics were nine panels. Students aged $21.9 + 1.2$ years said that their experience in using comics to prepare in-class digital material for children was meaningful. All students in our interview said that they achieved extra learning abilities and storytelling skills. Their emotional engagement for the creation of the story was high, and they had active engagement in the course material of teaching statistics to preschoolers. They used narratives or everyday life situations to create the short educational comics. All students tried to inspire children with the text in their comics. Additionally, they found simple graphical elements to present the mean, median, and mode were taken from everyday life situations.

Two of the students in this case study were male, and eight were female. The images included objects collected from everyday life activities. Students used pictures of common events to help children understand how statistical ideas appear in everyday activities. In the interviews following the completion of the comic, all students said that they were convinced to use comics to present statistical ideas in a digital storytelling form. A student said, "*It is very interesting to use digital comics and create a story to teach statistics to preschoolers.*" Other students said, "*There is no way to make a mistake when you use comics to deliver statistical notions to students,*" "*difficult ideas are transferred easily to children with comics,*" "*text and pictures make comics the best way for teaching statistics,*" and "*digital comics combine ICT tools in a learning by doing teaching method.*"

Students also reported about the ICT tools they used to create comics. Students said that "Pixton" (Fig. 2.) "*is an easy to use tool,*" "*it has bright colors,*" "*has many different characters but without a big variety of backgrounds,"* and *"the size of the objects remains unaltered in the free version, so we had limited ability for presentations."*

Fig. 2. "Pixton" comics.

Moreover, "Storyboardthat" (Fig. 3), was used by three students who said that *"it is an excellent tool for online comics,"* *"it has many pictures available for background, clear colors, and it is easy to use,"* and *"is character generator."*

Fig. 3. "Storyboardthat" comic.

Additionally, "MakeBeliefsComix" (Fig. 4) was an easy tool for comics for three students. Students said that *"the use of this tool is a great because has many cartoons"* and *"this cartoon has also many different sizes of balloons to write the text, so we can create unique narratives to teach statistical ideas."*

Fig. 4. "MakeBeliefsComix" comic.

Fig. 5. "Scratch" comic.

Lastly, one student used "Scratch" (Fig. 5) to create a comic for statistics. In this comic, there were fruits and bread to be collected from the child who later had to count them and find the mode. He said that "*Scratch is a good program for comics but we must have good knowledge of programming. … It is difficult for pre-preschool teachers to use it because they have little knowledge on scratch programming.*"

At the end, all students presented their comics in the classroom, discussed the creation of their comic, and replied to a questionnaire on the main characteristics of the digital comics they created (Table 1). Students had to tell us their opinions about the application of digital comics for teaching statistics to children. According to their replies, we found that all students felt positively for the application of comics in the teaching statistics. They supported the user contribution, cognitive effort, spatiality, collaboration with the classmates, virtuality of the comic, control of the comic, and engagement of children during teaching. We received only two negative replies on the logical structure of the comic, as they indicated that statistical notions are complicated for children so they need more frames and longer stories to offer a statistical reasoning. Additionally, we received three negative replies on the concreteness of the statistical ideas in digital comics. These students said that a series of more than three comics are needed to present properly a statistical idea to children. One student also said that only the comic in "Scratch" had interactivity for the children.

Table 1. Students' replies on the main characteristics of digital comics.

Characteristic	Positive replies	Negative replies
User contribution (role)	10	
Logical structure of the comic	8	2
Concreteness of the statistical ideas	7	3
Cognitive effort	10	
Spatiality	10	
Collaboration with the classmates	10	
Virtuality of the comic	10	
Control of the comic	10	
Interactivity	1	
Engagement of children during teaching	10	

4   Conclusions

The findings of this pilot study present the educational potential of use of comics to teach statistical ideas to children. Children in kindergarten learn by doing and learn easily using computers and online comics. Though it is difficult for all preschool teachers to use digital comics for teaching statistics in Greek kindergarten due to lack of infrastructure, comics remain as a valuable teaching ICT tool for preschool teachers. Comics combine active teaching strategies and structured learning. Although the generalizability is limited in this pilot study due the limited number the participating students, some conclusions obviously appear.

Students using online comics gain a unique opportunity to use technology for in-classroom activities, become interested in reading, and the comics keep them engaging. Moreover, preschool teachers must recognize the importance of digital storytelling and comics for teaching statistics and include them in their everyday school activities. In addition, comics increase student engagement on ICT tools, offer a rich learning visual environment, and help them to understand simple statistical ideas.

References

1. Pange, J.: Teaching probabilities and statistics to preschool children. Inf. Technol. Child. Educ. Annu. **1**, 163–172 (2003)
2. Lekka, A., Pange, J.: What ICT tools do undergraduate students use? In: 2015 International Conference on Interactive Mobile Communication Technologies and Learning (IMCL), pp. 386–388 (2015). https://doi.org/10.1109/IMCTL.2015.7359625
3. Rutta, C.B., Schiavo, G., Zancanaro, M., Rubegni, E.: Collaborative comic-based digital storytelling with primary school children. In: Interaction Design and Children IDC '20, June 21–24, London, United Kingdom, pp. 426–437. ACM, New York, USA (2020). https://doi.org/10.1145/3392063.3394433
4. Handler-Miller, C.: Digital Storytelling: A Creator's Guide to Interactive Entertainment, 3rd edn. Routledge (2014). https://doi.org/10.4324/9780203425923
5. Rizvic, S., Okanovic, V., Boskovic, D.: Digital storytelling. In: Liarokapis, F., Voulodimos, A., Doulamis, N., Doulamis, A. (eds.) Visual Computing for Cultural Heritage. SSCC, pp. 347–367. Springer, Cham (2020). https://doi.org/10.1007/978-3-030-37191-3_18
6. Wu, J., Chen, D.: A systematic review of educational digital storytelling. Comput. Educ. **147** (2020)
7. Çetin, E.: Digital storytelling in teacher education and its effect on the digital literacy of pre-service teachers. Think. Skills Creat. **39**(4) (2021). https://doi.org/10.1016/j.tsc.2020.100760
8. Kim, D., Li, M.: Digital storytelling: facilitating learning and identity development. J. Comput. Educat. **8**(1), 33–61 (2020). https://doi.org/10.1007/s40692-020-00170-9
9. Takemata, K., Minamide, A.: Poster: project based learning using digital storytelling: educational program for students before learning full-scale PBL practice. In: Auer, M.E., Rüütmann, T. (eds.) ICL 2020. AISC, vol. 1328, pp. 379–385. Springer, Cham (2021). https://doi.org/10.1007/978-3-030-68198-2_35
10. Tzifopoulos, M.: Digital comics in 21st century schools: preparing modern teachers. Pedagog. Rev. **65**, 160–178 (2018)
11. Tzifopoulos, M.: In the shadow of coronavirus: distance education and digital literacy skills in Greece. Int. J. Soc. Sci. Technol. **5**(2) (2020)
12. Manno, M.: Comics in the Classroom: Teaching Content with Comics (2014). https://teach.com/blog/teaching-content-with-comics/. Last accessed 06 February 2022
13. Lewkowich, D.: Talking to teachers about reading and teaching with comics: pedagogical manifestations of curiosity and humility. Int. J. Educ. Arts **20**(23) (2019). https://doi.org/10.26209/ijea20n23
14. Seelow, M.: Using Comics to Teach the 4 Cs (2020). https://www.edutopia.org/article/using-comics-teach-4-cs
15. Saputra, V. H., Pasha, D.: Comics as learning medium during the Covid-19 pandemic. Proc. Int. Conf. Sci. Eng. **4**, 330–334 (2021). http://sunankalijaga.org/prosiding/index.php/icse/article/view/681
16. Digital comic from Wikipedia (2022). https://en.wikipedia.org/wiki/Digital_comic. Last Accessed 12 February 2022
17. Rakhmayanti, F.: Comic maker app for enjoyable learning. J. Phys. Conf. Ser. **1987**(1) (2020). https://teach.com/blog/teaching-content-with-comics/
18. Nguyen, T.D., Cannata, M., Miller, J.: Understanding student behavioral engagement: importance of student interaction with peers and teachers. J. Educ. Res. **111**(2), 163–174 (2018). https://doi.org/10.1080/00220671.2016.1220359
19. Azman, F.N., Zaibon, S.B., Shiratuddin, N.: Digital storytelling tool for education: an analysis of comic authoring environments. In: Badioze Zaman, H., et al. (eds.) IVIC 2015. LNCS, vol. 9429, pp. 347–355. Springer, Cham (2015). https://doi.org/10.1007/978-3-319-25939-0_31

20. Azman, F.N., Zaibon, S.B., Shiratuddin, N., Dolhalit, M.L.: Evaluation of production model for digital storytelling via educational comics. In: Piuri, V., Balas, V.E., Borah, S., Syed Ahmad, S.S. (eds.) Intelligent and Interactive Computing. LNNS, vol. 67, pp. 513–524. Springer, Singapore (2019). https://doi.org/10.1007/978-981-13-6031-2_45

21. Azman, F.N., Zaibon, S.B., Shiratuddin, N., Dolhalit, M.L.: Strategies for learner-generated comic production in classroom: a comparative analysis. Int. J. Human Technol. Interact. **3**(1) (2019). https://journal.utem.edu.my/index.php/ijhati/article/view/5118

22. Psomos, P., Kordaki, M.: Analysis of educational digital storytelling environments: the use of the "dimension star" model. In: Lytras, M.D., Ruan, D., Tennyson, R.D., Ordonez De Pablos, P., García Peñalvo, F.J., Rusu, L. (eds.) WSKS 2011. CCIS, vol. 278, pp. 317–322. Springer, Heidelberg (2013)

23. Cabrera-Solano, P., Gonzalez-Torres, P., Ochoa-Cueva, C.: Using Pixton for teaching EFL writing in higher education during the Covid-19 pandemic. Int. J. Learn. Teach. Educ. Res. **20**(9) (2021). https://doi.org/10.26803/ijlter.20.9.7

24. Cabrera, P., Castillo, L., González, P., Quiñónez, A., Ochoa, C.: the impact of using "Pixton" for teaching grammar and vocabulary in the EFL Ecuadorian context. Teach. Engl. Technol. **18**(1), 53–76 (2018)

25. İlhan, G.O., Kaba, G., Sin, M.: Usage of digital comics in distance learning during COVID-19. Int. J. Soc. Educ. Sci. **3**(1), 161–179 (2021)

26. Kohnke, L.: Make beliefs comix. RELC J. **51**(2), 321–323 (2020). https://doi.org/10.1177/0033688218765301

27. Wahjuningsih, E., Santihastuti, A., Kurniawati, I., Arifin, U.M.: "Storyboard that" platform to boost students' creativity: can it become real? IOP Conf. Ser.: Earth Environ. Sci. **485** (2020). https://doi.org/10.1088/1755-1315/485/1/012095.2020

28. Zhang, L., Nouri, J.: A systematic review of learning computational thinking through scratch in K-9. Comput. Educ. **141** (2019). https://doi.org/10.1016/j.compedu.2019.103607

29. Lekka, A., Toki, E., Tsolakidis, C., Pange, J.: Literature review on educational games for learning statistics. In: 2017 IEEE Global Engineering Education Conference (EDUCON), pp. 844–847 (2017). https://doi.org/10.1109/EDUCON.2017.7942945

30. Fuentes, D., Grimes, N.: Creating Google classrooms using Bitmoji and Google slides: an early pandemic pedagogical response. In: Langran, E. (eds.) Interactive 2020 Online Conference, pp. 114–119. Online: Association for the Advancement of Computing in Education (AACE) (2020). https://www.learntechlib.org/primary/p/218128/

Creating a Powerful Employee Experience: Lessons from Product Management

Jimmy Pearson[1]([envelope]) and Donna Murdoch[2]

[1] Google, New York, NY 10011, USA
[2] Columbia University Teachers College, New York, NY 10027, USA

Abstract. "The Great Resignation" poses significant challenges to retaining employees, particularly mid-career professionals in innovative industries like technology and healthcare [1]. But significant opportunities for the learning and development function accompany this challenge, as employers increasingly recognize how robust learning can support retention. But are we ready to fully realize this opportunity? Fortunately, human resources departments—particularly in technology companies—can look to Product Managers' expertise driving user engagement for innovative employee retention approaches. Reimagining a career as a "product" and the employee as a "user" can unlock new ways of thinking about the employee experience.

Keywords: Employee experience · Learning & development · Employee retention

1 Introduction

According to the Product Development and Management Association (PDMA), "Product design … allows firms to create, develop, and maintain competitive offerings to customers … Product managers need a thorough understanding of the development tools and techniques to bring successful and meaningful design to the consumer" [2]. Just as product managers seek to create value for their companies through the design of products for customers, the Human Resources function creates value through the design of employee experience. This value can be created through increased employee retention and the reduction of accompanying recruiting and hiring costs; increase in productivity per worker delivered by learning and development programs; increased organizational efficiency achieved through internal mobility programs that help employees find their highest and best use; the design of compensation and other incentive programs to drive employee behavior; or in myriad other ways that Human Resources helps employees thrive while mitigating talent-related risk.

Yet while product managers have well established design processes that consider their customers' user experience holistically, Human Resources teams often work in silos with minimal coordination. As "The Great Resignation" poses significant challenges to retaining employees, particularly mid-career professionals in innovative industries like technology and healthcare [1], the cost of failing to deliver excellent, well-integrated

D. Guralnick et al. (Eds.): TLIC 2022, LNNS 581, pp. 290–295, 2023.
https://doi.org/10.1007/978-3-031-21569-8_27

employee experiences is high. But so is the up-side: companies that invest now in getting the employee experience right, integrating all of its elements to help employees grow and thrive, will win the battle for top talent and emerge from "The Great Resignation" a step ahead of competitors. This includes integrating all elements of employees' "user journeys"—including learning and development—into a coherent experience that meets their needs and drives business outcomes:

- Recruitment: Only 5 in 250 applications are ever seen by a human (TopResume). The rest are filtered out by artificial-intelligence-driven Applicant Tracking Systems (ATS). What tradeoffs are we making in candidate experience in the pursuit of greater efficiency?
- Learning and development: Employees need to know the career benefit of learning. Learning experiences should map to competencies that support advancement and internal mobility, and employees should have the autonomy to explore them.
- Flex assignments: Opportunities to gain new skills must be paired with opportunities to demonstrate them in project-based work (e.g., short-term assignments, "20% projects"), supporting employees' pursuit of advancement or internal mobility.
- Internal mobility: New Talent Marketplace Platform software allows employees to explore new opportunities without giving up their existing roles. Employers find "hidden talent" within their organizations, and employees can reimagine their careers in a lateral and linear fashion [3].
- Job To Be Done: Users - or employees - may be motivated not just by growth opportunities, but also the desire or need to complete the job they're currently engaged in (and its constituent tasks.) Employers can improve employee satisfaction and retention by enabling employees to complete their "job to be done" efficiently and effectively.

To fully realize the opportunities posed by the Great Resignation, learning and development functions must thoughtfully integrate with other elements of employees' user journeys to deliver a best-in-class employee experience.

2   Recruitment

The employee experience begins before a candidate is even hired. Candidates begin forming opinions of a company when first applying, and the experience delivered by recruiters can not only land top talent for an organization, but also set them up for success through the rest of their journey as an employee. Conversely, a poor recruiting experience risks mismatching candidates to roles, inadvertently reflecting negatively on organizational culture, or failing to attract talent in the first place. The analogous user experience in product design management is conceived as a user funnel: a series of steps that a user goes through from the moment they become aware of your product to the moment they begin engaging with it directly. Yet too often, recruitment processes shortsightedly optimize for efficiency and fail to focus on candidate experience.

Pandemic-related burnout has left the workforce generally less engaged than in previous years. Companies must set themselves apart by cultivating an inclusive environment for existing employees. Importantly, this process begins with job seekers, too, and is dependent upon organizational culture around learning and development. But too many job requisitions still include terms like "self-starter," "thrive in ambiguity," and "hit the ground running": buzzwords that can effectively screen out desirable candidates. Job seekers might draw quick conclusions from terms like these, which can convey certain messaging about low employee investment, not only in learning and development, but across the board, in compensation and benefits.

As ATS rely more heavily on automation, job seekers have become savvy about keyword inclusion to trigger a prescriptive algorithmic match. But a job seeker being skilled at beating the machine doesn't indicate a good fit for the team nor, importantly, that they'd want to learn more. Ultimately, a resume is not a person. Instead, employers must work to identify lifelong learners: candidates who have proven success in another role are most likely great learners and will welcome the opportunity to develop in a new or adjacent space.

Hiring managers interested in humanizing their approach to talent or their tone in recruiting should investigate loosening the guardrails on applicant tracking software. These tools that algorithmically filter and select aren't necessarily yielding desired results. Chasing the dream candidate who knows everything and is ready for anything, as determined by an algorithm, may end up being a toxic addition to any team. Instead, it's worth Human Resources teams' time to really look at job seekers, no matter how tedious the process may seem. This thoughtful approach will also demonstrate the care that the organization treats its employees with to candidates, making them more likely to accept an offer and begin with a positive impression of organizational culture.

3 Learning and Development

After a user is "converted"—that is, completes the user funnel and begins engaging with a product—their interaction with a product is represented by a user journey map. User journey maps document how a user flows through a product or site, often deepening their engagement over time. The most well designed products do this organically, without a user even realizing it: consider the last time you scrolled through a social media feed, engaging with content through likes and comments, for over an hour without realizing it. But a lot of design goes into the nudges and signposts that direct a user through their journey map. In employee experience, learning and development can serve this function. Sometimes overtly, as in new hire onboarding programs, and sometimes subtly within the flow of work, learning opportunities help employees explore new domains and build new skills that prepare them for the next phase of their journey: the next project, promotion, or internal transfer.

Indeed, being intentional about retaining the employees that they have means companies must implement learning and development programs that help them navigate the employee experience. There's an important connection between employee retention and on-the-job training, as reflected in this stat from the LinkedIn 2021 Workplace Learning Report: "82% of L&D pros report that engaged learners are also more likely to participate in internal mobility programs." If the user journey at their current employer isn't clear, employees will seek clarity elsewhere, and the expense of hiring, outsourcing hiring, and helping new hires learn their role and the company culture and function can be substantial. An organization driven by thrift can lose sight of the importance of retention, and there's even more at stake when the situation involves legacy knowledge born from continuous learning and development.

4   Flexible Work Assignments

An important element of successful user journeys is their flexibility; user journeys that are too rigidly defined are likely to lead to attrition, as users are unable to customize their experience or discover novel product features that keep them engaged. So too, an employee experience must allow for employee experimentation and flexibility. This flexibility helps foster innovation within the organization; reinforces new skills and knowledge gained through learning and development; and provides valuable experience and work artifacts that can help employees pursue other internal opportunities (see below).

Google cites the need for innovation as a key motivator behind its own "20% time" flex work program, which it borrowed from 3M and allows employees to use company time and resources to explore their own ideas [4, 5]. Other employers have experimented with models including short term "bungee" or "stretch" assignments that only last a matter of weeks or months. While these employers implemented these programs in the name of innovation, and not employee experience, the truth is that the two are mutually reinforcing. Employees are attracted to meaningful, interesting, and novel work experiences. The more they get to innovate in those experiences, the more attractive they are. And the more employees learn and apply new knowledge, the more likely they are to foster further innovation.

5   Internal Mobility

If learning and development provides the guidance and signposts that help an employee navigate their user journey, then navigating between projects and roles constitute the steps of that journey. Importantly, these steps can be linear or lateral. Linear progression through a user journey includes taking on greater responsibility within role or function, or promotion. Lateral moves include transferring between functions to find the best fit between the role and the employees' skills and interests. Lateral moves are important because they allow employees to find their highest and best use—both to deliver value to the company and to thrive professionally, boosting employee satisfaction and retention. In product design, it's taken as a given that no one solution will satisfy all users of a product. That's why the design process includes the creation of user "personas," allowing designers and product managers to think about creating a variety of specific experiences that feel tailored to each user's needs. These personas are akin to different functions or job ladders. Building robust internal mobility programs allows employees to discover the user journey that best matches their persona, delivering greater value as a result.

There are several new software programs, called Talent Marketplace Platforms, that several companies (Unilever, Schneider Electric, IBM, and many others) use to facilitate internal mobility. Employees can keep their existing roles and learn about another. While they help companies find the hidden talent within their organizations, they enable employees to reimagine their own careers in both a lateral and linear fashion [3]. The vertical career path is only one option. Internal mobility should be about talent and skills, and professionals often want to learn new things or adjacent skills. The investment it takes—of the platform, manager buy-in, etc.—makes employees feel they are valued and are working toward something bigger.

These are days when employees have a new mindset. They often want a portfolio of careers, rather than a linear career. While it may seem counterintuitive, by giving them the opportunity internally, they are likely to engage and stay longer. While they may not stay as long as people stayed in the past, they will stay longer to learn and develop. Adult learners are more likely to pursue learning and growth internally when the personal or professional benefit—the "What's in it for me"—is clear. They learn because they want to learn for themselves or for something they care deeply about. By surfacing new opportunities that are available from pursuing learning, internal mobility can make the return on investment of learning more tangible to the employees and make it more likely that they will engage.

6   Job to Be Done

The elements of user experience discussed thus far have been goal-agnostic. User funnels, journey maps, and personas apply equally to an employee looking to advance to the C-suite, achieve optimal work-life balance, or have a meaningful impact on their industry (not necessarily mutually exclusive goals). But, perhaps obviously, the goal an employee is trying to achieve through their work matters a great deal. The Job To Be Done approach to product design provides a powerful way to define these goals and design an employee experience to help employees achieve them.

Simply put, the Job To Be Done approach starts with figuring out what a user (or employee) is trying to achieve, instead of the solution to their (often assumed) problem. An often cited example is a quote attributed to Henry Ford: "If I had asked people what they wanted, they would have said faster horses." Ford's innovation was identifying his customers' Job To Be Done—getting from place to place—and helping them do it more easily.

There's no single answer to employees' Job To Be Done. Not every employee wants to get promoted, earn more pay, nor have greater business impact. Yet the employee experience, by design or by accident, is often optimized for a single Job To Be Done. Up-or-out talent models, for example, may assume employees' desire to advance. Learning and development opportunities, likewise, emphasize skills employees will need in their "next" role. The ability to leverage employee experience for linear or lateral career mobility is discussed elsewhere in this paper, but other potential Jobs To Be Done exist.

For example, during the COVID-19 pandemic, many employers recognized the need to help employees cope with stress, build resilience, and achieve a healthy work-life balance, and provided additional wellbeing benefits and training opportunities to help them achieve these Jobs To Be Done. But while the pandemic may have helped employers recognize these Jobs To Be Done, they likely existed before and are unlikely to fade as the pandemic (hopefully) does; indeed, as the great resignation continues, many employers are finding that attracting and retaining top talent means continuing to design employee experiences that take these Jobs To Be Done into account.

Likewise, many employees are motivated not by advancement opportunities but simply the satisfaction that comes with completing their work and achieving impact. For these employees, learning and development focused on advancement not only fails to advance their Job To Be Done, but it may actively impede it (by detracting from time for job-related tasks). For these employees, increasing organizational efficiency and effectiveness may improve employee engagement and retention, supplemented by training - especially delivered in the flow of work—that directly helps them complete their job more efficiently.

7   Conclusion

For those in workplace management, the pandemic talent marketplace may have come too hard and fast to pivot their entire workplace culture, but Human Resources professionals who are aware of the needed change and open to learning from their product management colleagues can move toward organic capture of avid learners disguised as job seekers, the creation of meaningful learning to help guide them through their professional journey, flex assignments that foster innovation and professional growth, internal mobility programs that help them thrive, and support systems that help them effective and efficiently complete their Jobs To Be Done.

Acknowledgments. We'd like to thank the Education Entrepreneurship program at University of Pennsylvania's Graduate School of Education, where we began our collaboration and friendship.

References

1. Explainer: What's driving 'the Great Resignation'? (2022). https://www.weforum.org/agenda/2021/11/great-resignation-career-change-mental-health-covid/. Accessed 15 March 2022
2. Product Design & Development Tools (2022). https://community.pdma.org/knowledgehub/bok/product-design-and-development-tools. Accessed 15 March 2022
3. Bersin, J.: Talent management platforms explode into view (2020). https://joshbersin.com/2020/07/talent-marketplace-platforms-explode-into-view/. Accessed 15 March 2022
4. Guide: Foster an innovative workplace (2022). https://rework.withgoogle.com/guides/foster-an-innovative-workplace/steps/introduction/. Accessed 15 March 2022
5. Von Hippel, E., Thomke, S., Sonnack, M.: Creating breakthroughs at 3M. Harv. Bus. Rev. **77**, 47–57 (1999)

Peer-to-Peer Learning at Google and Peloton: The Power of Internal Experts

Jimmy Pearson[1(✉)] and Thansha Sadacharam[2]

[1] Google, New York, NY 10011, USA
[2] Peloton, Toronto, ON M6K1X1, Canada

Abstract. The traditional professional development model involves spending thousands of dollars on conferences, training, seminars and courses. Though this spending provides employees with an opportunity to learn and continue to grow professionally, it has its pitfalls. Often, these opportunities are pre-built, untailored, and require a large time and financial commitment. This paper explores the benefits of developing internal experts through peer-to-peer learning programs, drawing on the experience of innovative learning organizations like Google and Peloton. Peer-to-peer learning can:

- ensure learning content is hyper-relevant and organizationally specific,
- increase learner engagement and enjoyment by leveraging deep understanding of workplace cultures
- improve employee experience by providing powerful opportunities to grow through facilitation, with ancillary benefits for employee mobility and retention, and
- provide a more scalable, cost-effective means of achieving organizational learning objectives by highlighting internal peer experts for employees to learn from.

Keywords: Peer-to-peer learning · Employee retention · Employee mobility

1 Introduction

As competition for top talent has intensified amidst the great recognition, many leaders have gained a renewed appreciation for learning and development as both a key differentiating factor in attracting top applicants, and as a means of growing talent for senior roles from within the organization. To fully realize learning's potential, learning and development professionals must look beyond the traditional professional development model of spending thousands of dollars on conferences, training, seminars, and courses. Though this spending may provide employees with opportunities to learn and continue to grow professionally, it has its pitfalls. It's increasingly common for employers to invest in these kinds of learning opportunities, and it's unlikely to differentiate a company from competitors in the market for talent. More importantly, these opportunities are pre-built, untailored, and require a large time and financial commitment. An alternative means of

D. Guralnick et al. (Eds.): TLIC 2022, LNNS 581, pp. 296–301, 2023.
https://doi.org/10.1007/978-3-031-21569-8_28

building and sustaining a robust organizational learning culture is through peer-to-peer learning programs, leveraging existing employees as internal experts to lead learning experiences for their peers.

At Google, employees actively volunteer to share their knowledge and passions through g2g (Googler-to-Googler). This vibrant peer-to-peer learning program empowers employees to deepen their own skills and help build the skills of others, while scaling learning for Googlers across the globe. g2g accounts for 80% of all formal learning activities, driven by a volunteer network of over 12,000 Google employees - known internally as "g2g'ers"—who volunteer a portion of their work time to develop their peers by facilitating instructor-led training, providing one-on-one coaching, or leading team-development activities.

Peloton's SME Network was created to power all internal technical learning programs, providing engineers the opportunities and tools required to pair, coach and teach each other. The group is made up of a core group of 40 engineers and since its creation in 2021, the group has been critical to the delivery of technical onboarding and continuing education programs at Peloton.

Learning and development teams at Peloton and Google have used peer-to-peer learning to:

- ensure learning content is hyper-relevant and organizationally specific,
- increase learner engagement and enjoyment by leveraging deep understanding of workplace cultures,
- improve employee experience by providing powerful opportunities to grow through facilitation, with ancillary benefits for employee mobility and retention, and
- provide a more scalable, cost-effective means of achieving organizational learning objectives by highlighting internal peer experts for employees to learn from.

2   Ensuring Relevance of Learning

"One of the things that makes a lot of g2g programs effective is that they can draw on examples from Google. And those make the lessons that they're teaching both more pointed and more memorable."

Benjamin Treynor Sloss

Engineering Vice President, Google

A perennial problem for learning organizations is how to provide learners with the right learning content in their time of need. Potential approaches fall on a spectrum: building curricula based on precise learning assessments may help us directly meet our learners' needs, but such assessments are difficult to conduct and risk missing important skills or competencies. Alternatively, we may be tempted to provide learners with access to vast content libraries maintained by external vendors, but they may be unable to navigate the daunting learning landscape on their own and have trouble finding needed resources. Between these extremes lie attempts at thoughtful curation of learning content libraries, paired with narrowly focused needs assessment and content development.

Peer-to-peer programs provide another approach to ensuring that learning content is relevant: employees can exercise autonomy not only in the learning they consume, but also the learning that they provide. Learners may still have many learning opportunities to choose from, but those that are delivered—and in many cases, developed—by their peers are presumed to be helpful and relevant. The hyper-relevance of peer-to-peer learning is difficult to match with externally sourced learning programs or centrally managed curricula because of the importance of organizational context; the opportunities it provides to uncover cross-functional learning opportunities; and its ability to respond in real time to learners' evolving needs.

We believe that the most relevant learning to employees of any company is organizationally specific. For example, the learning need for new engineers coming into a company is not learning how to code—they likely were hired because they already have that skill, and in any event, there are many options to further develop those skills outside of their learning organization. What is most relevant for new engineers at a company is how engineering at that specific company actually works: What are the tools used? What are the norms for reviewing code? This type of content can only be developed and delivered internally. To build Tech Onboarding at Peloton, more than 40 engineers contributed by writing content, pairing with new engineers or teaching in the program. The content would have been impossible to create without them.

In a pre-pandemic world, companies spent millions of dollars creating office spaces that allowed for moments of serendipity, to spark connections between employees who do not regularly collaborate in the flow of work. In a remote—or even a hybrid—world, impromptu opportunities for cross-functional collaboration become less frequent and far more difficult to create. Peer-to-peer learning programs help address this need by providing opportunities for learners and facilitators to interact with fellow employees they otherwise might not, and to pursue learning that, while not directly related to their role, can inspire and spark innovation. Peloton's Tech Talks is a program that features talks from engineers across the business twice a month to increase opportunities for collaboration and serendipity. In total, 92% of Tech Talks are presented by someone outside of the attending engineers' team.

Leveraging peer-to-peer learning also enables businesses to stay ahead of the learning curve: by empowering subject matter experts within the business to teach, a company can create an elite group of trainers within the business. As the people within the business learn, they are now empowered to share that knowledge with their peers creating a flywheel for a culture of constant learning that is far more agile than formal needs assessments and centrally-managed curricula. Google's peer-to-peer learning program provides an elegant alternative; since many g2g courses are open to all Googlers, the learning and development team can determine what Google's learning needs are by observing what sessions Googlers sign up for. Instead of asking Googlers what they need, or trying to derive learning needs from perceived business challenges, Googlers can "vote with their feet" and the learning and development function adjusts curriculum accordingly. For example, in the first three months of the COVID-19 pandemic, sessions of a course on the "Fundamentals of Mindfulness" increased 379% compared to the three months prior, even as overall instructor-led training sessions dropped 60% as Googlers transitioned to fully remote work. The learning and development function was able to respond by investing in related topics, rapidly developing and launching a resilience curriculum to help Googlers in their time of need.

It's important to note that peer-to-peer learning and a free market of training opportunities for employees are not dependent on one another; it may be possible to maintain centralized control of the curriculum each employee has access to while utilizing peer facilitators, or to provide open access to a wide variety of training content offered by learning vendors. But we strongly recommend implementing a peer-to-peer model where employees have open access to a wide variety of learning experiences to fully realize its benefits.

3 Learner Engagement and Learning Effectiveness

Learning and development professionals are often trepidatious about peer-to-peer learning because of a perceived risk that training quality will suffer when facilitation is entrusted to volunteers. Indeed, when g2g was first launched at Google, potential facilitators were subjected to several quality control measures before they could contribute. The process could take up to a year to complete. As Google was rapidly growing in size, g2g needed to grow with it and the quality control measures proved to be a significant impediment to program growth. As a result, the learning and development function removed the quality control measures as an uncontrolled "experiment": all general participation prerequisites were removed. As a result, facilitator participation doubled in six months. And surprisingly, quality remained high. Indeed, based on facilitator ratings provided by learners, quality actually increased as the program tripled in scale between 2012 and 2018, and has remained steady since:

Table 1. g2g Growth 2007–2019

	2007	2012	2015	2018	2019
Number of g2g facilitators	N/A (Program Launch)	3,209	6,709	9,000	10,500
Number of Google employees	16,805	53,861	61,814	96,000	115,000
Average g2g'er rating	N/A	4.0/5	4.5/5	4.7/5	4.7/5

While the data above focus on learner engagement (in the form of facilitator ratings), peer-to-peer learning can effectively deliver measurable business impact as well. As noted in the previous section, Peloton's Tech Onboarding was made possible through the contribution of engineers at Peloton. Not only was the content hyper-relevant to the new engineers coming into the business, it was incredibly effective at achieving the program goal of reducing "time to first pull request" for new engineers by 83%. Engineers at Peloton deploy in many different code bases, and depending on the team and the domain expertise of the engineer (front end, backend, mobile, etc.) the onboarding exercise that makes them the most effective is different. By working with over 40 subject matter experts who deploy code everyday into these various codebases we were able to create a library of onboarding exercises that helped new hires deploy into the right code base from them within their first week.

4   Learner Engagement and Learning Effectiveness

While there are many benefits for creating internal experts to those in a learning and development function, there are also benefits for the employees who are participating by teaching. Writing, mentoring, pairing, coaching or speaking through internal learning and development programs puts employees on a stage and gives them credibility as an expert. This allows employees to build a wider professional network and professional brand within the company.

These intuitive observations are borne out by participants' experience: in a survey of peer-to-peer facilitators at Google, 86% either agree or strongly agree that participating in the program helps them develop skills that are useful in their work. Importantly, more than half also agreed that their participation also helped them develop skills that have had a positive impact on their performance ratings and promotion prospects, helping encourage continued participation and delivering tangible benefits for an explicitly "volunteer" activity.

5   Scalable and Cost Effective Solution

Prior to peer-to-peer learning programs, Peloton employees filled learning gaps with external learning opportunities. Teams spent time and budgets on costly conferences, external classes, and learning platforms: Professional conference registrations are on the order of $1000 and Couresea licenses start at $398 USD/learner/per year [1]. This cost multiplied over hundreds or thousands of employees across a company can result in a significant investment for a business. However, participants returned with feedback that while some of these experiences were interesting, they weren't as relevant to their work as they had hoped. Ultimately these options ended up being costly and ineffective ways to fill hyper-specific learning needs from the business. Peloton's SME Network was created in response to this. Creating a peer-to-peer learning program allowed Peloton to create content that was hyper-relevant to the business, and by empowering engineers to teach each other Peloton created a model that scaled across the varying learning needs of the business.

The cost effectiveness of a peer-to-peer learning model will depend on the size of your organization and the operating model you choose to implement it. An organization of Google's scale (see Table 1), for example, can afford to invest in a small team of full time employees to administer the program and technological tools to enable it, and still spend far less than they would on available alternatives on a per-learner basis. Alternatives such as Coursera licenses may be cost-competitive for smaller organizations, or they may seek to implement a peer-to-peer program without standing up a dedicated team and/or with less sophisticated technical infrastructure.

6 Conclusion

In this paper, we've explored how two different companies at two different stages of the company lifecycle leverage peer-to-peer learning. Founded in 1998, Google employs over 100,000 people. Peloton was founded 10 years ago and employs under 10,000 people. Both companies, however, have successfully leveraged peer-to-peer learning to drive innovation and create hyper-relevant and organizationally-specific learning experiences at scale. By leveraging employee's deep understanding of workplace culture, these learning experiences have improved learner engagement and employee experience by providing both teaching and learning opportunities to employees. All of this has allowed both companies to stay ahead of the learning curve required to drive innovation.

Acknowledgments. The work presented in this paper is reflective of the support and energy that everyone at Google's g2g program and Peloton's Tech Learning team has placed in peer-to-peer learning.

Reference

1. Coursera Plus. https://www.coursera.org/courseraplus?action=enroll

Promoting Social Inclusion in Vocational Training Students with Disabilities: An Experience of Museum Education

Maria Rosaria Re[✉] and Mara Valente[✉]

University of Roma Tre, Rome, Italy
{mariarosaria.re,mara.valente}@uniroma3.it

Abstract. The present contribution aims to illustrate the results of a pilot experience conducted at the National Etruscan Museum of Villa Giulia with the participation of CFP Simonetta Tosi in Rome, a vocational training centre addressed to people with disabilities. The educational path, realized within the pilot experience, aims to promote well-being, analytical skills, and the use of digital technologies in museum education contexts, and it is addressed to adult users with problems of social inclusion. The achievement of the aims of the pilot experience is pursued using inclusive and innovative learning methodologies: Object-based Learning (OBL) and Digital Storytelling (DST). OBL is increasingly adopted in both formal and informal education contexts, especially in terms of well-being and transverse skills promotion. The focus on the museum object facilitates the involvement of users and supports communication, analysis, and argumentation skills. DST allows people to express, understand, and articulate everyday experiences in a creative way. Through DST, museum users can connect with the territory in which they have situated, identifying different types of stories and telling them through digital devices. Moreover, DST is not simply a vehicle for increasing digital literacy, but also a learning methodology aimed at overcoming social barriers and increasing understanding between generations, ethnicities, and displaced groups. The results of the pilot experience underline a good level of well-being at the end of the learning activities, an improvement of sense of community and digital and basic skills promotion within participants.

Keywords: Well-being · Museum education · Social inclusion · Object-based Learning · Digital Storytelling

1   Museums Education, Well-Being, and Social Integration

In the last few years, the consideration given by museum education to visitors, investigating different social goals and customised teaching methods [1], has delineated a new user experience at museums and offers an innovative orientation of contemporary study in the field. Moreover, the definition of learning is no longer applicable to the field of acquired knowledge, but especially the competences developed and usable throughout life (lifelong learning). This supported the development of research methods and approaches on museum learning that consider "education" as a wide, multifaceted, and social process. In fact, the skills acquired play a more important role in society, helping people to become more aware of the reality in which they live, providing them with a critical citizenship attitude. UNESCO has also recognised the role of culture in achieving the goals of sustainable development. In particular, it reiterates the importance of the need for cultural integration and the overcoming of poverty, also of a cultural type, in two goals, the 3 and 10 [2]. Specifically, in the goal 3, which is *Good health and wellbeing: specific target of promoting mental health wellbeing,* culture is defined as "who we are and what shapes our identity" and contributes to "poverty reduction and paves the way for human-centred, inclusive and equitable development" [2].

The role of museums in social inclusion is built by imagining the museum space as a place of education, when each social group succeeds in interacting with it (regardless of differences in age, cultural level, social status) and developing skills such as critical thinking, communication, collaboration and interpreting information [3–6].

As the Council of Europe (CoE) has emphasized, soliciting cross-sectional skills in a democratic cultural context improves the well-being of pupils and individuals, which implies being active, responsible, connected, resilient, valued, respected, and conscious [7]. In addition, well-being is seen by the CoE as the ability to participate in an active community and culture. To promote social cohesion and economic growth by reducing the gap between rich and poor, it is necessary to promote a people centred approach to heritage, that benefits all levels of society. When people get involved, learn, value, and promote their cultural heritage, they can contribute to social and economic development of the communities they are part of. An inclusive way of working that involves individuals and communities through heritage can lead to greater social well-being, including the 2030 Agenda for Sustainable Development [2]. In this way, heritage becomes a source of sustainability, a means "to celebrate the past in today's evolving world" [8].

Museums have a close connection to the history of the area where they are located; however, some social groups (such as refugees, first- and second-generation migrants, and people with disabilities) are often excluded by the cultural and artistic life of the territory in which they live and, as an effect, do not cover a dynamic role in creating and sharing of a collective social memory. Therefore, there is a need to reflect on and realize inclusive learning pathways in the museum context: individual and social rights, participation in the community, and equal opportunities cannot remain mere declarations of principle [9], but they should be the final and essential aims of any educational activity [10].

Consequently, the participation of formal educational institutions in museum education activities is necessary: the process of inclusion, starting from schools (at different levels), must be carried out through a synergic and common educational action, able at enhancing individual and group resources to ensure the personal growth of every student, particularly of learners with Special Educational Needs.

Starting from these assumptions, the Centre for Museum Studies (CDM), based at Department of Education of University Roma Tre, planned and implemented the pilot activity here described, within the "Inclusive Memory" project. The research questions linked to this pilot activity are the following:

- What kind of emotional and affective experience characterizes the students involved in the experience?
- Is there a relationship between OBL and DST activities and 4C Skills promotion? If yes, how is it configured in relation to social inclusion?
- Is there a relationship between 4C Skills and well-being development? If yes, how is it configured?

The didactic strategies used, the context in which the experience was carried out, the characteristics of the participants, and the didactic and research tools used were defined to provide reflections on the theme "Diversity, equity, and inclusion in learning" of the Learning Ideas Conference. The results obtained from the experience here described can give useful indications on how to design and can evaluate and implement integrated museum-vocational school paths, addressed to users with disabilities, in which transverse competences, which are essential for the workplace, are promoted.

2 The Inclusive Memory Project

The "Inclusive Memory" project, funded by University Roma Tre, fosters the construction of a shared and collective social memory through an inclusive system within the museum. The core of this project is a close connection between new teaching methodologies and the implementation of digital tools, to encourage the development of transverse competencies at all museum users' disposal. Thanks to strong cooperation between university, research institutes and the museum educational departments, the project aimed at creating educational paths to promote social inclusion and well-being, especially in democratic terms, within museum users.

The experience presented in this paper is one of the many educational activities stemmed from the theoretical framework of the "Inclusive Memory" project [6]. It takes into consideration three main aspects: the promotion of transverse competences (critical thinking and digital skills in particular) through heritage within vocational students with cognitive disabilities; the use of innovative learning methodologies for heritage education; and the promotion of well-being to improve social inclusion.

3 The Pilot Experience

3.1 Aims and Outcomes

The main purpose of the pilot experience is to combine the practice of innovative teaching methods and the digital resources in museum education setting to stimulate the development of transverse skills, social inclusion, and well-being within people with cognitive disabilities.

The general goals of the learning experience are the following:

1. to perceive and implement innovative, personalised and adaptive museum learning paths efforts to soliciting social integration of the target group;

2. to project and implement workshops in museum contexts directed at students with cognitive special needs, from one of the vocational training centres of the Municipality of Rome;
3. to enhance transverse skills within workshop participants, especially communication, critical thinking, collaboration and creativity; and
4. to design and produce digital products that record the experience and promote the creation of a common and shared memory, through the Digital Storytelling strategy.

The specific expected results are linked to an increase in emotional well-being, a sense of belonging to a community, and digital and basic skills development.

The achievement of the pilot experience aims is pursued through the adoption of Object-based Learning and Digital Storytelling as innovative learning methodologies.

3.2 Object-Based Learning and Digital Storytelling

Object-based Learning (OBL) is an active methodology to learn [11], a socio-constructivist approach that offers a tactile experience for learners, which challenges them to interrogate the object and conceptualise their thinking [12]. During Object Based Learning session, participants are asked to touch, handle, probe, and explore the exhibit or its 3D printed copy. They are encouraged to make observations on its shape, to derive meaning from it, to compare it to other objects, or to discuss its function. The exploratory investigation brings into play a series of skills and competencies that foster learning. Among these, especially observation and analysis skills, metacognition, critical thinking, communication, and problem-solving abilities are incentivized [13]. Moreover, since it encourages cooperation and interaction, it has proved to be a very useful strategy to engage students who have learning and/or relational difficulties with their peers/teachers, as it stimulates the development of transversal skills and, more generally, makes it possible to reach a condition of emotional wellbeing [14, 15] and using their previous knowledge [16]. OBL is increasingly adopted in both formal and informal educational contexts: the focus on the object facilitates the involvement of learners who do not respond well to written texts and seems effective in terms of well-being promotion [17].

Storytelling is a learning strategy through which people can express, understand, and articulate everyday experiences. Thanks to this process, persons can link with the world around them and recognise different kinds of stories. Therefore, Digital Storytelling (DST) is not only a way to expand the digital abilities but, above all, a learning method to break down barriers and increase understanding between generations, ethnic groups, and groups in a variety of ways [18]. In addition, DST is a powerful means of personal and creative expression, giving groups of people who are normally excluded from culture the opportunity to see and hear, and fostering the self-confidence needed to escape social exclusion and access, giving new opportunities that help improve the self-esteem and self-confidence needed to advance in life. DST workshops support the implementation of common stories and provide the incentive for addressing current social issues. The use of the DST in the context of heritage education supports the development of at least four of the eight competences that the European Reference Framework indicates as necessary to enable all European citizens to operate successfully in a "knowledge society" [19]:

communication in the mother tongue; digital skills; learning; and cultural awareness and expression, together with 4C skills [20]. Specific Research [21] has also confirmed, in the inclusive perspective of the DST, that this methodology is a powerful process for co-creation, as it allows marginalized people to script, film, and create videos of their own stories with support from an expert facilitator. In addition, digital storytelling has been tested to be an effective methodology to support the storytellers to process their mental health issues [22].

3.3   Methodology

The pilot experience was carried out in May 2019 with the participation of 6 students with mental disabilities of a vocational training centre of Rome. The museum selected for the realization of the pilot phase is the National Etruscan Museum of Villa Giulia.

Four educators from CDM and an archaeologist designed and implemented the pilot experience, which was realized with the supports of one teacher from the vocational training centre. The pilot activity was organised in two meetings, each of which focused on a specific learning methodology. Different evaluation activities were carried out before and after the pilot experience. The two meetings lasted 4 h each and were organized taking into consideration the following activities:

First meeting. Participants were introduced to the museum with a brief explanation of the Villa Giulia main building and garden history. The timeline of the activities was also introduced by the educators and participants took part in the first well-being evaluation activity. In the OBL workshop, one participant, engaged in the manipulation activities (hands-on), had to describe the assigned archaeological object to another participant (in pairs-activity). Then, the participants were engaged all together in the interpretation of the museum objects used during the OBL experience, through the support of an archaeologist.

Second meeting. The participants visited some rooms of Villa Giulia Museum with the support of an archaeologist and two museum educators: the exhibition of the museum was introduced together with the most popular types of museum objects (vases, jewellery, statues). A selected museum object ("Medea rejuvenates Jason with a spell" Etruscan Olpe) was collaboratively analysed by participants through the support of narrative texts selected for the experiences. After that, in the museum education room, the DST workshop started and the participants were involved in different activities: creation and realization of the story in a group inspired by the Etruscan Olpe; recording of the story with the support of a recorder; realization of the video through the use of a PC; and sharing of the video made with teachers and educators. The experience ended with a well-being outgoing evaluation activity.

4   Content Analysis of Workshop Recordings

The workshop activities were audio-recorded by CDM researcher and analysed through a specific content analysis. The purpose of this analysis is to investigate whether the teaching activity based on OBL and DST methodology has been a truly inclusive experience for the participants. The concept of inclusion can be operationalized in attitudes, external emotional states and opinions expressed.

The following categories and their codes have been provided (Table 1):

- Affective states (codes: interested, lively, motivated, positive, talkative, friendly; distressed, irritable, nervous, scared, unhappy, upset);
- 4C Skills (codes: communication, collaboration, critical thinking, creativity)
- Contents (codices: Renaissance architecture, Castellani, ancient civilizations, ceramics: technical-stylistic and/or functional aspects, history)

The recordings were fully transcribed, and the text was imported into *Atlas.ti* qualitative analysis software.

The transcript of the meetings was divided into units of meaning and these were codified. This allowed to carry out an analysis of the frequencies of the categories emerged during the discussion activities, and an analysis of the co-occurrences, useful to identify the coexistence of different emotional states and 4C Skills, and the educational contents offered. This procedure produces results in terms of percentage; in fact, the percentage of occurrence of a specific unit of meaning with respect to all the sentences expressed by the participants during the two meetings is presented in the following paragraph.

Moreover, it is important to underline that the codes are representative of group dynamics, not of the attitudes of individual students: thus, the evaluation data here presented refer to learning outcomes and results in terms of well-being promotion of the entire group of students participating in the activities, useful for an initial exploratory analysis of the educational model implemented within the experience described in the present paper.

4.1 First Results from OBL Workshop

From a general overview, the data analysis shows that participants express high levels of positive emotions in 28.26% of the cases; meanwhile, the codes classified as negative wellbeing are not present (Table 2).

In order to understand which emotions were reported by the students participating in the pilot experience, an analysis of the frequencies of the categories was carried out (Table 3). The high levels of positive emotions are mainly related to interest (15.22%), loquacity (5.59%) and friendliness (4.66%).

Students' interest in the activities is also confirmed by the observations made by the researchers during the pilot experience. As Table 4 shows, the interest is motivated by the possibility of communicating with others in 42% of the cases and by the characteristics of the museum objects (technical-stylistic and/or functional aspects) in 27% of the cases. These data underline that group discussion and object manipulation promote participants' interests by actively involving them in the learning experience. A good number of co-occurrences (18%) are found in the link between interest and critical thinking: this is in line with the reference literature that identifies curiosity as a necessary mental disposition for the good critical thinker [27].

The students participating in the pilot experience interact mainly with the mediators. An analysis of the co-occurrences between transverse competences (Table 5) shows that, in 22% of the cases, learners demonstrate that they stimulate critical thinking through

Table 1. Categories, codes and their definition/application.

Category	Code	Definition
Affective states	Positive well-being	Interested, lively, motivated, positive, talkative, friendly[23]
	Negative well-being	Distressed, irritable, nervous, scared, unhappy, upset [23]
4C skills	Communication	The ability to express and interpret concepts, thoughts, feelings, facts and opinions in written and oral form and to interact appropriately and creatively in a variety of social contexts, be they work-related, educational and of leisure [5]
	Collaboration	The ability to work together towards a common goal [24]. A process that leads to the actual achievement of desired individual and group outcomes" [25] *This code is applied when students, being given a stimulus, interact with each other to achieve the goal*
	Creativity	A new job considered at a certain point as sustainable or useful or satisfying by a group of people [26]
	Critical thinking	The ability to make a well-founded and self-regulated judgement that leads to the interpretation, analysis, evaluation and inference, as well as to the explanation of factual, conceptual, methodological, logical or contextual considerations on which the judgement is based [27, 28]
Contents	History, ancient civilizations, renaissance architecture, ceramic: technical-stylistic and/or functional aspects	Contents related to the Etruscan Museum collections and objects selected for the pilot experience

communication, which is carried out mainly with the mediators. These data highlight how the questions posed by the researchers and the organised activities of reflection and analysis prove to be effective in terms of critical thinking stimulation.

Table 2. Codes frequencies

Category	Code	Frequency (%)
Affective states	Positive wellbeing	28.26
4C skills	Communication	15.22
	Collaboration	1.86
	Creativity	0.31
	Critical thinking	6.83
Contents	Renaissance architecture	3.11
	Ceramic: technical-stylistic and/or functional aspects	10.56
	History	2.80
	Ancient civilizations	4.97

Table 3. Frequencies of positive wellbeing categories

	Friendly	Interested	Talkative	Motivated	Positive	Lively
Freq	4.66%	15.22%	5.59%	1.55%	1.24	0%

Table 4. Co-occurrences between positive wellbeing and other categories

	Friendly	Interested	Talkative	Motivated	Positive
Communication	10	42	18	5	3
Collaboration	3	4	4	3	2
Creativity	0	0	0	0	0
Critical thinking	5	18	3	4	2
Renaissance architecture	2	8	6	3	1
Ceramic: technical-stylistic and/or functional aspects	6	27	5	0	0
History	2	7	6	4	1
Ancient civilizations	4	14	7	3	1

Table 5. Co-occurrences between 4C skills

	Creativity	Communication	Collaboration
Critical thinking	1	22	6

5 Discussion and Conclusion

The present contribution aims to illustrate the results of a pilot experience conducted at the National Etruscan Museum of Villa Giulia with the participation of a vocational training centre addressed to people with cognitive disabilities. Six students participated into two webinars on OBL and DST strategies, designed and realised by the CDM research group. The pilot experience was analysed in terms of well-being and transverse skills promotion within participants, as indicators of social and cultural inclusion within museum education context.

The first results, from specific content analyses of the recordings made during the OBL meeting show that both transverse skills and positive well-being are stimulated during the learning activities. Participants show interest in the museum collection and activities (15.22%) and are friendly and talkative, especially with the mediators. OBL activities seem to promote communication (15.22%) and critical thinking skills (6.83%) within students, especially during moments of reflection and analysis on the museum objects selected for hands-on activity. Co-occurrence analysis shows that critical thinking is particularly evident in association with participant interest and communication, in line with studies in the field [27]. The relationship between positive well-being and transverse competences is particularly evident in relation to communication skills: 42% of the cases students show interest in the education experience while they communicate their thoughts.

The content analysis of the recordings of the DST workshop has yet to be concluded: the results may provide further indications in terms of creativity promotion and digital tools use within heritage education context.

The small number of participants and the way in which the experience was carried out do not allow generalisations; however, the results described provide first indications in terms of the use of active learning strategies to realise educational paths within museum context aimed at promoting social inclusion within users with cognitive disabilities.

Future research activities should include a larger number of participants in the experience and should include carrying out the activities in different museum and heritage contexts to test the degree of transverse competences and well-being promotion within users in different museum education contexts. It would also be useful to check whether participation in similar experiences, even over time, can improve the educational performance of participants and the promotion of technical and professional skills.

Authors' statement. The authors of the present paper contributed to the writing of this article as follows: M. R. Re (1, 4, 4.1 and 5), M. Valente (2, 3.1, 3.2 and 3.3).

References

1. Parry, R.: Museums in a Digital Age. Routledge, New York (2010)
2. UNESCO: Transforming Our World: The 2030 Agenda for Sustainable Development (2015). https://sdgs.un.org/2030agenda
3. Sandell, R. (ed.): Museums, Society, Inequality. Routledge, London and New York (2002)
4. Nardi, E.: Musei e pubblico. FrancoAngeli, Milano (2014)

5. Poce, A.: Il patrimonio culturale per lo sviluppo delle competenze nella scuola primaria. Cultural Heritage and the Development of XXI Century Skills in Primary education. FrancoAngeli, Milano (2018)
6. Poce, A. (ed): Memoria, inclusione e fruizione del patrimonio culturale Primi risultati del progetto Inclusive Memory dell'Università Roma Tre/Memory, Inclusion and Cultural Heritage First Results from the Roma Tre Inclusive Memory Project. ESI, Napoli (2020)
7. National Council for Curriculum and Assessment: Junior Cycle Wellbeing Guidelines. https://ncca.ie/media/2487/wellbeingguidelines_forjunior_cycle.pdf. Last accessed 31 May 2021
8. British Council: Cultural Heritage for Inclusive Growth (2018). https://www.britishcouncil.org/sites/default/files/bc_chig_report_final.pdf. Last accessed 31 May 2021
9. Chiappetta Cajola, L.: Indagine quantitative negli studi delle disabilità e dei DSA. Cult. Psychol. Stud. **9**, 311–346 (2014)
10. Booth, T., Ainscow, M.: Index for Inclusion. Developing Learning and Participation in the School. Centre for Studies on Inclusive Education (2002)
11. Freeman, S., et al.: Active learning increases student performance in science, engineering, and mathematics. Proc. Natl. Acad. Sci. U.S.A. **111**, 8410–8415 (2014)
12. Romanek, D., Lynch, B.: Touch and the value of object handling: final conclusions for a new sensory museology. In: Chatterjee, H.J. (ed.) Touch in Museums: Policy and Practice in Object Handling. Berg, Oxford and New York (2008)
13. Poce, A., Re, M.R., Valente, M., De Medio, C.: Musica e stampa 3D. Promuovere le competenze trasversali in alunni di scuola primaria tramite la riproduzione di strumenti musicali etruschi. In Antonella Poce (curated by), Veicolare l'inclusione attraverso il patrimonio. Alcuni risultati del progetto Inclusive Memory dell'Università Roma Tre, pp , 13–31. Edizioni Scientifiche Italiane, Napoli (2019)
14. Hannan, L., Chatterjee, H., Duhs, R.: Object Based Learning: a powerful pedagogy for higher education. In: Boddington, A., Boys, J., Speight, C. (eds.) Museums and Higher Education Working Together: Challenges and Opportunities, pp. 159–168. Ashgate Publishing, Farnham (2013)
15. Kador, T., Chatterjee, H.: Object-Based Learning and Well-Being: Exploring Material Connections. Routledge, London (2020)
16. Chatterjee, H., Hannan, L.: Engaging the Senses: Object-Based Learning in Higher Education. Routledge, London (2015)
17. Chatterjee, H., Noble, G.: Museums, Health and Well-Being. Routledge, London (2017)
18. Lambert, J.: Digital Storytelling, Capturing Lives, Creating Community, 2nd edn. Digital Diner Press. Routledge, London (2006)
19. European Commission: Proposal for a Council Recommendation on Key Competences for LifeLong Learning (2018)
20. Liguori, A., Bakewell, L.: Digital storytelling in culture and heritage education: a pilot study as part of the DICHE project. In: Poce, A. (ed.) Advanced Studies in Museum Education. Lectures, pp. 63–78. ESI, Napoli (2019)
21. Whitley, R., Sitter, K.C., Adamson, G., Carmichael, V.: Can participatory video reduce mental illness stigma? Results from a Canadian action-research study of feasibility and impact. BMC Psychiatry **20**(1), 16 (2020). https://doi.org/10.1186/s12888-020-2429-4
22. De Vecchi, N., Kenny, A., Dickson-Swift, V., Kidd, S.: How digital storytelling is used in mental health: a scoping review. Int. J. Ment. Health Nurs. **25**(3), 183–193 (2016). https://doi.org/10.1111/inm.12206
23. Thomson, L., Chatterjee, H: UCL Museum Wellbeing Measures Toolkit (2013). https://www.ucl.ac.uk/culture/projects/ucl-museum-wellbeing-measures. Last accessed 31 May 2021
24. Griffin, P., Care, E.: Assessment and Teaching of 21st Century Skills. Methods and Approaches. Springer Netherlands, Dordrecht (2015)

25. Kuhn, D.: Thinking together and alone. Educ. Res. **44**(1), 46–53 (2015)
26. Stein, M.: Creativity and culture. J. Psychol. **36**(2), 311–322 (1953)
27. Facione, P.A.: Critical Thinking: What It Is and Why It Counts. Insight Assessment, a Division of the California. Academic Press, San Jose (2015)
28. Poce, A.: Verba Sequentur. Pensiero e scrittura per uno sviluppo critico delle competenze nella scuola secondaria. Franco Angeli, Milano (2017)

Digitalization: Training University Professors and Students with Flashlearns

Anne-Dominique Salamin[(✉)] [iD]

University of Applied Sciences Western Switzerland, Delemont, Switzerland
adominique.salamin@hes-so.ch

Abstract. Microlearning offers an interesting training solution both for professionals and for university students. Short, brief, agile—the contents proposed in micro-learning formats correspond to current learning modes. However, they serve certain types of knowledge better, and learning methods, such as distance learning, are more favorable to this approach. This paper will briefly review the research in the field and will present a practical case of microlearning application at the university level. The University of Applied Sciences Western Switzerland (HES-SO) has indeed designed a digital concept based on micro-learning to train professors and students on issues related to the use of digital means in the classroom. Called *Flashlearns*, these short training units embed video and are structured and organized in a way that promotes learning. The paper describes the project and the instructional design, and it presents the results obtained. The paper concludes with some ideas for extending the project and opening new possibilities.

Keywords: Micro-learning · Digital competencies · Distant learning

1 Introduction

1.1 Context

Digitalization is impacting the tertiary educational sector by changing or modifying professor teaching methods, organization, coaching, and lecturing. The COVID-19 has forced most Universities worldwide to teach remotely, and revealed shortcomings, difficulties, ineffective habits.

The University of Applied Sciences Western Switzerland (HES-SO), the second largest university in Switzerland, is not different in this respect from any other tertiary institution: the challenges set by the digital technology, their impact on the university system, and career Microlearning offers an interesting training solution both for professionals in the field and for university studies. Short, brief, agile—the contents proposed in micro-learning format correspond to current learning modes. However, they serve certain types of knowledge better than others, and certain learning methods, such as distance learning, are more favorable.

This paper will briefly review the research in the field and will present a practical case of microlearning application at the university level. HES-SO has indeed designed a

D. Guralnick et al. (Eds.): TLIC 2022, LNNS 581, pp. 313–322, 2023.
https://doi.org/10.1007/978-3-031-21569-8_30

digital concept based on micro-learning to train professors and students on issues related to the use of digital in the classroom.

Called *Flashlearns* (FL), these short training units embed videos and are structured and organized in a way that promotes learning. The paper describes the project and the instructional design, and it presents the results obtained. The paper concludes with some ideas for extending the project and opening new possibilities, which may come unaware to either teachers or students. To modify mindsets and practices, the HES-SO has established in 2019 a Digital Competence Center (CCN). The CCN operationalizes the rectorate digital strategy and, among other objectives, contributes to train students and professors to address the actual impact of digitalization on learning and teaching.

A CCN specific workgroup dedicated to Digital Learning, designed a distant learning concept, based on micro-learning units (ML). FL training modules are based on short videos and animations and provide brief but rich insights on topics such as "student profession in the digital era," "critical approach to conspiracy theory," or "the educational scenario in the digital age."

This paper firstly questions the potential and features of ML, the reasons for which this notion, theorized some years ago, is back in the spotlight to emerge as a strong trend in training. Secondly, we will present a case study, the FL project, along with its institutional and epistemological components.

1.2 Definition and Features

The ML notion was formalized in 2005 [1] in connection with life-long learning approaches, and the growing need to train collaborators quickly and efficiently at their workplace. ML is linked to formal and informal learning, in distant or blended learning situations [2, 3].

Among the numerous definitions of ML, we adopt the Nikou and Economides one [4, 5]: "Micro-learning refers to a learning approach based on small learning units and short, focused learning activities. In micro-learning, learners obtain various micro-contents, including definitions, brief video segments, and micro-assessments so that their knowledge can be evaluated without requiring special testing arrangements."

In 2010, Buchem and Hamelmann [18] completed this definition by adding the notion of micro-content focusing on a single and defined topic.

The key interest with ML lies in its capability of sustaining self-study by letting the learners select their topics, and the moment they feel it is pertinent for them to access these topics [6].

The specific characteristics of ML [18] can be summarized as follows:

- a short duration (5–15 min),
- content divided into knowledge nuggets, chunks,
- dealing with restricted subjects, simple problems,
- generally based on video, and
- sometimes offering an accreditation of ML units to obtain recognition.

By its variety, its small scale, and the use of rich media, ML is engaging, and it allows the learner to keep up to date on notions proposed with a rapid flow. ML is motivating because the learner can choose the subject of his interest instead of remaining passive. However, ML has its limits. It is not suitable for acquiring and learning complex skills, processes, or behaviors. Reading something or even following steps listed in a micro-lesson is different from the process of "learning" complex knowledge [6].

1.3 New Habits Regarding Information Consumption

In recent years, the use of ML has extended to the academic training field, as societal habits are modifying teaching and learning.

In fact, the daily use of smartphones and tablets have impacted consumption patterns for knowledge, in two main ways.

First, people are used to get immediate gratification, through notifications from mobile applications, and with the massive use of social networks, users expect instant answers to their questions or concerns [23]. Second, consumption patterns for accessing information, have gradually changed. News, newspaper articles, television series, and 6-s "bumper ads" on YouTube-type videos are structured into brief units of information intended to keep the users' attention. This has an impact on the quantity, form, volume, and mediatization of the information, and therefore shapes the knowledge to be taught.

Thus, steadily, the continuous attention span has been declining in the last decade, in addition to the fact that the human brain shows a natural propensity for distraction [22].

For students belonging to the Z generation, mostly attending undergraduate courses at university level, the concurrent use of several screens (watching a series and chatting on their mobile) is natural, although the effectiveness of this approach for memorizing facts has not been proved.

These new habits of split attention have an impact on how knowledge is consumed. The use of ML, organized in small fractions of knowledge, which can be mobilized quickly and at the desired time, meets both the expectations and the observed behavior of the learners.

1.4 Cognitive Effectiveness of Microlearning

Is learning in small batches effective from a learning perspective? Two questions arise: Can knowledge be broken down into units so small that they can be learned without effort? Furthermore, can the content learnt in this way, be integrated, and be transferable to other situations, or does retention remain superficial, reducing the content to be learned to a mere noise among others emitted by social networks, news, applications?

We will distinguish two aspects: pedagogical engineering and assimilation.

Kerres [21] notes that ML constitutes a challenge for instructional design. In traditional instructional design, knowledge is organized vertically from the curriculum to the lesson or sequence and is broken down into rigorously organized knowledge elements. ML flattens this verticality and reduces this organization to its smallest element, to "assets." The author concludes, in this context, that instructional design is not dispensable but even becomes more complex, since the task of instructional design would imply to provide an arrangement of contents and tools that can be interwoven with the personal workspace of the learner. It is therefore important to consider not only the units of knowledge and the tools used to disseminate them, but also their fluidity of use, since learners will use them freely without following a path strictly prepared for them.

Can we learn with ML? Stohr et al. [19] indicate that ML has a positive influence on learning and implies an overall satisfaction for the learner with the learning object. Zhang [20] notes that the use of small chunks of resources offers the learner the opportunity to make effective use of his fragmented time to engage in learning. Hug et al. [1] indicate that learning takes place in micro-steps, the ML corresponds well to the micro-steps that form the basis of successful learning with an important level of sustainability. Furthermore, these microlearning steps facilitate the process of deep understanding and creation of profound knowledge, if the microlearning process is embedded in an appropriate learning design/setting [17].

However, several authors note that the integration of knowledge does not take place if it is not embedded in a broader, well-defined learning concept. Hug [1] notes that "microlearning does not consist in transferring knowledge nuggets from A to B," and that the challenge of learning consists in the construction of knowledge. Peschl [17] proposes a theoretical framework that starts from observation, to build relationships between knowledge to create new realities and to question MLs to make sense.

The ML facilitates the process of deep understanding and creation of profound knowledge, if the microlearning process is embedded in an appropriate learning design/setting. Peschl [17] also points out that it is not a particular feature that causes the learning process, but the explicit combination of these features related to the learning typology that makes the paradigm powerful. In other words, the ML must be conceived in a broader perspective, which can be divided into small units, each with its own and combined utility.

We have thus seen that the term ML covers a brief resource that enables rapid learning, close to the field, corresponding to current information consumption habits, that ML is effective when integrated in a general training concept, and that it can be deployed in university curricula, with the aim of clarifying theoretical knowledge. We present, in the following point, how this concept has been implemented at the HES-SO.

2   Case Study

2.1 Introduction

The CCN of the HES-SO, in its effort to kick-start the digitalization of the institution, would have had to train more than 30,000 people (professors, students, and administrative staff) spread over 28 campuses. It was clear from the start that setting up face-to-face training would be a logistical and human challenge. The choice was therefore made to offer a distance learning program. Although the concept of Massive Open Online Courses (MOOCs) was initially discussed, this approach was abandoned for three reasons.

First, the time factor was considered an impediment. As the Centre's activities are subsidized for a period of four years, and the production of several MOOCs covering different topics related to digitalization, would have taken several months or even years, it was difficult to implement the concept during the time available. Second, while the subject of digital technology is of wide interest, it was unclear what level of involvement would arise for long and demanding courses. Third, the Center was interested in developing a more agile, flexible approach without prejudicing the quality of the content.

Microlearning was therefore chosen. In this first period of the Center's existence, with the support of the HES-SO's Educational Department, the training effort was focused on the global topic of "teaching in the digital age," with the aim of explaining to teachers and students how digitization impacts both teaching and learning. Moreover, we intended to provide teachers and students with the key concepts of a notion, the vital minimum to know. We estimated that, as professional of their field, and of learning and teaching, they would be able to make rich and numerous links between their practice and these new concepts.

2.2 Approach

We decided to develop, in a first phase, a batch of 16 topics such as, for the teachers, "learning spaces," "scripting your digital course," or "using your smartphone to make an educational video" and, for the students, "the student profession in the digital era," "learning independently," or "social media addiction." The objective was to entrust the realization of these topics to experts specialized in the digital field. The project took place between February 2020 and September 2021.

Expert. From the start, we wanted to involve members of the HES-SO who could bring their experience and knowledge to the project. The experts chose the subject of which they considered to be the specialists and developed it from the pedagogical conception to the mediatization of the content. We estimated the time required to implement a FL to 40 h, which is equivalent to one week of full-time work, which corresponds to a cost of $4,300 ($75,000 for the 16 FLs).

We trained these experts, supported them when they were designing the instructional scenario, helped them to elaborate pedagogical objectives, as well as train them to realize the technical aspects. We then accompanied them throughout the realization by answering their questions and to follow the planning of the project.

Diffusion. once the resources were created, they were organized and distributed by a technical team on the platform dedicated to digital training at the HES-SO (https://numerique.hes-so.ch). Technically, this is a fork of Moodle, where some processes were automated, and new ones implemented. resources were then aggregated by using H5P/Moodle.

Each completed topic was assessed by a few final users, modified according to their remarks and propositions, and finally published on the platform. The summary of the topic, its pedagogical objectives, syllabus, and a teaser are then proposed to the learner, as shown in the image below (Fig. 1).

Instructional Design. The global duration of a microlearning resource varies according to their authors. For some authors, an FL should last seconds [24], for others less than 15 min [12], or between 5 and 10 min [1]. We found it difficult to produce learning units of such a short duration while guaranteeing the comprehension of the subject. We therefore applied an identical structure to each topic: four themes, each illustrated by four resources—i.e., 16 resources in all per FL (Fig. 2). The consumption of each topic lasts about 30 min in total, or an average of 5 min per resource. We called these training

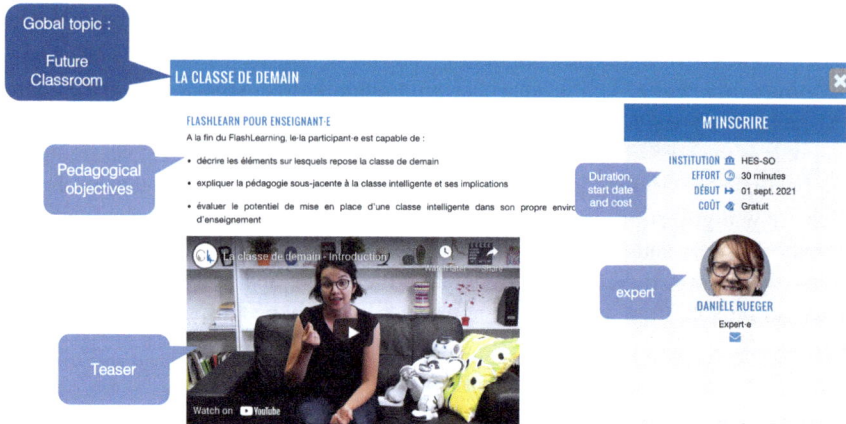

Fig. 1. Access to a specific topic.

entities FL, a term we believed to be engaging for our audience. We paired these FLs with a tagline: "30 min at the heart of training in the digital age."

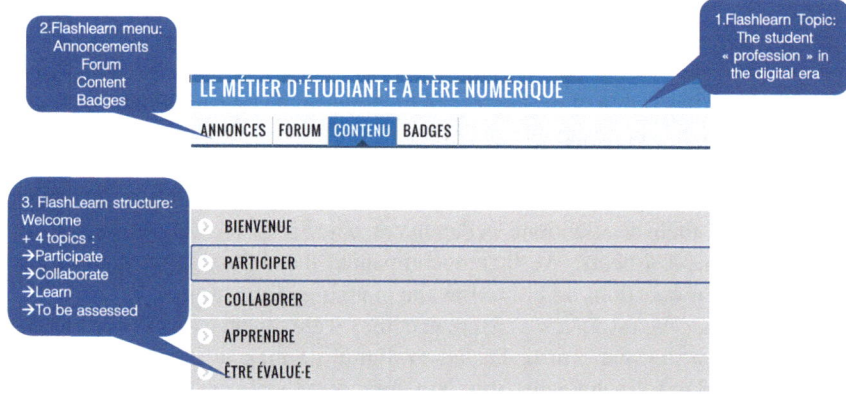

Fig. 2. FL global structure.

Moreover, as Glahn [24] points out, "Self-regulated learning means that learners use provided information to assess and control their learning processes." The aim was therefore to propose at least one self-assessment for the knowledge acquired on a given subject.

Each topic of the subject thus includes three video resources and one quiz, as shown in the image below (Fig. 3).

The first type of resource provided is a PowerPoint integrating the expert's voice. This resource offers a clear and illustrated review of the topic. The second type shows an animation including tips and tricks, which enhances or summarizes the theme. The third type of resource features an actor roleplaying the FL learner. She formulates questions

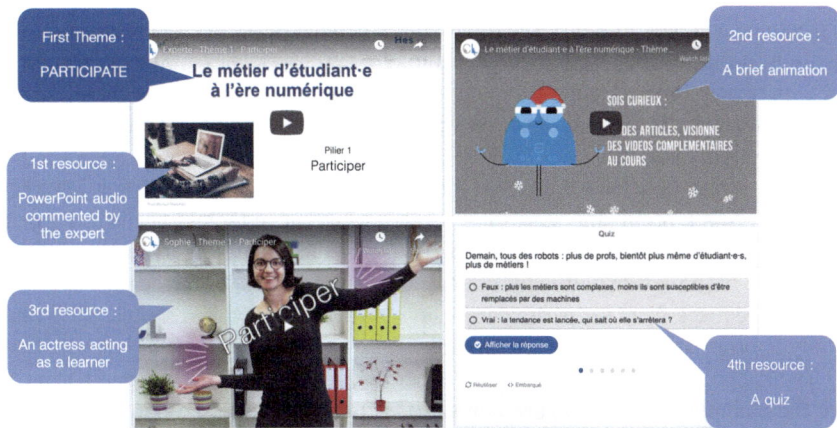

Fig. 3. Four types of resources.

that the learner might have about the theme, and summarizes the content of the other two resources, while adding a few "field" complements. Finally, the fourth type of resource consists in a short self-evaluation quiz on the three previous resources.

It is therefore both a precise subdivision of knowledge, and an organization which, for each theme, sheds light on the knowledge to be acquired in a variety of means, while avoiding pure repetition. Indeed, as Glahn [24] summarizes, "if learners repeat learning activities after their initial attempt, then forgetting takes longer and retention is improved." Regarding FLs, instead of watching the same resource several times, the knowledge to be acquired is displayed in several various means, each viewing dealing with the same topic from a slightly different angle.

2.3 Validation Concept

To encourage the use of FLs and thus support the process of digital transformation of teaching in the institution, a validation concept has been developed with the help of the Professional Development Center of the HES-SO (Devpro). This Center trains teachers on various aspects (technological, pedagogical) to support their career evolution and their teaching practice. After 15 days of training, either face-to-face or at distance depending on the courses, the teacher obtains a didactic certificate.

We have considered that 10 FLs are equivalent to one day of Devpro training. Each completed FL earns a digital badge; showing ten badges to Devpro validates 1/15 day of training for the didactic certificate. Regarding students, the recognition for their learning takes place through the integration of the FL into one of the courses they attend.

3   Discussion

The main difficulties faced in this project were threefold. First, the time availability of the experts, often involved in a number of projects and committees: they were often overwhelmed and had difficulty keeping to the schedule. Secondly, the exercise of simplification of the selected topic, while retaining a certain richness, proved difficult for some experts, who tended either to oversimplify the content or to produce resources over the time limit. Finally, special attention was given to the communication of the project to reach the right audience.

To date, 202 people have registered for the published FLs, and 74 badges have been awarded. This is a promising start for a project which nevertheless needs to be better embraced by its future users. Some Devpro courses in the flipped-class format use the FL as a resource for browsing before the course, as a starting point for further discussions in class. The feedback from the participants is positive regarding this resource: 67% found it totally appropriate, and 33% appropriate.

The FL concept has since spread to other entities of the HES-SO apart from the CCN, for example to develop micro-courses on ecology in the classroom, or to integrate the idea of sustainability in the teaching.

As we can see, the beginnings are small but progressive. With busy faculty members, it will take time for the project to reach everyone on the ground. The HES-SO spread over seven states, 28 campuses in western Switzerland, providing various cursus, with various culture approaches, ranging from Social Work, to Health, Engineering or Music, slows *de facto* the diffusion and adoption of the FL. Nevertheless, previous experiences have shown that it takes time for new training approaches to be integrated in the institution, and this project will certainly follow the same course of acceptance and use.

4 Conclusion

According to Glahn [24]: "The concepts of micro learning are useful to enrich the learning experiences and broaden the learning environment where conventional macro learning solutions are unsuitable." The needs of the HES-SO fit perfectly into this framework. The production of FL thus proved its interest and feasibility.

We have seen above how the format is interesting for presenting new concepts, not so well-known by learners. Thanks to its brevity and the use of rich media resources, microlearning acts on both the engagement and the motivation of learners [5], a major concern when the proposed courses are not compulsory. We have also shown how these formats, used for professional training, are appreciated by teachers.

The next step of the project consists in integrating some of these resources into the regular courses attended by undergraduate students in a bachelor's degree program, for example the FL "student's profession in the digital era" and to measure the retention rate of the information presented in this way.

The CCN is entering its second period of existence (2021–2024). A budget of $65,000 per year has been allocated to the FL project. The objective of this second phase is to deploy the system on a larger scale, while developing other digital topics. The CCN's Digital Learning working group has been renewed and is currently considering the expansion of the concept. It should be considered to better involve learners in the FL process right from the start, to avoid an excessive vertical approach from experts to users. For instance, users could participate in focus groups and ask questions about the given topics, or even design some FLs themselves: from the learner to the learner, according to a peer-teaching logic, since "teaching is learning twice" [25].

References

1. Hug, T., Lindner, M., Bruck, P.: Microlearning: emerging concepts, practices and technologies after e-Learning. In: Proceedings of Microlearning 2005, Learning and Working in New Media (2005)
2. Herrington, J., Oliver, R.: An instructional design framework for authentic learning environments. Educ. Technol. Res. Dev. **48**(3), 23–48 (2000)
3. De Vries, P., Brall, S.: Microtraining as a Support Mechanism for Informal Learning. eLearning Papers, 11 (November) (2008)
4. Nikou, S.: A micro-learning-based model to enhance student teachers' motivation and engagement in blended learning. In: Society for Information Technology and Teacher Education International Conference, Association for the Advancement of Computing in Education (AACE), pp. 509–514 (2019)
5. Nikou, S.A., Economides, A.A.: Mobile-based micro-learning and assessment: impact on learning performance and motivation of high school students. J. Comput. Assist. Learn. **34**(3), 269–278 (2018). https://doi.org/10.1111/jcal.12240
6. Jomah, O., Amamer, K., Xavier, P., Sagaya, A.: Micro learning: a modernized education system. BRAIN Broad Res. Artif. Intell. Neurosci. **7**(Issue 1) (2016). ISSN: 2068-0473 (print)
7. Giurgiu, L.: Microlearning an evolving eLearning trend. Sci. Bull. **22**(1), 18–23 (2017)
8. De Gagne, J.C., Park, H.K., Hall, K., Woodward, A., Yamane, S., Kim, S.S.: Microlearning in health professions education: scoping review. JMIR Med. Educ. **5**(2), e13997 (2019)
9. De Gagne, J.C., Woodward, A., Park, H.K., Sun, H., Yamane, S.S.: Microlearning in health professions education: a scoping review protocol. JBI Database Syst. Rev. Implement. Rep. **17**(6), 1018–1025 (2019)
10. Halbach, T., Solheim, I.: Gamified micro-learning for increased motivation: an exploratory study. In: 15th International Conference on Cognition and Exploratory Learning in Digital Age (2018)
11. Mohammed Wakil, G.S.K., Nawroly, S.S.: The effectiveness of microlearning to improve students' learning ability. Int. J. Educ. Res. Rev. **3**(3), 32–38 (2018)
12. Ning, M., Feilong, Z., Peng-Qin, Z., Jun-Jie, H., Lei, D.: Knowledge map-based online microlearning: impacts on learning engagement, knowledge structure, and learning performance of in-service teachers. Interact. Learn. Environ. (2021). https://doi.org/10.1080/10494820.2021.1903932
13. Chisholm, L.: Micro-learning in the lifelong learning context (foreword). In: Hug, T., Lindner, M., Bruck, P. (eds.) Microlearning: Emerging Concepts, Practices and Technologies after e-Learning, Proceedings of Microlearning 2005, Learning and Working in New Media, pp. 11–12 (2005)
14. Downes, S.: E-Learning 2.0, in ACM eLearn Magazine. http://www.elearnmag.org/subpage.cfm?article529-1&section5articles (2005). Accessed 7 Mar 2022
15. Huo, C.Q., Shen, B.G.: Teaching reform of English listening and speaking in China based on mobile micro-learning. Creat. Educ. **6**(20), 2221–2226 (2015). https://doi.org/10.4236/ce.2015.620228
16. Luhmann, N.: Die Gesellschaft der Gesellschaft. Bd.1, Frankfurt am Main u.a.: Suhrkamp (1998)
17. Peschl M.F.: Challenges for a microlearning-driven process of knowledge creation modes of knowing and creating knowledge in microlearning environments (On Microlearning). In: Proceedings of Microlearning, Learning and Working in New Media (2005)
18. Buchem, I., Hamelmann, H.: Microlearning: a strategy for ongoing professional development. eLearning Papers, vol. 21, no. 7, pp. 1–15 (2010)

19. Stohr, C., Stathakarou, N., Mueller, F., Nifakos, S., McGrath, C.: Videos as learning objects in MOOCs: a study of specialist and non-specialist participants' video activity in MOOCs. Br. J. Educ. Technol. **50**(1), 166–176 (2019). https://doi.org/10.1111/bjet.12623
20. Zhang, Q., et al.: Exploring the communication preferences of MOOC learners and the value of preference-based groups: is grouping enough? Educ. Technol. Res. Dev. **64**(4), 809–837 (2016). https://doi.org/10.1007/s11423-016-9439-4
21. Kerres, M.: Microlearning as a challenge for instructional design. In: Hug, T., Lindner, M. (eds.) Didactics of Microlearning. Waxmann, Muenster (2006)
22. Fiebelkorn, I.C., Pinsk, M.A., Kastner S.A.: Dynamic interplay within the frontoparietal network underlies rhythmic spatial attention. Neuron. **99**(4), 842–853.e8 (2018). https://doi.org/10.1016/j.neuron.2018.07.038
23. Wertz, J.: Why Instant Gratification Is the One Marketing Tactic Companies Should Focus on Right Now, Forbes (2018)
24. Glahn, C.: Micro-learning in the workplace and how to avoid getting fooled by micro-instructionists. https://lo-f.at/glahn/2017/06/micro-learning-in-the-workplace-and-how-to-avoid-getting-fooled-by-micro-instructionists.html. Accessed 7 Mar 2022
25. Whitman, N.A., Fife, J.D.: Peer teaching: to teach is to learn twice. ASHE-ERIC Higher Education Report No. 4 (1988)
26. Moussavi, A., Mander, J.: Global Trends among Gen Z. Global Web Index. https://assets.ctfassets.net/inb32lme5009/7wDIuSsLOnSxTUqPmRb081/603b8ffb77757549d39034884a23743c/The_Youth_of_the_Nations__Global_Trends_Among_Gen_Z.pdf (2019). Accessed 7 Mar 2022

Visual Storytelling and Interactive Iconography for the Museum of Zoology in Rome

Fernando Salvetti[1,2,3](✉), Giuseppe Amoruso[4,5], Sara Conte[4,5], Silvia Battisti[1,2,3,6], and Barbara Bertagni[1,2,3]

[1] Centro Studi Logos, e-REAL Labs, 10143 Turin, Italy
salvetti@logosnet.org
[2] Logosnet, e-REAL Labs, 6900 Lugano, Switzerland
[3] Houston, TX 77008, USA
[4] Design School, Politecnico di Milano, 20121 Milan, Italy
[5] INTBAU Italia, 40121 Bologna, Italy
[6] Università Cattolica del Sacro Cuore, 00100 Rome, Italy

Abstract. This article summarizes the concept of a new immersive and interactive setting for the Zoology Museum in Rome, Italy. The concept, co-designed with all the museum's curators, is aimed at enhancing the experiential involvement of the visitors by visual storytelling and interactive iconography. Thanks to immersive and interactive technologies designed by Centro Studi Logos, developed by Logosnet and known as e-REAL® and MirrorMe™, zoological findings and memoirs come to life and interact directly with the visitors in order to deepen their understanding, visualize stories and live experiences, and interact with the founder of the Museum (Mr. Arrigoni degli Oddi) who is now a virtualized avatar, or digital human, able to talk with the visitors. All the interactions are powered through simple hand gestures and, in a few cases, vocal inputs that transform into recognized commands from multimedia systems.

Keywords: Visual storytelling · Interactive iconography · Digitized experience

1 A New Immersive and Interactive Concept for the Zoology Museum in Rome

During 2021 and 2022, we designed and developed the concept of a new immersive and interactive setting for the Zoology Museum in Rome, Italy; then, we installed all the new solutions that are now part of the permanent exhibit.

The concept, co-designed with all the museum's curators, is aimed at enhancing the experiential involvement of the visitors by visual storytelling and interactive iconography. Through the promotion and organization of multimedia content, this invention is about the integration of various technologies to facilitate the use of cultural content by expanding the current exhibition space through the development of interactive solutions capable of increasing visitors' (of all ages) involvement.

D. Guralnick et al. (Eds.): TLIC 2022, LNNS 581, pp. 323–334, 2023.
https://doi.org/10.1007/978-3-031-21569-8_31

Thanks to immersive and interactive technologies designed by Centro Studi Logos, developed by Logosnet and known as e-REAL® and MirrorMe™, zoological findings and memoirs come to life and interact directly with the visitors in order to deepen their understanding, visualize stories and live experiences, and interact with the founder of the Museum (Mr. Arrigoni degli Oddi) who is now a virtualized avatar, or digital human, able to talk with the visitors. All the interactions are powered through simple hand gestures and, in a few cases, vocal inputs that transform into recognized commands from multimedia systems (Figs. 1, 2 and 3).

Fig. 1. Initial rendering of the Arrigoni degli Oddi room with the avatar of Mr. Arrigoni, founder of the museum, within his office.

To summarize, our intervention relates to

- The development of multimedia content and digital exhibits.
- The application of methodologies aimed at seeing the museum's heritage come to fruition.
- The realization of animal bones using 3D reconstructions through scanning-based technologies.
- The installation of immersive environments that allow a sensorial and gestural interaction with relevant contents (Figs. 4, 5, 6, 7, 8 and 9).

The main results are:

- Enhancement of the museum heritage, dissemination of its knowledge, and involvement of target audiences through visual and multimedia, tactile, audio, and olfactory communication tools.
- Use of new digital media to improve the museum's offerings.

Fig. 2. The interactive avatar of the museum's founder, Mr. Arrigoni degli Oddi, within his office.

Fig. 3. The interactive avatar of Mr. Arrigoni in detail. He is programmed to be a talkative digital human, able to share information about his life and the history of his zoological collection both in Italian and English.

Fig. 4. Skeletons room transformed into a hybrid reality setting: real skeletons, interactive animals (e-REAL technology) and 3D reconstructed footprints for tactile exploration.

Fig. 5. A representative 3D reconstructed footprint.

- Improved accessibility to the museum's collection, which has increased through the promotion and organization of multimedia content (Figs. 10 and 11).

The entire project was made possible thanks to the contribution awarded by the European Regional Development Fund (POR FESR Lazio 2014/2020, Action 3.3.1 b), relating to the public notice *L'Impresa fa Cultura*. The contribution awarded was 181,000 euros. The total investment supported by Centro Studi Logos, an Italian company (based in Turin and Rome) from Logosnet, was 227,850 euros [1].

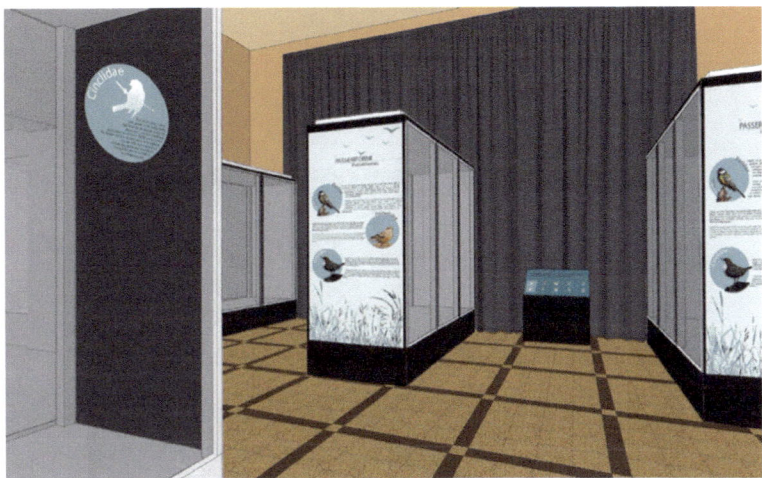

Fig. 6. Initial rendering of the birds' room: printed information, 3D animations and audio traces to recreate the sounds of the birds and to give them a new virtual life.

Fig. 7. Initial rendering: whale room transformed into a hybrid reality setting with real elements displayed jointly with a multimedia storytelling made interactive by gesture shaping (e-REAL technology).

Fig. 8. Initial rendering: a fully mirrored room with 360-degree projections.

Fig. 9. Augmented reality mirror (MirrorMe) to learn human evolution by mirroring the interactive skeletons of the human body and the chimpanzee.

Fig. 10. Details from the skeletons room: enhanced written and visual communication.

Fig. 11. The poster introduction to the new immersive and interactive system aimed at experiential involvement and the contents' diffusion by visual storytelling and interactive iconography modalities.

2 Digitalization and Cultural Heritage 4.0: Ways of Interactivity in Museums and Science Centers

Cultural heritage is a limitless source of innovation where traditions meet cutting-edge technology, mainly from the ongoing 4.0 digital revolution. Digital heritage, as well as science museums, are very interesting domains because the contemporary audience expects both stability and flexibility from museums, which should increase their attractivity without losing credibility. Technology and exhibition design can help in the creation of new spaces and innovative solutions to grant the audience the enjoyment of a living, memorable experience.

Digitization has a primary role to play in the conservation and promotion of cultural heritage, mainly by enhancing real-life experiences, rather than replacing them. Digital is not only a way to dematerialize our cultural heritage, but—mainly if associated to visual storytelling—also a powerful way to enhance the human capacity to generate engaging content and memorable experiences [2].

Social interaction and collaboration are critical to our experience of museums and galleries. Curators, museum managers, and designers are exploring ways of enhancing interaction and in particular using tools and technologies to create new forms of participation, with and around, exhibits. It is commonly assumed that it is the exhibit that is interactive. We speak of 'interactive exhibits,' 'interactive experiences,' and even of 'interactives' as something that can be designed by specialists, tested rigorously, and their outcomes measured. In all these cases, it is the object that is assumed to be interactive—something that can be touched, felt, or manipulated is claimed to be more 'interactive' than something that cannot.

Interactivity is normally used to mean physical interaction with an object or exhibit—a 'hands-on' experience. Most people, when they think of interactive exhibits at all, think of the experience of the Bernouilli blower, the ball bouncing gaily on a jet of air, making soap bubbles, or making bridges out of blocks—all commonly found exhibits in today's science centers. Limiting the notion of interaction to merely physical manipulation has been challenged for years, although most proponents still consider hands-on manipulation indispensable.

Richard Gregory, founder of Britain's first hands-on science center, the Bristol Exploratory, speaks of 'minds-on' exhibits and uses illusions to show the workings of the human mind. Jorge Wagensberg, Director of the Museum de la Ciencia in Barcelona, speaks of 'hearts-on' exhibits, which he uses to describe exhibits with a large affective dimension. In these cases, interaction seems to indicate a particularly tangible engagement with the exhibit. Even so, the notion of interaction itself—whether hands, minds, or hearts-on—does not give any real indication of the quality of the experience. Interaction is too vague a term to use precisely enough to be helpful.

Physical interaction is not a prerequisite for interaction, nor is the visitor obliged to publicly interact. It is, however, no coincidence that interactive exhibits and the corresponding educational theories that place interactivity at their core, stemmed from the science center movement. Bereft of objects, science centers had as their challenge to render phenomena visible, which almost by definition involved inviting the visitor to participate in the process of creating rainbows, making waves, and mixing colors.

Moreover, the cultural discourse that would have us believe that the experience of art is unmediated, is conspicuously absent in the world of science and technology - no one pretends that a steam engine explains itself, or that a chemical reaction can be appreciated without some small understanding of what is going on.

But is all as it seems? A closer look at the science center gives quite another impression. On closer inspection, the much-vaunted interactivity often masks experiences which in fact close down the visitor's ability to explore, and limit the ways in which they can direct their own discovery.

What if we started to look at interactivity as the property of the visitor and not of the exhibit? What if we looked at the exhibit as a tool that, if properly conceived, conferred the property of interactivity onto its user? What would this interactivity look like? Is interaction different in museums of fine art compared to those of applied art or design?

We would have to answer, no. The nature of the engagement in any informal setting is potentially the same, subject to the way in which the museum chooses the user-languages it employs and the degree to which the museum reduces the barriers that prevent the user from engaging with the material.

Where museums do differ, however, is in their deliberate use—or avoidance—of specific user-languages. Science centers were among the first to be forced to explore the user-languages of observation and variables, as both are proper to the natural sciences. Given their history, they were also among the first to explore the user-languages of problems and games. This is not to say, however, that fine art museums cannot make equally good use of these user-languages. Joaneath Spicer at the Walters Art Gallery in Baltimore, Maryland, turned her entire museum into a resource to solve an art historical puzzle and the new British Galleries at the Victoria and Albert are rich in exhibits which employ the user-languages of puzzles and games.

What is important, we believe, is not the nature of the museum's content, but the degree to which we make explicit use of particular user-languages in order to actively engage our visitors in the pleasure that comes from actively exploring and constructing the world in which we live in all its variety [3, 4]. All the above considerations are part of the vision behind our concept for the Museum of Zoology in Rome.

3 Digital Technology for Knowledge, Design and Experiential Education for Culture

The learning society represents a new human condition linked to contemporary social phenomena, a society where men and women live, work, organize themselves and utilize know-how and knowledge as a new form of capital. This vision lays the structural foundation for economics and social development: starting from Donald Schon's paradigm, "learning, reflection, and change" is translated into the promotion of creativity at all levels, addressing a critical and civic awareness and inducing a process of social change.

Design, considered as a whole set of disciplines in the universe of industrial design, deals with designing the value of processes, goods, environments and services, of increasing it and imparting this to society and citizens. Experiential design proposes a system of mediation between the territorial context and the cultural heritage system or the widespread heritage (memory, history, landscape) and the reference community intended as the final user. This makes it possible to have multiple forms of representation of goods and legitimizes their differentiated values, access, use and appropriation, whether directly or by using technology. Bearing this vision in mind, design does not solely restrict itself to designing the experience of use of goods (economy of experience), but introduces an innovative vision of systems and a shared vision of cultural heritage in all its forms; it also makes it possible to start upon a participatory and inclusive learning path and social well-being, which makes its diffusion in the community sustainable and cost-effective (from the institution to the cultural operator, to the different categories of users).

The service economy in recent years has shown considerable potential by creating an innovative system with a social nature, based on a particular type of economic performance. Goods and services are no longer sufficient as economic products; a new need has been created: through a design process, an integrated fruition project can be created, that is to say, the words, giving a sensorial and psychological form to experience.

Knowledge technologies are recognized as opportunities in terms of conservation, study and communication of heritage, but also of creating culture and awareness that is expressed in the contemporary forms of sharing and dissemination. Learning, in the different seasons of life, should therefore be considered as the source of an increasingly innovative economy that becomes sustainable and has an impact if it reaches a substantial and diversified number of users and social subjects.

Design for cultural heritage includes theories, methodologies and enhancement techniques that have the cultural heritage system understood in its cognitive, social and symbolic dimensions as their application sphere. The disciplines of representation interact with the multiple disciplinary specializations of design, proposing the definition of interpretative models for the analysis and representation of the historical, cultural, aesthetic and environmental values of a cultural asset as well as its material and immaterial meaning. The value enhancement strategy produces advanced visualizations as well as computer and multimedia modelling. Moreover, the experiential value, with its emotional imprint and fruition, is emphasized through immersive and interactive technologies. The applications make it possible to have a structured and flexible knowledge process including the simulation of forms of innovation and an increase in the social value of the transmission and sharing of cultural contents. In fact, in order to fulfil their educational mission, the spaces of culture need to go beyond the tangible and common sensorial dimensions in order to communicate and share a heritage, understood also as a process of appropriation and as such also linked to the intangible dimension. It is in this direction that the Convention for the Protection of Intangible Cultural Heritage (Paris, 2003) goes. It defines the intangible cultural heritage as "the practices, representations, expressions, knowledge, skills—as well as the instruments, objects, artefacts and cultural spaces associated therewith—that communities, groups and, in some cases, individuals recognize as part of their cultural heritage".

Within the framework of UNESCO, there is the "Recommendation Concerning the Protection and Promotion of Museums and Collections, Their Diversity and Their Role in Society (Paris, 2015), which underscores the importance of technologies in assisting museums in their task of educating and encouraging continuous learning. Technologies are therefore changing the relationship between users and cultural content in museums, libraries and places of learning. The environments must be imagined and transformed by also considering their virtual extension and allowing a range of customizations linked to the selection of contents. Participation and sharing mediated by the user can also create new cultural content by blazing a path to new forms of active and participatory learning. Among the cultural actions that are related to new media and their language, the creation and sharing of information and knowledge are included, as well as the accessibility to heritage through digital artefacts that represent ideas, identities and values of belonging. To these, Manovich also adds the interactive cultural experience, the opportunity to enjoy the experiences and cultural products by visitors, as well as ways to recreate the displayed objects, textual, vocal and/or visual communication and participation in a type of information that "ecologically" regenerates knowledge and its diffusion. Knowledge technologies offer multiple opportunities and challenges to cultural and scientific practitioners; the challenge of involvement and experience is not only one of technology and design, but also, and perhaps more importantly, a mental and imaginative one [5].

4 The Instructional Design for the Museum of Zoology in Rome

The instructional design for the Museum of Zoology in Rome is summarized by 3 keywords: Visualization, interaction, immersion. An effective visualization is the key to help untangle complexity: the visualization of information enables visitors—that are learners—to gain insight and understanding quickly and efficiently. Examples of such visual formats include sketches, diagrams, images, objects, interactive visualizations, information visualization applications and imaginary visualizations such as in stories. In such a way, visualizations show relationships between topics, activate involvement, generate questions that learners didn't think of before and facilitate memory retention. So visualizations act like concept maps to help organize and represent knowledge on a subject in an effective way.

Half of human brain is devoted directly or indirectly to vision and images are able to grab our attention easily. Human beings process images very quickly: average people process visuals 60,000 times faster than text. This is why we, as humans, are confronted with an immense amount of images and visual representations every day: digital screens, advertisements, messages, information charts, maps, signs, video, progress bars, diagrams, illustrations, etc. If we have to warn people, symbols and images are excellent: they communicate faster than words and can be understood by audiences of different ages, cultures and languages. Images are powerful: people tend to remember about 10% of what they hear, about 20% of what they read and about 80% of what they see and do [6].

Mainly the e-REAL and the MirrorMe technologies submerge learners in an immersive and interactive reality. Multi-surface environments, like the ones we created within the museum's rooms involved by our intervention, require users to be "physically" engaged in the interaction and afford physical actions like pointing to a distant object with the hand or walking towards a large display to see more details. Based on a body-centric paradigm, the e-REAL setting is well-adapted to device- or eyes-free interaction techniques because they account for the role of the body in the interactive environment.

References

1. https://centrostudilogos.info/visual-communication-and-interactive-experiences-for-the-zoo logical-museum-in-rome/
2. Scuderi, A., Salvetti, F. (eds.): Digitalization and cultural heritage in Italy. Innovative and cutting-edge practices. Angeli, Milan (2019)
3. Bradburne, J.: Ways of interactivity in museums and science centers. In: Scuderi, A., Salvetti, F. (eds.), Digitalization and Cultural Heritage in Italy. Innovative and Cutting-Edge Practices. Angeli, Milan (2019)
4. Salvetti, F., Bertagni, B. (eds.): Learning 4.0. advanced simulation, immersive experiences and artificial intelligence, flipped classrooms, mentoring and coaching. Franco Angeli, Milan (2018)
5. Amoruso, G.: Digital technology for knowledge, design and experiential education for culture. In: Scuderi A., Salvetti F. (eds.) Digitalization and Cultural Heritage in Italy. Innovative and Cutting-Edge Practices. Angeli, Milan (2019)
6. Salvetti, F., Bertagni, B.: Virtual worlds and augmented reality: the enhanced reality lab as a best practice for advanced simulation and immersive learning. Form@re **19**(1). University of Florence (2019)

The GW Community Medi-Corps Program: A Mobile Mixed-Reality Immersive Learning Center

Fernando Salvetti[1,2,3]([⊠]), Teri L. Capshaw[4], Linda Zanin[4], Kevin C. O'Connor[4], Qing Zeng[4], and Barbara Bertagni[1,2,3]

[1] Centro Studi Logos, e-REAL Labs, 10143 Turin, Italy
salvetti@logosnet.org
[2] Logosnet, e-REAL Labs, 6900 Lugano, Switzerland
[3] Houston, TX 77008, USA
[4] The George Washington University School of Medicine and Health Sciences, Washington, DC, USA

Abstract. The Community Medi-Corps Program—designed and implemented by the George Washington University (GW) School of Medicine and Health Sciences (SMHS) faculty with Growth and Opportunity Virginia funding (GO Virginia)—is aimed at leveraging the power of community, educational institutions, mentors, industry, and business partners to close the opportunity gap, transform student learning, and enrich the regional workforce. This program transforms educational experience through innovative virtual reality, augmented reality, and a mix between the two that is the enhanced reality (e-REAL). Students will be better prepared in the pathways they choose for high demand health and life sciences industry jobs that will help grow the economy.

Keywords: Immersive experience · Interactive visualization · STEM education

1 The Community Medi-Corps Program

Talent is everywhere, but opportunity is not. The Community Medi-Corps Program—designed and implemented by the George Washington University (GW) School of Medicine and Health Sciences (SMHS) faculty with Growth and Opportunity Virginia funding (GO Virginia)—is aimed at leveraging the power of community, educational institutions, mentors, industry, and business partners to close the opportunity gap, transform student learning, and enrich the regional workforce.

The Community Medi-Corps Program is a $1.6 million project made possible by a $700,000 grant from GO Virginia. The program will augment curricula and enhance health sciences education in the Alexandria City (Virginia) Public Schools (ACPS), Arlington (Virginia) Public Schools (APS), Fairfax County (Virginia) Public Schools (FCPS), and Loudoun County (Virginia) Public Schools (LCPS) The centerpiece of

Medi-Corps is a mobile Immersive Learning Center (ILC)—a 45-foot-long class-room/lab on wheels that provides students with cutting-edge technology, simulation, and immersive virtual reality and augmented learning experiences [1] (Figs. 1, 2 and 3).

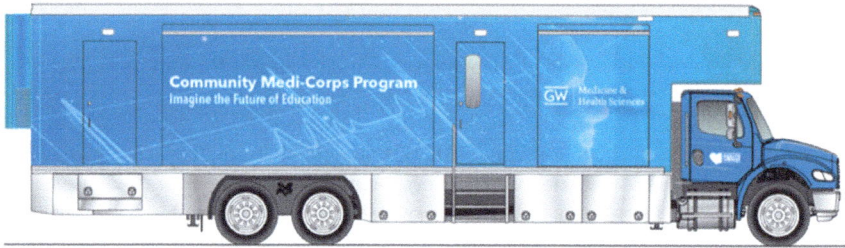

Fig. 1. Rendering of the mobile immersive learning center.

Fig. 2. Internal view of the mobile immersive learning center.

Medi-Corps and the ILC allow students to engage and work together to problem-solve, use virtual technology, and interact with experts in the life and health science fields. Faculty and staff members from SMHS and the four partnering school systems are supporting the project. The ILC incorporates the latest immersive learning technologies to support critical thinking and applied learning, and to maximize student learning and engagement. Our vision is for this initiative to serve as a best practice for other areas in Virginia and the region. We feel strongly that this innovative model, linking secondary education and four-year institutions, will benefit students in numerous ways and better prepare them in health sciences.

Although virtual reality (VR), augmented reality (AR), and mixed reality (MR) simulation training has gained prominence, review studies to inform instructors and educators on the use of these technologies—usually grouped under the name of extended reality (XR)—in science, technology, engineering, and mathematics (STEM) are still scarce. We found interesting references in Pellas, Dengel, and Christopulos that analyzed

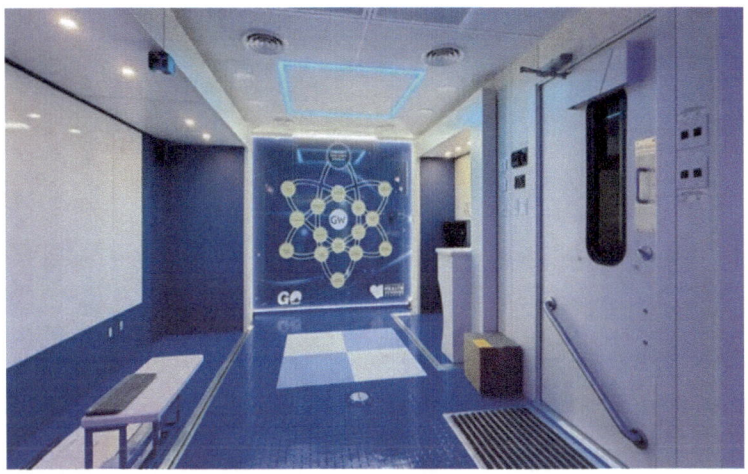

Fig. 3. Another perspective on the mobile immersive learning center.

various VR-supported instructional design practices in K-12 (primary and secondary), as well as higher education, in terms of participants' characteristics, methodological features, and pedagogical uses in alignment with applications, technological equipment, and instructional design strategies [2]. For the design of both the mobile ILC and the learning experiences we decided to implement the guidelines from Salvetti, Bertagni, Wieman, Waldrop, and Brenner, that are summarized by the title of the book Learning 4.0: Advanced simulation, immersive experiences, flipped classrooms, mentoring, and coaching [3] (Fig. 4).

Our project emerges from a large-scale collaborative effort supported by a large number of professionals and institutions. Our colleagues provided support for this initiative at multiple levels. The commonly shared goal is to inspire youth to explore and connect as we create tomorrow's next generation of health sciences leaders. Students are expected to be able to engage and collaborate, thanks to state-of-the art opportunities for students to learn STEM-H subjects interactively.

The Medi-Corps program is aimed at further bridging the gap between academics and the workforce by offering internships and mentorships with experienced health professionals.

The Medi-Corps team envisions the project to serve as a best practice that can be replicated in other areas in Virginia and the region. The innovative model, linking secondary education, community colleges, and four-year institutions, will benefit students in numerous ways and better prepare our future workforce in health sciences.

Medi-Corps is based at the Governor's Health Sciences Academy (Academy) at Alexandria City High School. Academy, APS, FCPS, and LCPS students are expected to experience interactive learning in the ILC, starting in the 2022–23 school year. Community events that showcase immersive learning and opportunities in the health and life sciences fields are also planned.

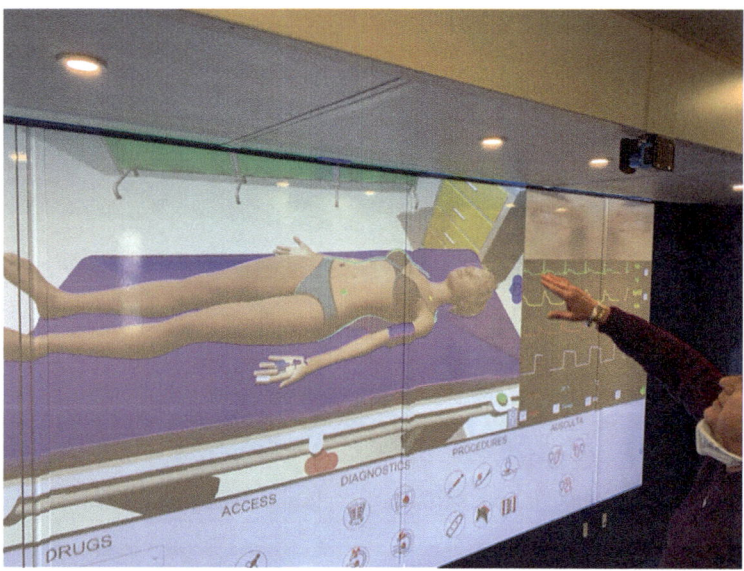

Fig. 4. An e-REAL interactive content: virtual patient.

2   The Mobile Immersive Learning Center and the Vision Behind the Program

The ILC was designed by the GW SMHS Community Medi-Corps grant team in collaboration with LifeLine Mobile, a vendor from Columbus, Ohio, and with the Logosnet Instructional Design Team, and the e-REAL Technology Team. The ILC features a 45-foot mobile immersive learning center that bridges the gap between the classroom and the field (Fig. 5).

The ILC transforms educational experience through innovative virtual reality, augmented reality, and a mix between the two that is the enhanced reality (e-REAL). It raises student aspirations and attracts students to postsecondary education, provides summer programs, job fairs, and community outreach events.

The main issues addressed by the Medi-Corps Program include the following:

- Limited diversity in the life sciences and health professions workforce has significant consequences for access to health care services, health outcomes, and health equity, especially for underrepresented minority patients and underserved communities.
- Youth who identify as racial or ethnic minorities are less likely to be exposed to and less prepared for a range of STEM-H careers.
- A long-term demand exists for skilled and credentialed health and life science workers: the field is growing at a faster rate than others with 7–26% growth forecast through 2028.

Fig. 5. Exterior view of the mobile immersive learning center.

The Community Medi-Corps program strives to:

- Provide summer programs, job fairs, and community outreach events.
- Engage professionals in STEM-H fields to serve as student mentors.
- Shape future leaders essential for a healthier society.
- Champion equitable excellence and robust academic opportunities.
- Enrich the diverse communities it serves.
- Promote equity, diversity, and inclusion in practice.
- Influence future workforce needs for high demand jobs.
- Contribute to the region's economic growth and resilience.

To learn more about the program and the STEM mobile lab please view this video: https://youtu.be/1UL51zFq_bM.

Experts in postsecondary education agree that the center can provide students with hands-on learning experiences and encourage them to stay in STEM and health care fields after graduation.

According to William Corrin, the director of K-12 education at MDRC, an organization that researches social policy, high school and university partnerships are beneficial because they smooth the transition for students, making it less of a gap and more of a bridge. "Those transition points are usually the places where there's the greatest risk for students to experience some kind of disruption to their educational trajectory," Corrin said.

Max Milder, the director of research at EAB, an education research organization, said partnerships between higher education institutions and high schools can create a pipeline of leading new students to the institution. "Universities are always interested in how they're going to continue to attract future students or enrollments in the coming years," Milder said. "There's a part of this that is getting George Washington University in front of high school students as early as possible, even before they're going into that decision-making process for enrollment."

Milder said exposing high school students to high-level technology that is common in medical education familiarizes students with what they'll be using throughout their medical careers. "Experiential learning is really critical," he said. "And that's true in K-12. That's true in higher education, as well. And so part of the effort here is to bring some of these scientific or medical concepts to life and do so in a way that is really engaging and hopefully fun for the students as well" [3].

The most innovative virtual, augmented, and mixed reality technologies are on board into the mobile ILC. Reality in the digital age is becoming more and more virtual, augmented, and mixed. These technologies offer options to improve learning methods. Sharing and mixing up the latest trends from digitization and virtualization, neurosciences, artificial intelligence, and advanced simulation allows us to establish a new paradigm for STEM-H education.

3   The Learning Setting and the Main Educational Outputs

The learning setting of the mobile Immersive Learning Center is designed according to the STEAM approach: It's the extension of an acronym that originally stands for science, technology, engineering and math, with the arts added because STEM alone misses several key components that many employers, educators, and parents have voiced as critical to thrive in the present and rapidly approaching future. The STEAM approach refers to a movement that has been taking root over the past several years and is surging forward as a positive mode of action to truly meet the needs of a 21st century society.

STEAM uses science, technology, engineering, the arts and mathematics as access points for guiding learner inquiry, dialogue, and critical thinking. The end results are learners who take thoughtful risks, engage in experiential learning, persist in problem-solving, embrace collaboration, and work through the creative process. STEAM is a way to take the benefits of STEM and complete the package by integrating these principles in and through the arts. STEAM takes STEM to the next level: it allows learners to connect their learning in these critical areas together with arts practices, elements, design principles, and standards to provide the whole pallet of learning at their disposal. STEAM removes limitations and replaces them with wonder, critique, inquiry, and innovation [5].

Designing a program that includes active learning requires more content knowledge, not less, than teaching in the classic lecture mode. If a teacher uses active learning techniques, he/she/they is still telling students information; but it's in response to their questions, their needs to solve a problem, and so they learn much more from it [6]. So, a teacher has to work hard to use active learning in the class and has to carefully structure problems and activities to get students to think like a scientist, mathematician, or, in our case, as a healthcare professional.

In active learning methods, students are spending a significant fraction of the time on activities that require them to be actively processing and applying information in a variety of ways, such as answering questions using electronic clickers, completing worksheet exercises and discussing and solving problems with fellow students.

The instructor designs the questions and activities and provides follow-up guidance and instruction based on student results and questions. Also, good active learning tasks simulate authentic problem solving and therefore teaching with these methods typically demands more instructor subject expertise than does a lecture [6].

The setting of the mobile Immersive Learning Center is designed around 3 keywords: Visualization, interaction, immersion. It is a fully immersive and multitasking environment, designed to present challenging situations in a group setting, engaging all participants simultaneously. The e-REAL instructional design and technology make possible teaching and learning with motion pictures, as well as with 3D visualizations and augmented reality tools. These tools are fully interactive and "talkative"; avatars or digital humans are a key-component of the setting.

Effective visualization is the key to help untangle complexity: the visualization of information enables learners to gain insight and understanding quickly and efficiently. Examples of such visual formats include sketches, diagrams, images, objects, interactive visualizations, information visualization applications, and imaginary visualizations in scenarios [7]. Visualizations within e-REAL show relationships between topics, activate involvement, generate questions that learners didn't think of before, and facilitate memory retention. So visualizations act concept maps to help organize and represent knowledge on a subject in an effective way.

Half of human brain is devoted directly or indirectly to vision and images are able to grab our attention easily. Humans process images very quickly: on average a person processes visuals 60,000 times faster than text. This is why we, as humans, are confronted with an immense amount of images and visual representations every day: digital screens, advertisements, messages, information charts, maps, signs, video, progress bars, diagrams, illustrations, etc. If we have to warn people, symbols and images are excellent: they communicate faster than words and can be understood by audiences of different ages, cultures, and languages. Images are powerful; people tend to remember about 10% of what they hear, about 20% of what they read, and about 80% of what they see and do [8].

Also, contextual factors have tremendous importance because they are key to learning. Learners practice handling realistic situations, rather than learning facts or techniques out of context. Context means "related factors," that can be influential and even disruptive. The most effective learning occurs through being immersed in context. Experience is lived and perceived as a focal point and as a key crossroad [9]. Much like being immersed within a videogame, people are challenged by facing real cases within complex scenarios that present a more than real wealth of information. This is because the many levels of the situation are made available simultaneously, by overlaying multisource information on the projected walls and inside a number of augmented reality displays made available within the setting.

The e-REAL setting and technology submerges learners in an immersive reality where the challenge at hand is created by sophisticated, interactive computer animation in three dimensions and holographic projections. Multi-surface environments require users to be "physically" engaged in the interaction and afford physical actions like pointing to a distant object with the hand or walking toward a large display to see more details. Based on a body-centric paradigm, the e-REAL setting is well-adapted to device-

or eyes-free interaction techniques because they account for the role of the body in the interactive environment. Very large interactive wall displays do not lend themselves to use with traditional interaction modalities such as mice and keyboards. It is a multi-surface environment that encourages users to interact while standing or walking, using their hands to manipulate objects on multiple displays. Within the e-REAL setting, the body itself is used for input: Users can interact by moving the body, or with a flick of the hands and some other gesture [10–12] (Figs. 6, 7, 8 and 9).

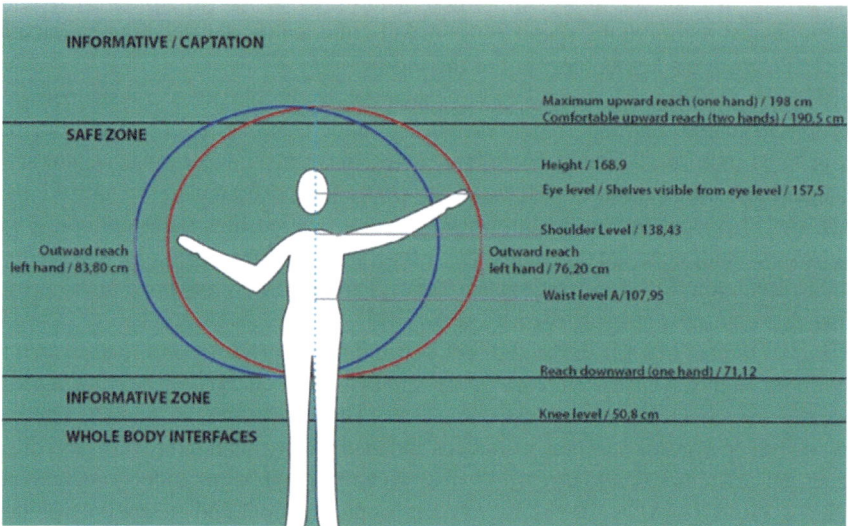

Fig. 6. Ergonomic reference table for e-REAL gesture shaping and body interaction.

Using the body enhances both learning and reasoning and this interaction paradigm has proven effective for gaming [13], in immersive environments [14], when controlling multimedia dance performances [15] and even for skilled, hands-free tasks such as surgery or emergency medicine [16]. Smartphones and devices such as Nintendo's Wii permit such interaction via a hand-held device, allowing sophisticated control. However, holding a device is tiring [17] and limits the range of gestures for communicating with co-located users, with a corresponding negative impact on thought, understanding, and creativity [18].

Advances in sensor and actuator technologies have produced a combinatorial explosion of options that do not require hand-held devices. The e-REAL interaction design team, since 2011, has tested and selected various options in order to combine them in a coherent, powerful way based on specific guidelines (like the ones displayed in Fig. 6). A few simple and intuitive gesture options are the solution, enabling the learning experiences within the mobile Immersive Learning Center. In such a way, learners are physically engaged in the interaction and afford physical actions like pointing to a distant object with the hand or walking toward a large display to see more details [19], listening to and interacting with one or more digital humans (Figs. 10, 11, 12, 13 and 14).

SENSOR'S SCANNING RANGE

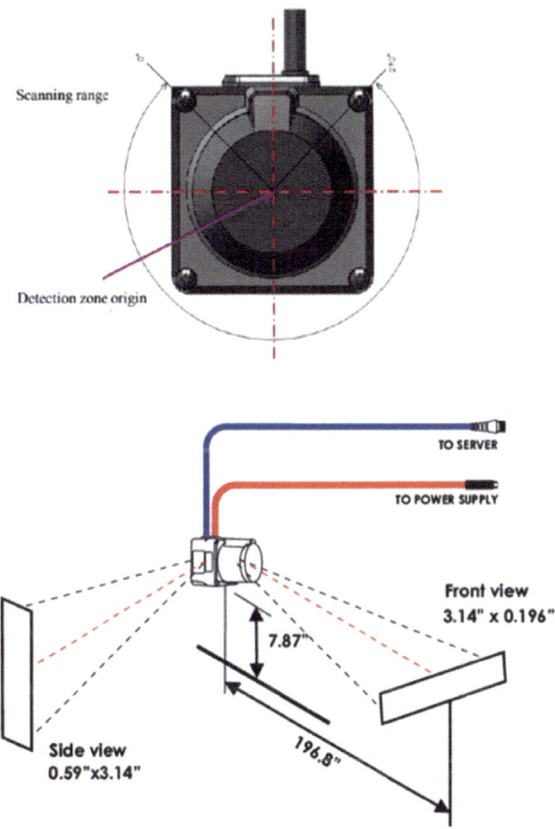

Fig. 7. The e-REAL sensor enabling gesture shaping and body interaction.

The e-REAL learning setting encourages students to learn by doing with the help of simulation or game-based tools. This setting allows learners to experience abstract concepts in three-dimensional space based on visualization, enhancing at the same time an active learning mindset by encouraging cooperative work among students. Visual storytelling techniques are essential to represent a realistic context where learners are proactively involved to analyze scenarios and events, to face technical issues, to solve problems. The most effective learning occurs when being immersed in a context: realistic experience is lived and perceived as a focal point and as a crossroad [20].

A context related experience within an e-REAL setting is similar to being immersed within a video game with our entire bodies. Characteristics of games that facilitate immersion can be grouped into two general categories: those that create a rich mental model of the game environment and those that create consistency between the things in that environment. The richness of the mental model relates to the completeness of multiple channels of sensory information, meaning the more those senses work in alignment,

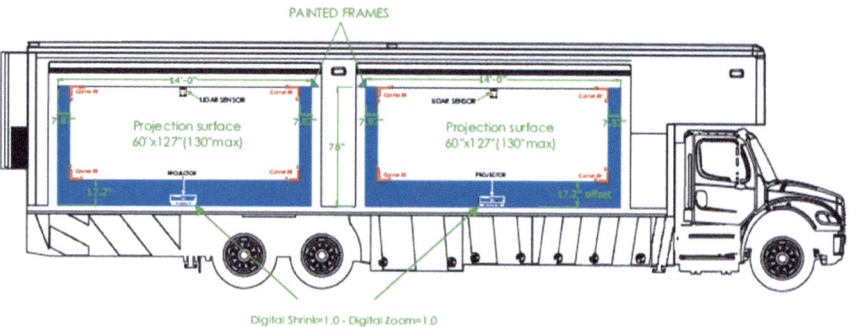

- Projection rapport: 16:9
- Projection surface: 60"x127"-130"
- Projection area will be surrounded, onto the same wall, by a frame painted like the side walls
- Projectors' settings (Corner Fit, Digital Shrink and Zoom) are displayed into the above image.
- Projections' height from the floor is expected to be 85": the visible projection's height will be reduced to 78" from the floor by an electronic "black strip" with parametrized height (approx 7"), adjustable at runtime.
Please notice that there is a 3% tolerance regarding projection's expected performance, due to optical component variations. It is recommended to physically test the projection size and distance before permanently installing the projector.

Fig. 8. Projection surfaces and details about the e-REAL immersive setting.

the better. The richness also depends on having a cognitively demanding environment and a strong and interesting narrative. A bird flying overhead is good. Hearing it screech is better.

Cognitively demanding environments in which players must focus on what's going on in the game will occupy mental resources. The richness of the mental model is good for immersion because if brain power is allocated to understanding or navigating the world, it's not free to notice all of its problems or shortcomings that would otherwise remind them that they're playing a game. Finally, good stories—with interesting narratives, credible because intrinsically congruent as much as possible—attract attention to the game and make the world seem more believable. They also tie up those mental resources. Turning to game traits related to consistency, believable scenarios, and behaviors in the game world means that virtual characters, objects, and other creatures in the game world behave in the way in which learners expect.

The process of learning by doing within an immersive setting, based on knowledge visualization using interactive surfaces, leaves the learners with a memorable experience. From an educational perspective, learners are not assumed to be passive recipients and repeaters of information but individuals who take responsibility for their own learning. The trainer functions, not as the sole source of wisdom and knowledge, but more as a coach or mentor, whose task is to help them acquire the desired knowledge and skills. A significant trend in education in the 19th and 20th centuries was standardization. In contrast, in the 21st century, visualization, interaction, customization, gamification, and flipped learning are relevant trends. In a regular flipped learning process, students are exposed to video lectures, collaborate in online discussions, or carry out research on their own time, while engaging in concepts in the classroom with the guidance of a mentor.

Fig. 9. A learner facing an e-REAL interactive image showing a brain cancer, divided in 8 pieces, during an experiment aimed at determining whether cognitive retention improves when visualization is broken into multiple smaller fragments first and then recomposed to form the big picture.

Critics argue that the flipped learning model has some drawbacks for both learners and trainers. A number of criticisms have been discussed with a focus on the circumstance that flipped learning is based mainly on video-lectures that may facilitate a passive and uncritical attitude towards learning, in a similar way to didactic face-to-face lectures, without encouraging dialogue and questioning—within a traditional classroom.

The e-REAL setting is a further evolution of a flipped classroom, based on a constructivist approach. Constructivism is not a specific pedagogy, but rather a psychological paradigm that suggests that humans construct knowledge and meaning from their experiences. From our constructivist point of view, knowledge is mainly the product of personal and interpersonal exchange. Knowledge is constructed within the context of a person's actions, so it is "situated": it develops in dialogic and interpersonal terms through forms of collaboration and social negotiation. Significant knowledge—and know-how—is the result of the link between abstraction and concrete behaviors.

Fig. 10. e-REAL representative avatars programmed to perform as digital twins of the learning facilitators.

Knowledge and action can be considered as one: facts, information, descriptions, skills, know-how and competence—acquired through experience, education and training. Knowledge is a multifaceted asset: implicit, explicit, informal, systematic, practical, theoretical, theory-laden, partial, situated, scientific, based on experience and experiments, personal, shared, repeatable, adaptable, compliant with socio-professional and epistemic principles, observable, metaphorical, linguistically mediated. Knowledge is a fluid notion and a dynamic process, involving complex cognitive and emotional elements for both its acquisition and use: perception, communication, association, and reasoning. In the end, knowledge derives from minds at work. Knowledge is socially constructed, so learning is a process of social action and engagement involving ways of thinking, doing and communicating [21].

The Community Medi-Corps Program is currently at its beginning and we will be able to research for and analyze its educational inputs in the coming years. So far, we can say that the program's first execution into the mobile Immersive Learning Center is reaching the expected outcomes.

Fig. 11. e-REAL representative digital human, programmed to perform as an athlete student expected to be injured and to start a healthcare rehabilitation program.

Fig. 12. e-REAL avatar programmed to perform as an injured lady, able to call for help and to interact dialogically with learners and simulation instructors.

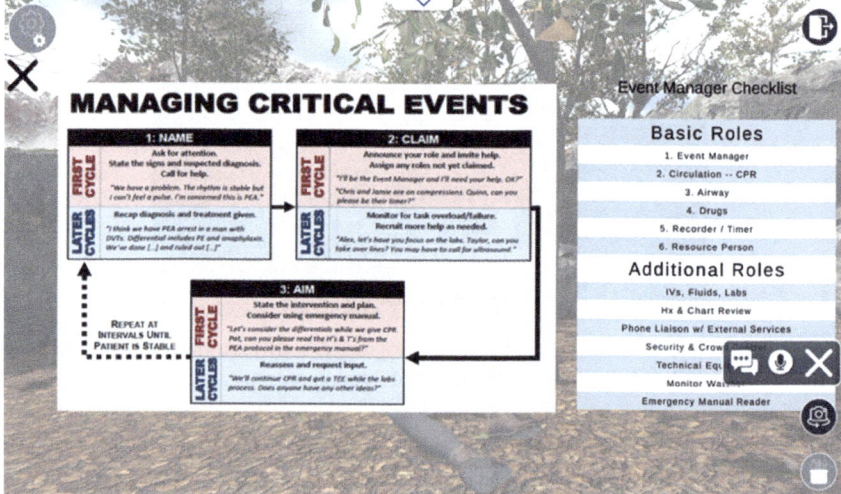

Fig. 13. Name-Claim-Aim ©: A mnemonic and checklist developed by faculty at the Center for Medical Simulation (Boston, Massachusetts), that encompasses a strategy to help health care professionals effectively organize a team for managing critical clinical events.

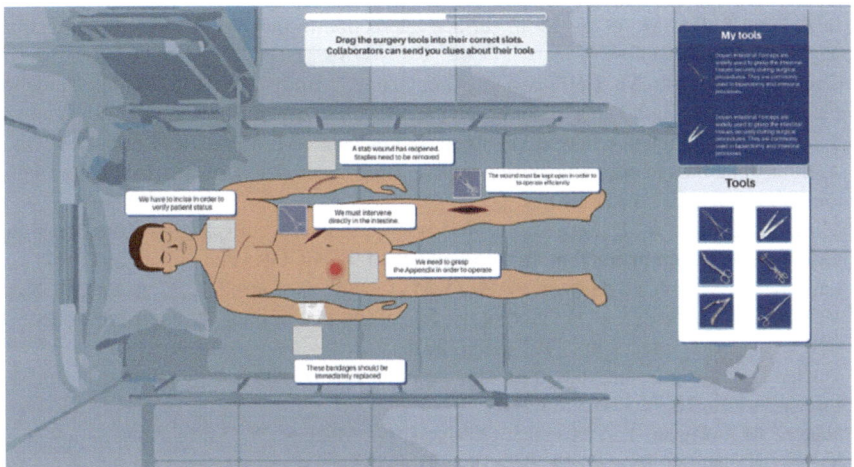

Fig. 14. e-REAL representative gamified healthcare activity designed to challenge the learners with a cooperative game aimed at understanding selected surgical procedures and to place surgery tools in the correct slots.

References

1. https://hssp.smhs.gwu.edu/news/community-medi-corp-program-expands-academy
2. A scoping review of immersive virtual reality into STEM education. IEEE Trans. Learn. Technol. **13**(4), 748–761 (2020)
3. Wieman, C.: STEM education: active learning or traditional lecturing?; Waldrop M.: Teaching science by active learning; Brenner, T.: Active learning tip sheet; Salvetti, F., Bertagni, B.: The past and the future in teaching and learning STEM. Salvetti, F., Bertagni, B.: Hololens, augmented reality and teamwork. In: Salvetti, F., Bertagni, B. (eds.) Learning 4.0. Advanced Simulation, Immersive Experiences and Artificial Intelligence, Flipped Classrooms, Mentoring and Coaching. Franco Angeli, Milan (2018)
4. https://www.gwhatchet.com/2021/01/11/smhs-receives-grant-to-construct-mobile-learning-center-at-local-high-school/
5. Salvetti F., Bertagni B., Foreword: Learning 4.0 & STEAM Education. In: Salvetti F., Bertagni B. (Eds.). Learning 4.0. Advanced Simulation, Immersive Experiences and Artificial Intelligence, Flipped Classrooms, Mentoring and Coaching. Franco Angeli, Milan (2018)
6. Wieman, C.: STEM education: active learning or traditional lecturing? In: Salvetti, F., Bertagni, B. (eds.) Learning 4.0. Advanced Simulation, Immersive Experiences and Artificial Intelligence, Flipped Classrooms, Mentoring and Coaching. Franco Angeli, Milan (2018)
7. Salvetti, F., Bertagni, B. Salvetti, F., Bertagni, B.: Virtual worlds and augmented reality: The enhanced reality lab as a best practice for advanced simulation and immersive learning. Form@re (University of Florence) **19**(1) (2019)
8. Arnheim, R.: Visual Thinking. University of California Press, Berkeley and Los Angeles, CA (1969)
9. Guralnick, D.: Re-envisioning online learning. In: Salvetti, F., Bertagni, B. (eds.) Learning 4.0. Advanced Simulation, Immersive Experiences and Artificial Intelligence, Flipped Classrooms, Mentoring and Coaching. Franco Angeli, Milan (2018)
10. Shoemaker, G., Tsukitani, T., Kitamura, Y., Booth, K.: Body-centric interaction techniques for very large wall displays. In: Proceedings of the NordiCHI, pp. 463–472 (2010)

11. Beaudouin-Lafon, M., et al.: Multi-surface interaction in the WILD room. IEEE Comput. **45**(4), 48–56 (2012)

12. Gaiani, M., Zannoni, M., Dall'Osso, G., Garagnani, S.: Body-Centric Interaction Guidelines for e-REAL. Research Paper under development, University of Bologna (2022)

13. Shotton, J., Fitzgibbon, A., Cook, M., Sharp, T., Finocchio, M., Moore, R., Kipman, A., Blake, A.: Real-time human pose recognition in parts from single depth images. In: Proceedings of the CVPR, pp. 1297–1304 (2011)

14. Mine, M., Brooks, Jr., F., Sequin, C.: Moving objects in space: exploiting proprioception in virtual-environment interaction. In: Proceedings of the SIGGRAPH, pp. 19–26 (1997)

15. Latulipe, C., Wilson, D., Huskey, S., Word, M., Carroll, A., Carroll, E., Gonzalez, B., Singh, V., Wirth, M., Lottridge, D.: Exploring the design space in technology-augmented dance. In: CHI Extended Abstracts, pp. 2995–3000 (2010)

16. Wachs, J., et al.: A gesture-based tool for sterile browsing of radiology images. J. Am. Med. Inform. Assoc. **15**(3), 321–323 (2008)

17. Nancel, M., Wagner, J., Pietriga, E., Chapuis, O., Mackay, W.: Mid-air pan-and-zoom on wall-sized displays. Proc. CH **I**, 177–186 (2011)

18. Goldin-Meadow, S., Beilock, S.L.: Action's influence on thought: the case of gesture. Perspect. Psychol. Sci. **5**(6), 664–674 (2010)

19. Wagner, J., Nancel, M., Gustafson, S., Huot, S., Mackay, W.E.: A body-centric design space for multi-surface interaction. In: CHI'13: Proceedings of the 31st International Conference on Human Factors in Computing Systems. ACM (2013)

20. Salvetti, F., Gardner, R., Minehart, R.D., Bertagni, B.: Enhanced reality for healthcare simulation. In: Brooks, A.L., Brahman, S., Kapralos, B., Nakajima, A., Tyerman, J., Jain, L.C. (eds.) Recent Advances in Technologies for Inclusive Well-Being. ISRL, vol. 196, pp. 103–140. Springer, Cham (2021). https://doi.org/10.1007/978-3-030-59608-8_7

21. Salvetti, F., Gardner, R., Minehart, R., Bertagni, B.: Effective Extended Reality: A Mixed-Reality Simulation Demonstration with Intelligent Avatars, Digitized and Holographic Tools. Research paper for The Learning Ideas Conference, New York (2022)

Time Traveling Towards a Climate-Neutral Society: An Interactive and Immersive Experience

Fernando Salvetti[1,2,3(✉)], Cristina Cavicchioli[4], Marco Borgarello[4], and Barbara Bertagni[1,2,3]

[1] e-REAL Labs, Centro Studi Logos, 10143 Turin, Italy
salvetti@logosnet.org
[2] e-REAL Labs, Logosnet, 6900 Lugano, Switzerland
[3] Houston, TX 77008, USA
[4] RSE S.p.A, 20121 Milan, Italy

Abstract. The European Union aims to be climate-neutral by 2050—an economy with net-zero greenhouse gas emissions. RSE carries out research in the field of electrical energy with special focus on national strategic projects supported through the Fund for Research into Electrical Systems. The activity covers the entire supply system with an application-oriented, experimental, and system-based approach. The e-REAL's time traveling immersive experience is based on RSE's know-how and allows interactive and immersive experiences that are powerful ways to use relevant communication and foster people's awareness about the European Union targets.

Keywords: Immersive experience · Interactive visualization · Communication of science

1 Time Traveling Towards a Climate-Neutral Society

The European Union aims to be climate-neutral by 2050—an economy with net-zero greenhouse gas emissions. This objective is at the heart of the European Green Deal and in line with the EU's commitment to global climate action under the Paris Agreement. The transition to a climate-neutral society is both an urgent challenge and an opportunity to build a better future for all. All parts of society and economic sectors will play a role—from the power sector to industry, mobility, buildings, agriculture, and forestry.

The European Union is leading the way by investing in technological solutions, aligning actions in key areas such as industrial policy, finance, and research, while ensuring social fairness for a just transition. Italy, as a leading European country, has been working towards ensuring the widest possible use of instruments that, together, serve to enhance energy security, environmental protection, and the affordability of energy, thus contributing to European objectives relating to energy and the environment. To reach these results, the Italian company RSE—a publicly controlled entity headquartered in

Milan—co-designed with the Instructional Design Team from Logosnet and the e-REAL Multimedia Graphic and Software Engineering Teams an immersive and interactive experience based on a time travel "machine" created by an e-REAL portable lab towards a climate-neutral society, in order to foster different communication campaigns.

RSE carries out research in the field of electrical energy with special focus on national strategic projects supported through the Fund for Research into Electrical Systems. The activity covers the entire supply system with an application-oriented, experimental, and system-based approach. The e-REAL time machine is based on RSE's know-how and allows interactive and immersive experiences that are powerful ways to use relevant communication and foster people's awareness about the European Union targets.

This work has been financed by the Research Fund for the Italian Electrical System under the Contract Agreement between RSE S.p.A. and the Ministry of Economic Development - General Directorate for the Electricity Market, Renewable Energy and Energy Efficiency, Nuclear Energy in compliance with the Decree of April 16, 2018.

2   Communication of Science and Immersive Experiences Within e-REAL

Communication of science is a process based on distilling technical information about science-related topics into understandable messages and stories for public consumption. It is a field concerned with bridging the gap between scientists and a general audience thanks to a multi-faceted form of communication that spans scientific fields such as the hard sciences, physical sciences, technology, health, environmental science, and more.

The e-REAL science communication professionals leverage their understanding of complex scientific topics, along with strategic communication and storytelling principles, to craft compelling and informative content about science and related disciplines. This combination of industry knowledge and practical communication skills advances the public's understanding of quite complex topics and issues.

To enhance the communication of science, Logosnet developed e-REAL®, that is both a phygital technology for immersive and interactive experiences and a cloud based platform for interactive communication [1, 2]. Since 2011, Logosnet has developed a number of projects in advanced simulation, virtual reality, online and lifelong learning, and interactive communications. e-REAL develops 3D scenarios, interacting with visitors in a natural way, with no need of helmets, glasses, or any kind of visors. Thanks to proximity sensors, visitors interact with virtual elements of the scenario with just a flick of the hands or by gesture shaping. Eye-catching interactive infographics disclose new ways to cascading science and technology related information to the general public (Figs. 1, 2, 3, 4, 5, 6 and 7).

Developing a communication based on the time traveling metaphor, the audience is engaged with an immersive and interactive experience in the years 2020, 2030, and 2050, and along 3 axes: Energy consumption trends, European Union and Italian public policies, and available technologies to reduce the carbon footprint. The audience may take the decision to travel from 2020 till 2050 or to explore only one or two time's periods. Another choice to take by the audience, within each time's period, is about the

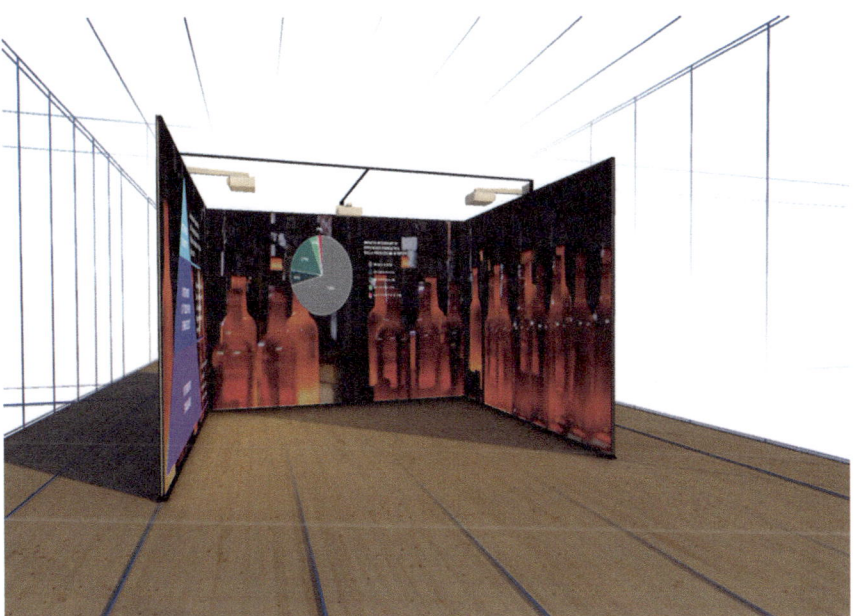

Fig. 1 Rendering of the e-REAL portable pop-up for the RSE immersive and interactive experience regarding time traveling toward a climate-neutral society: Infographics are interactive and responsive by gesture shaping.

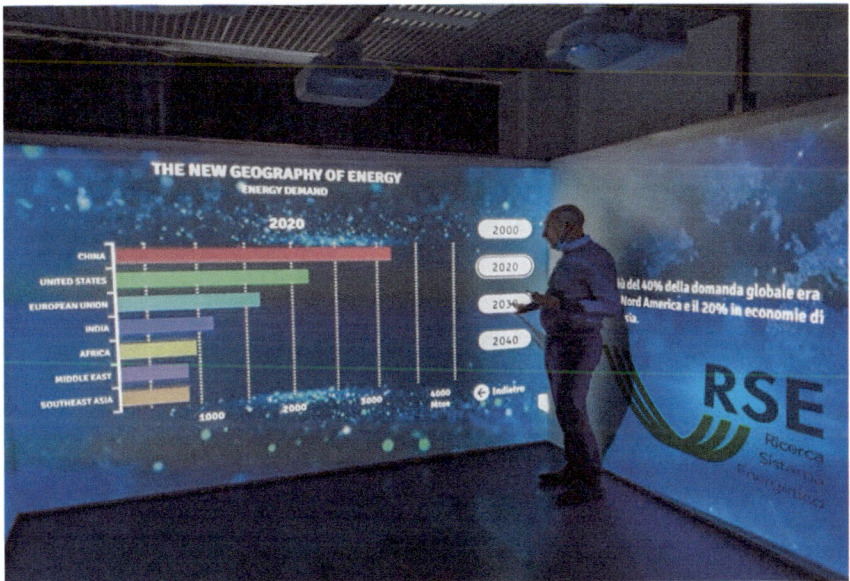

Fig. 2 Immersive and interactive e-REAL infographics, distributed on 3 walls.

Fig. 3 Another e-REAL infographics, immersive and interactive.

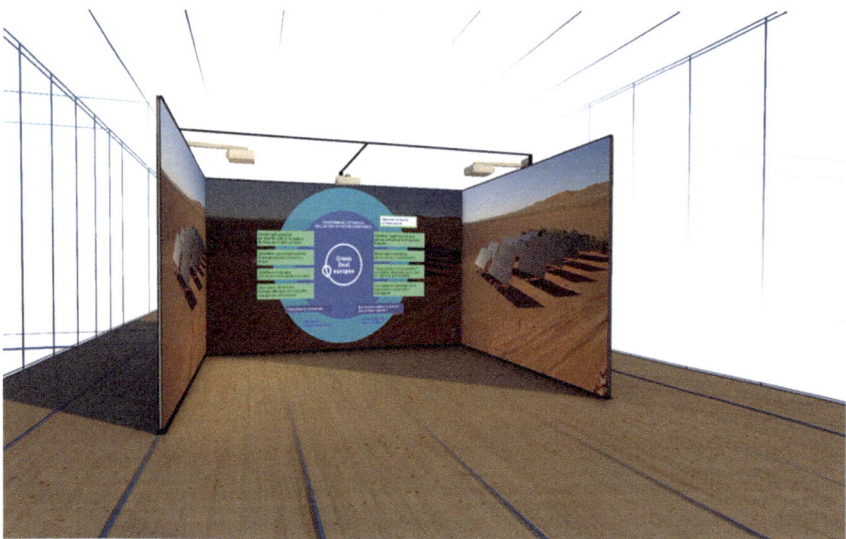

Fig. 4 Rendering of an interactive infographics with immersive videos on the side walls.

focus: energy consumption and/or public policies and/or technologies and production processes.

The setting of the portable e-REAL pop-up is designed around 3 key-words: Visualization, interaction, immersion. It is a fully immersive and multitasking environment, designed to help untangle complexity: the visualization of information enables the audience to gain insight and understanding quickly and efficiently. Examples of such visual

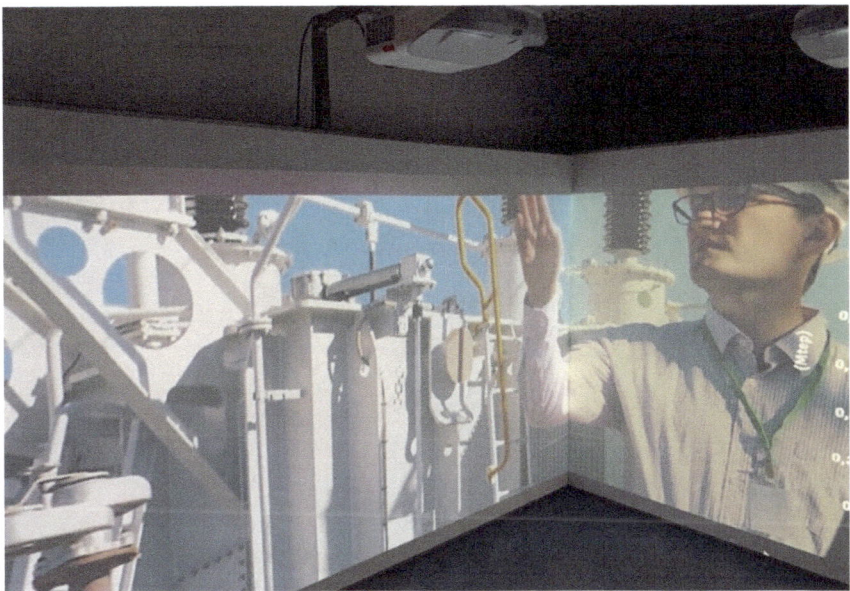

Fig. 5 Immersive video with content displayed on the three e-REAL walls.

Fig. 6 Interactive e-REAL puzzle about the carbon footprint.

formats include sketches, diagrams, images, objects, interactive visualizations, information visualization applications and imaginary visualizations such as in stories. Visualizations within e-REAL show relationships between topics, activate involvement, generate

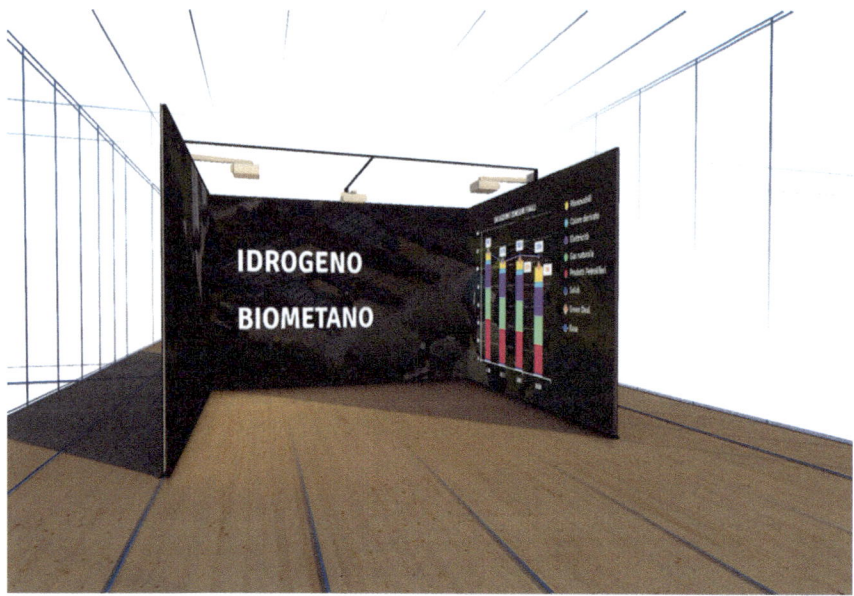

Fig. 7 Rendering of an e-REAL immersive video with an infographics on a side wall.

questions that learners didn't think of before and facilitate memory retention. So visualizations act like concept maps to help organize and represent knowledge in an effective way [3–9].

From a visual storytelling perspective, the main scenarios related to this time traveling adventure are the following (Figs. 8, 9, 10, 11, 12, 13, 14, 15 and 16):

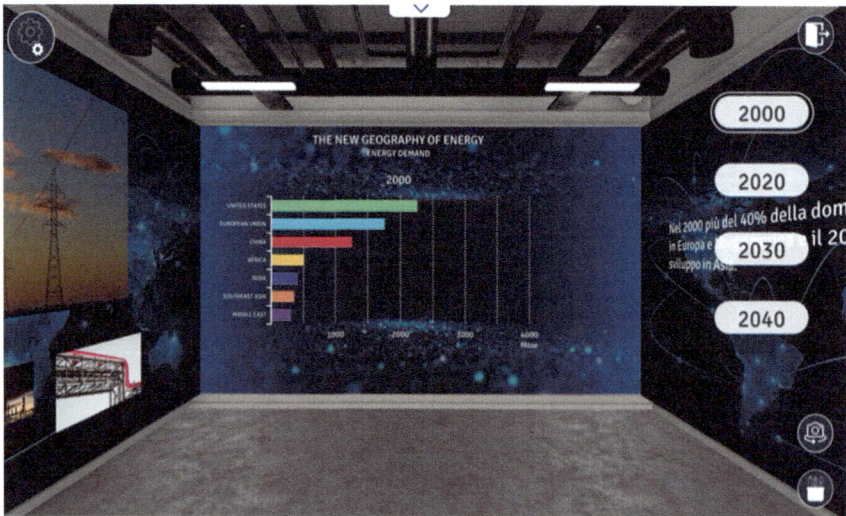

Fig. 8 Energy consumption trends in 2020: e-REAL interactive infographics.

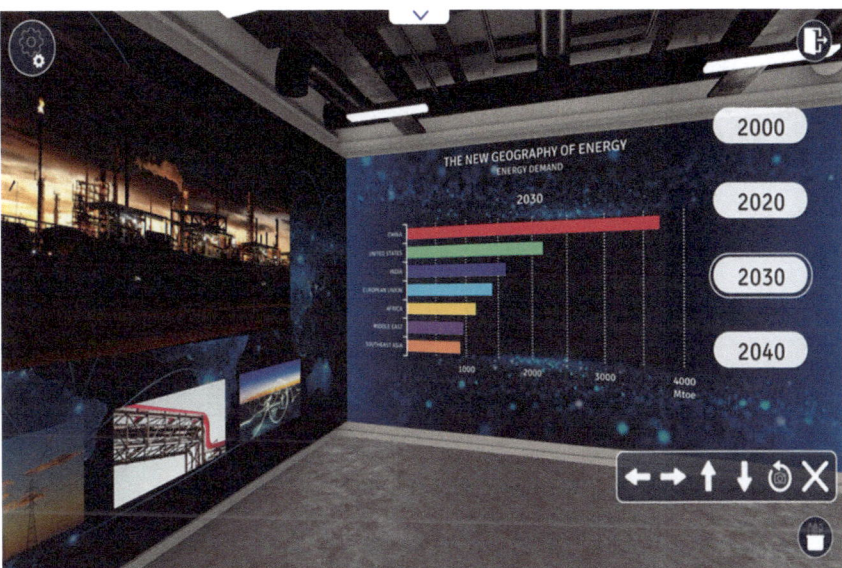

Fig. 9 Energy consumption trends in 2030: e-REAL interactive infographics.

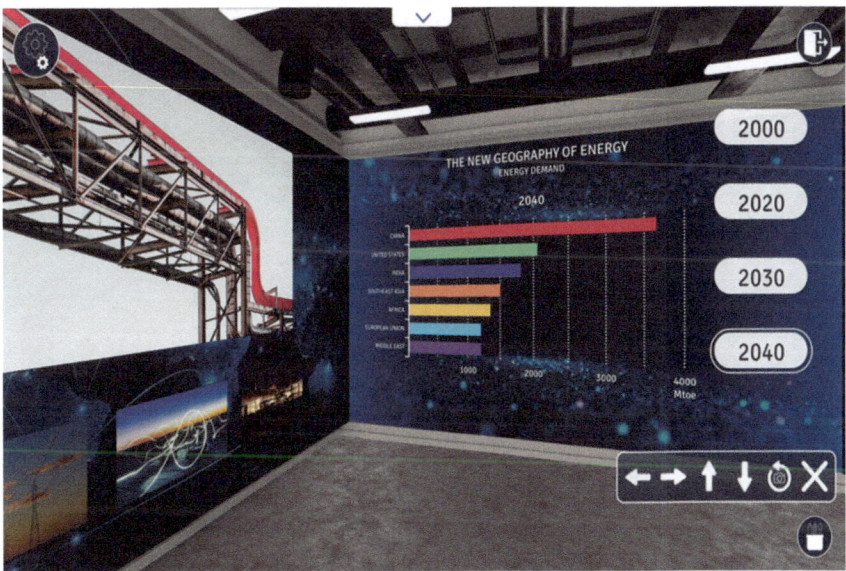

Fig. 10 Energy consumption trends in 2040: e-REAL interactive infographics.

All the scenarios are designed according to the visual thinking paradigm that, according to Rudolph Arnheim, implies that all thinking—not just thinking related to art—is basically perceptual in nature, and that the dichotomy between seeing and thinking, or perceiving and reasoning, is misleading [4]. Furthermore, all the scenarios are complaint

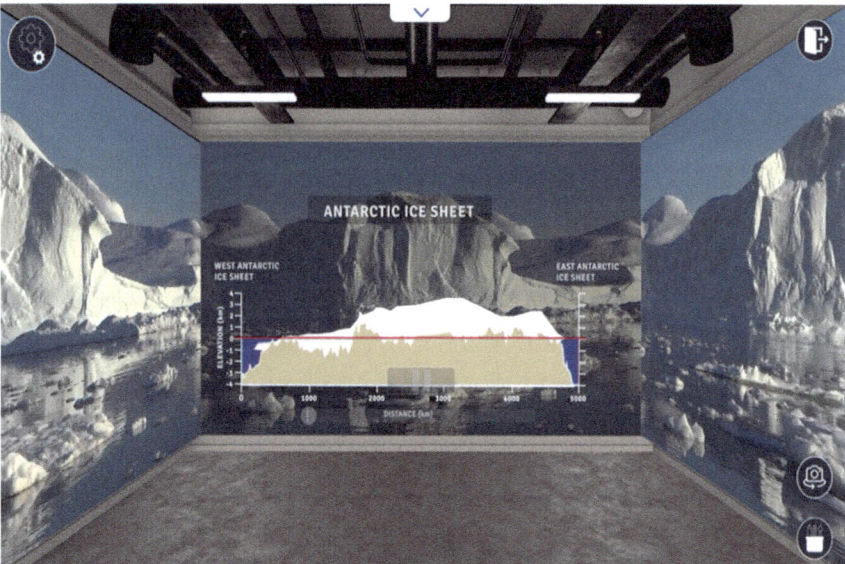

Fig. 11 Consequences of climate change: e-REAL interactive infographics.

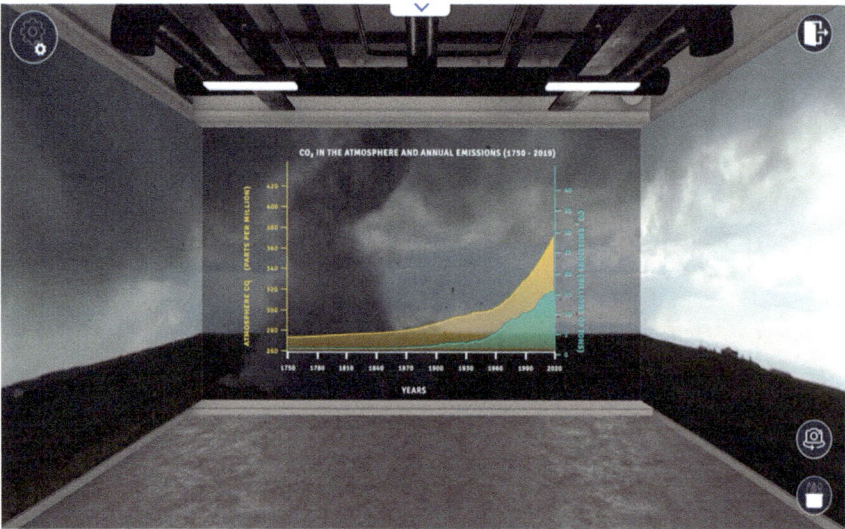

Fig. 12 CO_2 trends: e-REAL interactive infographics.

with the guidelines shared by the experts from the multimedia graphics and computer vision fields, that are interdisciplinary areas that deal with digital images or videos, audio clips, photographs, 3D models, volumetric data. Last but not least, all the scenarios are designed to provide an advanced simulation experience that is a highly realistic imitation of a real-world object, process or system. This type of experience encourages audience

Fig. 13 Available technologies: e-REAL immersive video.

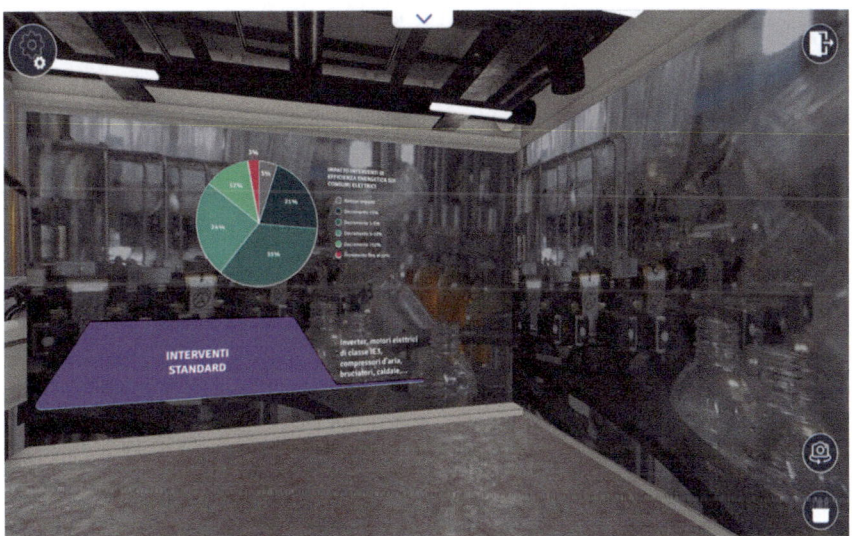

Fig. 14 European Union and Italian policies: e-REAL interactive infographics.

to cross conceptual and theoretical boundaries with the help of simulation or game based tools. It is one of the most promising methods for STEM education and communication of science and technology.

The e-REAL setting offers a unique user experience, a combination of visual communication and direct interaction with the content—by gesture shaping or spoken commands—immersing the audience in an entirely interactive ecosystem. Visual storytelling

Fig. 15. 4.0 digital revolution and 5.0 society trends: e-REAL immersive video.

Fig. 16. e-REAL interactive infographics about the main trends regarding renewable energies, energetic efficiency, emissions.

techniques are part of the simulation scene, to represent a realistic context where learners are proactively involved to analyze scenarios and events, to face technical issues, to solve problems. Effective visualization is the key to help untangle complexity: the visualization of information enables learners to gain insight and understanding quickly and efficiently. The most effective learning occurs when being immersed in a context: realistic experience is lived and perceived as a focal point and as a crossroad [10].

The richness of the mental model relates to the completeness of multiple channels of sensory information, meaning the more those senses work in alignment, the better.

The richness also depends on having a cognitively demanding environment and a strong and interesting narrative. Cognitively demanding environments in which players must focus on what's going on in the game will occupy mental resources. The richness of the mental model is good for immersion, because if brain power is allocated to understanding or navigating the world, it's not free to notice all of its problems or shortcomings that would otherwise remind them that they're playing a game. Finally, good stories—with interesting narratives, credible because intrinsically congruent as much as possible—attract attention to the game and make the world seem more believable. They also tie up those mental resources.

Turning to game traits related to consistency, believable scenarios and behaviors in the game world means that virtual characters, objects, and other creatures in the game world behave in the way in which the audience expects. Usually game developers strive for congruence among all the elements. The audience is challenged both cognitively and behaviorally in a fully-immersive and multitasking learning environment, within interactive scenarios that usually present also a wealth of information. The many levels of the situation are made available simultaneously, by overlaying multisource—words, numbers, images, etc.—within an environment designed by AR techniques based on the overlaying of multiple information. e-REAL submerges the audience in an immersive reality where the challenge at hand is created by sophisticated, interactive computer animation. Importantly, the system includes live and real time interaction on a peer-to-peer bases. Thus, it adds a very important social component and leaves the audience with a memorable experience.

References

1. Salvetti, F., Bertagni B. (Eds.).: Learning 4.0. Advanced Simulation, Immersive Experiences and Artificial Intelligence, Flipped Classrooms, Mentoring and Coaching. Franco Angeli, Milan (2018)
2. www.e-real.net
3. Salvetti, F., Bertagni, B., Salvetti, F., Bertagni, B.: Virtual worlds and augmented reality: the enhanced reality lab as a best practice for advanced simulation and immersive learning. Form@re 19(1) (2019). University of Florence
4. Arnheim, R.: Visual Thinking. University of California Press, Berkeley and Los Angeles, CA (1969)
5. Shoemaker, G., Tsukitani, T., Kitamura, Y., Booth, K.: Body-centric interaction techniques for very large wall displays. In: Proceedings of NordiCHI, pp. 463–472 (2010)
6. Beaudouin-Lafon, M., Huot, S., Nancel, M., Mackay, W., Pietriga, E., Primet, R., Wagner, J., Chapuis, O., Pillias, C., Eagan, J., Gjerlufsen, T., Klokmose, C.: Multi-surface Interaction in the WILD Room. IEEE Comput. 45(4), 48–56 (2012)
7. Gaiani, M., Zannoni, M., Dall'Osso, G., Garagnani, S.: Body-centric interaction guidelines for e-REAL. In: Research Paper Under Development. University of Bologna (2022)
8. Shotton, J., Fitzgibbon, A., Cook, M., Sharp, T., Finocchio, M., Moore, R., Kipman, A., Blake, A.: Real-time human pose recognition in parts from single depth images. In: Proceedings of CVPR, pp. 1297–1304 (2011)

9. Mine, M., Brooks Jr, F., and Sequin, C. Moving objects in space: exploiting proprioception in virtual-environment interaction. In: Proceedings of SIGGRAPH, pp. 19–26 (1997)

10. Guralnick, D.: Re-envisioning online learning. In: Salvetti, F., Bertagni, B. (eds.) Learning 4.0. Advanced Simulation, Immersive Experiences and Artificial Intelligence, Flipped Classrooms, Mentoring and Coaching. Franco Angeli, Milan (2018)

Medical Simulation in the Cloud: Learning by Doing Within an Online Interactive Setting

Fernando Salvetti[1,2,3]([⊠]), Roxane Gardner[4,5,6,7,8], Rebecca Minehart[4,7,8], and Barbara Bertagni[1,2,3]

[1] e-REAL Labs, Centro Studi Logos, 10143 Turin, TO, Italy
salvetti@logosnet.org
[2] Logosnet, E-REAL Labs, 6900 Lugano, Switzerland
[3] Houston, TX 77008, USA
[4] Center for Medical Simulation, Boston, MA 02129, USA
[5] Brigham and Women's Hospital, Boston, MA, USA
[6] Children's Hospital, Boston, MA, USA
[7] Massachusetts General Hospital, Boston, MA, USA
[8] Harvard Medical School, Boston, MA 02114, USA

Abstract. The process of learning by doing within an online interactive setting—accessible by a single and simple click of the mouse, without downloads of software or other technical procedures—is highly effective and leaves learners with a memorable experience, if the experience is multiplayer, highly cooperative, and glasses-free. That is the case with the e-REAL Online experience introduced in this article, which revolves around a case of multiple injuries acting in an alpine environment. Within this scenario, the learners are challenged to recognize a situation requiring rapid intervention, communication, knowledge sharing, decision-making, and management of an unforeseen event—while taking into consideration critical contextual factors such as a lack of time, scarcity of resources and tools, and a multitude of additional impactful factors (weather conditions, broadband availability, etc.). The entire experience is based on the visual exploration of an alpine environment and on a dialogue with the patient, which is an avatar appearing as a female that was injured during a hiking activity performed alone and was found by chance by an interprofessional rescue team.

Keywords: Medical simulation · Cooperative learning · Glasses-free

1 Learning by Doing Within a Glasses-Free Online Interactive Setting

The process of learning by doing within an online interactive setting is highly effective and leaves the learners with a memorable experience, if the experience is multiplayer, highly cooperative, and glasses-free. Such a hypothesis was tested during five online synchronous experiences of medical simulation, delivered into the e-REAL Online cloud platform with students from different medical and nursing schools based in Italy and Switzerland.

D. Guralnick et al. (Eds.): TLIC 2022, LNNS 581, pp. 363–369, 2023.
https://doi.org/10.1007/978-3-031-21569-8_34

The experience revolves around a case of multiple injuries placed in an alpine environment, co-designed by the instructional design team from Logosnet (Houston, Texas, USA, and Turin, Italy) and the subject matter experts from the Center of Medical Simulation (Boston, Massachusetts, USA). In this scenario, the learners are challenged to recognize a situation requiring rapid intervention, communication, knowledge sharing, decision-making, and management of an unforeseen event—while taking into consideration critical contextual factors such as a lack of time, scarcity of resources and tools, and a multitude of additional impactful factors (weather conditions, broadband availability, etc.).

The entire experience is based on the visual exploration of an alpine environment and on a dialogue with the patient, that is, a female who was injured during a hiking activity performed alone and was found by chance by an interprofessional rescue team [1] (Fig. 1).

Fig. 1. Alpine environment within the e-REAL® Online platform in the cloud: A case regarding multiple injuries, codesigned by the instructional design team from Logosnet (Houston, Texas, USA, and Turin, Italy) and the subject matter experts from the Center of Medical Simulation (Boston, Massachusetts, USA).

The e-REAL Online platform submerges learners in an immersive reality where the challenge at hand is created by sophisticated, interactive, computer animation. The platform allows for live and real-time interaction among peers and the medical simulation instructor—as well as with the injured patient, which is an avatar programmed to dialogically interact with the participants [2] (Fig. 2).

Dialogues with avatars can be automated and based on artificial intelligence, as was the case during the five online synchronous experiences, or they can be based on a real discursive interaction with a patient animated by a simulation instructor—as is the case in other experiences at the Center for Medical Simulation (Figs. 3, 4 and 5).

Fig. 2. The avatar is programmed to perform as an injured female patient within the e-REAL® Online platform: The avatar is a digital human enhanced with artificial intelligence, able to call for help and to interact dialogically with learners and simulation instructors.

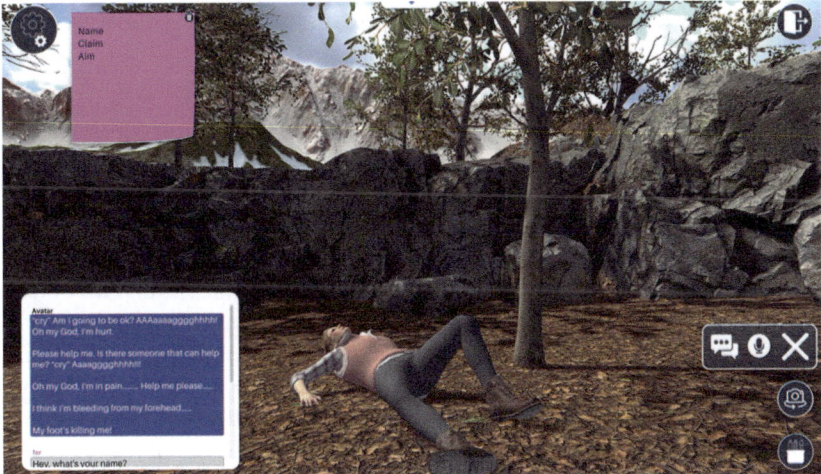

Fig. 3. Representative dialogue with the avatar, programmed to interact autonomously with the learners.

The e-REAL simulation setting is highly immersive, and learners may interact together in real time among themselves, as well as with the patient's avatar. Learners can also take notes, overlay key concepts, rotate virtual objects, or move them point to point.

This solution allows simulations in a virtual environment that display challenging situations; unlike other VR solutions, e-REAL allows users to experience full immersion without the need for glasses or goggles. And it is easy to use: an Internet connection and

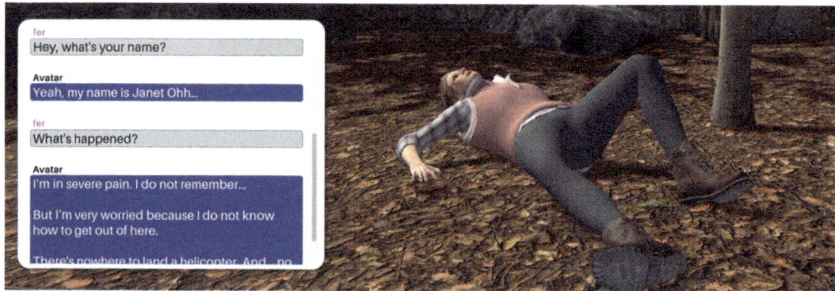

Fig. 4. Representative dialogue with the avatar, programmed to interact autonomously with the learners.

Fig. 5. Representative dialogue with the avatar, programmed to interact autonomously with the learners.

a browser are the only requirements, because the website hosting this experience is into the cloud and is accessible by a single and simple click of the mouse, without downloads of software or other technical procedures (Fig. 6).

The e-REAL teaching and learning approach is designed to have the learners working on tasks that simulate an aspect of expert reasoning and problem-solving, while receiving timely and specific feedback both from fellow students and the instructor. These elements of deliberate practice and feedback are key for actively involving the learners and making the learning experience effective [3]. Another key factor is the use of a mnemonic known as Name-Claim-Aim, developed by faculty at the Center for Medical Simulation, which encompasses a strategy to help health care professionals effectively organize a team for managing critical clinical events [4]. Last but not least, a key take away from this experience is the ability to have the complex decision making, an artificially intelligent avatar and communications with other learners in an online format easy and intuitive, that doesn't require downloads and doesn't limit the ability for collaboration (Fig. 7).

1.1 Boosting Teamwork, Verbal Communication, and Crisis Resource Management

Principles of crisis resource management, as described in 1992 by Howard, Gaba, and Fish, encompass those cognitive and interpersonal skills required to facilitate dynamic decision-making and effective teamwork during a crisis situation in complex environments. Effective verbal communication ensures a shared mental model and taps into

Fig. 6. Representative time management, writing and overlaying features available in the e-REAL Online platform.

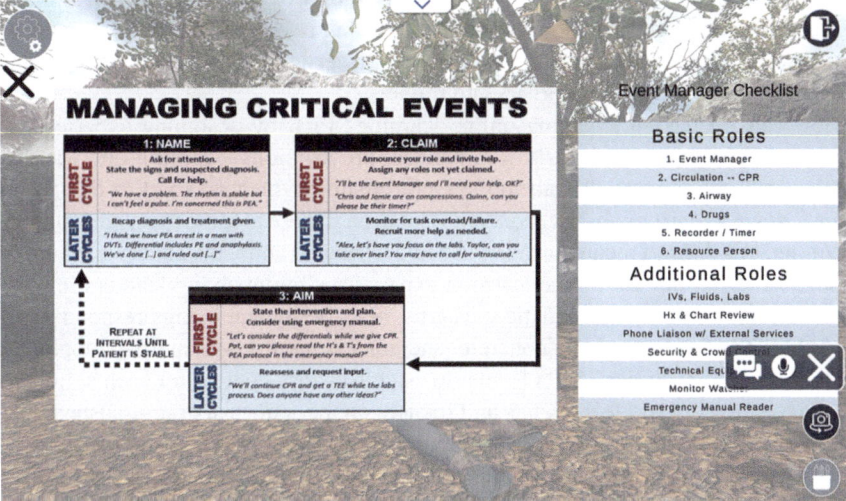

Fig. 7. Name-Claim-Aim ©: A mnemonic and check-list developed by faculty at the Center for Medical Simulation (Boston, Massachusetts, USA), that encompasses a strategy to help health care professionals effectively organize a team for managing critical clinical events.

the cognitive processes of the entire team [5]. Poor teamwork and communication are common factors contributing to errors and adverse events, especially while managing critical clinical events [6]. Health care professionals often find it challenging to effectively organize a team and establish explicit team leadership when managing critical events. When a crisis has been identified, an abrupt rise in cognitive load is triggered by the autonomic nervous system when even mild acute stress occurs simultaneously with increased tasks [7].

Opportunities to train and practice managing critical clinical events in simulated environments have existed since the early 1990s. The emergence of virtual reality-based training in procedural and surgical simulations was described in 1997 [8]. Since then, the use of virtual, augmented, or mixed reality environments has grown exponentially [9].

The e-REAL Online platform is a teamwork and communication booster, able to provide a memorable experience with a robust learning outcome. The process of learning by doing within the e-REAL Online interactive setting is highly effective as a multiplayer experience, highly cooperative, and glasses-free: 100% of the learners involved into our activity agreed with our hypothesis after having been involved in the above described activity and in two other different learning settings in order to compare the results (first alternative control setting: VR head-mounted displays; second alternative control setting: online, glasses-free, individual, and not cooperative).

From an educational perspective, within the e-REAL Online experience, learners are not assumed to be passive recipients and repeaters of information but to be individuals who collectively take responsibility for their own learning [10]. The feedback we received most often—both by the learners and by the simulation instructors invited to coordinate and facilitate the online sessions—was that they didn't expect such easy access to an intensely engaging and meaningful experience. This is critical feedback because our efforts are aimed at designing solutions that empower learning with a strong focus on the user experience, both for learners and trainers.

How can we design engaging online learning? First by designing experiences to connect with, and not "to-do-lists" or "medical prescriptions" working one way from the teacher to the student. It's all too easy for a subject matter expert to distill "core content" down to a list of abstract concepts, but even if there is someone who memorizes everything, this doesn't mean that he or she knows how to apply the concepts in real life. We suggest designing the online learning experiences by involving learners in stories that address key issues and realistic concerns, because human beings respond well to things that are relevant to them, concrete, and context related. We suggest adopting the situation-impact-resolution (SIR) format to establish story context for each simulation. This derives from Aristotle's *Poetics* and focuses on the sequence drama, suspense, and resolution—to use with epistemological acumen, within a systemic paradigm [1].

How can we make online learning effective? By encouraging learners to learn by doing and allowing them to cross conceptual and theoretical boundaries within a simulation's setting. In an online learn-by-doing simulation, the learners are asked to play specific roles and must make decisions in realistic situations. It's critical that learners then receive feedback as to the results of their decisions.

References

1. Salvetti, F., Gardner, R., Minehart, R., Bertagni, B.: Digital learning: healthcare training by tele-simulation and online cooperation. In: Guralnick, D., Auer, M.E., Poce, A. (eds.). Innovations in Learning and Technology for the Workplace and Higher Education. Proceedings of 'The Learning Ideas Conference'. Springer, Heidelberg (2021)

2. Salvetti, F., Gardner, R., Minehart, R., Bertagni, B.: Enhanced reality for healthcare simulation. In: Brooks, A.L., Brenham, S., Kapralos, B., Nakajima, A, Tyerman, J., Jain, L. (eds.) Recent Advances in Technologies for Inclusive Well-Being: Virtual Patients, Gamification and Simulation. Springer, Heidelberg (2021)
3. Rudolph, J., Simon, R., Raemer, D., Eppich, W.: Debriefing as a formative assessment: closing performance gaps in medical education. Acad. Emerg. Med. **15**, 1010–1016 (2008)
4. Salvetti, F., Gardner, R., Minehart, R., Bertagni, B.: Advanced medical simulation: interactive videos and rapid cycle deliberate practice to enhance teamwork and event management— effective event management during simulated obstetrical cases. Int. J. Adv. Corp. Learn. **12**(3), 70 (2019)
5. Howard, S., Gaba, D., Fish, K., Yang, G., Sarnquist, F.: Anesthesia crisis resource management training: teaching anesthesiologists to handle critical incidents. Aviat. Space Environ. Med. **63**(9), 763–770 (1992)
6. Brindley, P.G., Reynolds, S.F.: Improving verbal communication in critical care medicine. J. Crit. Care. **26**(2), 155–159 (2011)
7. Arnsten, A.: Stress signaling pathways that impair prefrontal cortex structure and function. Nat. Rev. Neurosci. **10**(6), 410–422 (2009)
8. Hoffman, H., Vu, D.: Virtual reality: teaching tool of the 21rst century? Acad. Med. **72**, 1076–1081 (1997)
9. Salvetti, F., Bertagni, B. (eds.): Learning 4.0. Advanced Simulation, Immersive Experiences and Artificial Intelligence, Flipped Classrooms, Mentoring and Coaching. Franco Angeli, Milan (2018)
10. Auer, M., Guralnick, D., Uhomoibhi, J. (eds.): Interactive Collaborative Learning: Proceedings of the 19th ICL Conference, vol. 1. Springer, Heidelberg (2017)

Effective Extended Reality: A Mixed-Reality Simulation Demonstration with Digitized and Holographic Tools and Intelligent Avatars

Fernando Salvetti[1,2,3]([✉]), Roxane Gardner[4,5], Rebecca Minehart[4,6], and Barbara Bertagni[1,2,3]

[1] e-REAL Labs, Centro Studi Logos, 10143 Turin, Italy
salvetti@logosnet.org
[2] e-REAL Labs, Logosneet, 6900 Lugano, Switzerland
[3] Houston, TX 77008, USA
[4] Center for Medical Simulation, Boston, MA 02129, USA
[5] Brigham and Women's Hospital/Children's Hospital/Massachusetts General Hospital and Harvard Medical School, Boston, MA 02114, USA
[6] Massachusetts General Hospital and Harvard Medical School, Boston, MA 02114, USA

Abstract. How can we design engaging and effective medical education both online and onsite? Extended reality (XR) is a term referring to all real-and-virtual combined environments and human-machine interactions generated by computer technology and wearables. It includes representative forms such as virtual reality (VR), augmented reality (AR), or mixed reality (MR), and the areas interpolated among them. MR is a domain of particular interest today. It takes place not only in the physical world or in the virtual world, but is a mix of the real and the virtual. Metaverses can be enabled by MR wearable augments. Glasses-free MR is another very interesting dimension: e-REAL®, as a MR environment for hybrid simulation and medical education in general, can be a stand-alone solution or even networked between multiple places through a link to a special videoconferencing system. Digital humans and human-sized holograms are part of the e-REAL scenarios, making this solution unique, rich, and diversified.

Keywords: Extended reality · Demonstrations · e-REAL

1 Extended Reality, Online and Onsite, for Medical Education

How can we design engaging and effective medical education both online and onsite? Opportunities to train and practice in simulated environments have existed since the early 1990s. The emergence of virtual reality-based training in the medical education field, first using procedural and surgical simulations, was described in 1997 by Hoffman and Vu [1]. Since then, the use of virtual reality (VR), augmented reality (AR), or mixed reality (MR) environments has grown exponentially. Extended reality (XR) is a term referring to all real-and-virtual combined environments and human-machine interactions generated by computer technology and wearables. It includes representative forms such

D. Guralnick et al. (Eds.): TLIC 2022, LNNS 581, pp. 370–376, 2023.
https://doi.org/10.1007/978-3-031-21569-8_35

as AR, VR, and MR, and the areas interpolated among them. XR is a superset which includes the entire spectrum from the complete real to the complete virtual in the concept of reality–virtuality continuum introduced by Milgram, Takemura, Utsumi, and Kishin [2].

The levels of virtuality range from partial sensory inputs to immersive virtuality, which is usually enabled by head-mounted displays. Mixed reality is a particularly interesting domain today. A mixed reality is a merging of real and virtual worlds where objects in the actual physical surroundings play a direct functional role within the virtual environment simulation [3].

Imagine a setting like the one displayed in Fig. 1 where a physical space is converted into a "phygital" place, with sensors embedded inside some components, like medical tools and patient simulators.

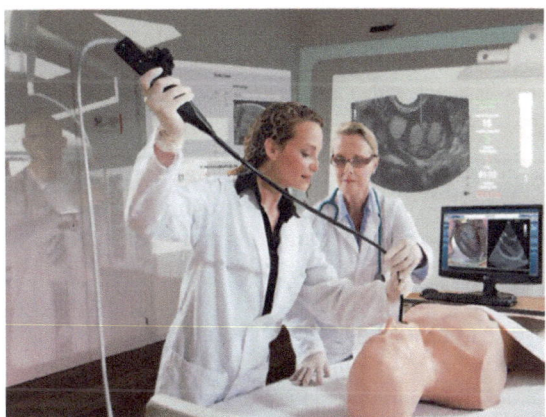

Fig. 1. e-REAL phygital classroom at the International Red Cross "Luigi Gusmeroli" Learning Center in Bologna, Italy: Learning medical procedures using a skilled trainer during a hybrid simulation within an environment enhanced by interactive medical imagery and a realistic avatar able to interact with the learners.

Sensors and embedded systems work together to provide one of the most important aspects of the Internet of Things (IoT): detecting changes in an object (device or asset) and/or the environment, allowing for capture of relevant data for real-time and/or post-processing. Sensors are used for detection of changes in the physical and/or logical relationship of one object to another(s) and/or the environment. Physical changes may include temperature, light, pressure, sound, and motion. Logical changes include the presence/absence of an electronically traceable entity, location, and/or activity. Within an IoT context, physical and logical changes are equally important. Sensor types are a number: a short list include acoustic, ambient light/optical, electric/magnetic, force/pressure, chemicals/gas/radiation, humidity, leakage/level/flow, locked/unlocked, motion/acceleration, temperature. By one or more sensors, MR is enabled and provides an interactive and immersive experience.

Now imagine another mixed reality implementation from a dramatically different angle, as shown in Fig. 2. As opposed to having the users interact with physical surroundings while viewing precisely correlated imagery within a fully immersive display, AR displays can be used to place computer-generated objects within the user's real-world surroundings [4].

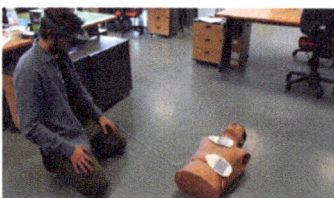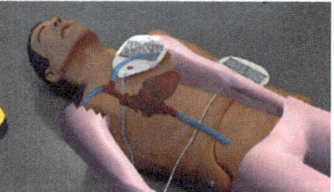

Fig. 2. A Ph.D. student from the Polytechnic School of Turin, Italy, is wearing AR head-mounted displays and experiencing a MR training onsite, where the half-manikin used for a training module about basic life support and defibrillation is digitally reconstructed within the glasses—that are also providing a multilayer vision with a schematic representation of the heart.

MR takes place not only in the physical world or in the virtual world, but is a mix of the real and the virtual. We define MR as a hybrid reality or the merging of real and virtual worlds to produce new environments and visualizations where physical and digital objects co-exist and interact in real time. It's the "phygital" that is an emerging key-word, concept, and trend. In a time where the digital world is so omni-present, it's important to not lose sight of the relevance and presence that physicality can provide.

For example, in team-based training, using TV monitors in portrait mode with interactive videos as a stand-in for a real team member [5]. Another example from the same team-based training field include experimenting with health care tools for telemedicine within a MR cooperative setting enabled by a "metaverse"; that is another pretty new word that stands for a network of 3D virtual worlds enabling social connection and collaboration from different places. The term "metaverse" has its origins in the 1992 science fiction novel *Snow Crash* as a portmanteau of "meta" and "universe" [6] (Fig. 3).

In recent years, various metaverses have been developed for popular use such as virtual world platforms like *Second Life* or videogames, and a number of new ones are announced or even advertised presently. What really matters, from our perspective, is that some metaverse iterations involve integration between virtual and physical spaces. These are the most interesting ones for an effective XR medical education to be delivered in real time both online and into phygital places.

2   e-REAL: Glasses-Free Mixed Reality

What makes MR so interesting today for medical education, and simulation in particular, is the glasses-free solution like that we developed under the name of e-REAL® [7]. In e-REAL, digital and physical objects co-exist in the real world, not within a headset, making e-REAL unique. e-REAL, as a MR environment for hybrid simulation, can be a stand-alone solution or even networked between multiple locations, linked by a

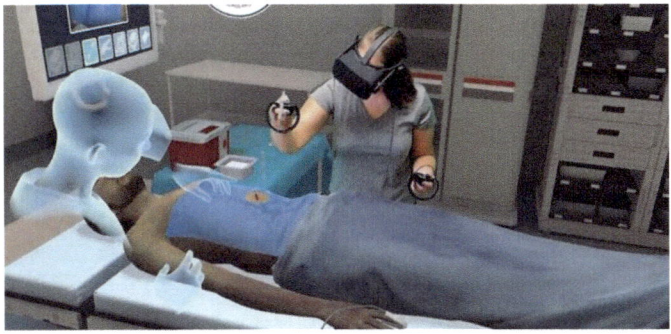

Fig. 3. Representative metaverse based on MR technology enabling visualization and some collaboration from different places.

special videoconferencing system optimized to process operations without perceivable latency. This connectivity allows not only virtual objects sharing (like medical imagery, infographics, etc.) in real time, but also remote cooperation [8] (Fig. 4).

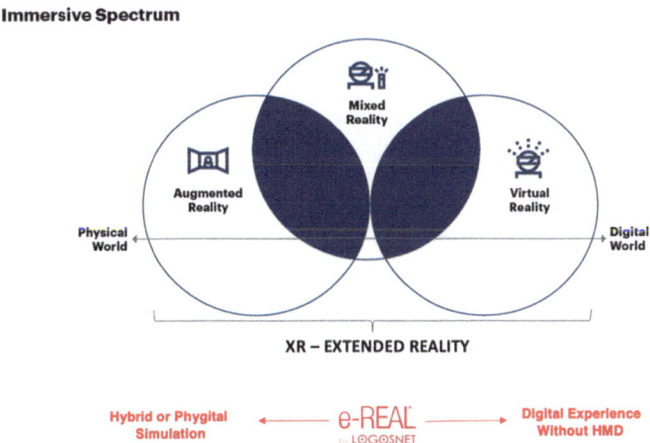

Fig. 4. The immersive spectrum of XR and e-REAL as a solution enabling glasses-free learning experiences, both onsite MR and online VR.

e-REAL is both an online solution for glasses-free VR and an onsite (or phygital) MR solution. The phygital configuration is a glasses-free solution that uses ultra-short throw projectors, gesture shaping lidars, and cameras to turn blank walls and empty spaces into immersive and interactive environments. It is designed as an easy, user-centric, and cost-effective solution compared to the old CAVE environments, which are too rigid, difficult to be managed, and expensive. The online solution is a glasses-free platform, available as both desktop or cloud-based, enabling remote cooperation in real-time. Both the e-REAL solutions are at the forefront of the extended reality options (Fig. 5).

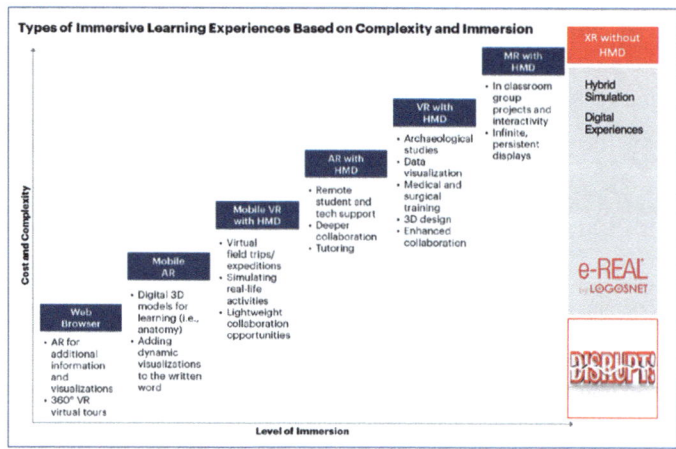

Fig. 5. Types of immersive learning experiences available today.

Similar to being immersed within a videogame, learners are challenged by facing cases within multifaceted medical scenarios. Digital humans, or avatars, are part of the solution both online and in the phygital settings (Figs. 6 and 7).

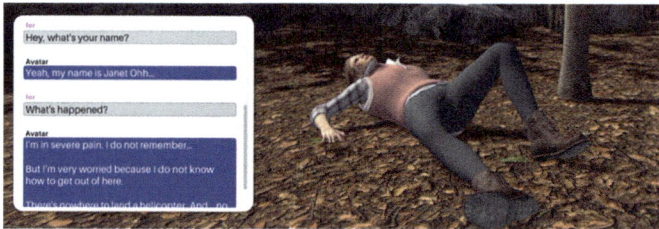

Fig. 6. An e-REAL digital human, enhanced by dialogic artificial intelligence.

Holograms may be part of the e-REAL phygital and online settings as well, with the learner utilizing wearable augments such as AR head-mounted displays. Also, human-sized holograms can be reproduced within the e-REAL phygital setting. Those specific types of holograms may be pre-recorded or may even be live (Fig. 8).

e-REAL digital holograms allow people to fully view parallax 3D images from every angle. These holograms show details that can't be viewed in traditional holograms that are horizontal parallax only.

e-REAL holograms are cutting edge, captivate audiences, are glasses-free, and offer an unparalleled degree of interactivity. Learners can see around a hologram from all angles, gaining rich insights and profound educational results.

e-REAL holograms are autostereoscopic and fully parallax, so cognitively they are directly translated to a mental image—while traditional holograms, as well as 2D textbook handouts, require 3D reconstruction within the working memory [9].

Fig. 7. Representative time management, writing and overlaying features available in both the e-REAL online VR platform and in the phygital MR classroom.

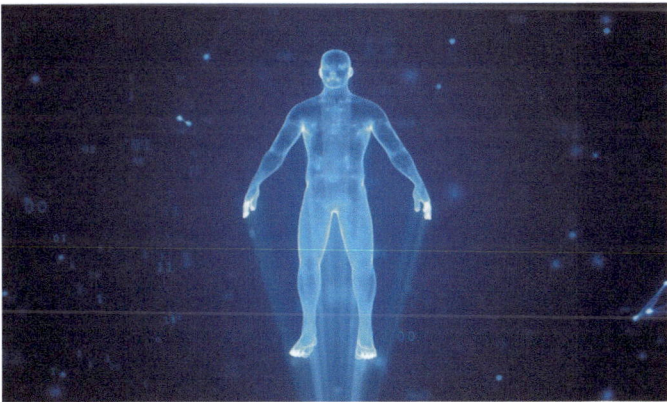

Fig. 8. Representative human-sized hologram. e-REAL holograms are autostereoscopic and fully parallax, so cognitively they are directly translated to a mental image.

Human-sized holograms and digital humans are part of the e-REAL scenarios, making this solution unique, rich, and diversified.

References

1. Hoffman, H., Vu, D.: Virtual Reality: Teaching Tool of the 21rst Century? Acad. Med. **72**, 1076–1081 (1997)
2. Milgram P., Takemura H., Utsumi A., Kishino F. Augmented Reality: A class of displays on the reality-virtuality continuum. In: Proceedings of SPIE—The International Society for Optical Engineering, Vol. 2351 (1994)
3. Aukstakalnis, S.: Practical Augmented Reality. Addison-Wesley, Boston (2017)
4. Salvetti, F., Bertagni, B.: Virtual worlds and augmented reality: the enhanced reality lab as a best practice for advanced simulation and immersive learning. Form@re **19**(1) (2019); Aukstakalnis, S.: Practical Augmented Reality. Addison-Wesley, Boston (2017)

5. Palaganas, J., Maxworthy, J., Epps, C., Mancini, M. (eds.): Defining Excellence in Simulation Programs. Society for Simulation in Healthcare, Wolters Kluwer (2014)
6. Stephenson, N.: Snow Crash. Bantam Books, New York (1992)
7. www.e-real.net
8. Salvetti, F., Gardner, R., Minehart, R.D., Bertagni, B.: Enhanced Reality for Healthcare Simulation. In: Brooks, A.L., Brahman, S., Kapralos, B., Nakajima, A., Tyerman, J., Jain, L.C. (eds.) Recent Advances in Technologies for Inclusive Well-Being. ISRL, vol. 196, pp. 103–140. Springer, Cham (2021). https://doi.org/10.1007/978-3-030-59608-8_7
9. Salvetti F., Bertagni, B. Interactive tutorials and live holograms in continuing medical education: Case studies from the e-REAL experience. In: Proceedings of the ICELW Conference, Columbia University, New York, NY, pp. 1–8 (2016)

Improving Scalability of Software Engineering Courses

Sigrid Schefer-Wenzl[(✉)] and Igor Miladinovic

Computer Science and Digital Communications, University of Applied Sciences Campus Vienna, Favoritenstr. 226, 1100 Vienna, Austria
sigrid.schefer-wenzl@fh-campuswien.ac.at

Abstract. University degree programs in software engineering often see an increase in student applications. Consequently, class sizes in this area are steadily increasing. There is a general perception that large classes can negatively affect student motivation and engagement. On the other hand, the workload for lecturers increases. At our university, we have mitigated these negative effects of large classes by using selected techniques to create a "small class feeling." Most importantly, we have improved the scalability of software engineering courses by reducing the load on our main lectures by outsourcing much of the non-scalable tasks to teaching assistants. In this paper, we describe this approach and our experience applying it in two software engineering courses. The findings of this paper represent our recommendations for the division of the tasks between main lecturers and teaching assistants and for creating a "small class feeling" in a large class.

Keywords: Higher Education · Large Classes · Software Engineering Education

1 Introduction

The demand for skilled software developers is increasing worldwide. Trend analyses from the United States Bureau of Labor Statistics predict that the growth of computing careers is about to continue through 2030 and that various computing skills, including software development, will be in strong demand for the foreseeable future [1]. Accordingly, we have seen a steady increase in the number of applicants to our computer science degree programs at our university in recent years. For example, in our computer science Bachelor degree program, the number of applicants has more than doubled from 2016 to 2021, with continuous growth each year. Despite selection through the admissions process, this has resulted in a higher number of students enrolled in our courses. To ensure a similar level of student support as in small groups, we were faced with several challenges. For example, the effort required to assess individual student performance increased linearly with the number of students. Furthermore, the number of groups for the practical tutorials increased as we work with a maximum of 25 students in a tutorial group. These and other challenges resulting from large class setting motivated us to look for solutions to improve the scalability of our courses without compromising quality.

We analyzed the tasks involved in delivering our courses and found that different levels of experience were required to complete the various tasks. Therefore, we divided the tasks into different groups, depending on the level of experience required. Certain tasks should be performed by the main lecturers. Other tasks may be performed by teaching assistants, such as less-experienced instructors or even by outstanding students of a higher semester. Outsourcing these tasks to teaching assistants allows for better scalability of courses from a main lecturer's perspective. The main lecturer can still focus on didactics and content selection in a similar amount as in a smaller group, whereas teaching assistants provide a small-group feeling for the students.

In this paper, we will present our approach and recommendation for tasks division between main lecturers and teaching assistants with a special focus on software engineering courses.

2 Teaching Elements for Small-Group Feeling in Large Classes

To address the challenges with large student classes, we have selected and adopted several techniques. In this section, we will describe these techniques.

2.1 Dividing Tasks Between Senior and Junior Staff

One disadvantage of larger class sizes is that interaction between students and instructors decreases and instructors are less accessible for students [2]. In our courses, we have improved the availability of the main lecturers by dividing tasks into two groups, according to the responsibility for a course as well as to the level of experience required.

There are tasks that should be performed by the main lecturers. Examples of these are the selection of the content to be taught, the selection of appropriate didactic methods, the delivery of lectures, the preparation of exams, the delivery of oral exams, and the definition and grading of tutorial exercises. These are the tasks that have the greatest impact on the nature of a course and should therefore remain in the hands of the "authors" of a particular course.

There are other tasks in the context of each course that may be performed by other, less-experienced instructors or even by outstanding students from a higher semester. Examples of these tasks include providing personalized feedback on smaller exercises, supervising and responding to discussions in online forums, and providing programming support in tutorials. Those tasks are supportive for students to better understand the course contents and, when done properly, can significantly increase student satisfaction with a particular course. They help create the feeling of a small class. However, in large classes, these tasks are very time-consuming for the lecturers.

Outsourcing these tasks to others allows for better scalability of courses from the perspective of a main lecturer.

2.2 Involving Students as Teaching Assistants

Involving experienced students as teaching assistants has repeatedly been shown to be beneficial, both for the teaching assistants and for the students in the course. For teaching assistants, teaching activities can stimulate a higher level of information processing [3] as they answer questions, proofread lecture material, support students with the more hands-on exercises, and sometimes hold tutorial sessions in preparation for exams. In our courses, we select outstanding students from higher semesters to serve as teaching assistants. Therefore, being a teaching assistant is also a kind of reward for our students.

For the students in the course, barriers to asking questions are lowered as students from higher semesters can bring in the students' perspective to the teaching activities and are more likely to understand students' problems students with the course contents. We also aim to select male and female teaching assistants to further reduce barriers.

When teaching assistants provide personalized feedback on small exercises, the typical process is for them to enter their comments into the learning environment. These comments remain invisible until they are approved by a junior or senior lecturer of the course.

2.3 Active Learning Activities

One of the biggest challenges with larger classes is the students' activation during lectures. This is because larges classes tend to cause passivity among students, due to anonymity, acoustic reasons or missing discussion opportunities [4]. Without activations, lectures mutate to frontal lectures, where it is hardly possible to keep the focus of students for the duration of the lecture [5].

In our software engineering courses, we usually activate students regularly, by alternating frontal lecture with active learning activities every 20–30 min. Examples for active learning activities in large groups are: live polls, quizzes, discussion in small sub-groups, thing-pair-share exercises, and little programming exercises. During exercises in small groups, we are visiting some of the groups and support them.

In tutorial parts of our courses, we divide our tutorial groups (usually up to 25 students) into small groups of 2–4 students working on the same exercises. We are usually accompanied by teaching assistants, who supervise the student groups together with the main lecturers.

2.4 Reflection and Personalized Feedback

We use learning diaries as a didactic tool to support, structure, and consolidate students' learning progress in different learning phases, as they are suitable to document students' written reflections of their learning progress [6]. In large-class settings, the main purpose of learning diaries is to facilitate individual comprehension and retention of complex issues and to encourage critical reflection on learning content [7]. It usually consists of several entries, also called learning protocols, that can vary in the extensiveness or in the degree of structure of the protocols. It is one of the few educational methods that can also be used to capture students' emotions, opinions, and attitudes as part of their learning process [8]. The main benefit for the lecturer is that a learning diary may provide valuable insights into the different learning processes and outcomes of participants in a course. Also, they help to diagnose possible misunderstandings. In addition, students are motivated to reflect on their learning activities and to take responsibility for their own learning process [9].

Especially in large groups of software engineering education, which are often characterized by very diverse levels of student knowledge, a high degree of group work and a considerable degree of informal learning from sources such as forums, videos, or reused code learning diaries offer an effective learning tool to develop students' autonomy and reflection capabilities.

Each learning diary entry gets personalized feedback and contributes to the final grade of the course. The personalized feedback is proposed by teaching assistants and approved by main lecturers.

2.5 Blended Learning

Blended learning provides a good learning setting as the benefits from both distance learning and in-class hours can be leveraged. When applying blended learning, in-class hours should be used for clarification of open questions, deepening knowledge of complex learning contents, or highly interactive exercises. It is also recommendable to provide ubiquitous functions to allow the full spectrum of advantages of mobile technologies and to enable students to learn anytime and anywhere.

We already have successfully applied mobile learning in different software engineering course [10, 11]. It supports easy repetition of contents and different learning speeds, and it is suitable to support peer learning, virtual collaboration, and teaching of distributed and large groups. For large groups, the mobile learning environment needs to provide corresponding features, such as a forum, chat, or other collaboration tools [12, 13].

In online or hybrid lectures, lecture casts are provided in order to support repetitions and also students who were not able to attend the lecture. Lecture casts and other eLearning material are designed to support consumption on mobile devices.

3 Scalability Improvements in Example Courses

In this section, we will outline our experience with two software engineering courses where we have applied the division of tasks between main lecturers and teaching assistants in order to improve scalability. The courses are in two subsequent semesters and complementary to each other.

The first course is "Software Engineering" in the third semester of our Bachelor degree program. Originally taught by one lecturer, the number of students gradually increased over the last five years, and this increase also increased the effort for this course. Today, the course is held by one lecturer and two teaching assistants. The course is composed of a lecture part and a tutorial part. The second course is "Mobile App Development" in the fourth semester of our Bachelor degree program. It can be seen as a deepening of the first course. This course also consists of a lecture and tutorial part but includes even more hands-on programming exercises with a subsequent project, which can be the basis for the bachelor thesis. The concept of both courses is presented in references [10, 11].

In both courses, we applied the task division as illustrated in Fig. 1. Some of the tasks require nearly the same effort for small and large classes. These tasks include the design of the didactic concept, selection of the course content, and organization of a hackathon even, which serves in both courses as the start of the tutorial part. In our courses, these tasks are mainly done by main lecturers.

The second group of tasks require increasing effort when increasing size of the class. However, this increase is less than linear. These tasks include: in-class exercises, pear learning, and preparation sessions for the exam. They are mainly (or even completely) performed by teaching assistants, as indicated in Fig. 1.

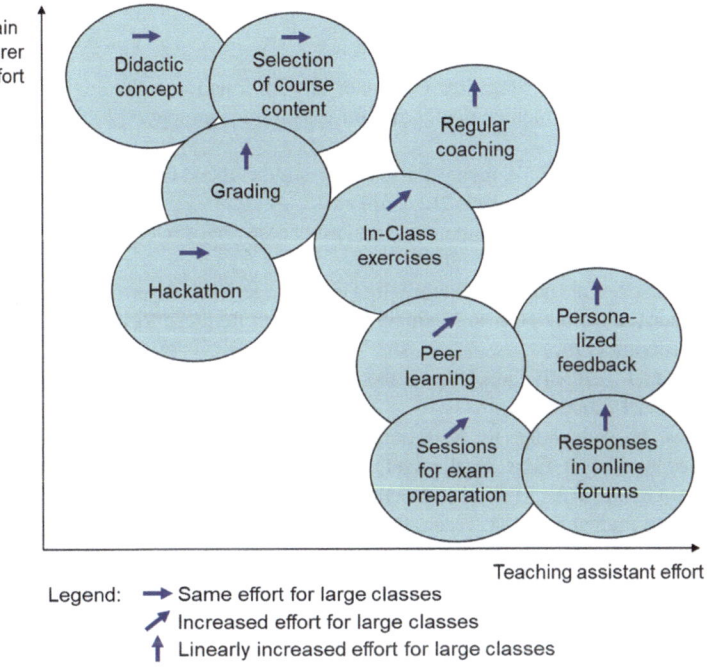

Fig. 1. Task division between main lecturer and teaching assistant with scalability indication.

The last group of tasks are activities requiring effort directly proportional to the number of students. These tasks are grading, providing personalized feedback, and management of online forums. In our courses, these tasks are performed by teaching assistants whenever possible. However, the final grading remains with the main lecturer, limiting scalability to some extent.

4   Conclusion

The rising number of students in the area of software engineering requires lecturers in this area to deal with ever increasing class sizes. This results not only in increased effort for some teaching tasks, but also often in distance and anonymity among students. This may lead to decrease motivation of both lecturer and students, and it may lead to a decreasing number of students reaching the learning outcomes of the lecture. At our university, we have developed several techniques to address these issues. The main idea is to systematically divide teaching tasks so that some of them can be performed by teaching assistants, who are less experienced than main lecturers. Usually, these are the tasks where the effort strongly increases with the number of students. In addition, these tasks can contribute to a "small-class feeling" in large classes. In our future work, we will apply our approach to courses beyond software engineering area, and evaluate the outcomes.

References

1. US Bureau of Labor Statistics: Computer and information technology occupation. URL: https://www.bls.gov/ooh/computer-and-information-technology/ (2017). Retrieved March 21, 2022
2. Lynch, R., Pappas, E.: A model for teaching large classes: facilitating a "small class feel." Int. J. Higher Educ. **6**(2), 199–212 (2017)
3. Ten Cate, O., Durning, S.: Dimensions and psychology of peer teaching in medical education. Med. Teach. **29**(6), 546–552 (2007)
4. Gibbs, G.: Control and independence. In: Gibbs, G., Jenkins, A. (eds.) Teaching Large Classes in Higher Education: How to Maintain Quality with Reduced Resources, pp. 37–59. Kogan Page, London (1992)
5. Carpenter, J.: Effective teaching methods for large classes. J. Family Consum. Sci. Educ. **24**(2), 13–23 (2006)
6. Nückles, M., Schwonke, R., Berthold, K., Renkl, A.: The use of public learning diaries in blended learning. J. Educ. Media **29**(1), 49–66 (2004)
7. Connor-Greene, P.A.: Making connections: Evaluating the effectiveness of journal writing in enhancing student learning. Teach. Psychol. **27**(1), 44–46 (2000)
8. Munezero, M., Montero, C. S., Mozgovoy, M., Sutinen, E.: Exploiting sentiment analysis to track emotions in students' learning diaries. In: Proceedings of the 13th Koli Calling International Conference on Computing Education Research, Koli Calling '13, pp. 145–152. New York, NY, USA, (2013)
9. Huisman, J., Wallenius, L.: Empowerment on-line collaborations: learning diaries as a sustainable learning tool. In: Proceedings of the 8th International Conference on the Future of Education, Florence, Italy (2018)
10. Schefer-Wenzl, S., Miladinovic, I.: Game changing mobile learning based method mix for teaching software development. In: mLearn 2017: Proceedings of the 16th World Conference on Mobile and Contextual Learning, Larnaca, Cypress (2017)
11. Schefer-Wenzl, S., Miladinovic, I.: Learning diaries—a valuable companion of mobile learning for higher education in software engineering. In: Interactive Mobile Communication, Technologies and Learning (2019)
12. Ally, M., Prieto-Blazquez, J.: What is the future of mobile learning in higher education? Revista de Unicersidad y Sociedad del Conocimiento (RUSC) **11**(1), 142–151 (2014)
13. Ramos, P. R. H., Penalvo, F. J. G., Gonzalez, M. A. C.: Towards mobile personal learning environments (mple) in higher education. In Proceedings of the 2nd International Conference on Technological Ecosystems for Enhancing Multiculturality (TEEM), Salamanca, Spain (2014)

A Conceptual Approach to an AI-Supported Adaptive Study System for Individualized Higher Education Services

Christian-Andreas Schumann(✉) 🆔, Claudia Tittmann 🆔, and Frank Otto 🆔

Faculty of Economic Sciences), West Saxon University Zwickau, Saxony, Germany
christian-andreas.schumann@fh-zwickau.de

Abstract. In the context of the digital transformation, the targeted implementation of AI-based or AI-supported technologies in "teaching & learning," as well as "administration & service," holds considerable potential for organizational change and quality enhancement for higher education institutions. The use of AI in higher education teaching and services lags behind the level in research. Therefore, holistic solutions must be planned and implemented in unity of teaching and research for the AI-based support of the stakeholders' inclusive administration, the further development or the establishment of new digital study programs and offers, as well as the prospective qualification of university staff in the field of AI. The applications that currently exist do not generally fit into a holistic concept. Therefore, they must be analyzed, systematized, and structured to generate a conceptual approach via an integrated architecture with adaptive services. A cross-university, transdisciplinary and modular system approach is being pursued to transfer AI methods as supporting technologies into the regular operation of teaching and associated services at the university in an interdisciplinary project over several years. From study orientation to the use of new study programs, the procedures at different process and structural levels of the university are made more flexible by means of AI and transferred into adaptive services for the individualization of learning and teaching. The paper explains the conceptual approach for the further development of the AI-based education system at a university and presents the first results of planning, developments, and applications. Selected case studies related to the characteristics of paths for study orientation and study specialization, as well as for special learning profiles serve as illustrations.

Keywords: Systematic Approach for AI-based Higher Education · Modular and Adaptive Services · Individualization of Learning and Teaching · AI-based Decision Support

1 Introduction

The support of a wide variety of systems by artificial intelligence (AI) will increase significantly and continuously. This is accompanied by uncertainties in evaluating the role of AI in society. The expectations for performance are very high and go beyond

D. Guralnick et al. (Eds.): TLIC 2022, LNNS 581, pp. 383–396, 2023.
https://doi.org/10.1007/978-3-031-21569-8_37

the currently validated results. The many possible applications of AI are faced with immense challenges, which is why it is important to formulate and adhere to fundamental guidelines for the use of AI. The use of AI should serve to support people and not create a dominance of technology over people. AI, its possible applications, and its acceptance must be discussed in society in a broad, transparent, and comprehensive manner. In a 2018 survey by the Association of German Engineers, AI was ranked first as a leading trend in IT by 56% of respondents. Together with automation and digitization, AI is one of the three basic technologies for autonomous, smart, and adaptive systems that take on complex tasks in cooperation with humans and support them in all areas of life. AI enables new solutions and offers potential for greater efficiency, flexibility, and quality of life. Large volumes of data are processed with high quality. By using AI methods, human values can be better communicated and implemented, and additional value added can be generated. AI will have a massive impact on life, education, and work processes, with AI systems serving as assistance systems for humans and not making human performance completely obsolete. Ethics have a special place in the use of AI. The relatively high level of development of AI, especially in technology-related fields, helps to transfer application experiences regarding methodology and tools to other areas of life such as education, considering the specifics of these application fields, to push the further dissemination of AI on a professional basis [1].

With the coordination and elaboration of the global consensus on AI in education under the aegis of UNESCO, a framework with the claim of educational planning in the AI area has been created as a guide for the use of AI under the specific conditions of knowledge transfer in schools and universities in 2019 [2].

Reference is made to UN's 2030 Agenda for Sustainable Development as a plan of action for people, planet, and prosperity in particular the Sustainable Development Goal SDG4, which as an education goal is to ensure inclusive and equitable quality education and promote lifelong learning opportunities for all [3].

The underpinning regarding the use of information and communication technologies (ICTs) to calibrate SDG4 by UNESCO's Quingdao Declaration in 2015 underscores the need to use emerging technologies to strengthen education systems, ensure access to education for all, ensure quality and effective learning, and ensure equitable and more efficient service delivery [4].

The 2019 UNESCO initiative on the application of AI in education, thus, serves to specify the overall UN and UNESCO sustainability and digitization goals of 2015. This finding is essential because it highlights the need to sustainably integrate AI into a general education framework, resulting in the challenge of complex solutions. The recommendations for action in the documents contain a clear message regarding a holistic and systematic approach to the use of AI in education, which includes: [5].

- Plan AI in education policies in response to the opportunities and challenges AI technologies bring, from a whole-government, multi-stakeholder, and inter-sectoral approach, that also allow for setting up local strategic priorities to achieve SDG 4 targets
- Support the development of new models enabled by AI technologies for delivering education and training where the benefits clearly outweigh the risks, and use AI

tools to offer lifelong learning systems which enable personalized learning anytime, anywhere, for anyone
- Consider the use of relevant data where appropriate to drive the development of evidence-based policy planning
- Ensure AI technologies are used to empower teachers rather than replace them, and develop appropriate capacity-building programs for teachers to work alongside AI systems
- Prepare the next generation of existing workforce with the values and skills for life and work most relevant in the AI era
- Promote equitable and inclusive use of AI irrespective of disability, social or economic status, ethnic or cultural background or geographical location, with a strong emphasis on gender equality, as well as ensure ethical, transparent, and auditable uses of educational data.

To address the dynamics of transformation and the complexity of the task in using AI in education, UNESCO is pursuing further development of the approach through multiple international events that ultimately led to the launch of the AI and the Futures of Learning Project, September 30, 2021. Participants at the International Forum on AI and Education debated "Ensuring AI as a Common Good to Transform Education" in late 2021 [6].

There is also an understanding in the EU, as well as in the countries of the community, that AI as an emerging technology must be understood systemically and has to be integrated into all areas via complex planning, models, concepts, and implementations. Topics currently being pursued in this context include Fostering a European approach to AI, Coordinating Plan on AI, up to a proposal for an Artificial Intelligence Act [7].

Experts commenting on the EU AI law proposal explicitly point out that education, just like, for example, critical infrastructure and security components, should be counted among the high-risk AI systems [8].

The high risk is derived from the complex effect of AI, which from a strategic point of view is one of the key policy areas, namely: High Performance Computing; Artificial Intelligence; Cybersecurity and Trust; Advanced Digital Skills; and Deployment and Best Use of Digital Capacities and Interoperability, listed in the context of establishing the Digital Europe Programme. Explicit reference is made to the fact that the five specific key topics are distinct but interdependent. The connection between digital transformation, which includes the application of AI, and education is made several times in the document in various relations [9].

Through the EU Commission, the integration of AI into education is linked to the updated Digital Education Action Plan [10].

2   AI Penetration in Higher Education

Recent European University Association documents explicitly address the impact of AI in the context of technological developments. Universities are encouraged to study and assess the impact of emerging technologies and prepare graduates for labor markets that are changing due to digitization and emerging technologies, especially AI, which will also transform the mode and way of working practiced by universities and their partners [11].

The Microsoft Education Transformation Framework (ETF) for Higher Education will be explicitly oriented toward AI applications in higher education. Through ETF, the complex approach of combining student success, teaching, and learning; academic research; and secure and connected campuses will be related to explore the current status of AI in Higher Education, categorize applications of AI in education, present a number of use cases, as well as discuss AI technologies supporting innovations [12].

The German Higher Education Forum on Digitization already took up this holistic approach in 2017 and placed AI as an emergent technology, especially in the form of machine learning, in the overall context of a digital turn for new ways of higher education in the digital age. Digital skills, digital teaching and learning, personalized learning, new understanding of roles and professions, academic program development, data security and privacy, and legal issues are mentioned as potentials and challenges for digital transformation. The focus should be on new business models, technologies, lifelong learning, internationalization, change management, organizational development, innovations in teaching, learning, and testing, curricular design and quality development [13].

Understanding the complexity of the task means that any intervention to establish AI in existing higher education systems requires a holistic approach. Punctual or insular concepts and solutions carry the high risk that the AI deployment will not be compatible with all other processes and components and will thus be viewed as a foreign body and isolated.

The "Digital Higher Education" funding priority of the Federal Ministry of Education and Research in Germany is used to develop various topics and application areas that are interrelated and encompass both practical knowledge for action and the development of framework conditions. The fields of application defined include experimental learning, digitized learning environments, and educational infrastructures and resources. Since 2021, the initiative "Artificial Intelligence in Higher Education" has been funding projects ranging from the development of new study programs and modules in the AI field to the development of AI-supported systems at universities, for example, using intelligent assistance systems or AI-based learning and examination environments. In order to do justice to the diversity and breadth of the higher education system and to achieve effective effects in studying and teaching, the following focal points are relevant: [14].

- Strengthening AI competencies in study and qualification programs.
- Improving higher education using AI.
- Sustainability, networking, and transfer between the former focal points.

Based on the described holistic approach of the use of AI in higher education, a project development for the establishment of an AI-based adaptive individualized study environment for students and university administration is currently underway, for which a conceptual design of an AI supported adaptive study system for the individualization of university services is required. Due to the complex impact of such a measure, the previous operating and process models of the university have to be analyzed, reengineered and optimized for AI support.

3   Process Organization of AI-Based System in Higher Education

Currently, many approaches to the use of AI in higher education are attempting to better address the complexity of the problem. In addition to the relatively narrow view of didactic, methodological and/or technological nature, proposals are made that include, for example, an entire AI campus [15].

Unfortunately, they usually reflect only on the academic sphere without sufficiently including the other tangential higher education and administrative processes. Even if such a holistic approach were to be taken, massive problems of transition to upstream and downstream educational levels outside the universities would again be created in the end, which could significantly hinder the freedom to choose among educational paths for years to come. For this reason, process models have been developed that encompass complex educational pathways from school education to university training and continuing professional development. They have already served in the past to create consistency and compatibility between different levels of education and degrees, including in the context of higher education (Fig. 1).

If service and administrative processes are integrated into the process models in addition to the educational pathways, the prerequisite for modeling and implementing processes and their AI support in a realistic manner is given [16].

Aspects of standardization, as well as individualization of educational modules, must be considered in the models. Processes can be controlled or regulated. When regulating educational processes, there are so-called feedback loops, whereby the processes can be better optimized. This approach is suitable for integrating AI support in a variety of ways. In particular, the adaptability of processes to the requirements of different stakeholders in higher education can be significantly improved by applications of machine learning, for example. Starting from integrated educational systems, flexible systems with adaptive control and assistance for users are created.

The further digitization and automation of processes goes hand in hand in principle with data science or AI, which also applies to higher education. The use of AI has two fundamental goals: Optimized and more efficient process management and user support for process operation. Assistance systems are also interesting in the field of higher education because they support users of different university groups (students, professors, and teachers, as well as administrative staff) directly or indirectly in the execution of actions. They provide information and, if necessary, make suggestions for decisions and actions. With the use of AI, these processes can be upgraded. In process control, AI is currently being incorporated primarily for process data analysis, process modeling, model use (for example, for design, process analysis, and control), and process intervention, which is also transferable to AI use in higher education, including pre-and post-processes [17].

The further development and optimization of process organization from the perspective of the use of AI in educational processes presupposes a fundamental revision of higher education models. Despite multiple efforts to consider the entire university,

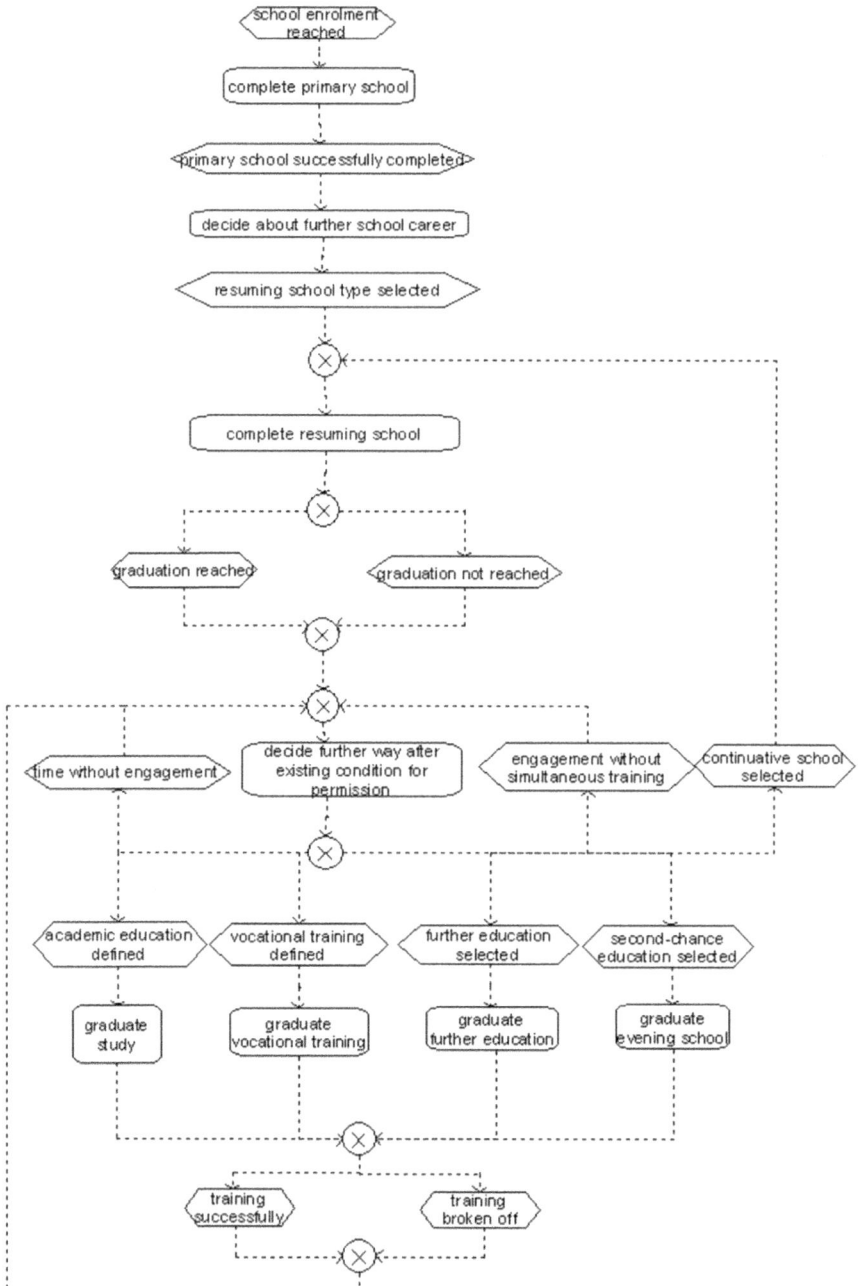

Fig. 1. Multi-stage and multi-optional educational process as an event-driven process chain.

including upstream and downstream or tangential activities, university operations are generally not based on system-oriented and systemic approaches. The permanently progressing individualization combined with the compulsion for permanent adaptation of all university components should be better mastered using AI. Therefore, fundamental reorientations of the educational institutions are necessary, which at the same time opens the possibility to rethink the whole system with all its processes and to replan it in a system-oriented way.

4   System-Oriented Approach for AI in Higher Education

The complexity of holistic, AI-based, and service-oriented higher education systems INCLUDING the required regular adaptation and adaptivity of systems to support individualized educational pathways requires the application of complexity reduction methods. One common and proven way is system decomposition. Structurally, it means describing the higher education institution through the interaction of many components, resulting in a building block structure. The new Modular System involving AI will be the basis of individualization with simultaneous standardization and unification for the new higher education.

Higher education systems are separated from the outside world by a service layer, which offers interested parties, students, lecturers, staff and other stakeholders the opportunity to use the university's own products and services. Depending on the use case, this service layer can be an interpersonal interaction facility (for example, during advising sessions between prospective students and student administration staff) or a human-machine interface (e.g., when accessing digital learning content).

Within this layer follows the assistance layer. This layer is responsible for providing the services offered by the university. In addition to administrative services, this primarily includes all services required for the proper execution of teaching and research. Basing on the assistance layer there is an orientation layer, whose purposes are:

1. the supportive provision of information for students and those interested in studying with regard to the courses of study and course content offered by the university (information layer), and
2. the provisioning of a learning path layer, whose primary task is to show students the possibilities of how to connect various study modules on the way to an individual and successful degree.

The module layer following the learning path layer, as well as the learning unit layer contained therein, are intended to enable students to access the individual modules and learning units and to support them in their daily tasks.

Figure 2 illustrates the relationships between the individual layers.

AI can be used in all of the named layers. However, a distinction must be made between the different ways in which AI can be used in institutions of higher education: As learning content within study modules, as a scientific method within research projects, and as a tool within administration. In the following, this chapter will focus on administrative processes, but the procedure described can be applied analogously to processes from the areas of teaching and research.

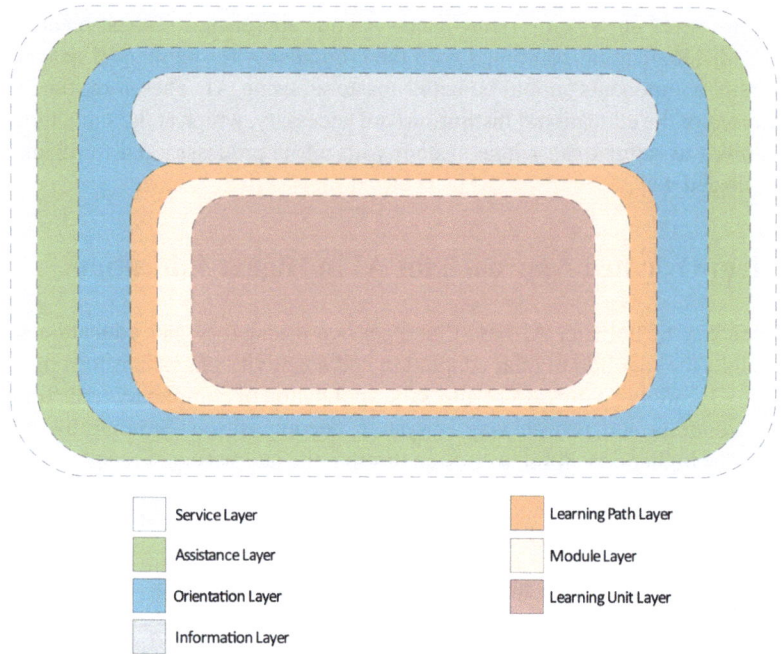

Fig. 2. Schematic illustration of the different layers of a university.

With regard to the use in administration, a further distinction must be made between the use of AI in:

1. general university administration processes, such as the administration of personnel matters or lecture rooms (Service, Assistance, Orientation and Information Layer),
2. student administration processes, such as AI-based decision support (Service, Assistance, Orientation, Information and Learning Path Layer), and
3. student recruitment processes (Service, Assistance, Orientation and Information Layer).

Users can interact with university members or the university's information systems by entering the assistance layer via a desk. This desk can be either physical (e.g., when speaking to staff members in their physical office) or a virtual access point (when interacting with some information system). When interacting at a physical desk, the AI supports the corresponding employee working on this desk, while at a virtual desk, the AI directly supports the user.

Figure 3 shows the layers already described for service, assistance, orientation, information and the learning paths. Within these layers exist different information sources and their connections, represented by the circles and lines. The black circles represent the desk described. For a clearer presentation, the desk has been drawn twice, because it describes the beginning and the end of the users' interaction.

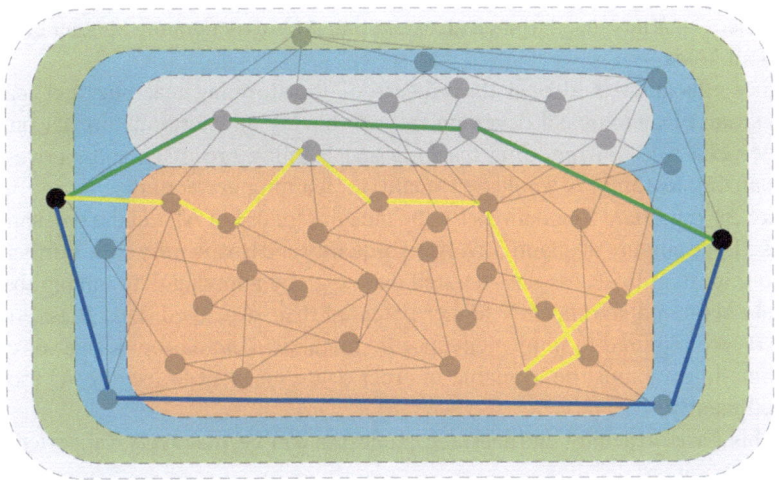

Fig. 3. Interaction possibilities of the users on administrative level.

When entering a desk (either virtual or physical) the user's intention is to get some orientation, information or specific content. If the user needs some orientation (e.g., on what to study at the university), an AI-supported assistance system could be a suitable approach. In this case the following options for AI-support are possible:

1. The sematic linking of distributed information,
2. A rule-based or deep-learning-based decision support system, and
3. A system, that provides speech recognition/language processing.

An example of a corresponding orientation path is shown as blue path in Fig. 3.

If a user is looking for information, the semantic linking of the existing, distributed information is also a possible way as well as a system with speech recognition or language processing provisioning. Figure 3 illustrates this option along the green path.

Another possibility is to support students on building their individual study path. In this regard, AI can also support by providing (semantic) linked information or by suggesting certain modules that suit the student's expectations (which have to be specified in advance) or that other students in analog situations have chosen and recommended. This process is represented by the yellow line in Fig. 3. As the yellow line indicates, hybrids between the layers for orientation, information and learning paths are possible.

5   Modular System for AI in Higher Education

5.1   Requirements for AI in Higher Education

AI has been an important tool in a wide variety of application areas for decades, including manufacturing, medicine, computer gaming, weapons and military, learning, and education. In recent years, technical and technological development—including computer performance—has advanced to such an extent that AI is experiencing another enormous boom in all fields.

Higher education is also in the process of transforming teaching and learning. The social and technological developments are having an enormous impact on it. AI plays a crucial role in supporting teaching and learning processes and, above all, in individualizing learning. A whole new quality of learning is emerging.

There are important foundations for the meaningful use of AI. On the one hand, this includes a large amount of quantitatively high-quality and trustworthy data. Furthermore, it is also important to access unstructured data, e.g., by applying data mining, because these data contain a significantly higher potential than structured data. Based on this, AI experts are required to implement specific methods. And on top of that, of course, human intelligence must be available in order to use or want to use the AI tools and results efficiently.

On this basis, teaching and learning processes can be massively supported in the following sub-areas:

Action, including language processing and knowledge representation,
Perception, including image processing, speech recognition, text recognition, face recognition,
Learning, including machine learning, deep learning, reinforcement learning, crowd sourcing, among others.

5.2 Our Experience from Projects

In our long-term project experience within our university in close cooperation with other educational institutions and companies, we have already done a lot of preliminary work, more or less intentionally aiming in the direction of AI.

Very intensively we have dealt with the topic and use of semantic networks in various contexts as well as the modeling of semantic networks. Semantic networks are formal models that represent entities and their relations as well as attributes. In the field of AI, semantic networks are used as a base of knowledge representation.

In parallel, we have been working on the modularization of learning content and its semantic connectivity. For example, we have developed models for determining content correlation of learning content and, based on this, learning concepts that exploit precisely these correlations in order to specifically close the knowledge gaps of learners. For example, this approach was used in a project to provide trainee mechatronics engineers with videos or learning content customized to a problem in their practice.

Other projects were more focused on data analytics, specifically learning analytics (LA). In more extensive studies on learning platforms, we tested appropriate methods. In the first stage, learning processes and learning scenarios were studied and enriched with LA methods. Subsequently, studies were conducted to determine the extent to which these methods add value for teachers and learners or are used at all.

5.3 How Can We Build on This as a Base for AI?

We have tested and evaluated the elementary technologies on which AI is based and examined their practicability.

Furthermore, we have dealt with the various stakeholders. By applying different methods, interviewing the stakeholders and evaluating the results, it is also possible to identify barriers and problem areas. Likewise, there are realizations from the analyses, which methods are fast convertible, which tools are more or less wanted, etc.

Based on this, the actual AI methods still need to be developed. In order to realize this, the processes in higher education as well as the higher education didactic requirements have to be considered in more detail. Also, important to be considered the perspectives of learners and teachers, who in their turn have different effects on the same process.

Higher education institutions will play a driving role in this AI process. AI-enhanced teaching and learning must become accessible to all and requires digital learning content.

Elemental to this strategy is the development of an AI competency among all stakeholders. It must be clear how AI works in order to achieve a high level of use and added value. In this context, data protection plays an important role. Participants must be able to decide for themselves how their data are used. This creates the need to define new roles and assign responsibilities.

6 Case Studies

6.1 Case Study I: Decision Support for Individual Curricula

Zwickau University is currently developing a multi-national double degree Master's program in the field of Management & Informatics. This is to be offered from the winter semester 2023/2024 in cooperation with several Asian partners. The idea is that the students will spend one year at the West Saxon University Zwickau and then another year at any partner university. Within this study program, the use of AI is planned on the one hand in the area of teaching in modules such as *Programming in Python, Machine Learning* or *Neural Networks and Deep Learning*. At the same time, a semantic network, which will ensure the exchange of data between the individual campus management systems of the partners, is planned.

In addition, a course-accompanying AI concept is to support students in their individual choice of study location and their individually favored study modules. In a first step, this will be done on the basis of a rule-based decision support system. The idea is that students provide the system with required parameters, such as language skills, previous degrees, desired specialization (specified with computer science, business informatics, management, and intercultural communication) or desired future career prospects, and as a result receive an individual module list with tailored elective modules. In a later step, a self-learning system can be created on the basis of the parameters received, the lists created, and the associated feedback from the students regarding the accuracy of fit of the proposed modules, which is intended to take over the task of the decision support in the medium term.

Currently, a prototype of the decision system exists, which, on the basis of the individual parameters provided by the students, desired place of study, desired specialization, and preferred form of examination—as well as the module descriptions and contents provided by the partner universities—outputs a sorted list of modules suitable for the student optionally over the entire course of study or alternatively filtered per semester. The individual parameters are weighted against each other. If, for example, modules A and B match the desired specialization, but only module B is completed with the preferred form of examination, both modules are in principle included in the list of results, but module B receives a better ranking.

In the next step, this prototype (which currently only functions locally) is to be published together with a website to be developed for the study program. In addition, the prototype will be extended, so that achievements already made are taken into account, which can lead to a recognition of individual modules if necessary and thus reduce the study effort.

6.2   Case Study II: Individual Learning Paths and Individual Knowledge Management

For many years, we have been working with the learning management system Moodle. Moodle, as well as all other learning management systems, allow students to follow individual learning paths.

Let's first clarify the question, what is an individual learning path. Learning paths are structured paths that—orientated on conditions—control and support individual learning.

There are many possibilities to enable self-directed learning, such as defining learning paths, opening branches for special interests, supporting heterogeneous groups with different levels of difficulty, monitoring learning progress.

Back to our LMS Moodle. Many teaching concepts say that modules build on other modules. However, the reality sometimes looks different: Course content from previous semesters is no longer present to the students, in master's programs students from different courses and universities come together, and their individual preferences for different topics. This makes a very heterogeneous community. Targeting students is almost impossible with larger groups, and there is not enough time for general repeating during the course.

But with some tools of Moodle, learning paths can be enabled according to individual needs, so that they can fill in their knowledge gaps and practice necessary techniques. Individuality is important so that no one gets bored or overwhelmed. Small online tests can be used to identify the subject areas of the course and teach the prerequisite knowledge. The self-assessment option gives participants the chance to identify their own knowledge gaps.

Based on the results achieved—e.g., the percentage of correct answers or the answers to individual questions—a targeted learning recommendation can be made. In any case, this can be implemented by AI in the future (learning videos, practice exercises, etc.). Furthermore, follow-up tests can only be unlocked after a certain learning level has been reached.

In addition, competition can be increased in a playful manner and motivation and ambition can be increased by defining and awarding badges.

Furthermore, competency frameworks can be defined and learning plans can be designed. This means that if a participant has worked through certain learning plans and successfully passed a test, he or she can be said to have acquired a competence.

Overall, it should be noted that the initial effort required to create digital learning content, the tests, and other elements is relatively high. But if these structures are set up wisely, they can be used later on.

Individual learning is therefore based on identifying knowledge gaps and closing them. This significantly improves the students' personal knowledge management.

References

1. Westerkamp, D., Bettenhausen, K.D., et.al.: VDI-Statusreport künstliche intelligenz (Artificial Intelligence—VDI Status Report). pp. 1–19. Düsseldorf (2018)
2. UNESCO BEIJING CONSENSUS on artificial intelligence and education. In: Outcome document of the International Conference on Artificial Intelligence and Education. Bejing (2019)
3. United Nations Transforming our World: The 2030 agenda for sustainable development. A/RES/70/1. pp. 21–22. New York (2015)
4. UNESCO Quingdao Declaration. In: International Conference on ICT and post-2015 Education. Seize digital opportunities, lead education transformation. ED/PLS/ICT/2015/01. pp. 2–4. Paris (2015)
5. UNESCO First ever consensus on Artificial Intelligence and Education (2019), https://en.unesco.org/news/first-ever-consensus-artificial-intelligence-and-education-published-unesco, last accessed 9 March 2022
6. UNESCO Artificial intelligence and the futures of learning. project on ai and the futures of learning (2021), https://en.unesco.org/themes/ict-education/ai-futures-learning, last accessed 9 March 2022
7. European commission ANNEXES to the communication from the commission to the european parliament, the european council, the council, the european economic and social committee and the committee of the regions. fostering a european approach to artificial intelligence. Brussels (2021)
8. Kop, M.: EU Artificial Intelligence Act: The European approach to AI. Stanford—Vienna transatlantic technology law forum, transatlantic antitrust and IPR developments, Stanford University, Issue No. 2/2021
9. European Parliament And The Council Of The European Union, Regulation (EU) 2021/694 oF the European parliament and Of the council of 29 April 2021. 2015/2240. Off. J. Eur. Union L 166/1 (2021)
10. European commission on artificial intelligence—a european approach to excellence and trust. White Paper. COM (2020) 65 final. pp. 7–8. Brussels (2020)
11. European university association universities without walls. A vision for 2030, pp. 4–13. Brussels (2021)
12. Papaspyridis, A., AI in Higher Education: opportunities and considerations. Microsoft Stories Asia. Higher Education, Asia Pacific Japan (2020), https://news.microsoft.com/apac/2020/03/26/ai-in-higher-education-opportunities-and-considerations/, last accessed 9 March 2022
13. Hochschulforum Digitalisierung, the digital Turn—Pathways for higher education in the digital age, Arbeitspapier Nr. 30, Berlin (2017)
14. Bundesministerium für bildung und forschung (German Federal Ministry of Education and Research), richtlinie zur bund-länder-initiative zur förderung der künstlichen intelligenz in der hochschulbildung, Berlin (2021)
15. Mah, D.-K., Hense, J.: Zukunftsfähige formate für digitale Lernangebote—innovative didaktische Ansätze am Beispiel einer Lernplattform für Künstliche Intelligenz. IN: Hochschulforum Digitalisierung. Digitalisierung in Studium und Lehre gemeinsam gestalten. Innovative Formate, Strategien und Netzwerke. pp. 617–631. Springer Wiesbaden (2021)

16. Schumann, C.-A., Töpfer, A., Weber, J.: eSystem integrated approach for regional development of human resources. In: Allan, J. et al., Making Knowledge Work. Proceedings of the PASCAL International Conference. pp. 353–361. Stirling (2005)
17. Both, P., Gaulocher, S., et.al.: KI-basierte Prozessführung. VDI Statusreport. pp. 3–21. Düsseldorf (2021)

Increased Student and Faculty Engagement Researched via an International Opportunity: A Case Study Overview

Barbara Schwartz-Bechet[(⊠)] and Colleen Duffy

Misericordia University, Dallas, PA, USA
bbechet@misericordia.edu

Abstract. As institutions of higher education have entered the 21st Century with several unexpected global crises, including the global pandemic and racial and social injustice events, the reliance upon standard interactions of old-fashioned face-to-face office hours and in-class engagement had to be sidelined. With change comes opportunity and an opportunity to engage in the building of an inter-university project inclusive of assignments, faculty collaboration and changing college environments was provided through an international experience to learn about global counterparts and cultural awareness.

Keywords: Student-faculty Engagement · Internationalization · Online Collaboration

1 Context

1.1 Faculty-Student Engagement in Teacher Education Programs

Specifically, in teacher education programs, there is a reliance upon active engagement in the use of modeled teaching behaviors, working in groups in instructional role play, the application of knowledge and skills under mentor guidance in field/clinical placements, and in process-oriented discourse. Engagement, as it affects learning outcomes, is necessary to understand and quantify variability and correlation in its impact on the students. Teacher education programs often rely upon engagement, limiting use of lectures and increased use of hands-on activities and student to student activities so as to play a focused role in the learning process, leading to greater engagement [27]. The American Educational Research Association (AERA), reported similar findings on pedagogical practices such as 1) micro-teaching and computer simulations based on a behavioral approach, 2) reliance on case studies, 3) reliance on video and hyper-media, 4) use of portfolios, and 5) involving students in practitioner research to increase engagement and improve learning outcomes [7].

Kuh, O'Donnell and Schneider [17] identified eight key elements at the intersection of high impact practices and engagement that could account for improved learning outcomes. These include: setting performance expectations at appropriately high levels; students investing effort over an extended period of time; interactions with faculty and peers about substantive matters; experiences with diversity, wherein students are exposed to and must contend with people and circumstances that differ from those with which students are familiar; providing frequent, timely, and constructive feedback; opportunities to discover the relevance of learning through real-world applications; public demonstration of competence; and structured and frequent opportunities to reflect and integrate learning. By design, Sharing Workplace Dilemmas, the course at the center of this international, inter-collegiate partnership, incorporates many of these essential elements, providing multiple avenues for increased faculty to student engagement.The engagement occurs in both face-to-face interactions- between student and student and student and faculty in the on-campus and clinical setting space and between student and student and student to faculty in the online learning space- making this partnership a blended learning experience. Blended learning, a combination of online and face-to-face learning, is increasingly of interest for use in both undergraduate and undergraduate education, especially as the pandemic begins to change how we teach and learn. In a study conducted by Morton, Saleh, Smith, et. al. [22] students indicated that the structure of building upwards from simple to complex ideas was logical and made competencies more connected, making for a clearer overview of the topic when engaged in blended learning situations whereby flipped classroom learning was most beneficial.

1.2   International Collaboration

Internationalization is "an intentional process undertaken by higher education institutions in order to enhance the quality of education and research for all students and staff, and to make a meaningful contribution to society" [20, p.12]. International partnerships can be formal agreements among and between institutions from different countries, and/or can be fluid in the developmental processes that encourage a variety of opportunities and processes to be tried in order to determine the best path to mutual success and cultural understandings. There is more of an affinity among global institutions of higher education to work together to develop cultural awareness and learning as all students will lead globalized lives, no matter the path that they will choose to pursue as a career. According to Harward [15], in classrooms where international educators work together and students interact with each other, there is an increase in open dialogue and an increased ability to listen and process differences and learn to better articulate subjectivities and objectivities in a safe and welcoming space. The classroom, although virtual in this case, allows for an immersive experience that may provide individual self-discovery. Learners will turn their own ordinary lives into lessons of global learning. The allowance of opportunity in shared safe spaces provides the impetus for the use of an effective pedagogy to be aligned between and among international higher education programs where students can attain and demonstrate successes, meeting their own goals and achieving state and national standards of practice

1.3   Current Institutions and Educational Programing

An opportunity to work with a university in the Netherlands was provided to a university in Northeastern Pennsylvania which occurred about 8 months prior to the COVID-19 global pandemic. With this ability to engage in online interconnected assignments, faculty discourse across institutions and programs, and stronger faculty-student discourse, a greater and more highly developed communicative interaction began, and an idea to investigate how the two possible variables that increased engagement—the international work, as well as the pandemic—was available. Three clusters of outcomes, identified by Cook-Sather, Bovill, and Felten [8] were investigated and discussed to include engagement, awareness, and enhancement. The research design sought to investigate if the outcomes support the framework of Cook-Sather, Bovill and Felton that identify how partnerships (partnerships will be defined relevant to this case) tend to make both students and faculty more thoughtful, engaged, and collegial as they go about their work and life on campus and through campus related events both virtual and face to face. Research has suggested that certain types of student-faculty interactions assist in greater content development as well as greater development of societal and cultural awareness and action. The case study documents how teacher education students from international partner universities were able to engage in joint assignments with faculty working with students and faculty across and between partner institutions.

Misericordia University is a private Catholic university located in northeast Pennsylvania. Founded by the Sisters of Mercy, it provides caring, motivated students with a challenging education. The university provides personal attention, support, and frequent opportunities for authentic application of knowledge and skills, resulting in student success across a variety of undergraduate and graduate programs. Service to others is woven throughout the mission and core values of mercy, justice, service, and hospitality. The University has been guided by these principles for over 95 years.

The Teacher Education Department (TED) in the College of Health Sciences and Education at Misericordia University offers programs in twelve certification areas and within a variety of graduate programs. It seeks to develop effective teachers who are masters of the content they teach and who teach in pedagogically-sound ways that inspire students to learn. Within a mastery model of learning, and using educational theory and methods courses with the liberal arts core as a base, the Teacher Education Department provides a variety of learning opportunities to facilitate student mastery of a thorough knowledge of human growth and development with an appreciation of diversity so that, as teachers, our graduates will understand, respect, and respond to the unique strengths, needs, and desires presented by individual students and their families.

Through coursework and field-based activities conducted in collaboration with partnered schools and agencies throughout Northeastern Pennsylvania, teacher candidates study, observe, and apply strategies to structure learning experiences and learning environments that are responsive to students' needs. Through teaching, supervision, and personal example, TED faculty establish the expectation that students will conduct themselves at all times in accordance with the highest standards of ethical practice and professionalism. The faculty strives to engage with students and develop in our graduates a commitment to on-going personal and professional growth.

NHL- Stenden University of Applied Sciences, a Dutch university, is firmly rooted in the northern part of the Netherlands while at the same time maintaining a strong international focus. NHL- Stenden works collaboratively to develop ways to successfully integrate education, research, and the latest developments in the professional world. The students, lecturers, and researchers work together in small teams on real-life assignments and share their innovative ideas in practice. Their courses incorporate all the latest trends and developments in the relevant sectors and industries, both at a national and an international level. The NHL has four different Faculties, one of which (Institute of Education and Communication) offers graduate (BA) and postgraduate (MA) teaching programs for Languages (Dutch, English, French, German, Frisian), Science (Biology, Physics, Maths, Chemistry), and Social Studies (Geography, History, Economics, Health Care). Besides content knowledge and professional training, there is a strong focus on educational technology and IT-didactics.

Staff and students of The Teaching Programs for Languages (English) at NHL have developed MySchoolsNetwork, the online platform site which hosts the course involved in this partnership, Sharing Workplace Dilemmas, to realize a variety of didactic and strategic goals: 1) to offer a virtual but authentic learning environment to students to practice their feedback, coaching and assessment skills; 2) to offer an attractive and authentic social platform to schools and their pupils to get to know other cultures and to practice their language skills; 3) to strengthen the relationship between the NHL and their network of professional development schools; 4) to offer a virtual "intervision" environment for students and expert teachers to engage in peer consultation activities (Intervision is defined as a community of learners of practice and will be defined further in the chapter); and 5) to foster cross-border learning activities and to promote the development of international learning communities. These ambitions are in line with recommendations of the European Council that education should aim at lifelong learning and the development of eight key competencies, of which global citizenship and mastery of foreign languages are examples.

1.4   Commonalities in Pedagogy

While differences do exist between the two preparation programs initially involved in the partnership, both are required to ensure their pre-service teachers meet competencies established by governmental regulatory bodies for their specialization areas and both rely on a clinical education model to provide effective teacher training.

Central in this training is a series of developmental field experiences that are designed and delivered for candidates to make explicit connections with content areas, cognitive development, motivation, and learning styles. Teacher candidates observe, practice, and demonstrate competencies under the supervision of education program faculty and the mentorship of certified teachers. As teacher candidates progress from observation to teaching small groups of students under the mentorship of a certified educator at the pre-student teaching level to the culminating student teaching experience; they rely on frequent engagement with faculty supervisors and mentor teachers to support them as they learn and demonstrate the complex competencies and responsibilities required by teachers.

Both institutions' teacher education programs rely upon seasoned faculty members who use active learning in their face to face classes to close the theory-practice gap. This results in future teachers who are able to meet the needs of diverse learners and it ensures familiarity with the pedagogy of active engagement. This face-to-face practice carries over to the virtual learning space where active engagement, student to student and student to instructor, is used in instruction and to create and strengthen communities of practice.

Relationship building in order to promote meaningful faculty to student interaction is a common tenet in both programs. Traditionally, faculty extend interactions beyond the classroom, meeting with students to promote the personal and professional development of teacher candidates. After the pandemic, these interactions primarily occurred via virtual office meetings between a student and an instructor. Held regularly to ensure engagement, the meetings provided an opportunity to re-emphasize the correct use of competencies in coursework as well as in classroom teaching in the field.

Because of its congruence with established learning theories and high impact practices, both programs rely on engagement, in various forms and settings, to prepare effective teachers. This includes faculty to student and student to student engagement in coursework and professional discourse and faculty to student and mentoring teacher to student engagement in field experiences. The onset of the pandemic threatened to greatly limit the opportunities for this level of frequent and substantive engagement. Programs were required to shift to online coursework and teacher candidates had limited access to faculty mentors and PK-12 classrooms for field experiences. Reduced in-person faculty to student and student to student interactions also threatened to reduce the amount and quality of collegial discourse, limiting the likelihood of teacher candidates to engage in meaningful reflection about their learning, their future practice, and their capacity to become teaching professionals.

2 Theoretical Underpinnings: The Three Clusters of Outcomes: Engagement, Awareness, and Enhancement

Cook-Sather, Bovill, and Felten [8] propose that partnerships produce similar outcomes for both students and faculty. They have clustered these outcomes from student-faculty partnerships into three themes:

Engagement. Partnerships enhance motivation and learning for students and faculty. They deepen students' learning, increase their confidence, and focus them on the process of learning rather than simply the product. For faculty, partnerships can produce new ways of thinking about, and understanding, teaching, enhanced enthusiasm in the classroom, and a reconceptualization of teaching and learning as a collaborative process.

Awareness. Partnerships also seem to lead students and faculty to develop meta-conitive awareness and an evolved sense of identity. Students become more reflective about their own roles in learning and teaching, and think in new ways about their own capacities as students. Faculty partners see both their teaching practice and their identity as teachers in new ways.

Enhancement. This set of outcomes emerges from the previous two, producing enhanced teaching and classroom experiences. As a result of partnership, students become more active as learners and take more responsibility for their own learning. Faculty have increased empathy for their students, including a better understanding of their experiences and needs and how to respond to them appropriately.

They assert that partnerships tend to make both students and faculty more thoughtful, engaged, and collegial as they go about their work and life on campus. In a later section, we will examine the outcomes of the current case study through the themes presented by Cook-Sather, Bovill, and Felten [8] with consideration given to outcomes enhanced or changed as a result of the instructional changes in response to the COVID-19 pandemic.

3   Application of the Theoretical Underpinnings

This new inter-collegiate partnership looked through a different lens, instituting a technology-based, transformational process to enhance learning. Rather than being teacher directed, it focused on the development of relationships and faculty moderated professional discourse. It began prior to the pandemic which allowed for the students to have previously been engaged in actual experiences within school buildings. Their discourse of issues and ideas stemmed from human interaction that was in person. So, when the transformation of virtual schooling metamorphosed around the world, the students had experience working in both a face to face and virtual classroom.

Students were enrolled in an asynchronous, online course, Sharing Workplace Dilemmas, offered on the MySchoolsNetwork site, during the spring and fall of 2020. Preservice teachers from Misericordia University learned alongside their international counterparts from NHL Stenden: University of Applied Science in the Netherlands. The partnership evolved to include additional partners including the Middle East Technical University, Turkey and an international cohort of Teacher Education students participating in the Erasmus Programme. Fall 2020 saw the addition of a cohort from Tampere University of Applied Science, Finland to the course.

3.1   Methods and Analysis

Quantitative methodologies, primarily pre- and post-survey results, were analyzed. While quantitative measures are often essential in educational research, they are not necessarily well suited as a sole measure for inquiry into partnership practices [16]. For this reason, we also interviewed faculty and students and analyzed their feedback using a constant comparison approach as outlined by Creswell [9] or elements of grounded theory [13]. Following analysis and open coding, key and emerging themes were identified.

The feedback received from the teacher candidates indicated that 93.3% felt they benefited from sharing and discussing classroom dilemmas with their international peers. The same percentage indicated they were actively involved in the experience by asking questions and giving advice to peers, and that the advice received was useful. When asked if sharing professional dilemmas caused them to reflect upon the essence of being a teacher, 83.3% agreed. Perhaps the most significant gain from pre- to post-survey items was the question that required students to indicate whether or not classroom challenges differed from country to country. The baseline survey indicated that 89% of respondents agreed that classroom challenges differed from country to country as opposed to only 30% of students agreeing with that statement on the post-survey, evidence of the development of a broader view of common themes in education. These findings indicate increased engagement in teacher preparation and a greater awareness of the teaching profession, which collectively translate to enhanced teacher training as a result of the international partnership.

Faculty from participating partner institutions identified similar benefits. Specifically, the observed development of students' digital literacy and global awareness. They reported that students engaged in professional discourse through active participation and, as a result, were able to identify common themes in education and teacher preparation. Faculty from the American university also reported on the value of the learning experience as related to student participation in diverse and cross cultural experiences.

Despite having established institutional and departmental goals to increase diversity, Misericordia University remains a predominantly white institution. Participation in Sharing Workplace Dilemmas gave the American students opportunities to participate in cross cultural experiences and increased opportunities to engage authentically with English Learners that they otherwise would not have been able to have. This enhanced learning experience counteracts the current trend of American teacher preparation programs failing to expose teacher candidates to diverse perspectives and experiences [19] and makes it more likely for those participating in the partnership to gain the knowledge and skills they need to effectively serve the diverse populations of students they will have in their future classrooms.

3.2   Phase 1. Sharing Workplace Dilemmas Course Landing Page on MySchoolsNetwork

The course required students to participate in an international community of practice. Teacher candidates' development as educational professionals occurred through the sharing of experiences and learning through peer consultation or "intervision", which is professional discourse centered on shared experiences and dispositions. Each student was required to upload eight professional dilemmas they encountered during their teacher training and respond to their classmates' dilemmas by asking probing questions and making context-appropriate, evidence-based recommendations to their peers.

The overarching goal of the course is both knowledge co-creation and empowerment of pre-service teachers through self-evaluation and critical thinking. Its theoretical underpinnings are grounded in high impact practices and the didactic design of the course is based on Gilly Salmon's [25] framework for online teaching and learning which comprises five essential components: Motivation, Socialization, Information Exchange, Construction of New Knowledge, and Metacognition.

Discussion forums were centered on relevant professional issues such as cultural/global learning, collaboration, online learning, assessment, classroom management, and mentoring. Each dilemma prompts students to reflect on their experience and post details about their dilemma to the forum without giving away the resolution or outcome. Then, their peers will read the dilemma and ask open, clarifying questions to gain a better understanding of the dilemma. Students then answer their peer's questions before entering the advice phase, in which students provide possible solutions to the dilemma by referencing background literature, teacher competencies, developmental and pedagogical theories, and differing perspectives. Importantly, the aim is not to identify a specific right answer, but to give and gain insight. Finally, students react by sharing the conclusion to the dilemma, identifying any advice that helped direct the decision and whether or not there was any resolution; indicating if they were satisfied with the result or if they would do anything differently.

3.3   Phase 2. Example of a Student Post to the Classroom Dilemma Discussion Forum

Throughout the process, the international faculty partners engage with students from the partnership programs by encouraging and moderating open, critical, and constructive dialogue. They identify areas for further reflection and elaboration and ask probing questions to steer student thinking and discourse as needed. All students, regardless of university affiliation, shared the same learning outcome, to participate actively in an international community of practice by sharing their own workplace dilemmas, experiences and dispositions and giving analysis and advice to peers. Respectful and reflective of the differences in programs and their related competencies, assessment of student performance varied by university (Fig. 1).

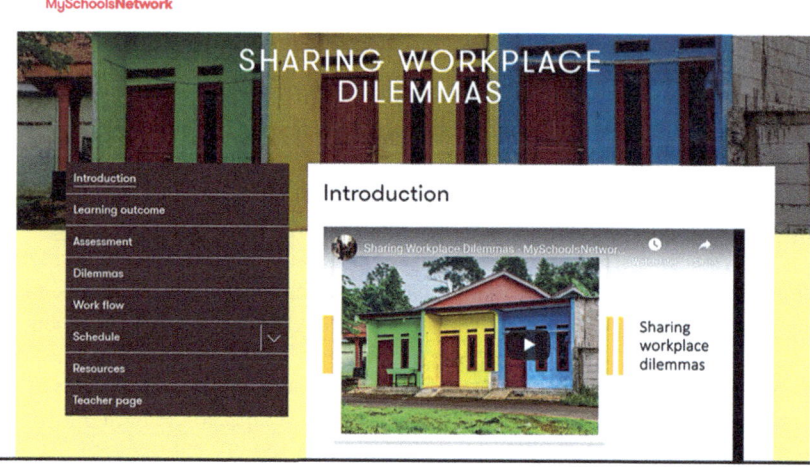

Fig. 1. Sharing workplace dilemmas course landing page on myschoolsnetwork

3.4   Post-pandemic Implications and Opportunities

Eight months into this international partnership a global pandemic, COVID-19, occurred forcing change in all aspects of life, including how we teach and learn. Face to face classrooms closed their doors, greatly limiting opportunities for student to student and student to faculty interactions. This restriction also posed great challenges to teacher preparation programs, like those involved in the partnership, that rely on access to PK-12 classrooms in order to afford teacher candidates the opportunity to apply their knowledge and skills in an authentic way and engage in professional discourse with mentor teachers in order to develop their identities as professional teachers.

Face to face coursework immediately shifted to remote synchronous or asynchronous format. This required faculty to reconsider course design and to utilize technologies to actively engage students. Being involved in the design and implementation of Sharing Workplace Dilemmas prior to the pandemic gave faculty involved in the partnership the experience needed to confidently adapt coursework and teaching strategies while preserving the efficacy of the teacher preparation programs. The pre-service teachers also had an advantage in this quick shift in modality as a result of their participation in the partnership. Specifically, familiarity with course navigation in various Learning Management Systems and comfort participating in discussion forums as both respondent and moderator lessened learner anxiety as all students found themselves limited to only virtual learning environments.

These skills also carried over to the pre-service teachers' practice as they completed field and student teaching placements in virtual and cyber classrooms, an unprecedented practice for many. In this way, the flexibility and responsiveness to student need enacted by the partnering faculty was serving as a model for best practice for the pre-service teachers who themselves were demonstrating great levels of flexibility and responsiveness in order to meet the needs of their PK-12 pupils who were also learning in unfamiliar ways and settings.

The partnership also provided an opportunity for faculty to respond to student experience by adapting course content. Since the course revolved around students sharing common experiences and challenges, a necessary and timely revision was to add forums centered on student experiences in online teaching and learning and E-mentoring. This afforded the opportunity for reflection, exploration, and sense-making about teacher candidates' current struggles while developing a deeper appreciation of the shared professional experiences they were facing.

Varying degrees of mentoring teachers' preparedness for online and virtual teaching also limited the time some of them had available to actively engage with the pre-service teachers assigned to their classrooms. Participation in Sharing Workplace Dilemmas gave the pre-service teachers increased access to inter-collegiate faculty who were able to provide E-mentorship while moderating open, critical, and constructive dialogue and engaging in professional discourse. This helped enhance teacher preparation by removing a limiting factor that would have been imposed by the pandemic, had the international partnership not existed.

Lived experiences with pupils in school settings of students both in the United States and internationally allowed and encouraged shared observations discussed on asynchronous boards. Several of the students provided discussion on similar dilemmas. Acknowledgement that there are shared experiences across cultures assists in the development of greater partnerships and understandings within a global perspective.

It is probative to consider that the pre-pandemic engagement in the classroom with students and teachers may have affected the outcomes of greater engagement in the online space once into the pandemic and may have generated a difference due to the timing of the partnership.

4   Results

Framing the outcomes of the current study within the three themes presented by Cook-Sather et al. [8] elucidates the value of this international, inter-collegiate student-faculty partnership with respect to increased engagement, increased awareness, and enhanced teacher training. Where appropriate, evidence includes reference to outcomes related to the impact of instructional shifts related to the COVID-19 pandemic (Table 1).

5   Conclusions

As the world learns to navigate best practices in engaging students in higher education, in online, hybrid, and face-to-face capacities, the basic research has demonstrated that whichever type of structure, successful engagement is essential to improve learning outcomes. How engagement is defined may lead to better student outcomes. Having the ability in this current case study to have worked with the students face to face prior to going virtual and then back to face to face may have been a variable associated with greater engagement and success of the students in terms of their development as future teaching professionals who are thoughtful, engaged, and collegial.

Implications of this research lends itself to investigating if students who are more active in working with and speaking with faculty will have a greater self-efficacy or if the interaction and engagement helps them to better develop their own self efficacy. It is the story of the chicken or the egg, which came first or which influenced the other. Additionally, internationalization and the fostering of global understandings may have been a variable that led to greater communication between the faculty-student dyad in the United States when clarification and cultural differences need to be explored and discussed. Engagement was identified between the faculty and students at Misericordia University as increasing over the time of the two institutions' project on shared dilemma discourse; it was perceived to shorten the distance between the two countries and attributed to recognizing the similarities of the profession across the ocean even though many students felt that the dilemmas were different in scope and function. The similarities of pedagogy that existed across the dilemmas included cognitive, behavioral, and emotional engagement of concepts, skills, and attitudes that affect engagement and could be identified as successfully achieved across most students who participated. If students are more likely to engage with their faculty when posed with challenges that are new to them, could this suggest a better way to develop pedagogy and would this type of learning be best instructed in a similar manner to be modeled when the students become teachers themselves? As more research emerges in the international classroom of joint experiences, the ability and agency of students to become joint presenters—or, shall we even say, instructors of content and designers of new pedagogy—may be part of the future of transnational global learning.

Additionally, conditions that allow for the interchange of ideas and learning are an additional important variable and, whether virtual or face to face, the appropriation of successful commitment to and by all members of the community of practice is essential. The class setting is a social domain. Civility and spoken as well as non-spoken rules need to be especially acknowledged in transglobal classes. The engagement is based upon trust and credibility between and among all people in the class setting. As a variable, the environment plays a critical role in the success of engagement as well as learning and developing the ability to facilitate new ideas and expressions of confidence from the students. Moving forward, the types of variables that allow for best practices under a variety of conditions/environments need to be researched and identified within the realm of instructional pedagogy that is provided as well the constructs to be mastered. Perhaps a framework that will guide international use of joint pedagogy to increase engagement in a sociocultural context needs to be developed through a meta-analysis of similar projects. Intentional effort can provide greater shared meaning and understandings, which are essential not only for intercultural understandings, but for increases in a collective construction of knowledge and learning and effective participation between the student-faculty dyad, leading to relevant engagement and growth.

Table 1. International partnership outcomes for students and faculty (adapted from [8]).

	Potential Outcomes as identified by Cook-Sather et al. [8]	Evidence of the outcome in this international, inter-collegiate partnership
Engagement outcomes		
For students	Enhanced confidence, motivation, enthusiasm	American students reported enthusiasm and excitement to collaborate with international students, a new feature in their program. Students who participated in the course reported being more confident to learn and teach in virtual and cyber classrooms after the COVID-19 instructional shift
	Enhanced engagement in the learning process	Post-survey indicates 93.3% of students were actively involved in the experience by asking questions and giving advice to peers; 93.3% indicated the advice was useful to current learning/future practice
	Enhanced responsibility for/ownership in learning	Students posted their original dilemmas but also assumed the role of forum moderator as they asked questions about their peer's dilemmas and offered evidence-based recommendations, opening the forum to faculty and students for professional discourse. By design, students had choice about which forums to post to and were able to choose which posts to question /expand upon. This created a sense of ownership and shared responsibility in the learning process
	Deeper understanding of, and contributions to, the academic community	Practical questions evolved from everyday interactions. The students engaged with one another and shared their field and classroom experiences and assisted one another in analysis and problem solving. After the COVID-19 shift, students gained valuable perspectives into the unprecedented challenges and existing inequities in education the pandemic brought to light. Post survey indicates 93.3% felt they benefited from sharing and discussing classroom dilemmas with their international peers
For faculty	Transformed thinking about/practices in teaching	By design, the partnership invited a team teaching approach into the teaching/learning space. International faculty were responding to students otherwise not in their charge and engaging in professional discourse with each other. The diverse perspectives enhanced learning for the students but it also expanded the notions of what and how faculty teach. This expanded view of teaching practice became valuable post-COVID as teacher candidates completed fields and student teaching in virtual and cyber classrooms using new and unfamiliar practices

(continued)

Table 1. (*continued*)

	Potential Outcomes as identified by Cook-Sather et al. [8]	Evidence of the outcome in this international, inter-collegiate partnership
	Changed understanding of learning and teaching by experiencing different viewpoints	Students and faculty brought diverse perspectives, often sharing differing interpretations of a dilemma being presented. This prompted rich discussion and substantive feedback that helped identify "teachable" moments and points for further discussion. The largest gain in pre- to post-survey items was the question that required students to indicate whether or not classroom challenges differed from country to country. The baseline survey indicated 89% of respondents agreed that classroom challenges differed from country to country as opposed to only 30% of students agreeing with that statement on the post-survey, evidence of the development of a broader view of common themes in education
	Reconceptualization of teaching and learning as a collaborative process	Teaching and learning goes both ways as discourse occurred between faculty and students. Students helped shape the learning experience (i.e., the addition of forums related to online teaching and learning following COVID-19)

Awareness outcomes

For students	Developing meta-cognitive awareness	Through the student-student and student-faculty dialog the partners articulated their experiences and understandings about their learning and experiences in PK-12 classrooms. This combined with the differences in shared perspectives prompted students to be more aware of their thinking and more intentional in their future actions
	Developing a stronger sense of identity	The dialogic interaction centered on common dilemmas between students and faculty changed the understandings and capacities of both students and faculty partners—making both sets of partners better informed teachers and learners. Post-survey results indicate that 83.3% agreed that sharing professional dilemmas caused them to reflect upon the essence of being a teacher
For faculty	Developing meta-cognitive awareness	Through the student-student and student-faculty dialog the partners articulated their experiences and understandings about their own teaching and learning. This combined with the differences in shared perspectives prompted critical analysis of their teaching practice and opportunities for reaffirmation or revision, as needed

(*continued*)

Table 1. (*continued*)

	Potential Outcomes as identified by Cook-Sather et al. [8]	Evidence of the outcome in this international, inter-collegiate partnership
	Developing a stronger sense of identity	The dialogic interaction centered on common dilemmas between students and faculty changed the understandings and capacities of both students and faculty partners— making both sets of partners better informed teachers and learners
Enhancement outcomes		
For students	Become more active as learners	In the forums, students had to seek out dilemmas to respond to. They were actively researching evidence-based practices in order to be able to make recommendations to their peers that were educationally sound. In doing so they were expanding or reinforcing their own knowledge base, making them more confidently prepared for future practice. Being an active learner in the asynchronous course proved helpful as they sought to engage their PK-12 pupils in cyber or online field/student teaching placements post-COVID
	Gain insight into faculty members' pedagogical intentions	Learning about the challenges the faculty partners had faced allowed the students to gain insight into why some teachers "do what they do." Being responsible for posting and moderating the forums under faculty guidance provided an avenue for increasing student capacity and agency as learners and future teachers
	Take more responsibility for learning	Students shared unique perspectives that, when shared in dialogue with peer and faculty partners' perspectives, raised awareness, increased engagement, improved teaching and learning for all, and created a culture of open communication about, and shared responsibility for, education and learning
For faculty	A deeper understanding of student experiences and needs	Faculty partners were able to use student input to identify and meet immediate learning needs. This resulted in a more reflective and responsive teaching practice. The flexibility and responsiveness modeled in the course became even more valuable to student partners as they pivoted in response to COVID-19 instructional shifts
	A reconceptualization of students as colleagues	Teaching and learning goes both ways as discourse is encouraged between faculty and students

References

1. Abdool, P., Nirula, L., Bonato, S., Rajji, T., Silver, I.: Stimulation in undergraduate psychiatry: exploring the depth of learner engagement. Acad. Psychiatry, **41**, 251–261 (2017). https://doi.org/10.1007/s40596-016-0633-9.pdf

2. Barrett, R., et al.: Social and tactile mixed reality increases student engagement in undergraduate lab activities. J. Chem. Educ. **95**, 1755–1762 (2018)

3. Beasley, S.: Student-faculty interactions and psychosociocultural influences as predictors of engagement among black college students. J. Divers. High. Educ. **14**(2), 240–251 (2021)

4. de Borba, G.S., Alves, I.M., Campagnolo, P.D.B.: How learning spaces can collaborate with student engagement and enhance student-faculty interaction in higher education. Innov. High. Educ. **45**(1), 51–63 (2019). https://doi.org/10.1007/s10755-019-09483-9

5. Brint, S., Cantwell, A., Hanneman, R.: The two cultures of undergraduate academic engagement. Res. High. Educ. **49**, 383–402 (2008). https://doi.org/10.1007/s11162-008-9090-y.pdf

6. Cavinato, A., Hunter, R., Ott, L., Robinson, J.: Promoting student interaction, engagement, and success in an online environment. Anal. Bioanal. Chem. **413**, 1513–1520 (2021). https://doi.org/10.1007/s00216-021-03178-x

7. Cochran-Smith, M., Zeichner, K.M. (eds.): Studying Teacher Education: The Report of the AERA Panel on Research and Teacher Education. Routledge, London (2009)

8. Cook-Sather, A., Bovill, C., Felten, P.: Engaging Students as Partners in Learning and Teaching. Jossey-Bass (2014)

9. Creswell, J.W.: Qualitative Inquiry and Research Design: Choosing among Five Approaches, 2nd edn. Sage, New York (2006)

10. Cuseo, J. (2018). Student-faculty engagement. New Dir. Teach. Learn. **154**, 87–97 (2018). https://doi.org/10.1002/tl.20294

11. DeBolle, S., Reddy, R.: Development of an academic surgical student program for enhancing student-faculty engagement. J. Surg. Educ. **76**(3), 604–606 (2019)

12. Gares, S., Kariuki, J., Rempel, B.: Community matters: student-instructor relationships foster student motivation and engagement in an emergency remote teaching environment. J. Chem. Educ. **97**(9), 3332–3335 (2020). https://doi.org/10.1021/acs.jchemed.0c00635

13. Glaser, B.G., Strauss, A.L.: The Discovery of Grounded Theory: Strategies for Qualitative Research. Aldine De Gruyter, New York (1967)

14. Giddens, J., Fogg, L., Carlson-Sabelli, L.: Learning and engagement with a virtual community by undergraduate nursing students. Nurs. Outlook, **58**(5), 261–267 (2010)

15. Harward, D.: Are higher education's efforts to advance global engagement, and global citizenship, un-American? In: Presentation at the AAC&U Network for Academic Renewal Global Engagement and Social Responsibility Conference. New Orleans, LA (2017). https://aacu.org/sites/default/files/files/global17/CS%205%20Presentation.pdf

16. Huber, M.T., Hutchings, P.: The Advancement of Learning: Building the Teaching Commons. Jossey-Bass, San Francisco (2005)

17. Kuh, G., O'Donnell, K., Schneider, C.G.: HIPs at ten. Change: Mag. High. Learn. **49**(5), 8–16 (2017)

18. Kuznekoff, J.: Online Video Lectures: The Relationship Between Student Viewing Behaviors, Learning, and Engagement, vol. 26, pp. 33–55. Association of University Regional Campuses of Ohio (20202). https://aurco.net/Journals/AURCO_Journal_2020/Harrison_Online_Video_2020.pdf

19. Marchitello, M., Trinidad, J.: Preparing Teachers for Diverse Schools: Lessons from Minority Serving Institutions. Bellwether Education Partners (2019). https://bellwethereducation.org/sites/default/files/Preparing%20Teachers%20for%20Diverse%20Schools_Bellwether.pdf

20. Marinoni, de Wit, H.: Is strategic internationalization a reality? Int. High. Educ. **98**, 12 (2019)
21. Matrin, F., Bolliger, D.: Engagement Matters: Student Perceptions on the Importance of Engagement Strategies in the Online Learning Environment. Online Learn. **22**(1), 205–222 (2018). https://files.eric.ed.gov/fulltext/EJ1179659.pdf
22. Morton, C., Saleh, S., Smith, S., Hemani, A., Ameen, A., Bennie, T., Toro-Troconis, M.: Blended learning: how can we optimize undergraduate student engagement. BMC Med. Educ. 16(195) (2016). https://doi.org/10.1186/s12909-016-0716-z
23. Park, J., Young, K., Salazar, C., Hayes, S.: Student-faculty interaction and discrimination from faculty in STEM: the link with retention. Res. High. Educ. **61**, 330–356 (2020). https://doi.org/10.1007/s11162-019-09564-w
24. Rodgers, T.: Student engagement in the e-learning process and the impact of their grades. Int. J. Cyber Soc. Educ. **1**(2), 143–156 (2008)
25. Salmon, G.: E-moderating: The Key to Teaching and Learning Online, 3rd edn. Routledge, New York, NY, USA (2011)
26. Seifan, M., Lal, N., Berenjian, A.: Effects of undergraduate research on students' learning and engagement. Int. J. Mech. Eng. Educ. **0**(0), 1–23 (2021). https://web-b-ebscohost-com.mis ericordia.idm.oclc.org/ehost/pdfviewer/pdfviewer?vid=8&sid=93f9ce2e-cf66-41cb-9fb0-e32cadd946e9%40pdc-v-sessmgr01
27. Schwartz-Bechet, B., Bos-Wierda, R., Barendsen, R.: Transatlantic online communities of practice. Int. J. Emerg. Technol. Learn. **7**(3), 50–53 (2012)
28. Snijders, I., Wihnia, L., Rikers, R., Loyens, S.: Building bridges in higher education: student-faculty relationships quality, student engagement, and student loyalty. Int. J. Educ. Res. **100** (2020). https://www.sciencedirect.com/science/article/abs/pii/S0883035519319068
29. Staines, Z., Lauchs, M.: Students' engagement with facebook in a university undergraduate policing unit. Australas. J. Educ. Technol. **29**(6), 792–805 (2013)
30. Thacker, M., Laut, J.: A collaborative approach to undergraduate engagement. Portal **18**(2), 283–300 (2018)
31. Trolain, T., Jach, E., Hanson, J., Pascarella, E.: Influencing academic motivation: the effects of student-faculty interaction. J. College Stud. Develop. **57**(7), 810–826 (2016). https://muse.jhu.edu/article/636338/pdf
32. Trowler, V., Trowler, P.: Student Engagement Evidence Summary. Higher Education Academy, New York (2010)
33. Webber, K., Laird, T., BrckaLorenz, A.: Student and faculty members engagement in undergraduate research. Res. High. Educ. **54**, 227–249 (2013). https://doi.org/10.1007/s11162-012-9280-5.pdf
34. Wenger, E., McDermott, R., Snyder, W.M.: Cultivating Communities of Practice, p. 304. Harvard Business School Press, Boston (2002)

Integrating Educational Components into the Metaverse

Tetyana Sergeyeva[1]([✉]) [iD], Sergiy Bronin[2] [iD], Natalya Turlakova[1] [iD], and Stanislav Iamnytskyi[1] [iD]

[1] National Technical University "Kharkiv Polytechnic Institute", Kharkiv, Ukraine
tv_sergeyeva@icloud.com
[2] Taras Shevchenko National University of Kyiv, Kyiv, Ukraine

Abstract. The prospect of integrating educational component into the "metaverse" has been analyzed, and the origin story of the idea is traced from the ancient religious, philosophical and psychological concepts. It is proposed: (1) to create a "metaverse" as a developmental ecosystem, taking into account the laws of personality development; (2) to involve not only figurative but also abstract thinking to expand the way of describing the "metaverse" through abstract schemes in addition to images; and (3) to introduce moral dimension into the assessment of possible "metaverse" impact on users' mind and body, as well as on the real world. It is assumed that modeling a digital learning environment based on the internal laws of human development will optimize the entire process. The built-in educational component will provide a virtual learning environment with developmental capacity vastly superior to simulated reality. An innovative Eco-Humanistic Developmental Model is proposed for integration. It focuses on human existential, social and career metasenses of self-development, synergistic interaction, and competencies development. In contrast to the situational approach, the process is modeled in its existential integrity. It is based on the idea of human development as a living self-developing system in interaction with the surrounding world as a developmental environment. In the process of balancing actual needs, personal abilities and conditions of environment, the synergetic development of all three components takes place. As a result, not only cognitive and personality development is optimized, but there is also an awareness of interdependence and mutual development of self and environment. The concept, structure, strategy, mechanisms and tools of innovative e-learning courses based on Eco-Humanistic principles are presented. To confirm the efficiency of the proposed approach, the results of research within the framework of a long-term developmental experiment are presented. The prospect of multidisciplinary research is revealed.

Keywords: Metaverse · Eco-humanistic technology of self-development · Human-environment synergetic interaction · Developmental eco-system · Existential model

1 State of the Art

Metaverse is becoming a mainstream concept of our time. It is tempting to use its potential for education, but at the beginning, it is important to understand what it is. So far, the metaverse refers to concepts that do not have clear definition, like energy or love. Everyone has their own idea of it. Mark Zuckerberg is convinced that the best way to understand the metaverse is to visit it. But it is difficult, because it does not yet fully exist.

One of the best ways to understand a phenomenon is to trace the history of its origin. We can trace the origin story of the idea from the Jewish interpretation of "Hochma," "Bina" and "Daat" as divine Sefirot (emanations) of wisdom, understanding, and aware-ness, and we can trace the ancient philosophical concept of "World of Ideas" through the psychological concept of "Cognitive Scheme" as an internal reflection of the real world to the concept of "Metaworld," using this as a simplified model of the world and as an interface of consciousness interaction with everything possible within the World Wide Web. This scientific way of analyzing sophisticated concepts will rather give an idea of the incomprehensible diversity and exceptional significance of the ideal world for humans. It is unlikely to advance us in understanding the metaverse, at least, at the current level of its development.

The other way that can lead to understanding is the analysis of the etymology of the word. The term "metaverse" was introduced in Neal Stephenson's seminal 1992 cyber-punk novel, Snow Crash [15]. In the book, it is a shared "imaginary place" accessible over the worldwide network where the developers can build buildings, parks and things that do not exist in Reality. They can visit areas where the laws of three-dimensional space are ignored, or zones where people go out to hunt and kill each other. Facebook's interpretation of the metaverse is quite different from Stephenson's description. Nev-ertheless, meta-rebranding over 30 years has inspired many followers. In the effort to create a metaverse, many online games have emerged that capture the most important concept without even using the term. Thus, the etymology of the word "metaverse" also did not give a clear understanding.

In the context of education, there are various entry points into understanding of the metaverse [10]. Most of the interpretations hold promise for the educational component of the metaverse [8]. These include the idea that metaverse is not a replacement for reality, but an attempt to make the virtual experience more realistic [1]. Zuckerberg's vision opens up new educational perspectives not achievable before. He presents the metaverse as more immersive and embodied web [7] where students can do everything they can imagine—meet friends and family, work, study, play, create—as well as brand new categories that do not fit into our understanding of computers or phones today [11]. They can enter a virtual environment through an avatar, sail through virtual worlds and not just watch it or their teachers and fellow students in cells on the screen. Students acquire the ability to teleport from one experience to another, as well as to act in the world of endless, interconnected virtual communities. According to Victoria Petrock, the metaverse is the next generation of communication where everything interacts in an invisible twin universe so that you live a virtual life as well as a physical one [4]. So, the metaverse can produce the developmental learning environment that is much richer and more diverse. It opens up completely new opportunities for cognitive and personal development within the framework of purposeful education [2]. Kyle Orland traced the metaverse history, both as a concept and as different online spaces [12]. He identified the elements that, together, can define the metaverse. Each of them has different degree of significance for education, but they are all applicable to one degree or another. If we summarize the following technical advances can invest into the education progress:

1. **Shared social space with avatars to represent users.** On a modern website or social media, students may be represented by a username or a small picture [6]. In the metaverse, they can be represented by a customized avatar that can move, speak, and/or perform animation actions. This is the building block of the metaverse as the more "embodied" Internet. Modern developments have really created amazing Avatars to interact with virtual objects. *The effect of presence can radically change the efficiency of students' development. In addition to the expected increase in their action awareness, students' personal involvement in a problem with the ability to track one's actions can significantly enhance the development of reflection as well as acquiring the most valuable metacognitive knowledge about oneself.*

2. **A continuously renewing "world"** for avatars **to inhabit and interact with.** There are a whole variety of worlds: (1) virtual world, which replicates the limitations of space and the imperfections of the real world; (2) a specially created world that is shared with the user for a particular game or time-sensitive event; (3) an ideal metaverse, where each user shares one virtual world, where property is retained between sessions. *From the point of view of education, the possibility of renewing the world is the chance to set various conditions for the developmental environment with which the student interacts. This is key condition for developing the ability to transfer knowledge into real life practice and as a result to automate actions.*

3. **Ability to create your own virtual property.** Allowing users to create content of their own materiality is an advantage both for users—who will be able to shape their virtual world at will as well as for the creators of the metaverse—who will not have to spend a lot of time and effort creating each individual virtual object from a sketch. Relatively simple building blocks can produce network effects and allow to create a whole variety of objects within the world. *In the context of education this co-creation of developmental environment with which the student interacts develops not only students' creativity but also helps to individualize the environment as well as to create new tools of training efficiency evaluation.*

Despite the temptation of the metaverse educational prospects it is clear that the creation of the metaverse is a complex task that requires solving the huge technical and economical problems. Much has been written on this subject in numerous articles [5, 9, 16]. But for us it is important to find the efficient way of educational component integration into the Metaverse.

1.1   Presentation of the Main Research Material

We put forward the idea of integrating the educational component into the meta-verse. One of the possible options is the transformation of the virtual reality into the developmental ecosystem. This transformation involves modelling existential conditions of self-development [14]. For clear presentation of the sophisticated concept to the multidisciplinary audience, we offer a format of generalizing conceptual scheme (see Fig. 1).

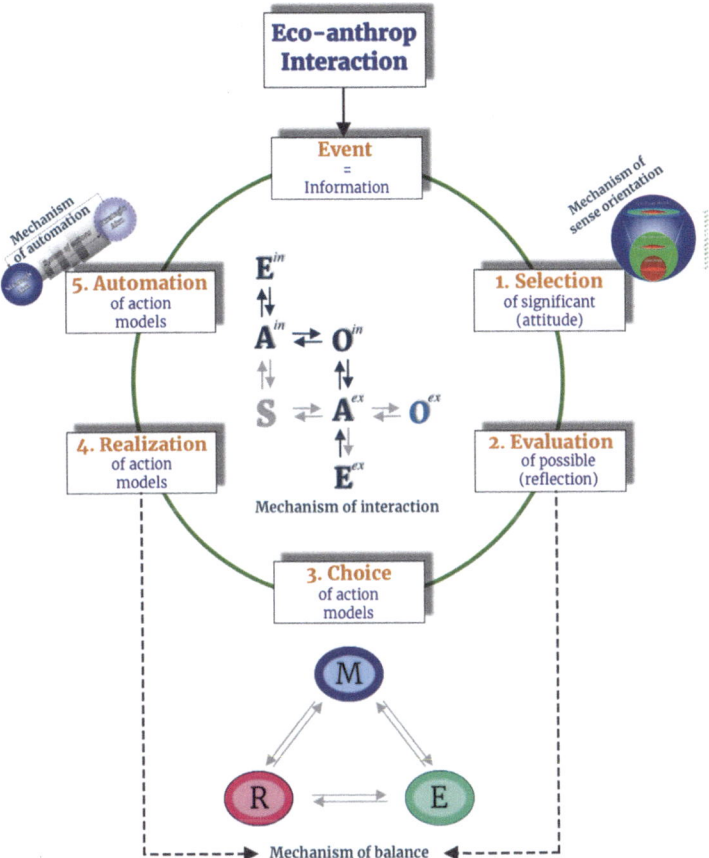

Fig. 1. Existential circle of self-development.

The proposed model reflects the cyclic process of self-development under condition of everydayness. Existential circle starts with an event that carries information. It is processed resalting action or inaction. The circle includes 5 stages, which are served by 4 internal mechanisms. It is noteworthy that the fundamental mechanism of internal-external interaction serves all stages. There are the following stages: Stage 1: 'Selection' of the event personal sense ('important' = action/'not important' = inaction'), internal mechanism of sense-cognitive orientation; Stage 2: 'Evaluation' of the possibility to realize personal sense under existing conditions ('possible' = action/'impossible' = inaction), mechanism of balance; Stage 3: 'Choice' of the action modes from the available resources for personal sense realization ('have a choice" = action/'no choice' = inaction), mechanism of balance; Stage 4: 'Realization' of the action modes with adoption to changing conditions; ('adapted' = action/not adapted = 'inaction'), mechanism of balance; Stage 5: 'Automation' of action modes by repeating them in case of success for sense realisation ('success' = action/'failure' = inaction), mechanism of automation.

Stages 1 and 2 deal with attitude and reflection, which are crucial. As a result, the automized actions integrate into internal resources as skills, abilities and qualities. Resources-enriched individual faces new events in the course of eco-anthropic interaction launching the existential circle of self-development again and again.

Really efficient integration of the educational component into the metaverse requires understanding internal processes that occur when a user interacts with the developmental ecosystem. The model of the existential process of self-development reflects internal mechanisms on which the whole process is based: 'Mechanism of interaction' reflecting consciousness-behaviour synergetic interaction that determines the developmental process flow and efficiency (Mechanism no. 1); 'Mechanism of balance' reflecting balancing personal senses – individual resources – external conditions that energizes the process (Mechanism no. 2); 'Mechanism of sense orientation' reflecting the synergy of sense-cognitive orientation as a determining factor of the process efficiency (Mechanism no. 3); 'Mechanism of automation' reflecting the system of actions organizations that optimizes the developmental process (Mechanism no. 4) (see Fig. 2).

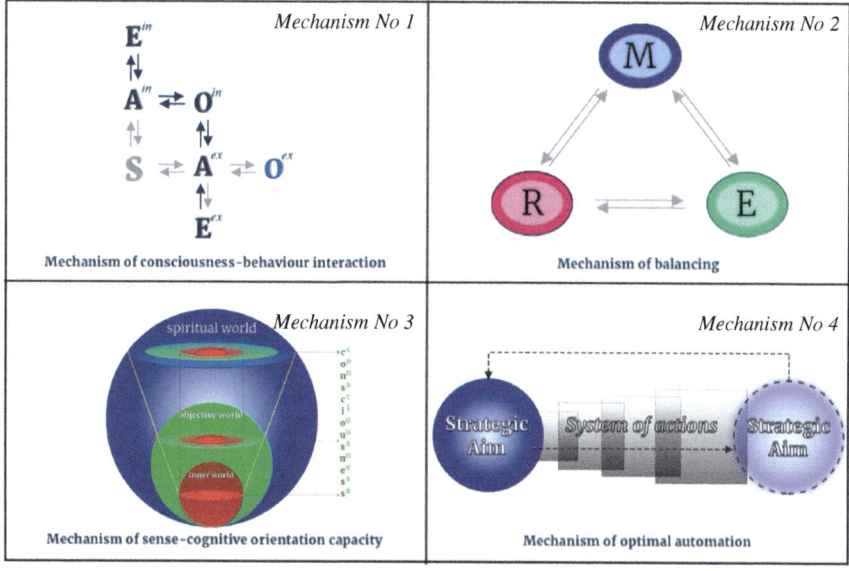

Fig. 2. Internal mechanisms.

Mechanism no. 1 is basic. User's consciousness-behaviour interaction (internal and external activity interaction) is a system-organizing factor. It is present at all stages of the existential circle of self-development. At that the 'selection of important', the 'evaluation of possible' and the 'choice of the activity modes' can be performed exclusively internally while 'realization' and 'automation' of the activity are performed both internally and externally.

The mechanism is working on the basis of positive feedback principle (see Fig. 2).

The user as the subject (S) of the activity performs both internal (A^{in}) and external (A^{ex}) activity appropriately aimed at internal (O^{in}) and external (O^{ex}) objects in the context of personal senses. It is important to note that A^{ex} is the essential content of O^{in} (the regulatory function of consciousness). In order to provide efficient activity, the user should consider both the nature of the O^{ex} as well as internal (E^{in}) and external (E^{ex}) conditions of the activity. While considering O^{ex}, E^{in} and E^{ex} characteristics (their perception, reflection and interpretation) the E^{in} (internal resources and personal senses) of the S are developed due to the principle of the positive feedback.

The mechanism reflects the dynamic system of mutually conditioned factors that influence the development of the user (its cognitive, regulative and communicative efficiency) as S of the activity interacting with the metaverse ecosystem. At that, the clockwise arrows point to S actions (the sphere of influence) and the counterclockwise arrows point to the feedback (the sphere of dependence). It is obvious that synergetic interaction of all factors is conditioned by both A^{in} directly and A^{ex} indirectly influencing the development of the S internal resources.

Realizing this mechanism on metacognitive level provides understanding: (1) once own role of the Subject of self-development; (2) the determining role of the activity itself, its aims as well as its internal and external conditions in the process of self-development; (3) *the necessity of **moral measurement** of one's own activity (the means of aim achievement). Mode of activity influences not only the O^{ex} but also the development of the S.*

Mechanism no. 2 is working at three stages of the existential circle of self-development: 'the evaluation of possible,' the 'choice of activity mode' and the 'realization of the modes of activity.' It reflects the driving force of human self-development energized by the ambivalent (differently directed) striving for development (as a condition of surviving in the changing dynamic environment) or desire for conservation (if the development is considered dangerous for surviving). This ambivalence is performed as balancing personal senses (M)—internal resources (R)—external environment (E) (see Fig. 2).

The nature of the balancing is such that the process itself constantly initiates a new unbalanced state as a result of producing new conditions caused by the effect of development of one of the components of the triad. As a result of the components' interaction, this process is starting again and again, gaining non-stop character. Balancing is perpetuum mobile that launches developmental process.

This mechanism works on the principle of positive feedback as well. The developmental process can be launched by any component of the triad in any direction. Going clockwise: the external environment (E) may require the development of resources (R), which in turn will stimulate the development of personal senses (M). Going counterclockwise: the environment (E) may initially stimulate the development of personal senses (M), which leads to a more efficient development of resources (R). Within this model, the case of the agents of changes development is interesting: if personality resources exceed the needs of the environment the removal of the braking factor may lead to the need of the environment development. *The integration of the mechanism into the metaverse developmental ecosystem will allow not only purposefully developing senses and resources of the user by the environment development, but also involving the users in the development of an existing or creation of a new reality.*

Realizing the system interdependence of the 'personal senses—internal resources—external conditions' on metacognitive level provides understanding: (1) the role of personal senses in the efficiency of self-development; (2) the role of strategies used for senses-resources-conditions balancing in the efficiency of self-development; and (3) the specific role of the subject of the environment development as a condition of self-development.

Mechanism no. 3 is working at the first stage of the 'existential circle. It reflects the factors that determine the efficiency of self-development (see Fig. 2).

The efficiency of development is achieved due to the capacity of the user's sense-cognitive orientation, which determines the diapason and quality of information processing, as well as the level of reflection. We consider that 'attitude—reflection—action' in their synergy determines this efficiency. The matter is that usually we divide verbally the factors that are indivisible existentially. Let's assume that attitude and reflection are represented in the consciousness as 'sense-cognitive scheme.' Its diapason determines the intensity and efficiency of actions (both at external and internal levels).

In fact, the wider is the sense orientation (to the 'surviving,' to the 'quality of life' or to the 'sense of life'), the bigger quantity of objects and links between them need cognitive development; in that way, the actions become more intensive for satisfying personal senses. The wider is the cognitive orientation in the reality ('phenomenological,' 'objective' or 'inner'), the bigger is the filter selecting important events from the flow of information that in its turn determines the character of performing the following activity aimed at satisfaction of the personal senses. The wider the diapason of the individual sense-cognitive scheme, the more intensive and efficient the actions and, as a result, the higher the potential of development.

Understanding the mechanism of sense-cognitive orientation in the context of the metaverse developmental ecosystem will allow to determine initially the user's cognitive potential based her/his sense orientation level and offer the developmental environment proportionate to her/his senses. However, we believe that it is more promising to introduce into the ecosystem initially a sense orientation, that will ensure the development of the user's senses as a condition of cognitive development efficiency.

Realizing the mechanism on the metacognitive level provides understanding the cognitive orientation to the essence as a condition of the efficient self-development—the sense orientation to self-development as the bases for existential success and a stabilizing factor as well as a condition of surviving within intensive transformations.

Mechanism no. 4 determines the optimization of the transforming activity means into operations, as well as their integration into the personality internal resources as skills, abilities and qualities. It corresponds to the fifth stage of the 'existential circle of self-development' (see Fig. 2). The integration of the activity modes that are used for personal sense realization into the internal resources is carried out in the process of automation of this action as a result of their frequent successful usage. Optimization of the developmental process is possible when the strategic/final aim is put forward, which prevents chaotic redundant actions, since they line up in an optimal system where the result of the previous action is becoming the means of realizing the next action. Automation is optimized thanks to excluding extra and not adequate actions.

The mechanism of optimal automation in the context of creating the metaverse developmental ecosystem can increase the developing effect of virtual reality. Bringing forward the ultimate goal ensures the awareness and consistency of the user's actions. Conscious systemic actions determine their efficient involuntary memorization. Sense orientation leads to understanding the sense of self-development as existential. Focusing on the existential sense of development in interaction with the environment, can lead to synergetic development of self and environment (see the mechanisms of balancing and sense-cognitive orientation). This conception allows developing user who is able to consciously manage both the process of once own development in synergy with development of a virtual environment. The initial metacognitive orientation can significantly enhance the sense orientation and create a synergistic effect of development. We managed to do this and experimentally verify within the framework of the natural learning process. It is assumed that the technological capacity of the metaverse developmental ecosystem will significantly increase the developmental effect of the virtual reality.

'Mechanism of automation' realization on metacognitive level helps understanding: (1) necessity of frequent use of the successful means of activities leading to their automation, transformation into personality skills, abilities and qualities integrating into individual internal resources; (2) the role of the 'strategic orientation' diapason that determines the awareness and frequency of the successful action usage for the efficiency of self-development; and (3) the role of the 'sense orientation' that stood ahead providing optimal conditions for automatizing the successful means of activity.

2   Experimental Verification

Our research is based on a prolonged developmental experiment within university training [14]. To carry out the experiment, the students were divided into 5 groups: 1 control (K) and 4 experimental (E1, E2, E3, E4). The experimental factor was the developmental environment modeled within the educational strategies ensuring the development of professional, social and existential competences on the bases of sense-cognitive orientation of different capacity. The dependent variable was the efficiency of personality self-development measured at three levels: sense level (development of Spr - professional, Ssc - social, Sek - existential personality senses), cognitive level (SKS - complexity of individual sense-cognitive scheme; ERZ - efficiency of problem solving) and activity level (efficiency, rate of automation and integration of operation modes into personality resources in the form of cognitive, communicative and regulative abilities for Csm-self-development, Cpr-decision-making, Cvz-synergetic interaction, as well as personal qualities of Mre-realism, Mpa-proactivity, Mav-autonomy, Mot-responsibility, Mkr-creativity, Mgb-flexibility and Mem-empathy) (see Fig. 3).

Systematization and generalization of empirical data were carried out with the help of correlation and factor analysis, which allowed to study the nature of the relationships between the studied indicators, as well as to identify and describe their regular combinations.

In the control group K, students acquired knowledge of the subject without strategic orientation. In the experimental groups, there were different strategic orientation/environment causing different senses/resources development: (1) senses - E1 - professional, E2 – social, E3 – existential, E4 - all three senses in synergy; (2) resources - ability E1 - for decision-making, E2 - for synergetic interaction, E3 - for self-development, E4 - all three abilities and corresponding qualities in synergy; and (3) environment - a system of tasks aimed at developing operation ways of decision-making (E1), synergetic interaction (E2); self-development (E3), all three in synergy (E4) activities. In group E4, the experimental conditions were as close as possible to natural. Professional and social senses were formed from the beginning in the context of the existential sense of self-realization.

The results of the empirical study are presented in the format of a generalizing matrix (see Fig. 3) and profiles (see Fig. 4) of self-development metacharacteristics.

GROUPS	INDICATORS	COGNITIVE DEVELOPMENT					PERSONALITY DEVELOPMENT												
		SKS complexity	problem solving				sense			abilities			qualities						
			velocity	accuracy	completeness	efficiency	professional	social	existential	decision-making	synergetic interaction	Self-development	objectivity	proactivity	autonomy	responsibility	creativity	flexibility	empathy
		SKS	Dsp	Dac	Dcm	ERZ	Spr	Ssc	Sek	Cpr	Cvz	Csm	Mre	Mpa	Mav	Mot	Mkr	Mgb	Mtm
K	Ks	0.37	0.72	0.36	0.73	0.35	0.29	0.43	0.41	0.38	0.37	0.36	0.37	0.34	0.34	0.35	0.35	0.34	0.34
	Kr	0.49	0.74	0.38	0.75	0.38	0.40	0.39	0.38	0.38	0.38	0.40	0.49	0.37	0.36	0.37	0.36	0.35	0.34
	ΔK	0.12	0.02	0.02	0.02	0.03	0.11	0.04	0.03	0.00	0.01	0.04	0.12	0.03	0.02	0.02	0.01	0.01	0.00
E1	E1s	0.37	0.81	0.36	0.83	0.35	0.32	0.43	0.42	0.40	0.38	0.38	0.42	0.32	0.37	0.37	0.37	0.33	0.36
	E1r	0.62	0.83	0.54	0.85	0.53	0.53	0.47	0.46	0.45	0.45	0.53	0.62	0.52	0.58	0.36	0.36	0.35	0.34
	ΔE1	0.25	0.02	0.18	0.02	0.18	0.21	0.04	0.04	0.05	0.07	0.15	0.20	0.20	0.21	0.01	0.01	0.02	0.02
E2	E2s	0.37	0.84	0.33	0.86	0.34	0.32	0.43	0.42	0.39	0.39	0.38	0.39	0.37	0.37	0.35	0.38	0.38	0.33
	E2r	0.64	0.88	0.50	0.90	0.51	0.52	0.55	0.53	0.55	0.56	0.52	0.64	0.62	0.35	0.37	0.50	0.50	0.51
	ΔE2	0.27	0.04	0.17	0.04	0.17	0.20	0.12	0.11	0.12	0.17	0.14	0.25	0.25	0.02	0.02	0.12	0.12	0.18
E3	E3s	0.37	0.91	0.35	0.93	0.35	0.31	0.43	0.42	0.39	0.38	0.37	0.35	0.39	0.38	0.36	0.34	0.37	0.37
	E3r	0.62	0.93	0.50	0.95	0.53	0.62	0.54	0.53	0.53	0.51	0.62	0.62	0.65	0.63	0.54	0.47	0.39	0.39
	ΔE3	0.25	0.02	0.15	0.02	0.18	0.31	0.11	0.11	0.14	0.13	0.25	0.27	0.26	0.25	0.18	0.13	0.02	0.02
E4	E4s	0.37	0.94	0.35	0.96	0.35	0.32	0.43	0.42	0.39	0.39	0.38	0.41	0.36	0.33	0.38	0.35	0.34	0.35
	E4r	0.79	0.98	0.65	0.99	0.69	0.69	0.67	0.65	0.64	0.65	0.69	0.79	0.70	0.67	0.58	0.62	0.57	0.55
	ΔE4	0.42	0.04	0.30	0.03	0.34	0.37	0.24	0.23	0.25	0.26	0.31	0.38	0.34	0.34	0.20	0.27	0.23	0.20

Fig. 3. Indicators of metacharacteristics of cognitive and personality development of the testee of the first (s) and second (r) sections in the control (K) and experimental (E1, E2, E3, E4) groups.

In group E4, there was a transition from external to internal regulation, first under controlled learning conditions, and then in reality. The transition was traced in the dynamics of development of motives from the basic motive of restoring balance, through the motive of development and cognitive motive to the triad of interconnected motives of self-cognition—self-development—self-realization. Initially, there was a strategy of external stimulation by creating a situation of imbalance between senses-resources-conditions. Then, in search of balance, each subsequent motive arose as a condition for satisfying the previous one. Qualitative shift occurred when the level of resource development allowed to manage the process of once own development. The motive of cognition was transformed into the motive of self-cognition, the motive of development into the motive of self-development, and the motive of restoring balance into the motive of self-realization. Self-cognition from the means of self-management of the developmental process was transformed into a means of strategic search for existential senses and self-identification. The motive of self-development as existential meta-motive has been realized as determining not only survival and quality of life, but also its sense.

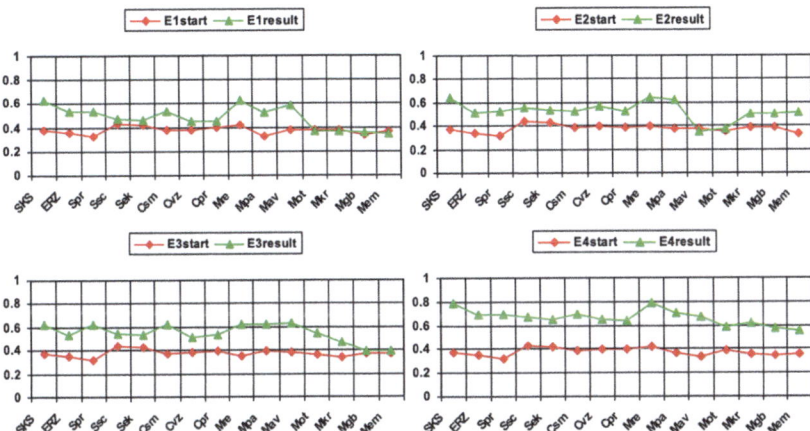

Fig. 4. Profiles of average values of metacharacteristics of experimental sample of the studied E1, E2, E3, E4 metacharacteristics of self-development of the 1st (start) and 2nd (result) sections.

Differences in the capacity of strategic orientation provided a different degree of systematicity, awareness and intensity of action, which, in turn, led to different developmental effects. In each subsequent group, the understanding of the developmental significance of one's own activity and its regularities was deepened in comparison with the previous one. Each group consistently answered three main questions: 1st (E1) focused on knowledge of the subject (WHAT?), 2nd (E2) - knowledge of the environment (HOW?); 3rd (E3) - knowledge of senses (WHY?); 4th (E4) integrated all knowledge, ensuring their synergy. Indicators of the E4 group can serve as criteria for the formation of personality, which is in the process of self-development. It is characterized by: cognitive orientation to knowledge of the essence and existential orientation to self-development (0.65); high complexity of individual sense-cognitive structure (0.79); high efficiency of problem solving (0.69); developed meta-abilities to decision-making (0.64), to synergistic interaction (0.65), to self-development (0.69) and developed meta-qualities of realism (0.79), proactivity (0.70), autonomy (0.67), responsibility (0.67), responsibility (0.59), creativity (0.62), flexibility (0.57) and empathy (0.55).

The synergistic effect of self-development appeared at the sense, cognitive and activity levels as a chain reaction. The synergy of sense and cognitive orientations appears directly in the synergy of modes of activity aimed at the realization of personal senses, and determines their various properties (individualization, intensity, awareness, system, ease of transfer and automation). These properties complement each other, which contributes to the appearance of their stability and synergy. It is proved that the synergy of personality qualities arises due to their meta-nature, which is reflected in the complexity, integrity, ambiguity and specificity. In the case of development simultaneously in several sense contexts, the complex nature is manifested integrally. The broader and more diverse the context of formation, the higher the synergistic effect.

It follows from experimental data that there exist prospects due to the integration of the educational component into the ecosystem of the metaverse. They can become subjects of a number of scientific studies. Here we note only one of the most tempting and probable prospects, which may appear due to the incredible coverage of the ecosystem of the metaverse as a developmental environment: the emergence of hitherto unattainable synergistic effect of development both of the user's personality at the sense, cognitive and activity level, and the virtual developmental environment itself.

3   Discussion of Results

It is assumed that in order to transform "metaverse" to really efficient developmental ecosystem it is necessary to take into account the laws of personality development. Modeling a digital learning environment based on the internal laws of human development will optimize the entire process. The built-in educational component will provide a virtual learning environment with developmental capacity vastly superior to simulated reality [3]. An innovative Eco-Humanistic Developmental Model [14] is proposed for integration, as it was experimentally tested and confirmed its high developmental efficiency. It is based on the idea of human development as a natural self-developing system in interaction with the surrounding world as a developmental environment. In contrast to the situational approach, the process is modeled in its existential integrity. It is focused on the development of human existential, social and career metasenses of self-development, synergistic interaction and competencies development.

The system-forming factor is the final goal (objective point), in the light of which the result of the previous task becomes a way to solve the next one. The regular transformation of an action as a goal into an action, as a means to achieve the next goal, ensures involuntary and efficient mastering of knowledge, skills and abilities. The development occurs through the conscious actions as ways to solve problems of a given class, which leads to the automation of knowledge and their integration into the user's internal resources in the form of skills and abilities. Thus, the content relevant to the target knowledge, entered through tasks, acquires educational characteristics.

There are innovative strategies and self-management toolkits, simulators and techniques for integration into the developmental metaverse ecosystem to provide targeted cognitive and personality development. The technique of superimposing competence profiles [13] allows users managing their own developmental process. There are competences profiles which: (1) the user wants to have; (2) the user really possesses; (3) are demanded by society at the present stage of its development; (4) are required in the existing labour market; and (5) can be developed within the appropriate metaverse ecosystem. Overlapping profiles make it possible to identify and even measure gaps that can be interpreted as zones of individual development, as optimal developmental program, as zone for adjusting the level of claims to real opportunities or conditions, etc. This approach allows the users consciously and purposefully choose a developmental environment within metaverse that corresponds to their personal senses.

Another entry point into the problem of choosing a developmental metaverse environment can be the complexity of the cognitive scheme that the environment offers. For this, the "Technique of superposition of sense-cognitive structures" [14] is used. Initially, based on testing and questionnaires, the complexity of the user's individual sense-cognitive structure is determined in the context of the content of the metaverse developmental environment. Then the individual structure is superimposed on the structure, which purposefully is developed within the framework of environment. Identified gaps are measured and interpreted as zone and program of the user's individual development.

The creation of a structured database of the capacities and content of the developmental environments of the metaverse ecosystem together with the proposed techniques for identifying the individual competence profile and the complexity of the individual sense-cognitive structure of the user will allow for a conscious choice of a virtual developmental environment that best suits senses, goals and needs of the user.

The integration of eco-humanistic technology for introducing an educational component into the metaverse ecosystem, on the one hand, will expand the functionality of the ecosystem, and on the other hand, will significantly enrich the educational component itself due to scale of impact and technological capabilities.

Thanks to the integration of the mechanisms of eco-humanistic technology into the metaverse developmental ecosystem, it will be possible: (1) to determine initially user's cognitive potential based on sense orientation level (Mechanism - M no.3); (2) to offer the developmental environment proportionate to user's senses (M no.3); (3) to introduce initially a sense orientation, that will ensure the development of the user's senses as a condition of cognitive development efficiency (M no.3); (4) to develop needed senses and resources by the virtual environment development (M no.2); (5) to increase the developmental effect of virtual reality (M no.4); (6) to involve the user in the development of existing or creation of a new virtual reality (M no.2); (7) to develop user who is able to consciously manage both the process of once own development in synergy with development of a virtual environment (M no.4); (8) to involve not only figurative but also abstract thinking to expand the way of describing the "metaverse" through abstract schemes in addition to images (Technique of superposition of sense-cognitive structure); (9) to create a synergistic effect of development (M no.4); (10) to introduce moral measurement of one's own activity (the means of aim achievement) on the bases of understanding and experimental testing the activity mode influence at the target object and at the development of the one who performs the action (M no.1); (11) to introduce moral dimension into the assessment of possible "metaverse" impact on users' mind and body as well as on the real world (Technique of profiles superimposing).

Each of these perspectives has innovative significance, but one of the most interesting is the possibility of educating agents of virtual reality development. The matter is that an organization of a developmental virtual environment with the integration of an educational component within the framework of the metaverse is possible in the format of targeted training, gaming and project activities. It should be noted that, in principle, any activity develops to one degree or another. There are users who choose environment and events that are relevant to their interests, but there are users who purposefully use any environment and events as an opportunity for self-development. For such users, the educational component is present in any activity within the metaverse. They represent a valuable group of potential agents for the development of the metaverse. If we assume a special status of the metaverse user as an agent of its development, then the educational component should be based on a metacognitive approach that optimizes developmental activities through knowledge about the process, about the environment, about one's role and opportunities in this environment. The development of such agents with special status could open up unforeseen educational prospects for metaverse.

Technological capacity of the metaverse developmental ecosystem may significantly increase the developmental efficiency of the educational eco-humanistic technology of self-development, that is: (1) to increase efficiency of students' development due to the 'effect of presence' provided by the metaverse and resulted in the development of action awareness; (2) to develop students' reflection and to acquire knowledge about oneself on the bases of personal involvement in a situation with the ability to track actions; (3) to develop students' creativity on the bases of co-creation developmental environment with which the student interacts; (4) to achieve powerful synergistic effect of development both of the user's personality at the sense, cognitive and activity level and the virtual developmental environment itself due to the incredible coverage of the metaverse ecosystem as a developmental environment; (5) to individualize the environment due to the possibility to co-create developmental environment with which the student interacts; (6) to develop the ability to transfer knowledge into real life practice resulted in actions automation speed increase due to the possibility of renewing the virtual world as the chance to set various conditions for the environment with which the student interacts; (7) to create new tools of training efficiency evaluation on the bases of the possibility to co-create environment with which the student interacts.

4   Conclusions

The integration of the educational component into the metaverse implies the organization of the ecosystem as a developmental one based on taking into account the patterns of personality development and the discovery of content relevant to target knowledge through a system of tasks. The development occurs in the process of user interaction with the virtual environment. The conditions of interaction determine the tasks/problems that user solves. Their solution determines both the internal (consciousness) and external (behaviour) activities of the user, which are connected synergistically. The solution of the tasks is carried out in the context of the final goal, the achievement of which occurs in the process of transition of intermediate actions from the status of the current goal to the status of the means of achieving a subsequent goal.

It is assumed that the integration of the educational component, built on the proposed eco-humanistic model of self-development, into the metaverse will not only provide the expected developmental effect, but also significantly enhance it due to opportunities hitherto inaccessible in the framework of traditional training and e-learning.

Immediate possibilities of metaverse enhance involvement in the developmental process not only at the cognitive, but also at the sense and emotional levels. Due to virtual possibilities, the set of developmental events expands almost unlimitedly, the scope of which can be limited only by fantasy. Access to development-supporting resources is also virtually unlimited. The creation of a powerful structured database of the capacities and content of the developmental environments of the metaverse ecosystem and the proposed techniques for identifying the individual competence profile and the complexity of the individual sense-cognitive structure of the user will allow for a conscious choice of an environment that best suits user's senses, goals and needs.

The range of prospects for the metaverse as a developmental environment is breathtaking, but whether the prospects turn into reality depends on what the founders' strategic goal is and how this goal is realized. We would like to hope that development as the existential sense of human life will be taken into account.

References

1. Alaman, X., Lasala, M.J., Jara, S.: Living in virtual and real worlds: a didactic experience. Proceedings **31**(1), 83 (2019)
2. Bronin, S., Kuchansky, A., Biloshchytskyi, A., Zinyuk, O., Kyselov, V.: Concept of digital competences in service training systems. In: Auer M.E., Tsiatsos T. (eds) Internet of Things, Infrastructures and Mobile Applications. IMCL 2019. Advances in Intelligent Systems and Computing, vol 1192. Springer, Cham (2021). https://doi.org/10.1007/978-3-030-49932-7_37
3. Bronin, S., Pester, A. (2021) Towards a national digital competence framework for Ukraine. In: SIST 2021 - 2021 IEEE International Conference on Smart Information Systems and Technologies. https://doi.org/10.1109/SIST50301.2021.9465946
4. Explainer: What is the metaverse and how will it work? The chronicle journal the newspaper of the northwest. https://www.chroniclejournal.com/business/national_business/explainer-what-is-the-metaverse-and-how-will-it-work/article_fecef5dd-acec-5da3-b9db-af76c73e0 bad.html. Accessed 28 Oct 2021
5. Former Oculus CTO John Carmack isn't a fan of Zuckerberg's Metaverse approach. https://www.pcmag.com/news/former-oculus-cto-john-carmack-isnt-a-fan-of-zucker bergs-metaverse-approach. Accessed 29 Oct 2021
6. Kurtz, G., Peled, Y.: Digital learning literacies—a validation study. Issues Informing Sci. Inf. Technol. **13**, 145–158 (2016)
7. 'Sandrine' Han, H.-C.: From visual culture in the immersive metaverse to visual cognition in education. IGI Global (2020)
8. Kerris, R.: The metaverse can revolutionize education, employment. CTECH by calcalist (2022). https://www.calcalistech.com/ctech/articles/0,7340,L-3926518,00.html
9. Kraus, S., Kanbach, D.K., Krysta, P.M., Steinhoff, M.M., Tomini, T.N.: Facebook and the creation of the metaverse: radical business model innovation or incremental transformation? Int. J. Entepreneurial Behav. & Res. **28**(9), 52–77 (2022)
10. Kye, B., Han, H., Kim, E., Park, Y., Jo, S.: Educational applications of metaverse: possibilities and limitations. J. Educ. Health Proffessions **18**, 1–13 (2021)
11. Mystakidis Stylianos: Metaverse. Encyclopedia **2**(1), 486–497, 12
12. Orland, K.: So what is "the metaverse," exactly. https://arstechnica.com/gaming/2021/11/eve ryone-pitching-the-metaverse-has-a-different-idea-of-what-it-is/. Accessed 11 July 2021
13. Sergeyeva, T., Iamnytskyi, S.: Multidisciplinary Approach to the Construction of Students' Competence Profiles, pp 14–18. Baltija Publishing, Leipzig (2019)
14. Sergeyeva, T.: Eco-humanistic Technology of Self-development. Blok, Kharkov (2010)
15. Stephenson, N.: Snow Crash. Publisher Bantam Books (US) (1992)
16. Xi, N., Chen, J., Gama, F., Riar, M., Hamari, J.: The challenges of entering the metaverse: an experiment on the effect of extended reality on workload. Inf. Syst. Front.: J. Res. Innov. 1–22 (2022)

BreakThrough Communication in a Hybrid World: Amplifying Interactive, Experiential Learning

Carol C. Shuherk, Susan R. Glaser[✉], and Peter A. Glaser[✉]

Glaser and Associates, Eugene, OR 97405, USA
theglasers@theglasers.com

Abstract. This paper describes the research foundation and instructional design of *BreakThrough Communication*, an evidence-based hybrid learning enterprise that builds organizational capacity by boosting communication skill in individuals and teams. BTC's tiered talent development program combines asynchronous, self-paced video instruction in foundational knowledge; live virtual workshops customized to engage learners in practicing new skills on organizational challenges they face; and change-sustaining follow-through to embed new skills through assessment, video reinforcers, and real-time practice sessions. A certificate program that prepares internal trainers to assume communication coach roles completes the process. The BTC learning system is grounded in the traditional pedagogy of the communication field, aligns with adult learning theory, and reflects recent neuroscience research on how the human brain most efficiently learns. Three distinct areas of communication skill comprise the program's subject material: interpersonal conflict, teamwork, and persuasive speaking. The communication models within each area have been validated in both experimental and field studies, shown to change behavior and affect outcomes.

Keywords: Competency-based instruction · Experiential learning · Hybrid learning · Virtual learning · Remote learning · Conflict resolution · Teamwork · Persuasion · Leadership

1 Performance Centered Learning

From the beginning, teaching in the field of Communication has centered attention on performance, on developing an individual's effectiveness in expressing themselves through speech. The oratory experts of ancient Greece and Rome, charged with preparing the offspring of the elite for their future roles as political leaders, devised methods that today would be called experiential learning. Every student engaged in multiple practice rounds of presentation and debate and received ongoing feedback from teachers and peers on their performance. The process is much like skills training for athletes or musical ensembles. Mastery is cultivated through individual study, collaborative group learning, public practice, and continuous feedback.

© The Author(s), under exclusive license to Springer Nature Switzerland AG 2023
D. Guralnick et al. (Eds.): TLIC 2022, LNNS 581, pp. 426–432, 2023.
https://doi.org/10.1007/978-3-031-21569-8_40

As the scope of the field expanded to include interpersonal and group communication and the role they play in relationships and organizations, the question of everyday communication mastery and whether it is teachable became pertinent. Research sought to identify the specific communication behaviors that constitute competency in one-to-one and group interaction and what instructional procedures increase the chances a newly acquired skill will transfer from the classroom to the communicative life of the learner.

1.1 Shaping Communication Competency

Compatible with communication's performance focus, behavioral competency approaches to skill development target specific linguistic and nonverbal behaviors associated with successful outcomes in a particular context, such as interpersonal conflict, team dynamics and persuasive messaging. A *shaping* process moves learners from intellectual understanding of a competency to closer and closer approximations of its use in real life [1]. Shaping follows a sequence: direct instruction in a target skill's component parts; identification of need for the skill in a learner's life; written scripting of personal word choices for each component; observation of models—effective and ineffective; practice with peers; and recurring performance feedback from instructors and peers.

Evaluation research of behavioral competency approaches has shown it to be effective in increasing communication skill as perceived by learners and those who interact with them. For example, in assessing a conversational skills training course for "communication apprehensive" adults—individual's for whom shyness was so debilitating it was adversely affecting their social relationships and work performance—researchers employed multiple measures: learner questionnaires and self-monitoring, lab samples of social interaction, and evaluation by confederates and coder ratings, to assess ability on completion of a program that targeted skills which previous research showed to be key: complimenting, agreeing, asking questions, describing shared experience, and expressing opinion. The findings showed clear support for the efficacy of this approach. The program produced greater improvements in comfort, conversation skill and positive impact on others for those who had completed the full learning sequence than for those in a control group who had received direct instruction only. Improvements were shown to have endured in a follow-up evaluation five months after the program's end [2].

Over the last two decades team-based work has come to dominate organizational life across sectors. With that has come increased demand for a constellation of interpersonal and group process skills. Team members need to be able to effectively express viewpoints, facilitate team process, build consensus, handle conflict, share leadership, and create safe work environments [3]. Research sponsored by a leading technology firm showed that with technical skills sets being equal, communication skills that equalize participation and create psychological safety for members are the two factors distinguishing the firm's most effective teams from all others [4]. Increasing organizational capability today means team members must learn new set of basic skills, for collaborative problem-solving, joint decision making and strong interpersonal relationships.

The pandemic and mass exodus to remote work that it triggered, has created further need for communication skills that support teamwork and talent development. A northeastern software developer now trains it managers to work with distributed teams, with an emphasis on conversations that establish team cohesion and build personal relationships.

A national insurance firm trained all its managers to facilitate career development for their employees, creating templates for conversations about skills and interests. They also developed a fully virtual four-week leadership course and made it available to employees at every level of the firm [5].

Field research in team development has shown that using the shaping process to teach concrete communication skills to intact teams can yield outcomes with long-term effects. Longitudinal study of a teambuilding intervention with a conflict-ridden public service organization, whose members were taught basic interpersonal skills and group problem solving methods, found lasting results. Three years after the initial intervention, comprehensive interviews with team members revealed four consistent impressions: increased ability to raise issues and manage conflict; increased mutual praise, support, and cooperation; clarified roles and responsibilities; and enduring commitment to teamwork and innovation [6].

A communication-based intervention in a large governmental organization treated five variables of organizational culture that take shape through communication practices and interaction rituals (supervision, teamwork, morale, involvement, information flow) in an initiative aimed at shifting the culture from hierarchical and authoritarian to participative and involved. One year after the intervention, organizational members at all levels and external public officials who interact with them indicated movement toward participative culture on five dimensions. Three variables—morale, involvement, and information flow—showed statistically significant change [7].

These studies, demonstrating the effectiveness of a behavioral shaping approach to developing communication competency, have also shown its applicability for learners of varying ages, in settings ranging from academic institutions to working organizations. The next logical step, given the nature of today's learning environments, is to apply this method of shaping skills incrementally to virtual and hybrid learning, developing solutions that are widely accessible, from anywhere. The learning system described further in this paper takes that step.

1.2 Teaching Adult Learners

Communication skills training in organizational environments requires a curriculum where the specific needs of adult learners are taken into account.

American educator Malcolm Knowles is widely known for making the term "andragogy," (Greek for "man-learning" in contrast to pedagogy, which means "child-learning") synonymous with adult education. He called it the "art and science of adult learning" [8]. He provides a platform of assumptions and principles for instructional design compatible with a competency-based approach.

Knowles' assumptions about adult learners: They are *self-directed*, and want learning experiences that give them autonomy and the freedom to draw their own conclusions. They bring a reservoir of *experience* to learning that will be diverse in terms of backgrounds, skill sets, and perspectives, and must be tapped to stoke engagement. Their *readiness to learn* is directly associated with the relevance of the material to their personal development in their current role. Adult learners want *immediate application*, subject matter that is going to solve problems they regularly face. And finally, they are *motivated* by engagement, active involvement with the learning material, and clarity on the reason behind every learning activity.

Four principles follow from these assumptions that underly curriculum design: involve adult learners in shaping the content of their instruction, employ activities that make direct experience the basis of learning, tie instruction directly to jobs or personal life, and make problem-solving the focus of course content [9].

Knowles published his theory of adult education almost forty years ago but its continuing relevance to instructional principles and practices is being demonstrated by cognitive science today. A 2016 review of neuroscience research on human learning found that the core assumptions of andragogy have a connection to the brain's neural networks related to memory and cognition [10].

2 Launching Experiential, Interactive Virtual Learning

As long-time practitioners of in-person, experiential learning, the key challenge in this initiative was to achieve the same engagement and quality outcomes in the virtual environment as we do live onsite. One central question ignited this work: How can we apply all we know about the art and science of teaching and learning to a hybrid environment? How can we drive experiential learning through every element of the virtual context, both synchronously and asynchronously?

Lessons being drawn from two years of remote work suggest some organizing principles that correspond to an outcomes-driven approach to virtual adult learning. Firms that managed remote work successfully took the time to compile important information in clear, easily digestible, flexible forms that employees could access from anywhere at any time [11]. This was particularly useful to new employees who were scaling the learning curve in their jobs and knowledge of the company. Successful firms were also very intentional about people dynamics, "scaffolding" the social dimensions of remote work. They established clear norms and etiquette around meetings, held town halls to reduce power distances, and ensured that all meetings were inclusive across time zones. They regularly sponsored virtual events that brought people together in remote coffee rooms, games, and happy hours, for interaction that kept them both physically distanced and yet in touch with each other's lives [12].

Finally, successful firms demonstrated awareness of diversity and equity issues. Among the employees most likely to prefer remote work are women and people of color, who even before the pandemic reported higher levels of underrepresentation and isolation at work. But being remote without conscious support and engagement could increase alienation and decrease opportunities for good assignments and career growth. Companies who managed remote work well put systems in place to ensure equal support and equal access for all [5].

These values of flexible, well-packaged information delivery; norms of structured yet informal interaction; and emphasis on equal engagement, are reflected in the content and structure of all three courses that comprise the BreakThrough Communication learning system. BreakThrough Conflict, Hardwiring Teamwork, and Persuasion & Influence all share a common goal: measurable increase in communication capability.

2.1   Each Course in Brief

BreakThrough Conflict—Navigate challenging conversations.

- CONVERT criticism from defensiveness to insight.
- RAISE issues that solve problems and strengthen relationships.
- CREATE habits of gratitude and recognition.

Hardwiring Teamwork—Build inclusive, productive meetings.

- TACKLE questions meaningful to the team's work.
- CRAFT action agreements from discord.
- CREATE meeting climates free from personal attack and wheel spinning.
- EQUALIZE input from dominating and quiet people.

Persuasion & Influence—Create powerful, convincing presentations.

- DELIVER with 5 dynamism capabilities.
- CONVERT nerves into productive enthusiasm.
- ORGANIZE with intention, clarity and impact

2.2   Each Course Has 3 Progressive Levels of Increasing Skill Mastery

Foundational Knowledge: **Self-paced** <u>Video Courses</u>. These asynchronous courses provide foundational knowledge on a schedule convenient for each learner, ranging in time from 1.5–2.5 h. They include engagement activities and <u>interactive digital workbooks</u> that move learners from knowing to doing, by applying skills as they are introduced.

- **Modular Learning**
  Each course is modularized, to build communication skill incrementally in its content area. Each module includes outlined learning objectives; text and audio overviews of core concepts; video lectures recorded before a live audience; audio/visual concept reviews; summary application exercises and interactive workbooks.
- **Engaged Learning**
  Learners actively engage with concepts and skills in two ways in each course. The application exercises built into course summaries provide timely tests of comprehension via multiple-choice reflection questions whose answers are provided immediately. Interactive workbooks make course material relevant to each learner's local circumstance with short answer questions that stimulate them to apply concepts to their life experience and a scripting tool that guides them through preparation for real-life interaction they are anticipating.
- **Autonomous Learning**
  The learning system is designed for easiest possible use. A click & listen navigation guide walks learners through each element of the program interface, showing them

how to control each aspect of course delivery, including starting, stopping and revisiting the overviews, video lectures, slides, and reflection questions for each module. Each course site is fully searchable and includes full transcripts of its contents.

Skill Practice: Live interactive Virtual Workshops. A serious challenge with asynchronous experiences is that they are solitary. To boost real-time engagement and collaboration, a synchronous virtual workshop is customized around scenarios taken directly from the organization's life, addressing learners' social and professional settings. Every person interacts, practices, and gets feedback in facilitated breakout rooms. Because practice enhances memory, this opportunity to perform new skills increases their "stickiness" and gives practice addressing actual challenges learners face in their daily lives. The social interaction of these breakout sessions also enriches learning by building a sense of community into the fabric of the experience.

Learning is further amplified with dramatic scenarios that are customized to capture the content and tone of challenging conversations from which the learners prepare. This is accomplished by trainers recreating compelling conflict scenarios directly from the work lives of the learners.

Within 30 d following training, key performance metrics are measured and participant learning evaluated using surveys, interviews, and follow up sessions.

Follow Through: Sustaining Change. Ongoing reinforcement embeds best communication practices and amplifies learning until Breakthrough Communication habits are formed. There are three follow-up initiatives:

- Micro reinforcement workshops: 45-min virtual practice sessions hardwire skills by practicing with new scenarios harvested from learners. In-house facilitators are trained to sustain this initiative.
- Mini learning capsules: 1 min video clips keep the skills in conscious awareness. A few titles from this library: Make Gratitude an Event, The Power of Apology, Be Curious not Furious, Never Weaponize Silence.
- Trainer training: a certification program that can be fully completed on-line. Participants work with coaches and then demonstrate mastery with application assignments and teach-backs.

References

1. Glaser, S.: Interpersonal communication instruction: a behavioral competency approach. Commun. Educ. **32**, 221–225 (1983)
2. Glaser, S., Biglan, T., Dow, M.: Conversational skills instruction for communication apprehension and avoidance: evaluation of a treatment program. Commun. Res. **10**(4), 582–613 (1983)
3. Glaser, S., Glaser, P.: Transforming organizational communication: changing how people resolve conflict and solve problems. In: International Association of Business Communicators Conference, pp. 1–25 (1998)
4. Duhigg, C., Graham, J.: What google learned in its quest to build the perfect team. New York Times Mag. (2016)

5. Robertson, K.: For Employees not in the office, the ladder gets trickier to climb. New York Times Bus. 1–6 (2022)
6. Glaser, S.: Teamwork and communication: a three-year study of change. Manag. Commun. Q. **7**(3), 282–296 (1994)
7. Zamanou, S., Glaser, S.: Moving toward participation and involvement: managing and measuring organizational culture. Group Org. Manag. **19**(4), 475–502 (1994)
8. Kansas State University Research and Extension: Chapter 6: A Word About Adult Learning Theory. Master Community Facilitator Notebook pp. 6–1 – 6–4 (2018)
9. Kearsley, G.: Andragogy (M. Knowles): the theory into practice database
10. Hagen, M., Park, S.: We knew it all along! using cognitive science to explain how andragogy works. Eur. J. Train. Dev. **40**(3), 1710190 (2016)
11. Choudhury, P.: Our Work-from-Anywhere Future: Best practices for all-remote organizations. Harvard Business Review Nov-Dec (2020)
12. Greer, L.: Why remote work makes teams (and leaders) better. Entrepreneur (2020)

The Power of Synchronous Sessions in Distance Education: Building Community and Resilience in the Age of COVID-19

Roxana Toma[✉] and Ali Ait Si Mhamed

State University of New York Empire State College, Saratoga Springs, NY 12866, USA
Roxana.Toma@esc.edu

Abstract. We draw on Bourdieu's work on the sociology of education and introduce the idea of building social capital and community in the often-misunderstood, one-sided narrative of online learning, which is seen both as an isolated and isolating experience. We also look at educational praxis, which is "informed, committed action," to address socially differentiated educational attainment—perceived to be more pronounced in online learning. To this end, we think that the field of distance education would benefit from a discussion of the significant value gained from adding synchronous sessions to online courses that are otherwise asynchronous, particularly for teaching research-based and analytical subjects at the graduate level. To investigate this, we perform a narrative analysis of qualitative data from student evaluations of five online courses taught within the past two years where we introduced regular synchronous sessions. Our findings indicate that synchronous sessions, especially during the pandemic, were perceived by students as cornerstone of a pedagogy of care. Further, these sessions work better than fully asynchronous courses for students who are prone to lower educational attainment due to prior conditions (e.g., SES, race) because of the added layer of support. Finally, our findings indicate that these sessions represent one, effective way for students to build social capital and community in courses that are otherwise fully asynchronous.

Keywords: Distance Education · Online Engagement · Community Building

1 Introduction

The COVID-19 pandemic has come to college with astonishing speed. In the fall of 2020, within weeks of reopening, many universities in the U.S. had reported clusters of cases, forcing schools to frantically backtrack on their plans of reopening face-to-face and suspend in-person classes. Meanwhile, enraged at paying face-to-face prices for education that was increasingly online, students and parents were demanding tuition rebates, increased financial aid, and reduced fees to compensate for what they felt was a diminished college experience and an indefinite period of "glorified Skype." While online courses have long been gaining credibility, many were arguing that with online classes, learning outcomes often disappoint, and that virtual instruction runs counter to

D. Guralnick et al. (Eds.): TLIC 2022, LNNS 581, pp. 433–445, 2023.
https://doi.org/10.1007/978-3-031-21569-8_41

the most important asset at a major university—personal interaction with your peers and with highly qualified experts. In this paper, our goal is to tackle and hopefully, refute these misconceptions. In doing so, we introduce the idea of online courses as a platform for building social capital, community, and resilience, which are especially important during the uncertain times we live in.

First, there is a difference between what has been practiced during the pandemic and online education. Findings from the largest cross-country research study undertaken during the pandemic suggest that the practices employed during this time can be defined as *emergency remote education* and this practice is significantly different from planned practices such as distance education, online learning, or other derivations [9]. The remarkable difference between *emergency remote education* and *distance education* is that the latter is an option, while the former is an obligation. Such an understanding is crucial because misconceptions in definitions would lead us to misconceptions in practices. Distance education "is not simply a geographical separation of learners and teachers, but, more importantly, is a pedagogical concept" [19, p. 22]. In contrast, the crash nature of emergency remote education inevitably results in its weakness in theoretical underpinning and is far from being a pedagogical concept in its own right. More importantly, the field of distance education has already proved its validity and value [31] and research indicates that there is no difference in learning outcomes between distance education and face-to-face education [25].

Second, in addition to the profound and global impact of the pandemic on our social, economic, and political lives, COVID-19 has also affected individuals both emotionally and psychologically. Surviving during the pandemic has required building support communities, sharing tools and knowledge, and listening to different voices. While it was advised that we keep our social distance, what was meant was keeping the spatial distance, not the transactional distance [20].

Waddingham [30] argued that "overwhelmed by the scale of things that are happening" (p. 104), we had to look after each other and make each other feel that nobody was alone in those traumatic times. Social media played an essential role by facilitating a space where educators were able to meet, share, and exchange their knowledge. While support communities were important for educators to collaborate and support each other, students similarly needed care, affection, and support. Although they were encouraged to self-organize to build community and support systems, from an educational perspective, providing support communities for students is vital because many people are psychologically overwhelmed, and in need of assistance from us, as we are theoretically better able to cope with the pandemic's implications in education.

Research has shown that emotions play a major role in the online learning experience [11] and that the online learning context is robust enough to allow for caring relations to emerge *at even a deeper level* than that experienced in face-to-face contexts [29]. As a result, several researchers have been investigating design elements and pedagogical practices that can enhance the emotional sensitivity and support the development of caring relations in online learning [10, 24, 28, 29].

With the uncertainty that characterized this period of human existence and the resulting anxiety and trauma that students and instructors have been experiencing, the theme of a pedagogy of care has surfaced within educational institutions. However, it is important to recognize that while the theme of care in education has been popularized during the crisis, it is a crucial element in learning that has always been needed, and that will continue to be essential long after COVID-19 [2]. Thus, we argue it is important to create online spaces where people can support each other, and our goal in the next sections is to show that adding regular synchronous sessions to otherwise asynchronous courses represents one, effective way to create that nurturing environment, especially for teaching analytic, research-based, and quantitative subjects.

2 A Pedagogy of Care

The emotional ramifications resulting from the trauma caused by this pandemic required intentional designs and practices that embody care, inclusion, compassion, and empathy as core values [32]. A care approach to education pushes educators to recognize and address the diversity of students' experiences and vulnerabilities, allowing them to be more receptive not only to the assumed needs of students, but also their expressed and individual needs. This requires structures and practices that go beyond academia and prioritize the emotional and psychological development and needs of students, especially during times of crisis. Concerned Academics [13] outlined a "Social Pedagogy"; an approach that is "consultative, inclusive, and sensitive to the contexts of students, teachers and their communities. It works toward a mutually supportive framework that will carry our pedagogic work through the current crisis, into a period of just recovery, and a more equitable future."

A key part of a pedagogy of care is listening to students and engaging in open and authentic dialogue—particularly marginalized and disadvantaged students who are struggling with the compounded effects of inequities that already exist in educational settings, which make them prone to lower educational attainment (e.g., low retention/lower graduation rates, longer times to degree completion)—and providing additional and stronger support to address these concerns and challenges [13, 21]. This involves understanding learners as individuals in their personal, social, economic, and political environments—beyond their role as a learner in a classroom/lecture hall (ibid).

Lambert's Six Critical Dimensions model, for example, incorporates learner diversity and agency into online learning processes as well as an understanding of students' skills, support, and learning materials that empower rather than reinforce existing inequalities [18]. In understanding the lived experiences of learners, education strategies need to be adapted to ensure that no learner is left behind or further disadvantaged (ibid.). Strategies and practices such as personal connections, reciprocity of caring, and students-centered design and teaching practices have shown potential in nurturing and maintaining a climate of care online [24, 28, 29]. This entails designing curricula that do not stop at content delivery and assigning tasks for assessment purposes, but that intentionally create spaces for learners to learn together in small groups (social constructivism) and to reimagine digital forms of informal social spaces (sometimes called third places) for connection similar to playgrounds and cafeterias [3] that help make school enjoyable for students and help build their social capital. To this end, we argue that adding synchronous sessions to otherwise asynchronous courses represents one, effective way to build students' social capital, sense of community and belonging, especially when teaching analytic, research-based, and quantitative subjects.

3 Building on BOUrdieu's Framework

To understand students' determinants of success in distance education, this study draws on Bourdieu's Theory of Practice [4] and his conceptualization of capital, habitus, and field. Bourdieu used the term *habitus* to describe an individual's unique characteristics; their tastes, perceptions, or ways of responding and thinking [6]. He was interested in the ways in which habitus both shapes and is shaped by practice. *Field* describes an area of practice characterized by an internal struggle for limited power or resources [4, 5, 6]. The distribution of power can be understood by considering the notion of *capital* or what is valued within the field [7]. Capital is an acquired form of power or influence, taking many forms, all of which are ultimately resources that can be exploited. The academic skills developed during the program of study, and the confidence in using them, represent forms of embodied cultural capital. Other relevant examples of capital include social networks (social capital); valuable objects and materials (objectified capital), such as the qualification itself; and the association of value through the reputation of an awarding institution (institutionalized capital), such as a university.

It follows, then, that changes in habitus convey changes in dispositions and capacities from which practices are adapted and developed and which, in turn, enhance the participants' agency within their fields. Learning, therefore, involves developing the habitus that is required to successfully operate within a particular field. As Schatzki [26, p. 29] noted, "the more the habitus is acquired, the better someone can proceed in these fields, and in a greater range of situations." However, it is important to recognize that these capacities and dispositions are not simply acquired by individuals, they are caught up in other forms of capital that are brought about by understandings of the institution and accreditation of qualifications (institutionalized capital), as well as the social connections generated through enrolment and engagement with the program (social capital).

Bourdieu's theory of practice emphasizes a dialectic relationship between habitus and field. As Bourdieu explains, "The habitus and the field maintain a relationship of mutual attraction, and the illusion (illusio) is determined from the inside, from impulses that push toward a self-investment in the object; but it is also determined from the outside, starting with a particular universe of objects offered socially for investment" [8, p. 512]. In other words, habitus does not just mirror an individual's social position, but is shaped by subjective (e.g., a student's intrinsic motivation or desires) and objective (e.g., societal structures like higher education systems) forces. These forces contribute to forming an individual's perception of possibilities and 'imposes a particular mode on desire' [8, pp. 512–513].

Bourdieu [4] conceived of people (in our case, students) as vying against each other within a *field* for different forms of capital and status. However, students can acquire capital within their field by using programs to learn about education, to obtain a relevant qualification, but also to generate valuable social networks. It is in this last issue that the greatest tension arises in Bourdieu's position: significant value comes from the development of supportive partnerships and networks, and participation in a community of learners. It is this aspect that we focus on—the social capital students can acquire through engagement with their program as a critical component to developing their habitus. Next, we introduce the idea of educational praxis and argue that we can use it in distance education to influence students' development of social capital, an important dimension for their habitus formation, but also an element that is critical for their resilience, especially during uncertain times like the pandemic.

4   Educational Praxis

Praxis is a Greek word that means moving back and forth in a critical way between reflecting and acting on the world. Because reflection alone does not produce change, Freire [14] advocated for the necessity of action based on reflection. Policy praxis involves inductive and deductive forms of reasoning. It also involves dialogue as social process with the objective of "dismantling oppressive structures and mechanisms prevalent both in education and society" [15, p. 383]. Critical, transformative leaders enter and remain in education not to carry on business as usual but to work for social change and social justice [1, 12, 22]. Unfortunately, Rapp, Silent, and Silent [23] found that 90% of educational leaders, both practitioners and professors, remained wedded to what Scott and Hart [27] call technical drifting—a commitment to emphasize and act on the technical components of one's work above the moral. Technical drifters fail to validate the cultural, intellectual, and emotional identities of people from underrepresented groups, they avoid situations where their values, leadership styles, and professional goals are challenged and dismantled, and they use their positions of power to reaffirm their own professional choices.

Given the devastating impact of this global crisis, prioritizing the issues of care, empathy, and emotional/psychological support to the classroom setting and targeted towards students became critical. Our goal as educators was to engage in *educational praxis*, which involves making "morally informed and committed action" that "helps to shape social formations and conditions" [17, p. 10]. In the midst of a chaotic and unprecedented time, we wanted to create more than an online class: we imagined a community of care that could transcend the limits of online teaching and the pandemic itself. We thought that adding regular synchronous sessions to our established online courses that were otherwise asynchronous, especially in graduate-level research, quantitative, and analytical subjects, will help in two ways:

1. Create social capital, community and resilience, ingredients that are hugely beneficial to students' *habitus* development, and implicitly beneficial to their operating in their respective fields.
2. Level out the field of socially differentiated educational attainment (due to preexisting conditions like students' socio-economic status and race) which is perceived to be more pronounced in online learning.

Further, we thought that if we can help instructors and students build social capital and community online, we could help further dispel the myths and misconceptions about online education.

5 Findings

We used a multiple case study approach to analyze online graduate-level courses spanning different student academic backgrounds from three masters programs (*Social and Public Policy, Community and Economic Development, Work and Labor Policy*) and one doctoral program (*Doctorate in Higher Education Leadership and Change*). We employed a narrative analysis of student perceptions about the newly added synchronous sessions that were drawn from course evaluations and discussion forums in five online courses taught between January 2020 and December 2021. For a list of those courses please see Table 1.

Table 1. Online courses by program.

Masters programs courses	Doctoral program courses
Research methods	Foundations of doctoral study
Public policy analysis	Principles of higher education leadership
Capstone project	

We followed the design of unstructured interviews to obtain data for this study [16]. We were interested to draw from the students' constructs embedded in their thinking and understand their perceptions by inductively helping them articulate their viewpoints such that they are understood clearly by the reader [16]. We asked them the following two open-ended questions: 1) *In your opinion, what was the best aspect of this course?* and 2) *What did you think about the recurrent synchronous sessions?* Students who wished to do so, provided their responses in short written reflections and shared them with us. Some of the data also came from course evaluations, which included the first question, and were anonymous.

For the number of students in each program by term of study and the response rate to our questions, please see Table 2. We organized our findings around the common themes identified in our narrative analysis. A few selected examples are included with each theme for illustrative purposes.

5.1 Common Themes

Student narrative evaluations demonstrated that participants had varied degrees of familiarity and understanding with synchronous sessions. Most of them indicated that they considered these sessions to provide a high level of supportive learning and understanding of difficult material. They also saw these sessions as a *social gap* filler because they provided them with an incredible sense of community and belonging. Due to their interactive mode, participants also thought of these sessions as empowering them to create an engaging environment. A very interesting and surprising finding is the fact that synchronous sessions were very helpful for students with learning disabilities. While we are not aware of the number of students with these disabilities in our courses, the data we

Table 2. Number of students, educational attainment, and response rate by term.

Academic term	Level of study	Number of students	Response rate
Spring 2020	Masters level	32	28.12%
Summer 2020	Masters level	20	25.0%
Fall 2020	Masters level	48	25.0%
Spring 2021	Masters level	74	31.08%
Summer 2021	Masters level	27	48.15%
Fall 2021	Masters level	49	36.73%
Fall 2021	Doctoral level	14	57.14%

collected relies on students self-identifying as such while mentioning this helpfulness. Moreover, our Programs serve a large population of students who are on financial aid, which may suggest that many of our students are likely to be prone to socially differentiated/lower educational attainment because of prior conditions like socio-economic status and race. To this end, our findings indicate that students who generally struggled more than others found these sessions the most helpful in their learning process and progress, because they added a layer of support to their independent, asynchronous learning.

Theme 1. High level of supportive learning and understanding of difficult material. Extracts from students' narrative comments reveal how the synchronous sessions fostered participants' intellectual competence to enhance and advance their learning of the course content. Student-led sharing of questions and discussions during these sessions displayed students' willingness to be more open and interested in engaging with others in the learning process.

Extract 1:

I wanted to thank you for the support over the past two semesters - you have been more supportive and responsive than any other professor and it is much appreciated. Truly - you're the only one who has offered the time for live classes and encouragement when, we as students, have been challenged and in need of that guidance with our education.

Extract 2:

The best aspect of this course were the online meeting sessions. The professor took an hour or more of their time to video chat the students and provide help with homework and other assignments. There was not one question [the professor] did not answer! [The professor] made me more confident with their continuous support.

Extract 3:

The synchronous sessions of [day and time specified] are essential to our learning, our professional development, and our ability to build community and connection, not to just each other but also to the material and institution. And, it makes sense… Research has demonstrated that the best predictor to being resilient, to overcome challenge is to surround yourself with community and connections that are grounded in trust, mutual respect, and support. The [day and time] sessions do this for the cohort of the Program. Having returned to the student role after [many years], this level of support is essential, especially during the first semester of the program. Special thanks to my colleagues and our faculty mentor for taking the time and spending the energy on creating and supporting this opportunity to keep us engaged.

Theme 2. Synchronous sessions as a social gap filler. Analysis of student narrative evaluations showed that some participants had articulated instances that curtailed their ability to develop their learning, including attempts to participate in all asynchronous activities to gain some interaction from their peers. More precisely, participants reported that the asynchronous mode lacked the social interaction they needed to make them feel they belong to a class and a "community" of learners. Although they were sometimes assigned group work, that, too, was completed asynchronously, in order to accommodate each other's schedules. Hence, students' narratives pointed to synchronous sessions as filling in their social aspect of learning and their lived experiences as students in a virtual community, which they indicated it would not have been possible without those synchronous sessions. Even though some had limited time to participate, they valued the quality of these synchronous sessions as social gap fillers.

Extract 1:

Much of great endeavors and worthwhile efforts, have their healthy, strong roots, in relationships built on mutual trust and support for each other. What makes the healthy yoke of enduring leadership model, is cultivating strong roots and enduring connections. I appreciate you recognizing this critical theme of our experience and allowing for the space in which it may grow.

Extract 2:

The asynchronous model of our program is what works best for many of us. But asynchronous can be lonely and isolating, and when going through a [name of the program] program you need support and companionship. The weekly optional synchronous meetings give us the opportunity to meet "face to face" and connect with our cohort and instructor. In a time where there is a lack of connection worldwide, these meetings allow us to develop relationships, feel supported and gain experience in a co-curricular setting. No matter how busy life gets, this one hour is something I look forward to each and every week!

Extract 3:

There is nothing that compares to a synchronous session when it comes to learning and collaborating with faculty and students in a course. It matters less if the session is in person or virtual, but more on the fact that it is synchronous and not asynchronous. The natural dynamic that occurs when two or more people are interacting live over the internet is not replicable in a videotaped, pre-recorded session. The ability to listen and react and pose questions during a synchronous session far strengthens the learning aspect because one gets an immediate response and the back and forth can actually flow the discussion in different ways, with the professor choosing the path forward in the discussion based on the audience, the areas in which the group wants to participate, and the connections made among the students on the spot. One learns from the next and so on, and the faculty member, either as a teacher or a moderator, can make the learning experience more practical and relevant to the class. I highly recommend synchronous sessions as part of an online course. Even if sessions don't focus on course content all the time, or even a majority of the time, the synchronous session bonds the online class in a common experience and likely leads to better outcomes and retention for future terms.

Theme 3. Empowering tool for creating an engaging environment. Another theme that emerged from the data was that although synchronous sessions may come to be considered a regular task, or maybe even a slight burden in the schedule for some students, most participants agreed that these sessions empowered them to create an engaging environment. The more they attended these sessions, the better they engaged in creating other environments in which they could interact.

Extract 1:

I cannot emphasize enough the importance your clear explanations in the live classes play in my comprehension of this material. The second review class was just as helpful and informative as the first one. I appreciate the level of care you demonstrate toward students. You promote learning while empowering the class. This is a rare talent in a professor, and a gift to students.

Extract 2:

Coming from an educator perspective – [professor's name] is a really great teacher! I loved the live discussions - this was a refreshing change from the written discussions in Moodle that other courses do - I wish all courses did this instead!

Extract 3:

This is by far the BEST class I have taken at [college name] yet. I thoroughly enjoyed [professor's name] teaching style and think the professor has nailed an effective format for asynchronous online learning. My favorite part was the live sessions the professor hosted reviewing the assignments and taking questions. Online school can feel isolating, and it was so good to get instant responses to questions, have something explained verbally, and hear from the other students in the course. It allowed me to realize that I was on the same track and that my fellow students had the same questions as I did.

Theme 4. Synchronous sessions were very helpful for students with learning disabilities. A particularly interesting finding was highlighted by only a few students but the value of the finding itself was invaluable for this research. Online programs are primarily being offered to meet the needs of non-traditional students. These students are busy with their personal and professional obligations and they are not able to add an even heavier schedule to their lives. Asynchronous programs appeal exactly to this kind of student population. However, incorporating optional synchronous sessions to these programs adds a "spice" to students' learning experience and drives the ideas home. Importantly, these sessions are deemed extremely valuable for students who self-identify with a learning disability.

Extract 1:

My personal favorite aspect of the course is the "flipped classroom" model. [Professor's name] posts slide shows and recorded lectures, then holds online group sessions where we discuss the concepts we have learned on our own. This is a fantastic learning format that I feel works best for me as an adult learner with some minor learning disabilities. This allows me to set my own pace while knowing my professor is available live at regular intervals for any questions or clarification.

Extract 2:

I really appreciated the live sessions you provided. Though I do prefer not taking courses in a physical classroom, listening to your explanations, and answers to questions, really helped me to understand the concepts more clearly.

Extract 3:

The video office hours reminded me of the value of live collective class discussion and the insight that can be achieved through verbal instruction. As someone with a learning disability who has always had to overcompensate to succeed academically, [professor's name] teaching style, clear and concise communication/organization and thoughtful manner during the live sessions have allowed me the opportunity to enjoy the learning process.

Theme 5. Synchronous learning is useful for struggling students. Struggling students find the synchronous sessions the most helpful for their learning because they provide an environment whereby conversations, discussions and learning happen in the live interactive mode.

Extract 1:

The online session was so informative and really set me on the right path! You do an awesome job of making difficult concepts understandable!!

Extract 2:

I appreciate your time during the live classes. I know you will say it is your job and you are correct, but you do your job with what I believe is true care and concern for your students.

Extract 3:

I really, really enjoyed the live classes. It was awesome to actually have the professor explain the material. It was very helpful that there were these opportunities.

6 Conclusion

Findings from this two-year experiment which consisted of adding regular synchronous sessions to otherwise fully asynchronous courses indicate that a combination of synchronous sessions with asynchronous work and discussions seems to work better for adult learners, for students with learning disabilities, and for students who are prone to lower educational attainment (prior conditions like SES, race, affecting retention/graduation rates and time to degree completion) because of the added layer of support students encounter in these sessions. Our findings also suggest that synchronous sessions empower students and provide them with an effective way to build community and social capital, something that students claim it cannot be achieved online through asynchronous work only. We have shown earlier that these elements of social capital and community-building are intrinsic to students' habitus development and how well they will operate in their respective fields. Finally, our findings indicate that for students, synchronous sessions are strongly connected with a pedagogy of care.

Thus, our position, upon reflection on the findings presented here, is that non-traditional students getting their degrees online can become more competitive in their fields by building social capital and community during their studies. It follows that programs should function as sites of nurture and mutual support, and that may be best facilitated by pedagogies that actively work towards lasting, useful relationships and the fostering of a collaborative culture. We argue that adding synchronous sessions to otherwise asynchronous courses is one effective way to do that, especially in analytic, research-based, and quantitative courses at the graduate level, which are more difficult to teach online than in traditional settings.

References

1. Ayers, W., Hunt, J.A., Quinn, T.: Teaching for Social Justice: A Democracy and Education Reader. Teachers college Press, New York (1998)
2. Bali, M.: Care is not a fad: Care beyond COVID 19 (2020). https://blog.mahabali.me/pedagogy/critical-pedagogy/care-is-not-a-fad-care-beyond-covid-19/. Last accessed 01 March 2022
3. Bali, M.: Literacies teachers need during COVID-19 (2020). Al-Fanar Media. https://www.al-fanarmedia.org/2020/05/literacies-teachers-need-during-covid-19/. Last accessed 10 March 2022

4. Bourdieu, P.: Outline of a Theory of Practice. Cambridge University Press, New York (1977)
5. Bourdieu, P.: The Logic of Practice. Stanford University Press, Stanford, CA (1990)
6. Bourdieu, P., Wacquant, L.: Symbolic capital and social classes. J. Classic Sociol. **13**(2), 292–302 (2013)
7. Bourdieu, P.: The forms of capital. In: Richardson, J. (ed) Handbook of Theory and Research for the Sociology of Education. Greenwood, New York (1986)
8. Bourdieu, P.: The contradictions of inheritance. In: Bourdieu, P., Accardo, A., Balazs, G., Beaud, S., Bonvin, F., Bourdieu, E., Bourgois, P., Broccolichi, S., Hampagne, P., Christin, R., Faguer, J-P., Garcia, S., Lenoir, R., Oeuvrard, F., Pialoux, M., Pinto, L., Podalydes, D., Sayad, A., Soulie, C., Wacquant, L.-J.D. (eds) The Weight of the World: Social Suffering in Contemporary Society, pp. 507–513. Polity Press, Cambridge (1999)
9. Bozkurt, A., Jung, I., Xiao, J., Vladimirschi, V., Schuwer, R., Egorov, G., Lambert, S. R., Al-Freih, M., Pete, J., Olcott, Jr., D. Rodes, V., Aranciaga, I., Bali, M., Alvarez, Jr., A. V., Roberts, J., Pazurek, A., Raffaghelli, J. E., Panagiotou, N., de Coëtlogon, P., Shahadu, S., Brown, M., Asino, T. I. Tumwesige, J., Ramírez Reyes, T., Barrios Ipenza, E., Ossiannilsson, E., Bond, M., Belhamel, K., Irvine, V., Sharma, R. C., Adam, T., Janssen, B., Sklyarova, T., Olcott, N. Ambrosino, A., Lazou, C., Mocquet, B., Mano, M., Paskevicius, M.: A global outlook to the interruption of education due to COVID-19 pandemic: Navigating in a time of uncertainty and crisis. Asian J. Distance Educ. **15**(1), 1–126 (2020)
10. Chng, L.K.: Learning emotions in e-learning: how do adult learners feel? Asian J. Distance Educ. **14**(1), 34–46 (2019)
11. Cleveland-Innes, M., Campbell, P.: Emotional presence, learning, and the online learning environment. Int. Rev. Res. Open Distrib. Learn. **13**(4), 269–292 (2012)
12. Cochran-Smith, M.: Teaching for social justice: Toward a grounded theory of teacher education. In: Hargreaves, A., Lieberman, A., Fullan, M., Hopkins, D. (eds.) The International Handbook of Educational Change, pp. 916–951. Kluwer Academic, The Netherlands (1998)
13. Concerned Academics: Public Universities with a Public Conscience: A Proposed Plan. https://drive.google.com/file/d/1tyiyKND-5xT1W2BNaYZ43yCJmWb7Y-vR/view?usp=embed_facebook. Last accessed 28 Feb 2022
14. Freire, P.: Pedagogy of the Oppressed, Rev Continuum, New York (1994)
15. Freire, P., Macedo, D.: A dialogue: culture, language, and race. Harv. Educ. Rev. **65**(3), 377–402 (1995)
16. Holstein, J, A., Gubrium, J. F.: The active interview. In: Siverstein, D. (ed) Qualitative Research: Theory, Method and Practice, pp. 140–161. Sage, Thousand Oaks, CA (2004)
17. Kemmis, S.: Research for praxis: knowing doing. Pedagogy Cult. Soc. **18**(1), 9–27 (2010)
18. Lambert, S.R.: Six critical dimensions: a model for widening participation in open, online and blended programs. Australas. J. Educ. Technol. **35**(6), 161–182 (2019)
19. Moore, M.: Theory of transactional distance. In: Keegan, D. (ed.) Theoretical Principles of Distance Education, pp. 22–38. Routledge (1997)
20. Moore, M.G.: The theory of transactional distance. In: Moore, M.G. (ed) Handbook of Distance Education, 3rd edn, pp. 66–85. Routledge (2013)
21. Noddings, N.: The caring relation in teaching. Oxf. Rev. Educ. **38**(6), 771–781 (2012)
22. Oakes, J., Lipton, M.: Teaching to Change the World. McGraw-Hill, Boston (1999)
23. Rapp, D., Silent, X., Silent, Y.: The implications of raising one's voice in educational leadership doctoral programs: Women's stories of fear, retaliation, and silence. Journal of School Leadership, **11**(4), 279–295. (2001)
24. Robinson, H.A., Al-Freih, M., Kilgore, W.: Designing with care: Towards a care-centered model for online learning design. Int. J. Inform. Learn. Technol. **37**(3), 99–108 (2020)
25. Russell, T.L.: The No Significant Difference Phenomenon: As Reported in 355 Research Reports, Summaries and Papers. North Carolina State University (1999)

26. Schatzki, T.: Practices and learning. In: Grootenboer, P., Edwards-Groves, C. (eds.) Practice Theory Perspectives on Pedagogy and Education, pp. 23–43. Springer, Singapore (2017)
27. Scott, W., Hart, D.: Organizational America: Can Individual Freedom Survive the Security It Promises? Houghton Mifflin, Boston (1979)
28. Sitzman, K., Leners, D.W.: Student perceptions of caring in online baccalaureate education. Nurs. Educ. Perspect. **27**(5), 254–259 (2006)
29. Velasquez, A., Graham, C.R., Osguthorpe, R.: Caring in a technology-mediated online high school context. Distance Educ. **34**(1), 97–118 (2013)
30. Waddingham, R.: COVID-19: how can we support each other (and ourselves)? Psychosis, 1–5 (2020)
31. Xiao, J.: On the margins or at the center? Distance education in higher education. Distance Educ. **39**(2), 259–274 (2018)
32. Zembylas, M.: Critical pedagogy and emotion: Working through 'troubled knowledge' in posttraumatic contexts. Crit. Stud. Educ. **54**(2), 176–189 (2013)

Implementation of a Talent Development Program in Higher Education

Vilmos Vass[(⊠)] and Ferenc Kiss

Budapest Metropolitan University, Budapest Nagy Lajos király útja 1-9, 11148 Budapest, Hungary
vvass@metropolitan.hu

Abstract. The context of the paper is, on the one hand, a growing trend of internationalization in higher education, especially focusing on creativity and innovation. Talent development programs play important role in this process. On the other hand, 50% of all employees will need reskilling by 2025, as adoption of technology increases, according to the World Economic Forum's Future of Jobs Report. Analytical thinking and innovation, creativity, originality, and initiative are on the Top 10 Skills of 2025. The purpose of the paper is to introduce some results of the implementation of a 3-years talent development program at the Budapest Metropolitan University focusing on creative and startup thinking. This program has transdisciplinary approach and modular structure from different domains, such as tourism, business, communication, and arts. This paper introduces some important phenomena and impacts of the program implementation, for instance formulating professional learning community of students and teachers, creating and evaluating business plans, and changing creative and startup thinking skill set at individual, team, and organizational levels.

Keywords: Talent development program · Creative and startup thinking · Transdisciplinary approach

1 Context

No doubt, higher education faces some challenges in the 21st century, because of globalization and internationalization, growing international competition, and closing the gap between work life and education. In fact, giving the answers to these challenges characterizes standardization, but more higher education institutions can find individual, creative, and innovative ways [3, 4]. Under the umbrella of the above-mentioned trends, the question of quality gets increased attention. Basically, quality of higher education is a controversial topic, especially from the concept of knowledge and learning. Traditionally, higher education has focused on academic, declarative knowledge, which is based on memorization and concentration. The growing needs on applied, procedural knowledge emphasizes project-based and problem-based learning. From this perspective, the question of quality focuses on talent development programs, as well. Personalized learning, tutoring, and mentoring system are significant quality indicators on this process in

D. Guralnick et al. (Eds.): TLIC 2022, LNNS 581, pp. 446–450, 2023.
https://doi.org/10.1007/978-3-031-21569-8_42

higher education. Parallel to personalization, talent development programs are based on networking and cooperative learning. The fundamental dilemma of talented education, especially in higher education, focus on the competency development. Basically, this is a domain-specific focus, which is related to chosen specialization in order to deepen knowledge. Firstly, the concept of domain-specific knowledge in higher education is based on academic, explicit knowledge. Deepening knowledge, in this meaning, is finding specialized topic in order to dig deeper and to analyse and synthetize the main research results. No doubt, analysing and synthetizing are more difficult than describing something, but this is one of the most fundamental characteristics of talented education. Secondly, domain-specific knowledge in higher education has strong consistency with creativity. As Mihály Csíkszentmihályi [7] defines, "…domain, which consists of a set of symbolic rules and procedures. Mathematics is a domain or at a finer resolution algebra and number theory can be seen as domains. Domain are in turn nested in what we usually call culture, or the symbolic knowledge shared by a particular society, or by humanity as a whole."

This is a complex and social-cultural concept of domain-specific knowledge, which can open the gate from the narrow specialization. This complexity has transdisciplinary phenomena, where different culture-based domains can enrich themselves in order to give opportunities to analyse and synthetize processes and results in a creative way. In order to rethink about talent development programs in higher education in this context, we need to cite Csíkszentmihályi's definition of creativity: "Creativity is any act, idea or product that changes an existing domain, or that transforms an existing domain into a new one" [7].

In this sense, changing or transforming the domain is the fundamental objective of talent development programs in higher education. But this is, only, one side of the coin.

The other side of the coin is the impact of fast-changing economy and word of work to changing and transforming domain. The Closing the Skills Gap: Key Insights and Success Metrics document [1] conceptualizes reskilling and upskilling. Why is the "reskilling revolution" so important? Because, according the World Economic Forum's Future of Jobs Report, 50% of all employees will need reskilling by 2025 [8]. The Forum estimates that by 2025, 85 million jobs may be displaced by a shift in the division of labour between humans and machines. Thus, they defined the top 10 skills: "Analytical thinking and innovation, Active learning and learning strategies, Complex problem-solving, Critical thinking and analysis, Resilience, stress tolerance, and flexibility, Creativity, originality, and initiative, Leadership and social influence, Reasoning, problem-solving, and ideation, Emotional intelligence, Technology design and programming."

Focusing on higher education, from the above-mentioned list and context of complex domain-specific talent development programs, analytical thinking and innovation, creativity, originality, and initiative deserve unique attention. The feasible coherency between social-cultural-based and economy-based domains have been changed the fundamental vision of this concept. Basically, domain is based on knowledge, as we see from uni- to transdisciplinarity. But growing complexity can block changing and transforming the domains, because of traditional vision of knowledge in higher education. This traditional vison of knowledge is factual, rigid and explicit. In order to understand the revision, we need to evoke tacit dimension of knowledge. As Michael Polányi wrote, "We can know more than we can tell" [6].

Parallel to the personal phenomena of knowledge, it is worthy our attention the "knowing how" characteristics. "Tacit knowledge is the personal knowledge resident within the mind, behaviour and perceptions of individuals. Tacit knowledge includes skills, experiences, insight, intuition and judgment. It is typically shared through discussion, stories, analogies and person-to-person interaction; therefore, it is difficult to capture or represent in explicit form. Because individuals continually add personal knowledge, which changes behaviour and perceptions, tacit knowledge is by definition uncapped" [2].

In fact, there are some methodological elements on this above-mentioned definition, such as personalization, discussion, and interaction, which are dominant on the talent development programs in higher education [5].

2 Practice

METU STARTUP LAB talent development program has been implemented for 3 years at Budapest metropolitan University, focusing on soft skills and competencies of digital entrepreneurship. The Budapest Metropolitan University has received non-refundable project support from the Ministry of Human Resources' National Talent Programme, University's Talent Management Committee. The project named "Digital company building – STARTUP LAB" has started.

The STARTUP LAB is based on Creative and Innovative Thinking Development Workshops. Building on these LABs, we have integrated this program into the University's life, bringing together students and lecturers to motivate them for talent development. We have focused on the development of digital skills, which is needed to become an e-entrepreneur, beside the important goal of developing creative thinking and startup mindset. The workshops have included 25 students with different fields of study in order to strengthen transdisciplinarity and collaboration among different faculties and institutions (Media and Communication, Business and Arts). The students have formed groups of 4–5 people practising techniques like cooperative learning, various project-based learning exercises and problem-solving. The workshops have modular-structured and have built upon each other. These modules covered the following topics:

- Competency based development of digital entrepreneurship
- HR questions of the startup ecosystem
- Creative thinking in Digital Marketing
- Digital business models from tourism
- The circular skill-development opportunities built on company

The purpose of the talent development program, on the one hand, is for the students to be able to understand the advantages and difficulties of startup thinking. On the other hand, students should be able to diagnose the innovative challenges and be able to focus on them. The learning outcomes are students should be able to compare and discuss different innovative challenges and potential startup ideas in an evidence-based professional way. Students should be able to make collaborative startup business plan (planning and development), be able to make decisions (from idea to plan) and evaluation about the progression of the startup idea and business plan.

At the end of the course, students should be able to cooperate, to work in collaborative project work. Students should be able to do collaborative presentations and evaluation on the startup idea and business plan.

Students should be able to reflect own work and reflect others work.

At the end of the course, students should be able to solve problems in a creative way, be able to listen and understand others' opinions, and make conscious decisions.

We have provided to the students:

- Specialists, experts in various fields who will mentor you to acquire the startup mindset
- Visit to some startup incubation centres where students can gather real life experiences
- An opportunity to take part in the Startup DEMO DAY event to meet and talk to successful start uppers

Preparing the talent development program, it is based on collaborative curriculum planning via mapping of the different modules and consistent project design. The next preparation phase has diagnostic assessment focusing on motivation letters, needs analysis and making individual competency web. The significant prioritization of the talent development program is mapping prior knowledge, mind mapping, brainstorming and making place mats. The outcomes of the program has business and project management plans. Regarding formative assessment, the talent development program has created new feedback culture with continuous evaluation following the students' progression using self-assessment, rubrics, personal portfolios and reflective essays in order to develop competencies effectively [9]. This continuous formative assessment and innovative feedback culture work among the teachers using teaching portfolios and professional discussions.

To sum, the main challenges from the teaching perspective is strengthening and maintaining students' motivation, especially focusing on intrinsic motivation via developing self-directed learning and startup thinking. The other challenge is collaborating with the colleagues in order to focus on collaborative professionalism. Collaborative curriculum planning and continuous cooperation, discussion, and reflection are innovative mindsets at the university. Last but not least, innovative feedback culture under the umbrella of formative assessment requires special portfolio-based teaching competencies, especially skills and attitudes, such as empathy, openness, tolerance, critical thinking, communication, and problem-solving.

Finally, student feedback gives evidences of the effectiveness on the talent development program. They emphasized strong pragmatism, experience-based learning, practice-oriented approach, and collaboration among the students and teachers on the program. They stressed the importance of discussions, questioning and collaborative planning and continuous feedback and reflections.

3   Conclusion

Transformation of tacit knowledge and practical application of the complex web of tacit and explicit knowledge are some of the biggest challenges on talent development programs in higher education. The first phase of this transformation is based on the innovation of consistency among curriculum planning, teaching methodology, and assessment. Collaborative curriculum design comes into prominence, which stresses soft skills or key competencies, flexibility, transdisciplinarity, and modular structure. The second phase is progressive teaching methodology, such as project- and problem-based learning, debating, questioning, and different forms of interaction. The third phase of this transformation is renewing assessment functions, namely diagnostic and formative assessment come to the fore. Concerning the effective implementation of the talent development programs, at the individual level, personalized learning and brand building are dominant. At the team level, professional collaboration and networking between students and teachers are important. Last but not least, at the organizational level, moving forward to the professional learning community is the biggest challenge in higher education.

References

1. Closing the skills gap: key insights and success metrics. White Paper. World Economic Forum, November 2020. http://www3.weforum.org/docs/WEF_GSC_NES_White_Paper_2020.pdf
2. Dampney, K., Busch, P., Richards, D.: The meaning of tacit knowledge. Australas. J. Inf. Syst. **10**(1) (2002). https://doi.org/10.3127/ajis.v10i1.438
3. De Witt, H., Gacel-Ávila, J., Jones, E., Jooste, N. (eds.): The Globalization of Internationalisation. Routledge, London (2017)
4. Maringe, F., Foskett, N.: Introduction: Globalization and Universities. In: Maringe, F., Foskett, N. (eds.) Globalization and Internationalisation in Higher Education, pp. 1–15. Continuum International Publishing Group, London (2012)
5. Mészáros, T., Vass V.: The transformation and impact of tacit knowledge to the concept of knowledge in the higher education. SZABAD PIAC: GAZDASÁG - TÁRSADALOM- ÉS BÖLCSÉSZETTUDOMÁNYI FOLYÓIRAT 4: 2, pp. 23–38, 16 pp. (2021)
6. Polányi, M.: The Tacit Dimension. Doubleday, Garden City, NY (1996/2009)
7. Csíkszentmihályi, M.: Creativity. Flow and the Psychology of Discovery and Invention. Harper Collins Publisher, New York, NY (1996)
8. The Future of Jobs Report 2020. World Economic Forum, October 2020. http://www3.weforum.org/docs/WEF_Future_of_Jobs_2020.pdf
9. Vass, V., Kiss, F.: The role of competency development in the implementation of portfolio-based education in higher education. In: Michael, E.A., Dan, C. (szerk.) Visions and Concepts for Education 4.0. ICBL 2020: Proceedings of the 9th International Conference on Interactive Collaborative and Blended Learning (ICBL2020), 571 pp., pp. 42–48, 7 pp. Springer International Publishing, Cham, Svájc (2021)

What if We Radically Reimagined Assessment? An Experimental Design for Participatory Assessment Practices and Learning Community Agreements

Jessica Walker[1]([✉]), Jeongki Lim[1,2], and Srikrithi Srinivasan[1]

[1] Parsons School of Design, The New School, New York, NY, USA
walkerja@newschool.edu
[2] School of Arts, Design, and Architecture, Aalto University, Espoo, Finland

Abstract. To develop assessment practices that can dynamically adapt to different learning environments, the latest learning research in science, and a constantly evolving economic landscape, we are examining a technological intervention that can expand participation in assessment selection and agreement practices. In this position paper, we propose Allgrade, an online collaboration platform that enables various stakeholders within a learning community to collaboratively create a unique combination of assessment methods using a visual and simple interface. The research is in the early stage of development, where we will implement and test the conceptual design with educators at Parsons School of Design at The New School and within the New York City Department of Education school system. In this paper, we will present the methodologies, theoretical frameworks, and a design proposal that can elicit constructive feedback toward further refinement of the prototype platform.

Keywords: Participatory assessment · Learning community agreement · Learning technology

1 Introduction

Despite recent advancements in learning science and educational technologies, assessment practices in typical K-12 and college learning environments have remained consistent with standardized approaches that were adopted by the US education system in the late 1800s [2]. Educators are often overextended and disempowered to acquire more efficient and equitable assessment methods that move beyond standardized models. The ever-present pressure to 'teach to the test,' coupled with a lack of planning time and resources, leaves educators in a position where they must rely on conventional assessment methods as opposed to more nuanced and adaptive approaches [4]. Reliance on standardized testing as the sole indication of a learner's progress and potential has been a contributor to inequality for learners as evidenced in the lower graduation rates among students of color from low-income backgrounds who engaged in standardized testing

D. Guralnick et al. (Eds.): TLIC 2022, LNNS 581, pp. 451–456, 2023.
https://doi.org/10.1007/978-3-031-21569-8_43

without other forms of assessment being considered [6]. Capacities including resilience and self-efficacy are required in higher education and career contexts; however, these are difficult to assess through standardized means alone.

We see a promising research domain where the objective is to develop a self-organizing assessment practice that can adapt to different learning environments within a landscape of ever-changing economic forces. In this article, we propose a conceptual design of Allgrade, an experimental assessment tool that challenges the dominant top-down approach to classroom assessment. Our model will center on participatory assessment practices such as community agreements and peer-review, as potential ways to consider how frameworks of inclusion promote resilience, self-efficacy, and belonging among all stakeholders within the learning environment.

This article serves as a position paper through which we will articulate our research, which is in an early stage of development. In the following sections, we will present the methodological and theoretical frameworks and an early design proposal that will be implemented in subsequent empirical studies. We will conclude by discussing the potential impacts such a tool can have on a wider community of education researchers and practitioners.

2   Methodologies and Theoretical Frameworks

As we are developing a pedagogical intervention in a real educational context, we are using research-based design and design-led research methodologies [3, 9]. Creating well-informed artifacts with multiple stakeholders is an essential part of the research. An iterative design approach built on continual testing of artifacts will result in research augmentation. Our theoretical framework will be based on cultural-historical activity theory, which will allow us to analyze the division of labor and collaborative dynamics among identified stakeholders in the learning environment [5]. In addition, we are using program planning and evaluation tools including a logic model and a theory of change to investigate potential impacts on a systematic level [13] (Fig. 1).

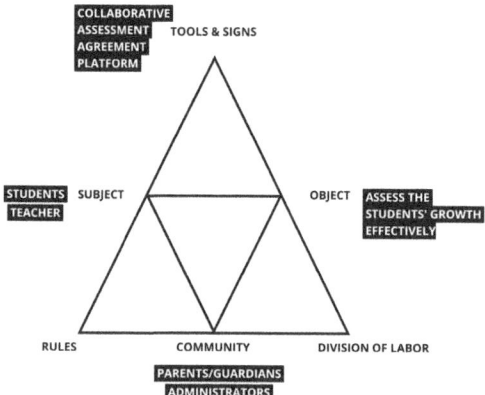

Fig. 1. An analysis of the learning environment using activity theory frameworks.

In the initial design of the tool, we are incorporating various research and pedagogical practices that help us reframe assessment practices to be inclusive of multiple stakeholders in the learning environment. The learner-centered assessment is considered an ideal pedagogical practice in higher education [12]. Participatory evaluation is considered to empower participants—especially for those who often lack decision-making power—by sustaining organizational learning [8, 10]. Setting active learning goals for students can result in sustained motivation [7]. In order to create a fair and equitable evaluation practice, there is a need to accommodate a range of social and cultural backgrounds represented by all students [11]. Lastly, we are examining the pedagogical practice of learning contracts that aid in a negotiation process that can lead to an assessment agreement between educators and students. This framework has been used in higher education settings for a specific subject matter and not frequently in K-12 contexts [1].

3   Design Proposal

Allgrade is an online collaboration platform that prioritizes inclusionary practices as the key driver for determining assessment methods and goals of a learning environment. Below is a walkthrough of the conceptual design of the platform.

First, multiple classroom stakeholders such as teachers, students, administrators, and parents are invited to form a learning community. The learning community is defined by the collective goals and shared social agreements that the group members set as initial ground rules (Fig. 2).

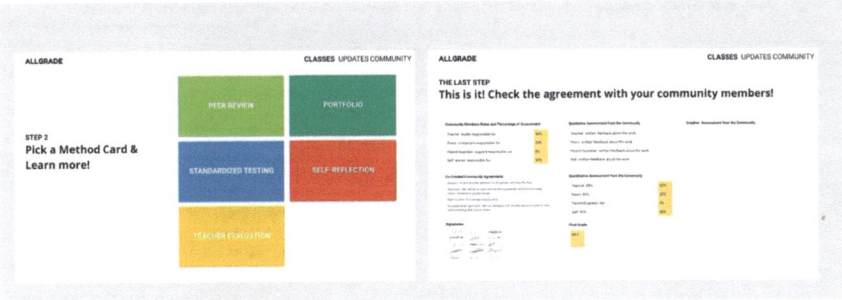

Fig. 2. Example prompts and collaborative interface of Allgrade.

After the learning community has been defined, the group will co-create a unique combination of assessment strategies to be used specifically for their learning environment. The group will be presented with a marketplace of possible assessment strategies that include standardized testing, peer review, self-reflection, teacher evaluation, and portfolio creation. Each method can be clicked, which opens additional information about that particular assessment method's functionality. Through a drag-and-drop builder, the group will co-construct their assessment recipe by combining the available choices. Once the assessment blocks are in place, options to apply different weighting and rubrics will be available to create a customizable plan. The final chosen combination of elements will essentially determine how grading will function within the classroom. After the assessment combination is created, each member of the learning community has an opportunity to sign off in agreement on the ground rules and assessment recipe, enacting a shared governance contract. This initiates the grading period.

During the grading period, each member of the learning community can see their own progress in relation to anonymized assessment data from other members of the group. Through the online collaborative functionality of the tool, peer reviews, tests, visual portfolios, and other assessment platforms are enabled. The learning community can update their ground rules and assessment combination to be responsive to unforeseen changes in the learning environment (Fig. 3)..

1. Participation in Allgrade
  · Regular engagement as a member of a learning community on the platform

2. leads to

3. these outcomes
  · More transparency about grading policies and standards
  · Leadership and decision-making equity for all stakeholders in the classroom
  · Higher interest to participate in learning environments that prioritize cooperative frameworks

4. which are represented by

5. Classroom Community Agreements
  · All stakeholders must discuss and agree on policies and standards enacting a shared governance model
  · Many learning styles can be voiced and valued
  · Metacognitive planning is central to the process
  · Community-based inclusionary pedagogies are prioritized

6. which enable

7. Assessment Combinations
  · Collectively built and curated recipes designed to meet the needs of the specific learning context
  · This is customizable with options to apply weighing and rubrics, which may change over time
  · Many assessment strategies (standardized testing, peer review, portfolios, etc.) can be easily combined

7 a. Assessment Recipe Artifacts
  · Documented outputs from completed recipes including test scores, reflective writing, peer reviews, teacher evaluations, portfolio images, etc.
  · Artifacts are unique to the individual but can be combined with others to gain insight into classroom trends

8. which contribute to
  · Sense of tangible achievement in classroom learning
  · Self-efficacy and understanding path to personal growth in relation to others
  · A sense of belonging in the learning community based on a shared governance model
  · Transferring extrinsic to intrinsic motivations leading to more resiliency for teachers and learners

Fig. 3. An iteration of Allgrade logic model.

We are currently iterating on the logic model of the assessment platform. To better understand the assumptions, inputs, and outputs of the model, we will be working with a group of higher education teachers and students to engage in interview-based activities and develop a variety of lo-fi prototypes. Through a participatory process, we will hone the design parameters that will later inform the functionality of the Allgrade prototype.

4 Discussion and Future Work

In this position, we propose that a self-organizing assessment practice can adapt to dynamic learning environments and economic landscapes. We propose an approach that invites and empowers different stakeholders in a learner's community to develop and implement assessment plans collaboratively. Allgrade is a technology platform that enables these activities in a learning environment.

Many aspects of the design require further consideration in this experimental phase. The long-term vision is for Allgrade to be supportive in a wide variety of learning communities, which may include formal and informal learning spaces for classes of all ages everywhere. While expanding participation is a key driver of increasing adaptability, the current model does not consider the unequal access to the internet and technology as well as the time constraints in the classroom. For example, educators, students, and other stakeholders may not have ample time to organize and develop a set of community agreements. Our technological approach has an implicit assumption that the design and functionality of the virtual tool will enable in-person participatory behaviors and facilitation activities in the classroom.

To address these issues in an evidence-based approach, we will establish school-based teams among our existing high school and university collaborators that prioritize the support of learners in under-resourced communities throughout New York City. A core group of educators, students, and families will be identified as the pilot design team. These key stakeholders will provide invaluable and ongoing feedback about the platform and processes we adopt. From this group, we will seek to build a teaching team with a range of academic subjects. Following a summer design sprint, we will build a working prototype of the Allgrade app that works with existing products available to educators and students. Throughout the year, we will host retreat sessions, classroom visits, community meetings, and interviews to learn how Allgrade is being incorporated into classroom learning and how we can adjust and improve the platform. Throughout the pilot phase, we will document our interactions and develop external resources and a guide for educators to implement Allgrade into their classroom teaching. These external materials along with key learnings from the in-progress pilot will position us to recruit additional classrooms to join the initiative. We hope these works will provide tangible benefits to the K-12, university, and adult learning educators nationwide and positively contribute to the learning science research community.

References

1. Anderson, G., Boud, D., Sampson, J.: Learning Contracts: A Practical Guide. Routledge, London (2013)
2. Association NE History of Standardized Testing in the United States | NEA. https://www.nea.org/professional-excellence/student-engagement/tools-tips/history-standardized-testing-united-states. Accessed 14 March 2022
3. Brown, A.L.: Design experiments: theoretical and methodological challenges in creating complex interventions in classroom settings. J. Learn. Sci. 2, 141–178 (1992)
4. Cooper, A., Klinger, D.A., McAdie, P.: What do teachers need? an exploration of evidence-informed practice for classroom assessment in Ontario. Educ. Res. 59, 190–208 (2017). https://doi.org/10.1080/00131881.2017.1310392

5. Engeström, Y.: Activity theory and individual and social transformation. In: Engeström, Y., Miettinen, R., Punamäki-Gitai, R.-L. (eds.) Perspectives on Activity Theory, pp. 19–38. Cambridge University Press (1999)

6. Froese-Germain, B.: Standardized testing + high-stakes decisions = educational inequity. Interchange **32**, 111–130 (2001). https://doi.org/10.1023/A:1011985405392

7. Grant, H., Dweck, C.S.: Clarifying achievement goals and their impact. J. Pers. Soc. Psychol. **85**, 541–553 (2003). https://doi.org/10.1037/0022-3514.85.3.541

8. Guijt, I.: Participatory Approaches: Methodological Briefs—Impact Evaluation No. 5 (2014)

9. Leinonen, T., Toikkanen, T., Silfvast, K.: Software as hypothesis: research-based design methodology. In: Proceedings of the Tenth Anniversary Conference on Participatory Design 2008. Indiana University, USA, pp. 61–70 (2008)

10. Odera, E.L.: Capturing the added value of participatory evaluation. Am. J. Eval. **42**, 201–220 (2021). https://doi.org/10.1177/1098214020910265

11. Scott, S., Webber, C.F., Lupart, J.L., Aitken, N., Scott, D.E.: Fair and equitable assessment practices for all students. Assess. Educ. Princ. Policy Pract. **21**, 52–70 (2014). https://doi.org/10.1080/0969594X.2013.776943

12. Webber, K.L.: The use of learner-centered assessment in US colleges and universities. Res. High. Educ. **53**, 201–228 (2012)

13. Wholey, J.S., Hatry, H.P., Newcomer, K.E.: Handbook of Practical Program Evaluation. John Wiley & Sons (2010)

Teaching in Times of COVID-19: The Effects of Course Standardization as Perceived by Lecturers

Daniela Waller[1] , Alexandre Woelfle[1], and David Rueckel[1,2]([⊠])

[1] Department Computer Science, University of Applied Sciences Technikum Wien,
Hoechstaedtplatz 6, 1200 Vienna, Austria
{daniela.waller,alexandre.woelfle,
david.rueckel}@technikum-wien.at
[2] Institute of Business Informatics - Information Engineering, Johannes Kepler University Linz,
Altenberger Straße 69, 4040 Linz, Austria

Abstract. The current study reports on a university's initiative to standardize courses and the effect of this on teaching during the COVID-19 pandemic. Based on the Bologna Process, which aims to harmonize academic programs that improve national and international comparability in European countries, the case institution started initiatives to foster the standardization of course didactics. At course level, the standardization provides a blended-learning approach, a close interlinking of self-study phases, and an application-driven approach during classes. Literature clearly shows the effects of standardized course infrastructure and didactics in general (e.g., comprehensive learning paths) and the impacts of certain didactic elements in particular (e.g., video quality). However, we identified a perceivable gap in finding effects and impacts of standardization during the COVID-19 pandemic and its associated challenges in teaching those academic contents online. Therefore, we conducted a qualitative study aiming to clarify those effects and impacts. Lecturers, teaching in both standardized and non-standardized courses, have been interviewed. Afterwards, the semi-structured interviews were transcribed and analyzed using qualitative content analysis. Preliminary results show inhibiting effects concerning flexibility and agile course handling on the one hand and unequivocally positive effects towards clarity and transparency in teaching standardized courses. To summarize, we were able to derive thoughtful insights into the field of tension between the need to react deftly to profound global interventions and the benefits of having a transparent, standardized, but rather rigid course didactic.

Keywords: Standardization · Courses · Pandemic · Qualitative research · Lecturer

1 Introduction

The Bologna Process provides a means by which higher education institutions (HEIs) can be encouraged to provide more attractive curricula for the younger generations whilst considering the broad range of engineering fields [1]. The major goal of the Bologna Process aims to harmonize academic programs that improve national and international comparability in European countries. This leads to both centralized (e.g., country-wide) and decentralized (e.g., university-wide) initiatives for fostering the standardization of course didactics [2].

Within the case institution—a university of applied sciences in central Europe—a project to standardize all bachelor courses was initiated in the winter of 2019/2020. Since then, all of the courses in 12 bachelor study degree programs have been standardized to harmonize the academic courses throughout the institution. The standardizing process of a course comprises, on the one hand, the creation of Moodle courses according to a unified format, which strengthened the constructivist learning and teaching approach, and on the other hand, the adaptation and development of didactic methods based on the constructivist learning and teaching approach, which is now in focus.

The case institution's teaching and learning center (TLC), which is a key development and support team of the standardization project, outlines the advantages of the standardization process from various perspectives [3]. A principal advantage for students is their spatial and temporal independence: i.e., the course contents can be freely edited and processed in a self-regulated manner within the set deadlines. Another advantage for students, in terms of learning content and performance requirements, is the consistent transparency of course design. Advantages for lecturers include the reusability or adaptability of learning and teaching content for further courses and contexts, thereby requiring minimal effort and saving them time in developing eLearning courses and lesson materials, as well as the sustainable use and further development of the learning management system (LMS) Moodle. Thus, the standardized process, as well as the standardized teaching and learning content of courses, and the comparable examination results within the case institution as well as with other universities, are major advantages of university-wide standardization for the case institution [3].

2   The State of the Field

Here, we briefly summarize the state of the field by introducing three key topics: (1) Constructivism in learning, (2) (applied) blended learning approaches, and (3) teaching during the COVID-19 pandemic.

2.1   Constructivism as the Learning Theory of Choice for Standardization/Harmonization

As mentioned above, the standardization of courses at the case institution strengthened constructivism. Constructivism is one of the learning theories that evolved from substantial studies of cognitive development (i.e., how thinking and knowledge develop with age) by Swiss psychologist Jean Piaget and Russian psychologist Lev Vygotsky. Vygotsky's theory was like Piaget's assumptions, but Vygotsky placed more importance on the social context of learning. While in Piaget's theory, teachers played only a limited role, in Vygotsky's theory, the teacher played an important role in learning activities by active engagement, inquiry, problem solving, and collaboration with others. The teacher is more of a guide, facilitator and/or coach who encourages learners to question, challenge, and formulate their own ideas, opinions, and conclusions, rather than just a knowledge giver [4]. Thus, constructivists understand learning as an explicatory, repetitive building process by active learners that interrelates with the physical and social world [5]. In contrast, as emphasized by Snelbecker [6], teachers cannot restrict themselves to only one learning theory as constructivism; they are urged to examine each of the basic science theories in the study of learning and to select those principles and conceptions which seem to be of value for one's particular educational situation [6].

The construction of the standardized courses at the case institution represents a blended learning approach for all study degree programs, which includes presence/contact phases as well as self-study phases to harmonize all courses. The idea, based on the explicit implementation of self-study phases in contrast to the previous distance-learning phases, was to strengthen the constructivist learning approach of students. This means that students must learn the "theoretical" knowledge by themselves during the self-study phases and apply the learned competence during the contact phase on further examples with higher complexity. Even if in the self-study phases the learning paradigm of constructivism was strengthened, mixed didactic methods, which also support other learning paradigms, have been added not only to the presence/contact phases, but also to the self-study phases.

2.2 Blended Learning Concepts

Before the case's institution-wide process of standardization started, the bachelor and master's degree programs had been set up according to three distinct blended learning concepts [7]. Full-time study degree programs have been set up according to the enrichment concept where classes/presence phases are only supported by self-study phases. In contrast, part-time study degree programs comprised a balanced proportion of classes/presence phases and self-study phases. The distance learning study degree programs were set up using the virtualization concept, wherein classes were mainly replaced by online phases supported by lecturers as "eCoaches" (Fig. 1).

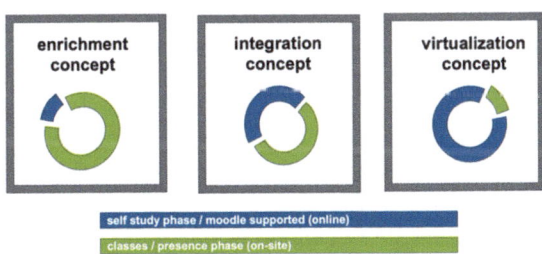

Fig. 1. Blended learning concepts (according to Ref. [8]).

As mentioned above, the part-time study degree programs have been set up using the integration concept; therefore, the standardization process less of a deep impact on these study programs than on the full-time study programs, wherein the enrichment concept and, therefore, the presence phases have been more dominant. The integration concept has been realized in all types of study degree programs, which, in full-time studies, leads not only to the reduction of presence phases but also to a switch in the learning paradigm from behaviorism to a more constructivist teaching and learning approach within the case institution. The standardized process led from the student point of view to more intensive self-study phases and less accompanied distance study phases. Thus, the responsibility for acquiring knowledge lies in advance, before the respective presence phase, at students' side and is reflected in the presence phase with the lecturer. From the student's point of view, this led to topic discussions at eye level with lecturers.

2.3 The COVID-19 Crisis and Switching to Online Learning

As the COVID-19 crisis hit the case institution by March 2020, the organization was forced to immediately convert its teaching from face-to-face to online. The case institution's faculty of Computer Science runs two bachelor programs and six master programs, offering more than 22,500 h of teaching in the field of Computer Science, especially in Software Engineering, DevOps, Digital Enterprise, UX, AI and Data Analytics, and Information Security. At this time, study programs were either designed as face-to-face or as blended learning programs. The main course types are integrated courses (45%) and exercises (20%), followed by distance learning (10%) and others (totalling 25%). The department consists of 38 permanent employees and about 200 external lecturers. Within the blended learning programs, around 30–60% of teaching already took place online before the COVID-19 lockdown came into effect. Shifting all teaching to online within days is a risky process. A complete online course requires an elaborate lesson plan design, teaching materials such as audio and video contents, as well as technology support teams [9, 10]. However, as there was no other choice, all teachers involved put enormous effort into rapidly shifting their individual courses into a suitable online format.

The main challenges in these distance studies have been the didactic planning of the courses, as well as ongoing support and motivation of the students during the eLearning phases, as also mentioned in Ref. [11]. In various funding projects, tools to support lecturers as well as students have been developed. These tools comprised, for example, a matrix to transfer didactical approaches of on-campus phases to eLearning phases, extensive training offerings in the application of eLearning tools such as Learning Management Systems (e.g., Moodle), Authoring tools (e.g., Articulate Storyline), and communication tools (e.g., Skype4Business), as well as certain licenses to use these software systems. Despite all these supporting instruments, there are many potential obstacles for lecturers and students during eLearning phases. Based on the shock which was created by the COVID-19 crisis, it took some time to check the reusability of these instruments in the in-person study degree programs [12].

To summarize, we derived the following research question: What effects concerning the standardized course environment were perceived by lecturers during the COVID-19 pandemic?

3   Methodology

To answer the defined research question, we initially conducted literature research to assess the current state of the field. This was followed by expert interviews at the case institution within the Computer Science department. After the expert interviews were done, we applied qualitative data analysis [13].

3.1   Literature Research

We reviewed the existing literature and assessed the current status of standardized learning documentation, allowing us to determine the current situation and related problems. We conducted a paper review on widely used scientific databases such as IEEE Explore, OECD Digital Library, Springer, and used search engines such as Google Scholar. We used keywords such as "standardized teaching," "standardized lecturing," and "covid 19 university." We narrowed down the results by excluding studies that focused neither on standardized learning nor on distance learning during the pandemic.

3.2   Expert Interviews and Transcription

We conducted semi-structured interviews with experts, following a guideline to ensure the rigor of our research. The main goal of the interviews was to obtain an expert review, relying on a set of experts to gain a broad evaluative view of our designed framework. We selected didactic experts working at the case institution, as well as internal and external lecturers, who gave both standardized courses and non-standardized ones. This procedure resulted in a total of nine interviews, which generated insights into various domains linked to online lecturing, as well as standardized and non-standardized courses, and enabled us to evaluate the framework from both academic and practical perspectives. We developed an interview guideline to gain a deeper understanding of the different types of courses (i.e., standardized, non-standardized, face-to-face, distance learning), in the context of the relevant domain. We revealed and explained the framework to the nine interviewed participants, who are internal and external lecturers at the case institution (see Table 1) and asked them nine questions to share their opinions, thoughts, and impressions about the mentioned subjects. The interviews were conducted in the native language of the participants and took place between February 17 and February 22, 2022. The interviews were digitally recorded, transcribed, and analyzed using MAXQDA software. The interviews' lengths ranged between 22 and 28 min. an average of 25 min.

3.3   Qualitative Data Analysis

To perform our qualitative data analysis, we opted for the qualitative content analysis method according to Mayring [13]. As described in Ref. [13], we followed the seven-step model of inductive category development. In the first step of the qualitative content analysis, we defined the research question as earlier proposed. In the second step, we defined the categories (criterion of selection) and levels of abstraction for inductive categories (see Table 2). The inductive categories were formulated step by step based

Table 1. Interviewed experts of the Computer Science Department at the case institution.

#	Participant	Length of employment	Area of expertise	Standardized course/s	Non-standardized course/s
1	External lecturer	>3 years	Communication skills, change management	Creativity and complexity	Moderation and problem solving
2	Internal lecturer	>3 years	Database fundamentals, big data infrastructures	Data management	Database systems
3	Internal lecturer	<3 years	Usability evaluation and user experience design	Selected chapters of computer science	User experience design
4	Internal lecturer	>3 years	Object-oriented programming	Structured programming lab	Structured programming lab
5	Internal lecturer	>3 years	Business process and software analysis, modeling	Introduction to business informatics	Fundamentals of ERP systems
6	External lecturer	>3 years	Enterprise resource planning systems	IT-based accounting	IT-based accounting
7	External lecturer	>3 years	Algorithm and data structures, artificial intelligence	Algorithm and data structures	Introduction artificial intelligence
8	Internal lecturer	>3 years	Web development, data analytics	Data ethics and open data	Smart data
9	External lecturer	>3 years	Business process and software analysis, modeling techniques	Introduction to business informatics	Software selection project

on the transcribed interviews. The more general upper-level categories were deduced by evaluating the framework, and the lower-level categories, which emerged from the raw material, followed an inductive approach allowing for open coding. We defined five categories: *lockdown online education, participation in course development, evaluation, standardized courses,* and *non-standardized courses.*

After our categories were set, we started coding via MAXQDA, subsumed some categories, and formulated subcategories. After coding approximately one third of the expert interviews, we revised the categories to avoid bias and ensure reliability. In the next step, we proceeded to code the remaining expert interviews. Once all the interviews had been coded, we conducted a consistency check through an intra-coder reliability check, as commonly recommended. At a later stage, an inter-coder reliability check was done based on the codings of two authors, using MAXQDA. In the final step, we

Table 2. List of categories and frequencies per category.

#	Main category	Sub category	Frequency
C1	Lockdown online education		
C1a		No change	8
C1b		Advantages	12
C1c		Disadvantages (organization, didactics)	35
C2	Participation in course development		
C2a		Yes	5
C2b		No	9
C3	Evaluation		
C3a		Feedback not derivable	8
C3b		Non-standardized courses	3
C3c		Standardized courses	23
C4	Non-standardized courses		6
C4a		Course implementation	14
C5	Standardized courses		7
C5a		Planning versus real implementation	16
C5b		Continuous improvement process	1
C5c		Course implementation	105
C5d		Preparation of the presence phases	34

interpreted the results of our qualitative analysis, which we describe in more detail in the following chapter.

4 Results

As our first step, we had to reference all the courses our interview participants are lecturing for. Thus, we identified whose standardized and non-standardized courses the interviewed expert had taught since the pandemic broke out in the summer semester of 2020.

4.1 Differences in Lecturers' Preparation of Face-to-Face Classes

For the first question of the interviews, we asked the lecturers what differences they have been noticing between standardized courses and non-standardized courses regarding the preparation of face-to-face classes. Many of the lecturers reported that switching to standardized learning made a significant difference for them as standardized courses offer them fewer options to choose from in terms of didactics and learning material, in comparison with non-standardized courses. While this makes class preparation require less effort from the teachers, it also gives them less flexibility in their teaching. Due to the more limited options for decision-making, some lecturers experienced feeling less important in the entire didactic process.

4.2 Participation in the Development of Standardized Courses

The amount of effort lecturers need to put into the preparation of face-to-face classes has been presented by some of them as depending both on the personal participation in the development of standardized courses as well as on the years of teaching experience in the corresponding courses. Those insights and findings led us to question which lecturers who took part in our interviews have been actively participating in the development of the standardized courses they are teaching themselves. Five out of nine lecturers confirmed that they have been involved in this development process, which means that more than half of the respondents had deeper insights and influence on the planned didactical methods and learning materials prior to giving those standardized lectures.

4.3 Advantages and Disadvantages of Giving Standardized and Non-standardized Courses

The next question aimed at creating a discussion about the advantages and disadvantages of giving standardized and non-standardized courses. Regarding the implementation of the courses, the main disadvantages mentioned by lecturers related to the didactical approach. Participant 2 mentioned the lack of flexibility of the teaching and learning materials as well as problems they experienced when articulating their knowledge within the requirements. Participant 5 raised the issue that it seems antagonistic that, even though it is based on a constructivist approach, everything in the standardized approach is quite linear, fixed, and rather inflexible. This seemed to be more observable for long-time lecturers with a lot of experience. Regarding organizational factors, lecturers further mentioned the lack of preparation from students during self-study time and a higher administrative effort than in non-standardized courses, as indicated by Participant 4. The advantages of standardized learning could rather be pinpointed on the students' side, as standardized courses are providing opportunities to students with autonomous learning skills. With such a learning structure, students need to develop new skills and think more by themselves, which can be of great help. Even if it is challenging, they are confronted with independent acting and thinking during the whole learning process, which is always rewarding one way or another.

Our results show that there is a clear differentiation that can be grasped in the advantages and disadvantages of the standardization process for students and lecturers. For the former, standardized learning constitutes challenges and growth opportunities, yet it is considered as too overwhelming and difficult for many students, especially the ones that have not already experienced self-study in a constructivist way, as in the standardized courses of the case institution. Lecturers see fewer advantages in standardized courses, and it is generally experienced as lowering the chance to involve individual aspects of teaching and personal involvement. Participant 1 further stated that his practical knowledge and competence in study field could hardly be turned into standardized courses.

Analyzing and gathering insights about non-standardized courses was particularly important in order to compare them with the findings of the standardized courses. Most participants of the expert interviews reported that non-standardized courses were more advantageous to them than standardized ones and showed better results in terms of course implementation. According to Participant 2, one major advantage is based upon the possibility of adapting the classes to the student's actual state of knowledge and to react rapidly to all kinds of organizational changes. Since lecturers see higher flexibility in non-standardized courses, it seems that not having to follow strict didactical guidelines reduces pressure for them. Participant 9 further mentioned that, while the students' preparation prior to the lectures is necessary for standardized courses, it can be considered as more optional for non-standardized lectures. In other words, in some cases, there is evidence of poorer preparation from students before classes in non-standardized courses. Moreover, the non-standardized courses require more effort in terms of learning content creation, thus the quality of the course depends on the amount of time spent by lecturers for all the preparations upstream. Lower-quality preparation by the lecturer leads to a lower-quality course in general, which is not necessarily the case for standardized courses as standardized courses provide lecturers with ready-to-use learning content, providing better quality assurance.

4.4 Completion of Planned Didactical Guidelines in Standardized Courses

Following our interview guideline, we aimed, with the next question, to assess whether lecturers have been able to follow the given didactical guidelines or if they had to deviate from the didactical path (and if they did, for what reasons). Our interview results demonstrated that lecturers almost followed the didactic requirements which were given for the standardized courses. The participants responses show that it was not always possible for them to follow those requirements but that they did whenever it was possible.

4.5 The Influence of Lockdowns and the Switch to Online Education

COVID-19 has been a major challenge during the last two years, in terms of both didactics and education. Therefore, we wanted to find out what the main challenges and pitfalls regarding online education were as well as its opportunities and advantages, especially for standardized courses.

Asking experts that gave lectures in online as well as non-online settings has been highly valuable for making some further comparisons. The interviews' results showed that their general satisfaction significantly depended on the type of classes the lecturers were involved in, as also shown in Ref. [14]. According to some lecturers, online education and classes have been an opportunity to consider new perspectives and have even helped the learning process, especially with the use of breakout rooms and the new ways of experiencing group work.

Some of the expert interview participants acknowledged that their classes have been more difficult to teach on the campus during the COVID-19 pandemic and that online education brought some significant benefits—for instance, using screen sharing when exercises had to be done within the software applications. Another aspect that we could point out from the interviews as being an advantage is the fact that technical issues could often be handled in an easier and quicker fashion online in contrast to a campus setting (mentioned by Participant 4). Moreover, it was quite interesting to notice that the online setting also offered an opportunity for students to ask questions with less pressure, especially when in smaller exercise groups.

One commonly mentioned disadvantage of online education is the social factor, which every participant confirmed as something truly negative. The human factor plays an important role in didactics and in the well-being of students as well as lecturers. The isolation due to online teaching and missing physical meetings have mainly been experienced as a pain point and sometimes led to a lack of motivation and attention difficulties on the students' side. Another disadvantage that has been mentioned by a few participants (for example, Participant 3) is that lecturers have difficulties seeing and following what the students are doing (such as if they are working on exercises) and obtaining students' feedback during lectures.

4.6 Feedback

In the final part of the interview, we asked the lecturers if they noticed significant differences between students' feedback for standardized and for non-standardized courses. Standardized courses received some mixed feedback, from neutral to negative. Most of the feedback concerning standardized courses were on the negative side of the spectrum, with effort and overload during self-study phases being the main factors of dissatisfaction. Furthermore, it is worth mentioning that some topics were too complex for students to be learning by themselves.

The constructivist learning approach was not originally designed or intended for students at universities of applied science, in contrast to standard universities which already worked in a more constructivist manner. Participants of the expert interviews noted that students frequently experienced feeling lost and not knowing what exactly to do and how. This new way of approaching studies and articulating both very strict guidelines while still having a certain freedom could lead, in some cases, to frustrating and overwhelming the students. Participant 7 also mentioned that some students had the impression that the learning material was no longer properly explained by lecturers and that they had to do almost everything by themselves. They did not feel as supported in the learning process as they might have needed. However, in accordance with what lecturers reported themselves, some of the students saw standardized courses as an opportunity to put more effort into learning collectively so that face-to-face lectures could work better (Participant 1). Working more autonomously requires more skills and resourcefulness but it can truly be an asset for the student's future, as independence, creativity and improvisation are some valued skills, both for studies and for work life.

5   Discussion

Our study provided us with various insights into standardized, non-standardized, and online lecturing—topics that have been significant during the COVID-19 pandemic. Talking to lecturers through expert interviews has been an opportunity to discuss and confront different perspectives and to compare the individual experiences of lecturers.

We discovered recurrences in the gathered feedback, as well as discrepancies. This can be partly explained by the fact that courses and programs cannot always be compared in terms of structure, organization, and requirements. Subjectivity also makes a difference regarding perception.

Standardized and non-standardized courses both offer opportunities and weaknesses. Standardized courses have been described as more rigid, often as less efficient, and as more difficult to implement by lecturers. Standardized courses also show a noticeable challenge for students because of the higher level of independence and personal initiative required in the learning process. The courses are based on a constructivist didactical approach but do not really offer students more flexibility in finding their own individual learning strategies and paths. To improve the standardization of courses and its constructivist learning approach, one recommendation could be to define the learning requirements less strictly while guiding students to a more individual and self-determined way of learning. For instance, students could be helped in finding their own methods of knowledge acquisition by providing them a pool of didactical methods and different tasks. Additionally, students may benefit from integrating more face-to-face classes when necessary. Whilst the number of face-to face lessons is working well for students and lecturers for some courses within the standardized environment, other courses with technical learning content of high complexity and a strong focus on practical application (e.g., programming, databases, modeling) would need more face-to-face lessons on campus and support during self-study phases. In this way, students could understand the learning essentials of their respective discipline and be able to apply this newly gained knowledge in practical examples. This is crucial for full-time students because the majority of students in these programs are much younger than the part-time students and therefore have less experience in the practical application of theoretical knowledge, and particularly in self-studying. A further recommendation would be to train the students by teaching organization techniques and methods of constructivist learning in various courses within the field of social competence. The COVID-19 pandemic, and especially the abrupt change to online learning, showed us how important it is for lecturers and for students to cope with online learning environments and settings without sacrificing quality.

References

1. Uhomoibhi, J.O.: The Bologna Process, globalisation and engineering education developments. Multicult. Educ. Technol. J. **3**, 248–255 (2009). https://doi.org/10.1108/175049709 11004255
2. The SAGE Handbook of International Higher Education. SAGE Publications, Inc., 2455 Teller Road, Thousand Oaks California 91320 United States (2012). https://doi.org/10.4135/ 9781452218397
3. Krizek, G., Langer, K., Lietze, S.: Standardisierung & Digitalisierung von Hochschullehre am Beispiel der Lehrveranstaltung "Grundlagenlabor Physik" (2020). https://doi.org/10.13140/ RG.2.2.23872.56328
4. Remmel, E.: Constructing cognition. Am. Sci. **96**, 80 (2008). https://doi.org/10.1511/2008. 69.80

5. VanTassel-Baska, J.: Book reviews. In: Fosnot, C.T. (ed.): (1996) Constructivism, theory, perspectives, and practice. Teachers College Press, New York. Gift. Child Q. **41**, 113–114 (1997). https://doi.org/10.1177/001698629704100308

6. Snelbecker, G.E.: Instructional design skills for classroom teachers. J. Instr. Dev. **10**, 33–40 (1987). https://doi.org/10.1007/BF02905309

7. Overview of blended learning: the effect of station rotation model on students' achievement. J. Crit. Rev. **7** (2020). https://doi.org/10.31838/jcr.07.06.56

8. Henrich, A., Sieber, S.: Blended learning and pure e-learning concepts for information retrieval: experiences and future directions (2008). https://doi.org/10.1007/s10791-008-9079-3

9. Paul, J., Jefferson, F.: A comparative analysis of student performance in an online vs. face-to-face environmental science course from 2009 to 2016. Front. Comput. Sci. **1**, 7 (2019). https://doi.org/10.3389/fcomp.2019.00007

10. Knapp, N.F.: Increasing interaction in a flipped online classroom through video conferencing. TechTrends **62**(6), 618–624 (2018). https://doi.org/10.1007/s11528-018-0336-z

11. Fojtík, R.: Problems of distance education. Int. J. Inf. Commun. Technol. Educ. **7**, 14–23 (2018). https://doi.org/10.2478/ijicte-2018-0002

12. Pucher, R., Holweg, G., Dolezal, D., Eckkrammer, F., Geyer, S., Redl, C., Salzbrunn, B., Waller, D.: Did COVID19 improve our teaching? In: ICERI2020 Proceedings. ICERI Proceedings 3981–3992 (2020). https://doi.org/10.21125/iceri.2020.0897

13. Mayring, P.: Qualitative Inhaltsanalyse: Grundlagen und Techniken. Beltz, Weinheim Basel (2015)

14. Bezic, H., Balaz, D., Buljat, B.: Harmonization of curriculum with the needs and requests of the Fourth Industrial Revolution: Case of Faculty of Economics and Business Rijeka. In: 2020 43rd International Convention on Information, Communication and Electronic Technology (MIPRO), pp. 694–699. IEEE, Opatija, Croatia (2020). https://doi.org/10.23919/MIPRO4 8935.2020.9245191

ALICE (Adaptive Learning via Interactive, Collaborative and Emotional Approaches) Special Track

Lessons Learned from Using Conversational Agents to Support Collaborative Learning in Massive Open Online Learning

Santi Caballé[(✉)] ⓘ, Jordi Conesa, David Gañán, and Joan Casas-Roma

Faculty of Computer Science, Multimedia, and Telecommunication, Universitat Oberta de Catalunya, Barcelona, Spain
{scaballe,jconesac,dganan,jcasasrom}@uoc.edu

Abstract. Although Massive Open Online Courses (MOOCs) have been reported as an efficient and important educational tool, there are issues related to their educational impact. More specifically, there are dropouts during a course, little participation, and lack of learners' motivation and engagement overall. This paper describes the results of the European project "colMOOC" that aimed to enhance the MOOCs experience by integrating collaborative settings based on conversational pedagogical agents to support both learners and instructors during a MOOC course. Integrating this type of conversational agents into MOOCs to trigger peer interaction in discussion groups was evidenced to considerably increase the engagement, commitment, and satisfaction of online learners and, consequently, reduce MOOCs dropout rate. This paper presents the experimentation and evaluation results of a series of pilot trials in a real largely participated MOOC course. Lessons learned and good practices are eventually provided for practical application to make the most of conversational agents to support collaborative learning in MOOCs.

Keywords: MOOC · Conversational pedagogical agents · Collaborative learning

1 Introduction

This paper reports on the results of a European project called colMOOC (https://colmooc.eu/) [1] that aimed to enhance the MOOCs experience [2, 3] by integrating collaborative settings based on Conversational Agents (CAs) in synchronous collaboration conditions. CAs guide and support student dialogue using natural language both in individual and collaborative settings [4] and have been produced to meet a wide variety of applications, and studies exploring the usage of such agents have led to positive results. Integrating this type of CAs into MOOCs is expected to trigger productive peer interaction in discussion groups and, therefore, to considerably increase the engagement and the commitment of online students, reducing, consequently, the overall MOOCs dropout rate. To achieve this, the project employed collaborative activities supported by CA [3, 4].

To achieve the above aims, the work methodology of the colMOOC project followed a four-step loop approach, including (i) the educational design, (ii) the system integration,

(iii) the MOOC evaluation, and (iv) community's dissemination (Fig. 1). This paper focuses on the third step of this work methodology and, in particular, the pilot MOOC named "Educational Technologies in the Classroom (2nd edition)", which is the MOOC used for the evaluation purposes reported in this paper.

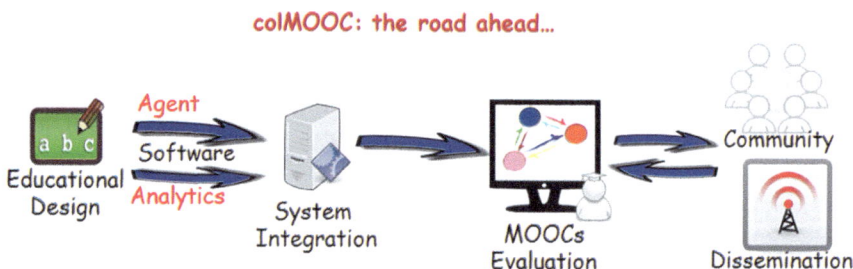

Fig. 1. colMOOC project main activities.

Considering the above, this paper aggregates the evaluation results of two educational experiences in a row of using CA to support collaborative learning activities in a real MOOCs with overall participation of 3,000 learners where CAs automatically mediated about 1,000 chat discussions in dyads (i.e., 2 learners). A full report of the first experience was published in Ref. [5], including a throughout methodology to conduct the evaluation process, which also served for this second accumulated experience, and where results were shown in terms of participation, performance, and satisfaction of both the CA-mediated activities and the MOOC overall. To avoid redundancy, we refer to Ref. [5] for the background, detailed methodology of the first study, and results while reporting here the methodology and results from the second study of using collaborative activities supported by CAs.

The remainder of the paper is structured as follows: Sect. 2 reports a detailed evaluation methodology of the pilot, while Sect. 3 presents the evaluation results from a descriptive statistical perspective. Section 4 provides the lessons learned from the evaluation results. Finally, Sect. 5 concludes the paper with final remarks and outlines the following steps in the research agenda.

2   Evaluation Methodology

In this section, we present the methodology for evaluating the CA component in the mentioned pilot MOOC. Following standard methodologies to report online education research [6], information about the participants, the apparatus used for the evaluation, and the procedure of the pilots are provided in this section. The results of the evaluation are reported in the next section.

2.1   Evaluation Goals

This study is intended to evaluate the effects of the CA-mediated activities in the participation, performance, and satisfaction behavior, as well as its connection with potential learning benefits at the MOOC and CA-mediated activities level.

2.2 Participants

Two trials were run as two editions of the same MOOC course on the MiriadaX platform[1]. The first one was run in January-March of 2020, and the evaluation results were reported in Ref. [5], while the 2nd edition was run between September and November 2020. The results of the latter are reported in this paper. This MOOC course was delivered in Spanish mainly to Spaniards and Latin-Americans.

About 1,000 people registered for the 2nd edition of MOOC with the following demographics (data refers to about 30% of registrants who answered a survey during the registration process):

- Gender of participants was balanced (43% men and 57% women).
- Age of participants ranged on average between 24 and 35 years old, and a great majority of them was between 25 and 54 (82%), while a marginal part was under 25 or older than 54.
- Academic profile of participants related to higher education included mostly teachers and researchers (43%), as well as people who had a university degree (31%), being the former group the targeting profile of this MOOC. A small part of participants were university students without a degree yet (17%).

Finally, MOOC participants were geographically distributed widely among 40 countries. The vast majority of them (88%) were located in Spanish-speaking countries (Spain and Latin-America), with Spain being the country with more participants (32%), followed by Peru (10%), Mexico and Colombia (9%), Ecuador (8%), and Argentina (5%).

2.3 Apparatus and Stimuli

The aim of the MOOC was to endow course participants with competencies and practical knowledge as for designing and implementing ICT-mediated collaborative learning and e-assessment activities. A complete description of the course is reported in Ref. [5] and a summary of the syllabus of the course is depicted in Fig. 2, highlighting the key CA activities conducted during the course.

2.4 Procedure

Overall, the five CA activities depicted in Fig. 2 were designed and implemented by the colMOOC platform throughout the course in order to support chat discussions. Each designed CA activity was implemented by the colMOOC Player tool (see Fig. 3) in the form of a chat where dyads mediated by an agent performed the proposed activity (see details of the colMOOC platform and how to create and configure a CA activity in Ref. [5]).

---

[1] Miríadax is the world's leading non-English speaking MOOC platform. https://miriadax.net/.

Start day (M0)	Week 1 (M1)	Week 2 (M2)	Week 3 (M3)	Week 4 (M4)	Week 5 (M5)	Final day (MF)
Introduction	Forum discussion	CA2 + Survey		CA4 + Survey		
Profile survey	Videos + Formative tests	Videos + Formative tests	Videos + Formative tests	Videos + Formative tests	Videos + Formative tests	Conclusive video
	CA1	CA3	Peer-review	CA5	Forum discussion	
Diagnostic test	Sumative test + Survey	Sumative test + Survey	Sumative test + Survey	Sumative test + Survey	Sumative test + Survey	Final Sumative test + Survey

Fig. 2. Course syllabus of the MOOC with the 5 CA activities highlighted in red.

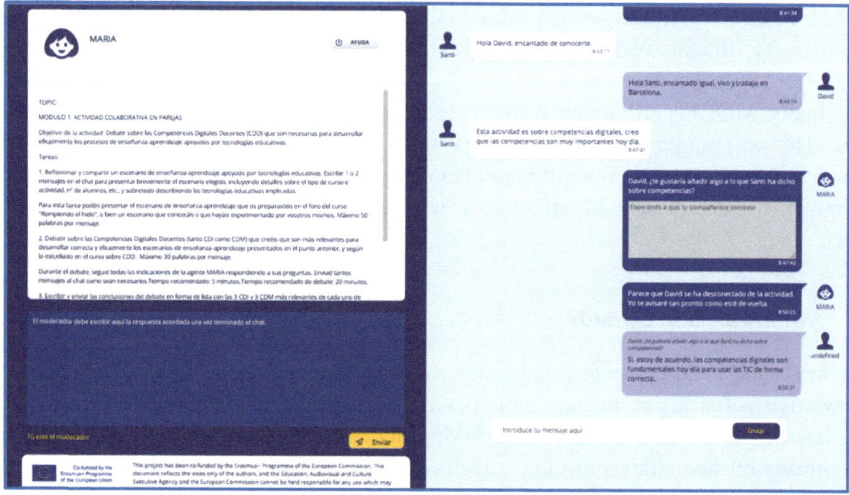

Fig. 3. Learner's view of the colMOOC Player showing a chat discussion between two learners mediated by an agent (see https://colmooc.eu/presentation-video).

3   Evaluation Results

This section presents the results of the evaluation at the MOOC and CA-mediated activities level along with a throughout discussion and interpretation of them.

3.1   MOOC Evaluation Results

The MOOC evaluation results are reported here in terms of participation and completion rate as well as general satisfaction and performance.

Participation and Completion Rate. From about 1,050 learners registered in the MOOC, 623 started the course (about 40% initial dropout rate) and 125 finished it, managing a 20% completion rate (27% the 1st edition of the MOOC and 24% on average for both editions). This is considered a fair result as the general completion rate of participants who start a MOOC is about 15% according to ample studies on MOOCs [7].

Performance. In line with the 1st edition of the MOOC [5], general performance results of the 2nd edition are positive as they show that the diagnostic test was participated by 537 learners and 69.6% passed the test, while the summative test was participated by 125 learners and 81.9% passed the test, which are quite similar results to the 67.9 and 83.9% obtained in the previous edition. This means that, as it happened in the 1st edition, the great majority of learners who performed the entire MOOC improved their cognitive state through the course's contents and understood the main concepts, thus achieving those concept-understanding objectives declared for the MOOC.

Satisfaction. As in the 1st edition, the overall satisfaction of the MOOC was measured in this 2nd edition by applying sentiment analysis on the learners' open comments in the survey conducted at the end of the course (see Ref. [5] for details of this survey). From 126 learners who submitted the survey, 71 (56%) expressed positive feelings towards the course and 13 (10%) expressed negative feelings, while the remaining 42 learners (33%) did not express any opinion with respect to the MOOC (Fig. 4). This result is consistent with the 1st edition [5], which showed 60, 9, and 31%, respectively. Therefore, the overall feeling towards the MOOC was mostly positive. The overall results are even clearer if the DK/DA responses are discarded, and then the overall satisfaction of the MOOC is evaluated in terms of responses containing positive feelings (85%), versus responses containing negative feelings (15%) for the 2nd edition (which are consistent with the 86 versus 14% results obtained for the 1st edition).

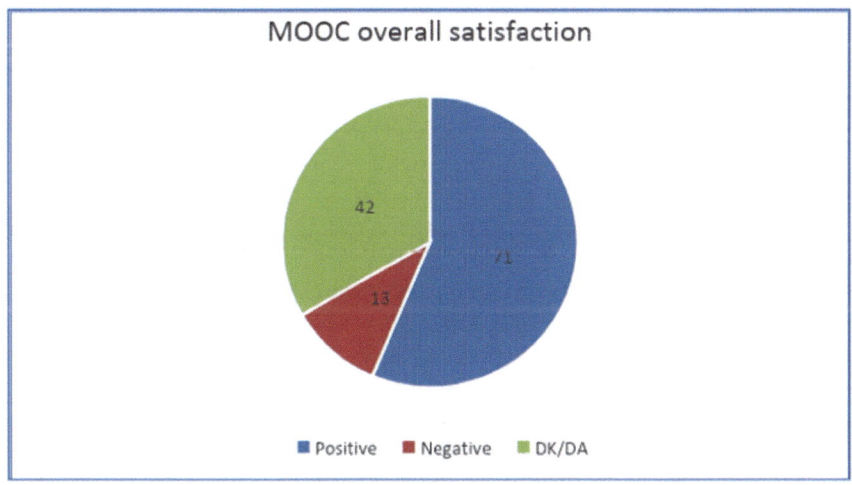

Fig. 4. Survey results on MOOC satisfaction in terms of positive/negative feelings.

3.2   CA Evaluation Results

The implementation results of the CA activities are reported here in terms of participation and satisfaction in this type of activities. Considering the 5 CA activities (see Fig. 2), about 180 chat discussions were mediated by CAs. It is worth mentioning that CA2 and CA4 were not mandatory and therefore the level of participation was lower than that of the mandatory CA1, CA3 and CA5 (see Fig. 5). Moreover, the participation in the CAs was in line with the decreasing number of learners attending the course.

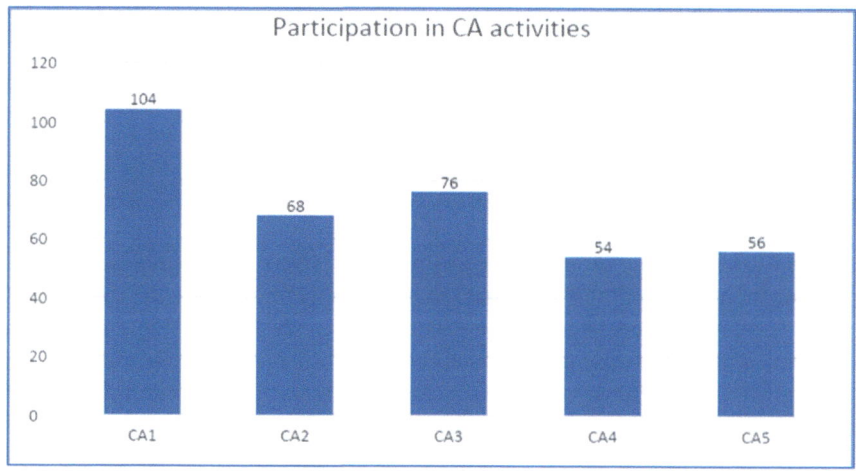

Fig. 5. Participation in each of the five CA activities of the course.

Regarding the overall satisfaction of the CA, most of the MOOC learners (54%) who participated in the research survey at the end of the course found the CA activities an interesting experience, while 17% of the learners felt indifferent to these activities. Only a marginal 7% of the learners felt negative towards those activities, as they found that the CA activities were a barrier to completing the course. Finally, 26 learners (22%) did not participate in any CA activity because of technical and organizational reasons (Fig. 6). This result is consistent with the 1st edition of the MOOC [5], which showed 61, 14, 9, and 16%, respectively, confirming the overall positive satisfaction with the CA activities.

When asking those learners who participated in the CA activities (77% of the respondents to the final survey) about what aspect they liked the most in this type of CA-mediated activities, most of them (61%) liked the exchanging of ideas with their partner and then managing to reach a common solution to the activity, while the rest of participating learners (16%) liked the interaction with the agent—either by receiving the agent's interventions or by answering to the agent's question (see Fig. 7). Again, these results are very much in line with those of the 1st edition of the MOOC (80, 65, and 15%, respectively) [5].

The previous results are supported by 3 questions addressed to the learners after each module with CA activities (for comparison, see results of the 1st edition in Ref. [5]):

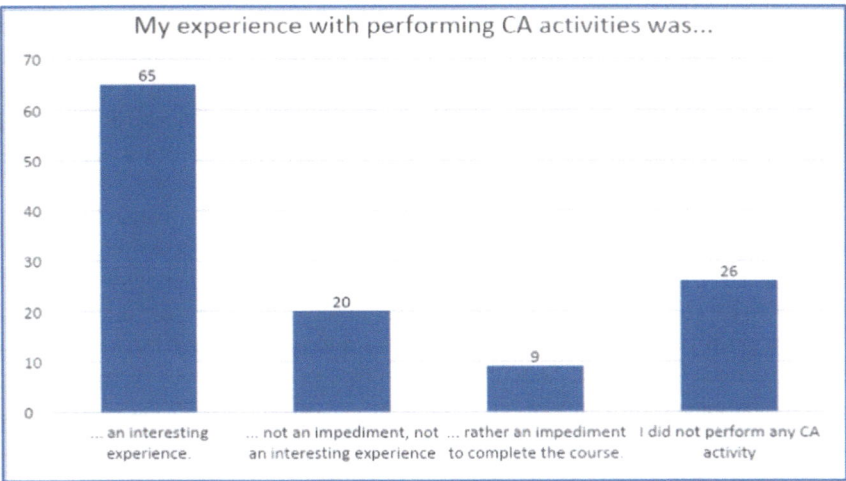

Fig. 6. Survey results showing the learners' experience with the CA activities.

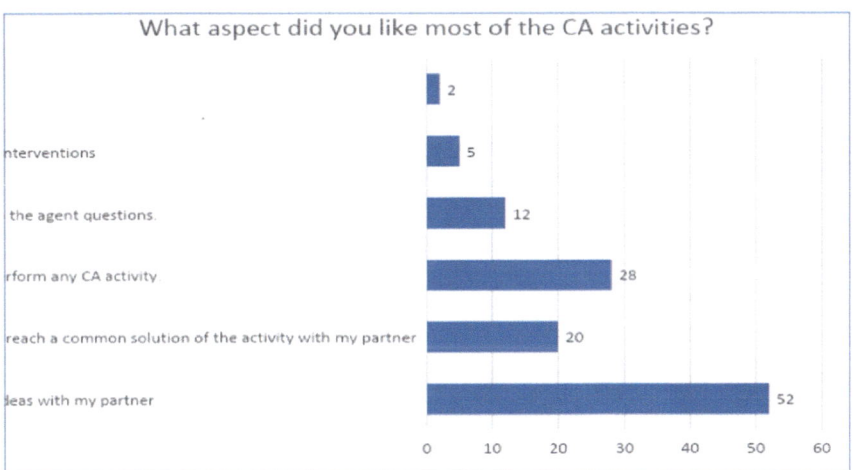

Fig. 7. Survey results showing the most interesting aspects of the CA activities.

- The first question was about whether the CA activities had been beneficial for their learning of the module. Most of the learners (between 60 and 70%) somewhat or totally agree with the question, while around 20 and 35% of them were indifferent. Only a marginal part of learners (about 10%) somewhat or totally disagreed (see left side of Fig. 8). These results are in line with the 1st edition of the MOOC with an improvement in the range of somewhat or totally agree (which was between 50 and 60%) and those who disagreed (between 10 and 20%).
- The second question asked learners about their satisfaction with the CA activities in terms of the discussions with their partners mediated by an agent. Similar to the previous question, between 60 and 70% of the learners somewhat or totally agreed

(featuring more learners who totally agreed, rather than somewhat agreed) while between 20 and 30% of them were indifferent to this question. The rest of the learners (about 10% and 20%) somewhat or totally disagreed (see right side of Fig. 8). These positive results are again in line with the equivalent results of the 1st edition, with a little improvement in the range of somewhat or totally agree (which was between 50 and 60%) and those who disagreed (about 15 and 20%) for both questions.

- The third question was new to this 2nd edition and asked the learners about their satisfaction with the chat-based CA technology to support the discussion activity with their partner. Unlike the second question, this was aimed at extracting the satisfaction level from a technological perspective. The results are depicted in the following figure, showing positive results similar to the previous question: between 60 and 70% of the learners somewhat or totally agreed (with more learners who somewhat agreed, rather than totally agreed), while between 15 and 25% were indifferent to this question. The rest of learners (about 10 and 20%) somewhat or totally disagreed (see Fig. 9).

Finally, related to satisfaction with the CA activities, the majority of them (64%) answered positively to this question, while 21% felt indifferent and only 15% answered negatively. These results and the reasons behind them are also very consistent with those of the 1st edition of the MOOC [5] (62, 19, and 18%, respectively). See Ref. [5] for the detailed justification of these results.

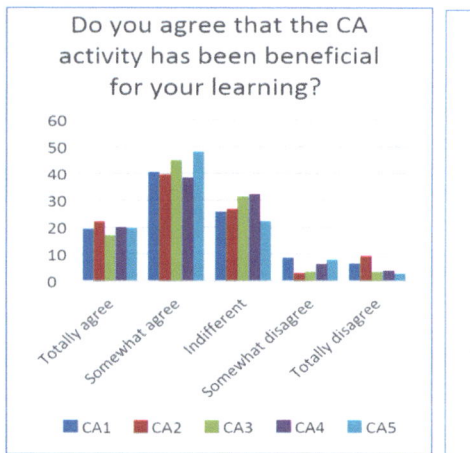 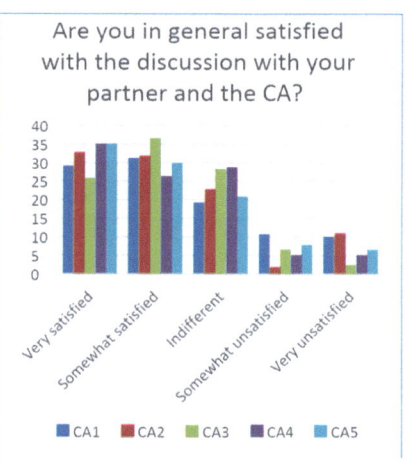

Fig. 8. Survey results on learning benefits (left) and discussion satisfaction (right) for each CA activity.

Overall, the previous results, which highlight different pedagogical and technological perspectives of the satisfaction dimension, confirm the positive perception of the CA activities by the learners of both editions of the MOOC.

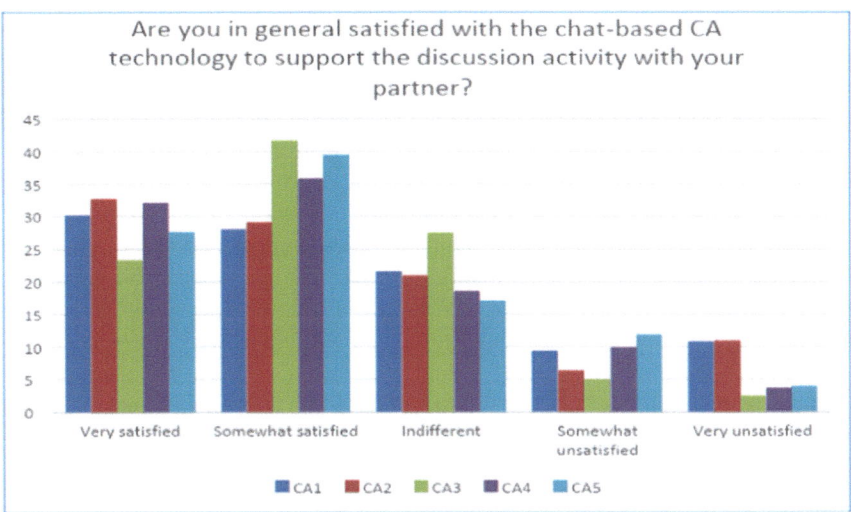

Fig. 9. Survey results on the satisfaction from the CA technological view to support the chat-based discussions.

4   Lessons Learned and Good Practices

This section summarizes the lessons learned from the CA activities during the two editions of the MOOC course while providing good practices as well as recommendations for similar experiences with teaching practices supported by these technologies. To this end, the following report is supported by the data and feedback collected in the research surveys at the end of each CA activity of the 2nd edition of the MOOC (i.e., CA1-CA5), which are informed here in terms of CA activities and agent interventions. The lessons learnt for each of these two items are compared to those reported for the 1st edition of the MOOC and evaluated overall [5].

4.1   CA Activities

The 2nd edition of the MOOC was supported by all the accumulated knowledge and experience in the 1st edition with the CA activities at pedagogical and technical levels. In particular, the design of the CA activities was carried out by the teaching team leveraging the good practices proposed after the previous trial, which highlighted as a key requirement the need to have the colMOOC platform (i.e., Editor and Player) fully implemented, tested, and integrated in the MiriadaX platform and ready to operate. The application of these good practices allowed the teaching team to conduct a smooth and effective design and implementation of the CA activities, thus avoiding the multiple technical issues faced in the 1st edition of the MOOC, as well as other shortcomings resulting from the lack of experience with both technological and pedagogical implications [5]. As a result, the satisfaction in the 2nd edition with the CA activities by learners improved about 10% the same results of the 1st edition, as reported in the previous section, and overall achieved high satisfaction levels reaching out 70%.

Moreover, the improvements achieved with regards to the satisfaction with the CA activities reported above confirms the need to carry out a sound preparatory work before running these activities in a real context of learning. In particular, the following preparatory steps are recommended: (i) the colMOOC platform and the related technologies are to be fully integrated and tested in the targeted learning platform; (ii) the teaching team involved in the design of the CA activities need to be well trained in the use of the colMOOC platform (i.e., Editor and Player) by testing different types of colMOOC configurations using fake learners' accounts in order to check and learn from the col-MOOC execution and performance; and (iii) the teaching team should test different CA designs either as part of the previous training, or separately, although in all cases it is key to receive both feedback from other teachers involved, as well as the outcomes of the colMOOC platform itself. Being a long process, these steps of preparatory work should be planned well in advance to the start of the CA activities, thus ensuring the best possible learning experience for the participants (teachers and learners).

4.2   Agent Interventions

In line with the 1st edition of the MOOC, in general, the participants of the 2nd edition were satisfied with the agent interventions because the agent had the role of a mediator between two unknown partners and allowed a reflection on the key concepts taught in this module. In addition, the CA activity provided more interactivity between the two learning partners. However, from the feedback received in the 1st edition, learners claimed a critical issue—namely, the agent interrupting the discussion flow in two different ways: content (i.e., unrelated questions with respect to the discussion) and time (i.e., too frequent interventions breaking the natural discussion flow) [5].

As good practices, different approaches were conducted in the 1st edition to overcome these issues and improve the agent interventions throughout the different CA activities of the course while iterating the process from collecting feedback from the learners and analyzing the data of the surveys after each CA activity. These improvements are related to the creation of a CA domain model with a few but very focused agent interventions and a good balance between the task objectives of the activity and domain model. From the positive results collected in the 1st edition about the time and content dimensions of the agent interventions (between 25 and 33% totally or somewhat agreed, between 23 and 33% were indifferent, and only between 2 and 11% totally or somewhat disagreed), the same approaches were followed and reinforced for each CA activity of the 2nd edition, which led to achieving similar positive result (see Fig. 10). In terms of content appropriateness, between 23 and 46% totally or somewhat agreed, between 14 and 15% were indifferent, and only between 2 and 13% totally or somewhat disagreed. In terms of time appropriateness, between 33 and 39% totally or somewhat agreed, between 13 and 23% were indifferent, and only between 1 and 19% totally or somewhat disagreed. These results confirm the benefits of this approach to support CA activities in terms of time and content of the agent interventions.

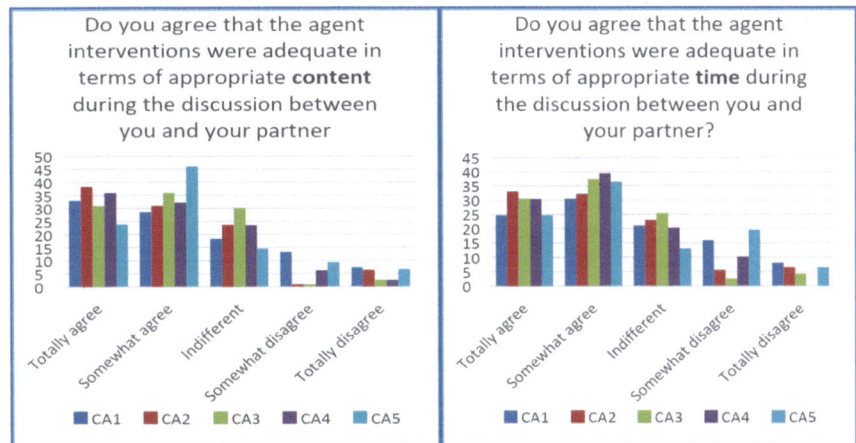

Fig. 10. Survey results on agent interventions in terms of content (left) and time (right) appropriateness for each CA activity.

5   Conclusions and Further Steps

This paper has presented detailed knowledge from a real experience with using CAs for supporting synchronous collaborative learning activities in the context of a MOOC with large participation. The main focus of the paper has been on a sound and effective evaluation with results that have been then discussed in terms of lessons learned and good practices. Overall, the valuable feedback and data received from learners along with the evaluation data analysis presented in this paper and the previous report published in Ref. [5] will serve as a primary data source for preparing next experiences with improved versions of the CA integrated in real MOOCs and other forms of online education in order to support peer discussions and sustain productive dialogues.

In addition, in next steps of our research we would like to explore the ethical dimension behind the application and evaluation of the technologies reported in this paper in education [8]. This includes considering how potential detrimental ethical effects can be foreseen via protective approaches and exploring the potential ethically beneficial opportunities that such technologies provide, as well as considering the educational environments in which they are deployed.

Acknowledgements. This research was funded by the European Commission through the project "colMOOC: Integrating Conversational Agents and Learning Analytics in MOOCs" (588438-EPP-1-2017-1-EL-EPPKA2-KA).

References

1. Demetriadis, S., Karakostas, A., Tsiatsos, T., Caballé, S., Dimitriadis, Y., Weinberger, A., Papadopoulos, P.M., Palaigeorgiou, G., Tsimpanis, C., Hodges, M.: Towards integrating conversational agents and learning analytics in MOOCs. In: Lecture Notes on Data Engineering and Communications Technologies, vol. 17, pp. 1061–1072 (2018)
2. Siemens, G.: Massive open online courses: innovation in education. In: Open Educational Resources: Innovation, Research and Practice, vol. 5 (2013)
3. Daradoumis, T., Bassi, R., Xhafa, F., Caballé, S.: A review on massive e-learning (MOOC) design, delivery and assessment. In: Proceedings of the Eighth International Conference on P2P, Parallel, Grid, Cloud and Internet Computing, pp. 208–213. IEEE Computer Society (2013)
4. Demetriadis, S., Tegos, S., Psathas, G., Tsiatsos, T., Weinberger, A., Caballé, S., Dimitriadis, Y., Sánchez, G.E., Papadopoulos, M., Karakostas, A.: Conversational agents as group-teacher interaction mediators in MOOCs. In: Proceedings of 2018 Learning with MOOCS (LWMOOCS), pp. 43–46. IEEE Education Society (2018)
5. Caballé, S., Conesa, J., Gañán, D.: Evaluation on using conversational pedagogical agents to support collaborative learning in MOOCs. In: 10th International Workshop on Adaptive Learning via Interactive, Collaborative and Emotional Approaches (ALICE-2020), Yonago, Tottori, Japan. October 28–30, 2020. Lecture Notes in Networks and Systems, vol. 158, pp. 199–210. Springer (2020)
6. Caballé, S.: A computer science methodology for online education research. Int. J. Eng. Educ. **35**(2), 548–562 (2019)
7. Reich, J., Ruipérez-Valiente, J.A.: The MOOC pivot. Science **363**(6423), 130–131 (2019)
8. Casas-Roma, J., Conesa, J.: Towards the design of ethically-aware pedagogical conversational agents. In: Lecture Notes in Networks and Systems, vol. 158, pp. 188–198. Springer, Cham (2021)

Natural Language Understanding for the Recommendation of Learning Resources Within Student Collaboration Tools

Nicola Capuano[1]([✉]) [ID], Luigi Lomasto[2,3] [ID], Andrea Pozzi[4] [ID], and Daniele Toti[4] [ID]

[1] University of Basilicata, Potenza, Italy
nicola.capuano@unibas.it
[2] University of Salerno, Fisciano, Italy
[3] Ministry of Education, Rome, Italy
[4] Catholic University of the Sacred Heart, Brescia, Italy

Abstract. Discussion forums are popular tools in Massive Open Online Courses (MOOCs), used by students to express feelings, exchange ideas, and ask for help. Unfortunately, the huge number of enrolled students hinders educational scaffolding activities, including the active participation of instructors in discussion forums. Therefore, students seeking to clarify the concepts learned may not receive the answers they need, reducing engagement and promoting dropout. This work presents a methodology for supporting learners within discussion forums, by analyzing conversations among students and providing them with recommendations in terms of relevant learning resources. The methodology involves several steps: the initial definition of an ontology that details the topics of the course, the real-time analysis of student posts within the discussion forums to extract different attributes including intent of the post, the concepts it is about, the sentiment, and the level of urgency and confusion. The extracted information is then used by a rules-based mechanism to assess whether the learner needs a recommendation. If so, the system suggests the most suitable learning resources. The paper also includes an initial evaluation of the proposed methodology.

Keywords: Natural language processing · Recommender systems · Massive open online courses

1 Introduction

In the latest years, Artificial Intelligence (AI), Machine Learning (ML) and Natural Language Processing (NLP) methodologies and techniques like Named Entity Recognition [1, 2] Sentiment Analysis [3, 4], Text Classification [5–7], and word, text, and document embeddings [8, 9] have provided a great opportunity to solve problems and improve resources within the context of e-learning. In this regard, Massive Open Online Courses (MOOCs) are a fertile ground upon which these methodologies and techniques are being applied to provide support both to learners and teachers.

Among the open issues lies the possibility of providing learners enrolled in MOOCs with an effective support when using course-related discussion forums, which are in principle a popular way for learners to convey their feelings, exchange ideas and ask for mutual help. Given the sheer size of the students usually involved and the corresponding scarce number of teachers available to help them, however, it is unsurprising that many questions are left unanswered and doubts on courses and topics often remain unclarified. This may lead to a reduced engagement and an increased drop-out rate for students as a physiological consequence of this situation.

The purpose of the present work is to try and fill this gap, by proposing a methodology based on AI and NLP meant to analyze conversations among students expressed in natural language, so that it may be possible to provide students with recommendations in terms of relevant learning resources in accordance with the topic, sentiment, and intent of the conversations themselves. This is carried out via several steps, including a preliminary definition and population of an ontology detailing the courses' topics. Elements in the ontology are then matched and associated with the corresponding relevant learning resources, via an automatic NLP pipeline that analyzes the resources according to their respective format.

Each conversation is then analyzed as well via linguistic techniques that identify the topics discussed and recommend the most relevant learning resources accordingly. This is done by matching the identified topics with those in the ontology both at the syntactic and the semantic level, via syntactic distances and embedding methodologies. For considering syntactical similarity and semantic difference, a pre-trained deep contextual language model is fine-tuned on the texts to provide finely grained contextual embeddings, and thus compare the vector representation of a conversation with that of the topics in the ontology.

The work is structured as follows. Section 2 discusses related work. Section 3 describes the methodology by detailing the steps it is made up of and the techniques and tools involved. Section 4 presents an initial evaluation of the methodology. Finally, in Sect. 5 conclusions are drawn.

2 Related Work

Several research works have addressed various aspects related to the problem of helping students learn in online courses. As explained in reference [10], one of the main critical points is represented by collaborative tools like discussion forums, where students post questions and problems about their learning activities, which are difficult to manage, especially in massive contexts like MOOCs due to the small number of instructors. To solve this problem, in such a work the authors propose a tool to label forum posts to simplify the answering tasks.

Another solution, proposed in reference [11], tries to solve the problem of the amount of the students' requests. Because of the large number of students and the small number of teachers, distinguishing posts that require rapid intervention is challenging. In this case, a pre-trained BERT model [9] is used for embedding representations to find the more urgent questions to be solved.

Conversational agents are another important solution for this kind of problem. In the colMOOC platform [12], a conversational agent was developed to help intervention activities to facilitate productive dialogue for chat-based collaborative learning. A conversational agent named COLA was also proposed in reference [13] with the aim of suggesting optimal educational paths.

The authors in reference [14] developed a conversational agent called CAERS whose architecture supports the construction of knowledge for students, tutors, and teachers through autonomous interference and recommendations of educational resources. Another approach to solve the question answering problem is shown by the authors in reference [15] where a ML model is proposed that can predict the best answer for a question by using historical forum data; other approaches show how it is possible to answer questions by relying upon an underlying ontology that semantically models the available information [16].

Solutions were also presented with which students can be supported for filling their learning gaps. The work in reference [17] proposes an agent-based recommender system that aims to help learners overcome their gaps by suggesting relevant learning resources. The system can recommend relevant resources to provide support to improve the learning experience. Another very important aspect to analyze in the e-learning process is represented by the feelings of the students with respect to the online courses. In reference [18], the authors aim to detect and analyze the involvement of course participants by considering their feeling, urgency, and confusion elements, via the use of ML on students' comments and posts.

Also, similar students may have similar preferences, and in platforms that offer a very high number of MOOCs like Coursera, ProfessionAI, and Udemy, it may be important to implement a collaborative system, to suggest relevant contents. The authors in reference [19] propose a method called Multi-Layer Bucketing Recommendation (MLBR) to recommend courses on MOOCs. In MLBR, student profiles are vectorized and distributed in a multidimensional space containing similar students with multiple courses in common. Moreover, in reference [20], the authors present good results for the collaborative filtering task applied to the MOOC context using deep learning and word embedding to catch semantic meaning of multimedia content.

3   Methodology

The proposed methodology is intended to support learners within discussion forums, by analyzing conversations and providing timely recommendations in terms of relevant learning resources. It improves the methodology presented in reference [18], where similar techniques were used only to assess the level of student engagement within an e-learning system. The methodology includes the following steps:

1. The initial definition and population of an ontology detailing the topics of the MOOC courses, by which the corresponding learning resources are cataloged.
2. The real-time analysis of the student posts within the discussion forums. From each post, the following cognitive and emotional information attributes are extracted: (i) topics (i.e., the learning concepts the post is about within the scope of the corresponding course), (ii) intent (the aim of the post), (iii) confusion (the level of confusion expressed by the post), (iv) sentiment (the affective polarity of the post), and (v) urgency (how urgently a reply to the post is required).

3. The extracted information is then passed to a rule-based mechanism to assess whether the student who posted a message needs a recommendation. If that is the case, the most suitable learning resources are suggested.

These steps are detailed in the remainder of this Section as follows. Section 3.1 describes how course topics are organized and matched with available relevant learning resources, while Sect. 3.2 describes the extraction of domain-independent attributes (i.e., intent, confusion, sentiment, and urgency) from the text of a forum post. Section 3.3 describes the rule-based approach that, given the extracted information, is used to decide when a student needs a recommendation. Finally, Sect. 3.4 details how the system finds the learning resources best suited to meet the students' needs and suggests them through an automatically generated post.

3.1 Ontology Definition and Conceptualization of Course Topics

An ontological schema for representing a given course along with its related information is defined beforehand by using the RDF [21], RDFS [22], and OWL [23] formalisms. The ontology includes the following set of main classes and properties:

- `Course`: this owl:Class represents a given MOOC course, and is the subject of the following datatype properties:
  - `hasTitle`, whose object is the textual title of the course.
  - `hasDescription`, whose object is the description or abstract of the course.
- `Module`: this `owl:Class` represents a given sub-module of a MOOC course, and is thus associated with a `Course` via the triple: `Course hasPart Module`. It has the following datatype properties:
  - `hasTitle`, whose object is the textual title of the module.
  - `hasDescription`, whose object is the description or abstract of the module.
  - `releaseDate`, whose object is the release date of the module.
- `Topic`: this owl:Class represents a topic of the course. It is the subject of the following datatype properties:
  - `hasName`, whose object is the actual textual name of the topic.
  - `hasRelevanceScore`, whose object is a floating-point value representing the semantic relevance of such a topic for the given course.
  - `hasTopic`: this object property connects a course with a topic, having Course as `rdfs:domain` and `Topic` as `rdfs:range`.

With the above schema defined, instances are created for each MOOC course, by instantiating the Course class and the Module classes along with their datatype properties accordingly. Each course then undergoes a conceptualization process, either in terms of its sub-modules or as a whole, where the associated textual elements (title and description), as well as those from the learning resources (whose text is extracted by means of mechanisms like Apache Tika [24]) of the course or of its sub-modules, are passed as input to CONCEPTUM [25], an advanced knowledge discovery system capable to extract semantically meaningful information from unstructured natural language texts. Here, the focus is on its ability to identify the most relevant concepts from a given text.

Such automatic conceptualization relies on linguistic analysis based on Part-of-Speech tagging, syntax parsing, and custom rules dependent on the language family.

This offline, preliminary conceptualization step works as follows. The textual resources of each course, as stated above, are first processed by CONCEPTUM. Its conceptualization mechanism, implemented via an NLP pipeline, proceeds along the following steps:

1. Language is detected (in this case, the focus is on the English language) and the macro-family of CONCEPTUM's supported languages (Anglo-Saxon/Neo-Latin) is assigned. As such, the corresponding pre-built models for these languages are considered to carry out the subsequent NLP tasks (sentence splitting, tokenization, POS tagging, etc.).
2. The text is split into sentences, each sentence is split into tokens, and tokens are tagged with their Part-Of-Speech value.
3. Aggregation rules based on the language family, the POS tagging, and the proximity of the tokens are applied to produce compound words made up of nouns, adjectives, and modifiers, which will be considered the concepts of the given text.
4. Acronyms are detected via CONCEPTUM's PRAISED module [26] and associated accordingly with their explanation, if found within the text.
5. For each concept, its term frequency is computed and used as a preliminary means for ranking the extracted concepts.
6. The whole text is then passed to an inner library able to carry out a wikification process: in accordance with the content and structure of the Wikipedia commonsense knowledge base for the detected language, this process returns the Wikipedia topics deemed most relevant for the given text. Such topics may be either variations of the actual concepts featured in the text or be something else entirely (and thus not explicitly featured in the text itself).
7. Concepts and topics are then matched via a combination of syntactical distances (Levenshtein/Jaccard) and synonym expansions (via WordNet), to re-rank the former and increase the semantic relevance of those concepts that are also matched, either fully or partially, with a topic from the wikification.

The resulting concepts extracted and ranked as described above are then stored in the ontology, as connected properties to the course or to its sub-modules, by creating instances of the Topic class and generating, for each of those instances, a triple whose subject is the Course/Module instance, whose predicate is the hasTopic object property and whose object is the Topic instance. The actual topic names (the concepts extracted) along with its score (the semantic relevance detected by CONCEPTUM) are then associated with the corresponding datatype properties to the given Topic instances.

3.2 Extraction of Domain-Independent Attributes of a Forum Post

The first, real-time step of the proposed methodology is meant to extract domain-independent attributes by using a corpus-based text classification approach that works on three phases. In the first phase, the word representations are learned from an existing corpus of annotated forum posts. In the second phase, the word representations are combined to produce document (post) representations. In the third phase, the document representations are classified. Four classification tasks have been addressed according to the following dimensions: intent (question, answer, opinion), confusion (low, medium, high), sentiment (positive, negative, neutral), and urgency (low, medium, high).

The defined methodology, which is an improved version of that presented in reference [27], uses a deep neural network combining an embedding layer that transforms each word into a dense numerical input vector, a convolutional layer that extracts local contextual features on neighboring words, a gated recurrent unit layer that combines extracted features considering the temporal dependencies and a fully connected output layer for classification. The model is trained on the Stanford MOOCPosts[1] data set, which includes about 30,000 anonymized forum posts from 11 Stanford University public online classes in three domain areas. Therefore, four trained models were obtained, one for each classification task.

During data set preprocessing, the HTML tags, hyperlinks, and words replaced by automated anonymization were removed. The polished posts were tokenized, and the 12,000 most frequent tokens were retained as features. Each post was then represented as 100-dimensional sequence of word indexes. Posts longer that 100 words were truncated while post shorter than 100 words where right-padded with 0. This last step is needed because the first two network layers only process tensors of fixed size and, therefore, it is necessary to choose an adequate length for most of the inputs but not too large to require training of too many parameters. Experiments were made with the input length varying between 25 and 250 words, obtaining increasing performances up to 100 words and substantially constant for greater lengths.

The *embedding layer* of the classification model was initialized with GloVe (Global Vectors for Word Representation)[2] 6B.50d, which includes 400,000 pre-computed word vectors in 50 dimensions corresponding to English tokens obtained from Wikipedia thanks to word co-occurrence statistics [28]. Each 100-dimensional input vector (representing a single post) was then transformed by the embedding layer into a 100×50 output matrix where the i-th row is the word vector associated to the i-th token of the post. During the training, the embedding layer was enabled to fine-tune word vectors to better characterize the classification task.

The *convolutional layer* consisted of 16 filters with a kernel size of 7 (number of subsequent words to be convolved) and depth of 50 (embedding dimension). The *rectified linear activation function* (ReLU) was adopted by the convolutional units. Such layer combines the representation of each word to its local neighbors, weighted by several kernels (one for each feature to be extracted). Since the same input transformation is performed on each word, a pattern learned at a certain position in a sentence can be subsequently recognized at a different position, making such transformation invariant for temporal translations [29]. The operation of this layer is summarized in Fig. 1.

Although convolution is useful for extracting local semantic patterns, to effectively classify documents, it is also necessary to consider long-term semantic dependencies— i.e., those that occur between words that are not close to each other in the text. A person who reads (and understands) a text does so incrementally, word by word, maintaining

---

[1] https://datastage.stanford.edu/StanfordMoocPosts/.

[2] https://nlp.stanford.edu/projects/glove/.

an internal model built from previous information that is constantly updated when new information comes in [30]. To this end a *gated recurrent unit layer* made of 16 units is added to the network. Such network model processes sequences of information element by element maintaining a hidden state (memory) containing information on what has been seen so far. Unlike standard recurrent neural networks, gated recurring units control how much past information is needed and how much can be forgotten, thus speeding up network training and operation [31].

To proceed with classification, a fully connected output layer made of three nodes with *soft-max* activation is used to classify the summative document representation in one of the three available classes (in fact each classification task has exactly three possible output classes). Dropout is used as regularization technique between network layers [32] to ignore randomly selected signals during training, thus leading to networks capable of better generalization and less likely to overfit training data. The model was trained separately for each classification task. The *root mean square propagation* variant (RMSprop) of the classical stochastic gradient descent algorithm was used for model fitting with categorical cross-entropy as loss function. The network was trained in 30 epochs with a mini-batch size of 256 items.

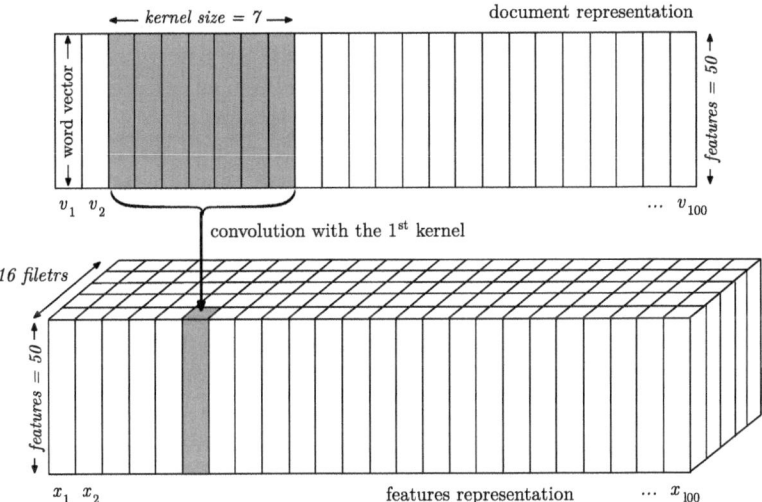

Fig. 1. Convolutional transformation of an input document representation.

3.3 Rule-Based Intervention Planning

The designed methodology adopts a rules-based approach to decide when the recommender intervention is needed during the students' conversation in the course forum [33]. The rules are intended to be completely customizable by the teacher and take the form of conditional statements *if <condition> then <consequent>* where the *condition* is a Boolean expression over the previously extracted attributes and the *consequent* is a request to the recommender to suggest a suitable learning resource. Some examples of rules are provided below.

- *IF intent = question AND (urgency = high OR sentiment = low) THEN recommend* i.e., if a student asks for something and a strong urgency or negative feeling is detected, then a recommendation is generated to prevent dropout.
- *IF intent = question AND urgency = medium AND idleTime > 30 THEN recommend* i.e., if a student asks for something, even if with medium urgency, but no one responds within 30 min, then a recommendation is generated.
- *IF intent = opinion AND confusion < > low THEN recommend* i.e., if a student posts a confusing statement, then a recommendation is generated.
- *IF intent(−1) = question AND intent = answer AND confusion = high THEN recommend(−1)* i.e., if a student provides a confusing answer to question, then a recommendation related to the question is generated.

The rules use the functions *intent, confusion, sentiment,* and *urgency* to get the attributes extracted from the forum posts. When no argument is given to these functions, the attribute refers to the last forum post. To consider a previous post, a negative argument referring to the forum history is given (i.e., -1 refers to the first previous post, -2 to the second previous, etc.). Other functions are provided to consider supplementary information such as the idle time (in minutes) since the last post (*idleTime*), the number of post views (*views*) and likes (*likes*). The basic Boolean logic is applied in the definition of the condition that support AND, OR, and NOT operators.

The *recommend* function invokes the recommender system. When no arguments are provided, a recommendation for the last post is generated. When an argument is provided, a recommendation is generated for a specific post in the forum history. When a list of arguments is provided, a recommendation is generated considering several forum posts. No information on course topics is provided to the recommender system that makes autonomously its evaluations based on the post content. It may happen that the recommender does not find any course topics connected to a post for which a recommendation has been requested. In this case the recommendation is skipped.

3.4   Recommendation of Learning Resources

Once a recommendation is deemed necessary for a post in the forum of a particular course, a classification process is employed to determine which of the course topics (also called course modules) in the ontology described in Sect. 3.1 can most likely be associated with the considered forum post. This classification process is necessary to recommend the appropriate resources to the student(s) who posted in the forum.

The process starts with the conceptualization of the forum posts to extract the most relevant semantic concepts from them. A vectorized representation of the extracted textual concepts is then obtained by exploiting the same pre-trained embedding layer discussed in Sect. 3.2. Finally, the retrieved embeddings are used to define a similarity score between each new student post and the available resources. This process is an expanded version of the one described in [34–36], relying upon the embedding mechanism that moved the NLP efforts significantly ahead from 2018 onwards.

The CONCEPTUM system is applied within the context of forum posts as well to conceptualize a given forum post and extract its relevant concepts, with the same mechanism described in Sect. 3.1. Basically, every time that a new post is written by a student, an online conceptualization is performed to obtain a set of concepts which semantically describe the considered post.

All the concepts extracted by CONCEPTUM are vectorized through a pre-trained embedding layer (see GloVe in Sect. 3.2). Then, the semantics of each topic in the ontology is assumed to be represented by the centroid of the vectors related to its most relevant concepts. More formally, given a topic $t \in T$, where T is the set of all the possible topics for the specific course, we consider $w_i^t \in R^n$ the n-dimensional vector associated with the i-th concept of the topic t, with $i = 1, 2, \ldots, n_t$, being n_t the number of the most relevant extracted concepts for the topic t (the relevance of a concept can be computed according to a tf-idf approach), and we define the topic representation as the following centroid:

$$\overline{w}^t = \frac{1}{n_t} \sum_{i=1}^{n_t} w_i^t \tag{1}$$

Similarly, the concepts extracted online from a new post p can be represented as n-dimensional vectors w_i^p with $i = 1, 2, \ldots, n_p$ where n_p is the number of concepts extracted from the post p. Furthermore, a vectorized representation of a post is assumed to be provided by the centroid of the vectors of its concepts:

$$\overline{w}^p = \frac{1}{n_p} \sum_{i=1}^{n_p} w_i^p \tag{2}$$

Then, a similarity score $S_C(p, t)$ is computed between the embedding of the post p and the representation of each of the course topics $t \in T$. We consider the *cosine similarity* measures, that, for a post p and a topic t, is defined as follows:

$$S_C(p, t) = \frac{\overline{w}^t \cdot \overline{w}^p}{\|\overline{w}^t\| \|\overline{w}^p\|} \tag{3}$$

where $a \cdot b$ represents the dot product between the vectors a and b, and $\|a\|$ is the norm of the vector a. The topic $t^* \in T$ which exhibits the highest similarity score with respect to the post p is then retrieved, where the match is obtained by solving the following optimization problem:

$$t^* = argmax_t S_C(p, t) \tag{4}$$

Finally, the resources related to the topic t^* are recommended to the student.

4 Evaluation

Several experiments were performed to evaluate the proposed methodology step by step. To evaluate the performance of the extraction of domain-independent attributes, the Stanford MOOCPosts data set was divided into 4 disjoint subsets of equal size and at each step, the k-th subset with $k \in \{1, \ldots, 4\}$ was used as the validation set, while the remaining subsets have been used for training (k-fold validation).

Performance on the validation set was measured in terms of accuracy (Acc), macro-averaged $f1\text{-}score$ ($F1_{M\text{-}avg}$) and weighted averaged $f1\text{-}score$ ($F1_{w\text{-}avg}$) [37]. While $F1_{M\text{-}avg}$ is the unweighted mean of the $f1\text{-}score$ calculated for each label, $F1_{w\text{-}avg}$ is the mean of the $f1\text{-}score$ calculated for each label, weighted by the support (i.e., the number of true instances for each label). So, while $F1_{w\text{-}avg}$ (as the accuracy) considers the label imbalance, $F1_{M\text{-}avg}$ does not, giving each class the same importance. The classifier performance, for each dimension, averaged between the 4 validation phases, is shown in table 1. The accuracy obtained is quite high for each classification task and comparable with the reference literature, reported in Sect. 2.

As it can be seen, the performance obtained from the proposed approach in terms of both accuracy and $F1_{w\text{-}avg}$ is satisfactory and in line to related literature. On the other hand, the performance in terms of $F1_{M\text{-}avg}$ is relatively lower than $F1_{w\text{-}avg}$, thus reflecting the imbalanced distribution of training items on the data set which hinders a fair and unbiased representation of the three classes for almost all the classification tasks. This can be overcome in the future by using a different data set or by applying specific techniques for unbalanced data sets such as resampling.

Table 1. Performance of the attribute extraction.

Dimension	Acc	$F1_{M\text{-}avg}$	$F1_{w\text{-}avg}$
Intent	0.87	0.73	0.86
Confusion	0.82	0.68	0.81
Sentiment	0.87	0.67	0.86
Urgency	0.87	0.76	0.86
Intent	0.87	0.73	0.86

Finally, it is worth emphasizing that the problems of topic conceptualization and resource recommendation have been addressed here only from a methodological point of view in Sect. 3.1 and Sect. 3.4, respectively. The validation of such techniques is beyond the scope of the present paper, and it will be part of future works.

5   Conclusion

This work has presented a methodology for addressing the problem of supporting learners in MOOCs when asking for help and support within discussion forums, given the huge number of students usually enrolled in these online courses and the limited number of human teachers available to answer their questions and clarify their doubts.

The proposed methodology takes advantage of a combination of state-of-the-art NLP tools and techniques for classifying the learning resources of a given course, analyzing the forum posts written by the students, assessing whether a recommendation is needed and providing relevant learning resources when deemed necessary. The results show that the proposed approach exhibits a satisfactory accuracy level in the extraction of domain independent attributes when compared to the existing literature.

Future works may involve the fine-tuning of a model based on the Transformer architecture on MOOCs' documents, with the aim of enhancing both the conceptualization capability and the recommendation process. To classify the students by also considering their semantic profile and preferences, models like BERT and GPT3 can be considered and tuned. The introduction of mechanisms for the summarization of the learning resources and contents may in principle be useful to offer students additional support and provide them with even more effective recommendations. Finally, methods and standards supporting the integration of the proposed system within existing e-learning and MOOC platforms and frameworks will be explored.

References

1. Shelar, H., Kaur, G., Heda, N., Agrawal, P.: Named entity recognition approaches and their comparison for custom ner model. Sci. Technol. Libr. **39**(3), 324–337 (2020)
2. Li, J., Sun, A., Han, J., Li, C.: A survey on deep learning for named entity recognition. IEEE Trans. Knowl. Data Eng. **34**(1), 50–70 (2022)
3. Zhang, L., Wang, S., Liu, B.: Deep learning for sentiment analysis: a survey. Wiley Interdiscip. Rev. Data Min. Knowl. Dis. **8**(4), 1253 (2018)
4. Yadav, A., Vishwakarma, D.K.: Sentiment analysis using deep learning architectures: a review. Artif. Intell. Rev. **53**(6), 4335–4385 (2019). https://doi.org/10.1007/s10462-019-09794-5
5. Kowsari, K., Meimandi, K.J., Heidarysafa, M., Mendu, S., Barnes, L., Brown, D.: Text classification algorithms: a survey. Information **10**(4), 150 (2019)
6. Lomasto, L., Di Florio, R., Ciapetti, A., Miscione, G., Ruggiero, G., Toti, D.: An automatic text classification method based on hierarchical taxonomies, neural networks and document embedding: the NETHIC tool. In: International Conference on Enterprise Information Systems, pp. 57–77. Springer (2019)
7. Ciapetti, A., Di Florio, R., Lomasto, L., Miscione, G., Ruggiero, G., Toti, D.: NETHIC: a system for automatic text classification using neural networks and hierarchical taxonomies. In: ICEIS 2019—Proceedings of the 21st International Conference on Enterprise Information Systems, pp. 284–294 (2019)
8. Mikolov, T., Sutskever, I., Chen, K., Corrado, G.S., Dean, J.: Distributed representations of words and phrases and their compositionality. Adv. Neural Inf. Process. Syst. **26** (2013_
9. Devlin, J., Chang, M.W., Lee, K., Toutanova, K.: BERT: pre-training of deep bidirectional transformers for language understanding. In: Proceedings of the Conference of the North American Chapter of the Association for Computational Linguistics: Human Language Technologies, vol. 1, pp. 4171–4186 (2019)
10. Capuano, N., Caballé, S.: Multi-attribute categorization of MOOC forum posts and applications to conversational agents. In: International Conference on P2P, Parallel, Grid, Cloud and Internet Computing, pp. 505–514. Springer (2019)
11. Khodeir, N.A.: Bi-GRU urgent classification for MOOC discussion forums based on BERT. IEEE Access **9**, 58243–58255 (2021)
12. Karakostas, A., Nikolaidis, E., Demetriadis, S., Vrochidis, S., Kompatsiaris, I.: colMOOC—an innovative conversational agent platform to support MOOCs a technical evaluation. In: 20th International Conference on Advanced Learning Technologies (ICALT), pp. 16– 18. IEEE (2020)

13. Penstein Rosé, C., et al. (eds.): AIED 2018. LNCS (LNAI), vol. 10947. Springer, Cham (2018). https://doi.org/10.1007/978-3-319-93843-1
14. Rossi, D., Ströele, V., Braga, R., Caballé, S., Capuano, N., Campos, F., Dantas, M., Lomasto, L., Toti, D.: CAERS: a conversational agent for intervention in MOOCs' learning processes. In: The Learning Ideas Conference, pp. 371–382. Springer (2021)
15. Jenders, M., Krestel, R., Naumann, F.: Which answer is best? Predicting accepted answers in MOOC forums. In: Proceedings of the 25th International Conference Companion on World Wide Web, pp. 679–684 (2016)
16. Toti, D.: AQUEOS: a system for question answering over semantic data. In: Proceedings—2014 International Conference on Intelligent Networking and Collaborative Systems, IEEE INCoS 2014, pp. 716–719. Institute of Electrical and Electronics Engineers Inc. (2014)
17. Brigui-Chtioui, I., Caillou, P., Negre, E.: Intelligent digital learning: agent-based recommender system. In: Proceedings of the 9th International Conference on Machine Learning and Computing, pp. 71–76 (2017)
18. Toti, D., Capuano, N., Campos, F., Dantas, M., Neves, F., Caballé, S.: Detection of student engagement in e-learning systems based on semantic analysis and machine learning. In: International Conference on P2P, Parallel, Grid, Cloud and Internet Computing, pp. 211–223. Springer (2020)
19. Pang, Y., Jin, Y., Zhang, Y., Zhu, T.: Collaborative filtering recommendation for MOOC application. Comput. Appl. Eng. Educ. **25**(1), 120–128 (2017)
20. Wu, L.: Collaborative filtering recommendation algorithm for MOOC resources based on deep learning. Complexity (2021)
21. W3C: RDF resource description framework (2014). http://www.w3.org/RDF/
22. W3C. RDF schema (2014). http://www.w3.org/TR/rdf-schema/
23. W3C: Web ontology language (2012). https://www.w3.org/OWL/
24. The Apache Foundation: Apache Tika (2007–2022). https://tika.apache.org/
25. Toti, D., Rinelli, M.: On the road to speed-reading and fast learning with CONCEPTUM. In: Proceedings—2016 International Conference on Intelligent Networking and Collaborative Systems, IEEE INCoS 2016, pp. 357–361 (2016)
26. Toti, D., Atzeni, P., Polticelli, F.: Automatic protein abbreviations discovery and resolution from full-text scientific papers: the PRAISED framework. Bio-Algorithms Med-Syst. **8** (2012)
27. Capuano, N., Caballé, S., Conesa, J., Greco, A.: Attention-based hierarchical recurrent neural networks for MOOC forum posts analysis. J. Ambient. Intell. Humaniz. Comput. **12**(11), 9977–9989 (2021)
28. Pennington, J., Socher, R., Manning, C.: GloVe: global vectors for word representation. In: Proceedings of the 2014 Conference on Empirical Methods in Natural Language Processing (EMNLP), pp. 1532–1543. ACL (2014)
29. Chollet, F.: Deep Learning with Python. Simon and Schuster (2021)
30. Capuano, N.: Transfer learning techniques for cross-domain analysis of posts in massive educational forums. In: Intelligent Systems and Learning Data Analytics in Online Education, pp. 133–152. Elsevier (2021)
31. Cho, K., van Merriënboer, B., Gulcehre, C., Bahdanau, D., Bougares, F., Schwenk, H., Bengio, Y.: Learning phrase representations using RNN encoder–decoder for statistical machine translation. In: Proceedings of EMNLP 2014, pp. 1724–1734. ACL (2014)
32. Srivastava, N., Hinton, G., Krizhevsky, A., Sutskever, I., Salakhutdinov, R.: Dropout: a simple way to prevent neural networks from overfitting. J. Mach. Learn. Res. **15**(56), 1929–1958 (2014)
33. Capuano, N., Gaeta, M., Miranda, S., Orciuoli, F., Ritrovato, P.: LIA: an intelligent advisor for e-learning. In: Emerging Technologies and Information Systems for the Knowledge Society, pp. 187–196. Springer (2008)

34. Capuano, N., De Maio, C., Salerno, S., Toti, D.: A methodology based on commonsense knowledge and ontologies for the automatic classification of legal cases. In: Proceedings of the 4th International Conference on Web Intelligence, Mining and Semantics (WIMS-2014), pp. 1–6. ACM (2014)
35. Capuano, N., Dell'Angelo, L., Orciuoli, F., Miranda, S., Zurolo, F.: Ontology extraction from existing educational content to improve personalized e-learning experiences. In: Proceedings of the 3rd IEEE International Conference on Semantic Computing (ICSC-2009), pp. 577–582. IEEE (2009)
36. Capuano, N., Toti, D.: Experimentation of a smart learning system for law based on knowledge discovery and cognitive computing. Comput. Hum. Behav. **92**, 459–467 (2019)
37. Sokolova, M., Lapalme, G.: A systematic analysis of performance measures for classification tasks. Inf. Process. Manage. **45**(4), 427–437 (2009)

Teaching Ethics in Online Environments: A Prototype for Interactive Narrative Approaches

Joan Casas-Roma(✉) ⓘ, Jordi Conesa(✉) ⓘ, and Santi Caballé(✉) ⓘ

SmartLearn Research Group, Universitat Oberta de Catalunya, Barcelona, Spain
{jcasasrom,jconesac,scaballe}@uoc.edu

Abstract. Ethics is an important part in the training of professionals. Most associations have their own code of ethical practices, and many public and private institutions highlight the importance of integrating ethical standards into their practice. Due to this, ethics already appears, in one way or another, in almost every syllabus in professional and higher education courses. Nevertheless, ethics is often taught in a descriptive fashion that refers almost exclusively to regulations, and thus ethics quickly becomes a matter of compliance. Learning ethics, however, should also lead to becoming skilled in the sort of ethical awareness, reflection and reasoning capabilities that is needed to anticipate, understand and react appropriately to current and future challenges that can seldom be reduced to a set of regulatory principles. In order to teach (and learn) ethics as a skill, a more direct experience to ethically relevant situations that prompt for subjective reflection is needed. In this regard, interactive narrative experiences can be used as a way to prompt subjective involvement. This paper presents the design of a prototype of an interactive narrative created to teach ethics. The prototype integrates insights from game design, such as attachments, meaningful choices, and the creation of spaces for subjective reflection, and is conceived as a complementary tool that could easily be integrated as part of a learning module. This work introduces the required background, presents the prototype and its design, and provides reflections on the relevant decisions behind the development of interactive experiences aimed at teaching ethics.

Keywords: Ethics teaching · Interactive narrative · Online learning · Design methodology

1 Introduction and Motivations

Actions have consequences that often affect individuals, collectives, and even the physical world around us, both in the short and the long term. Due to this, considering and anticipating the potential outcomes of a decision is key in order to decide what to do and how to act. These outcomes can often be classified as being "good" or "bad," or even "right" or "wrong" from an ethical point of view; that is, regardless of the observable (i.e., physical) changes that a decision might cause to the world, there is a layer of ethical considerations that can be used to evaluate whether that decision has been fair, or just, or whether the person that made that decision has acted with integrity, honestly, or in an egoistic way. Regardless of the observable changes that might happen into the physical world, the ethical dimension of the outcomes of a decision can have a profound impact, either for the better, or the worse, on the lives of potentially many individuals. This is why ethics (that is: understanding and reasoning about the good, bad, right and wrong dimension of events) is an important layer that concerns practically each and every action and decision we make. This is also particularly important in the context of professional practice, where actions and decisions might be more far-reaching and might affect a greater number of individuals, and in where there might exist potential conflicts of interest between what would be ethically desirable to do, and what would be profitable: this is why almost every professional community has its own code of ethical practices [1, 6, 11], as well as why there exist other, more overarching guidelines concerning the ethical dimension of professional practices [7, 10, 12].

Due to this, ethics is an important part of education, including the training of professional practitioners, an appears in many syllabi of professional and higher degree courses. Nevertheless, ethics has some particularities that often makes its teaching (and learning) present a series of challenges that are not found in other matters. In some cases, ethics is seen as a set of legal regulations that one must adhere to, and therefore ethics ends up being refurbished into compliance. This approach falls short in terms of learning ethics as a skill that has to be incorporated in professional everyday practice in order to both identify and mitigate potential ethically undesirable effects, as well as to recognize and exploit ethically beneficial ones. The way ethics is taught (and learned) as part of a professional profile should, therefore, integrate ethics in a way that is relevant throughout the whole process of a professional practitioner, rather than just a checklist to be looked at certain stages of the process.

This paper argues for the need to understand ethics as a skill that can be integrated actively as part of a professional practitioner's formation, and presents the design and implementation of a prototype of an interactive narrative created to complement the learning of ethics as an active skill. This work is structured as follows: the remainder of Sect. 1 introduces the background and motivations regarding ethical theories, the challenges of teaching and learning ethics as a skill, and the benefits of using fiction and interactive media to support such teaching and learning; Sect. 2 presents the main decisions around three key aspects behind the design of this interactive narrative; Sect. 3 introduces the technologies used in the implementation of the prototype; and finally Sect. 4 provides some concluding remarks and directions of future work.

1.1 Ethical Theories and Ethical Autonomy

Different ethical theories are based on different conceptions of what ethics (or acting in an ethically acceptable way) is. For example, *consequentialist* theories place their focus on the outcomes of a choice; utilitarianism, in particular, favors those actions that bring the greatest amount of happiness to the maximum number of people [13], and is based on "calculating" the ethically desirable choice in accordance to that. Conversely, theories like *deontology* focus on the rightness of the act itself, rather than on its outcomes [5], and are usually expressed as a compendium of rules that summarize what should, or should not be done.

A different approach comes from theories based on *virtue ethics*, such as the Aristotelian, Confucian, and Buddhist conceptions of ethics [18]. In a nutshell, those theories understand ethics as the personal growth associated with acting according to a set of virtues, or principles to strive towards. Some common examples of such principles could be honesty, integrity, or honor, to name a few. What makes this approach to ethics particular is that those virtues are not meant to be prescriptive commands according to which one must always act, but rather beacons that can guide one's decisions, but which might not always be the path that must be chosen: these virtues, therefore, need to be balanced out with what is usually called *practical wisdom*, which refers to the subjective ability to distinguish when to act according to a certain principle.

Conversely to consequentialist and deontological theories of ethics, which seek to "externalize" ethical reasoning into an "objectivable" procedure, theories of virtue ethics keep the subject central and relevant to the decision. Because virtues, or principles, are meant to act as a guidance, the subject is always required to understand and reason about what the ethically desirable course of action would be; in other words, the subject is required to retain their *ethical autonomy* to recognize the context, nuances and potential conflicts involved in an ethically relevant decision, instead of delegating their decision to an externally driven calculus, or set of rules. Although one could think that resorting to external procedures to decide on ethically relevant matters might be preferable, current and upcoming challenges in professional and social contexts will inevitably require both individual and collective involvement from an ethically informed position in order to recognize, understand, and reason about their ethical dimension.

1.2   The Challenge of Teaching (and Learning) Ethics

Nevertheless, teaching (and learning) ethics often presents a series of challenges that are not found in other topics where their learning is mainly based on either memorizing concepts, or practicing certain procedures. When reduced to a descriptive format based on a set of rules, ethics often turns into a shallow and legalistic approach to notions as complex as "goodness" or "rightness," which are often heavily context-dependent, and the nuances inherent to complex scenarios are often lost, or heavily stripped off. However, when ethics is understood as a skill that can be practiced and developed, it actually requires a combination of different abilities:

- An understanding of what the relevant *ethical principles* are in each context, as well as of what those principles are about.
- An *awareness* of, and a *sensitivity* towards ethically relevant situations in order to identify when the ethical dimension needs to be taken into account.
- An *empathetic recognition of the patients* (i.e., those agents that will likely be affected by the outcomes of a choice) and their particular situations, which might involve considering their drives, goals and needs.
- The *reasoning skills* needed to balance the ethical dimension of a decision with other, potentially conflicting interests and tensions.

Even though these abilities can be conceptually understood from a theoretical point of view, they need to be trained to be applied in a practical situation—just like any other skill. Nevertheless, training ethics in real-life situations where something is at stake might not be the most convenient (or pressure-free) way of training this skill.

Ethically relevant abilities may be strengthened through the use of examples that situate the relevant ethical features in different contexts [2, 15]. But, even when using those example cases, the spectator is still a sort of "back-seat driver": they are external to a scenario in which they have no agency. In order to further involve the spectator into the scenario one additional feature is needed: interactivity.

1.3 Fiction, Interactive Media and Subjective Involvement

Even if they are not real, stories of fiction have always been able to prompt a wide range of emotions in their audience through many different formats, like books, movies, or theatre plays, to name a few. Nevertheless, in this "traditional" forms of media, the spectator is still "separated" from the fiction and is kept outside of it, with no power over the events that happen; in other words, these forms of media lack interactivity.

There are, however, ways of making a fictional narrative interactive and thus give some degree of agency to the spectator. For example, branching narratives such as choose-your-own-adventure books, or narrative-driven games [3] (be them digital, or tabletop) provide the spectator (which, in those cases, could be already referred to as the "player") with the opportunity to provide their own input to determine how the story may move forward. This allows to potentially prompt a range of emotions that cannot appear in non-interactive media formats: *moral emotions* [16], such as guilt, or pride which require the person who experiences them to feel responsible for their decisions. Probably, the kind of media that offers the most interactivity are games (either digital, tabletop, or blended). As such, many narrative-intensive games feature some profound decisions that could prompt these moral emotions into their players [14, 16, 19]. Although there probably is no infallible formula that guarantees achieving such level of subjective immersion in the players, there are a few features that are necessary (although maybe not sufficient) to at least open up the path towards achieving that effect: *attachments* and *meaningful choices* [9].

Attachments refer to the ability that some works of fiction have to make the audience develop a sympathy towards their fictional characters --in other words, to make the audience "care" about them. Because, otherwise, why would it matter to the players what happens to the fictional characters, if they do not feel attached to them? *Meaningful choices* refer to the requirement of choices to have (or be perceived as having) actual consequences within the fictional setting, rather than being simply cosmetic choices. Because, if the players' choices do not actually affect the fiction and their characters in any relevant way, why care about those choices?

This latter feature clearly distinguishes interactive media from other types of media and allows to prompt the range of the aforementioned moral emotions, which are linked to the feeling of responsibility of the players. This reflection that usually accompanies this kind of emotions is, precisely, part of what the prototype presented in this paper aims for. Ideally, students who take part on this interactive experience should feel invested enough in the story, its characters, and the outcomes of the choices they make so as to prompt genuine reflection regarding those choices.

But this is not the only requirement that the prototype aims to fulfill. Ethics as a skill is not just about reasoning about the right choice, but also about identifying (i.e., becoming aware of) the ethical dimension and ethical principles behind our decisions. In this sense, the prototype also aims at prompting genuine subjective reflection about relevant ethical principles. In regards to this, there is the example of a digital game, *Ultima IV* [8, 19], that provides the right directions. Long story short, the player must act within that game according to a series of virtues that are part of the game's lore. What makes that game so distinct, in this regard, is that the player sees no explicit account of their progress (namely: the variables keeping track of the player's progress are hidden to them); in order to fulfill the game's goal, the player has to subjectively *understand* what each virtue is about, and act in ways that are in accordance to them. In this sense, Ultima IV asks from the player a level of genuine and subjective understanding of the game's fictional virtue system that requires their comprehension and reflection around those virtues.

Considering this, the use of an interactive narrative experience is proposed as a useful learning complement in order to help fill the gap between the theoretical understanding of ethically relevant matters and the need of interactive scenarios that prompt subjective reflection in order to train ethics as a skill.

2   Designing an Interactive Narrative to Teach Ethics as a Skill

Although there has been many relevant decisions during the design and implementation of the prototype, some of them are particularly relevant towards providing insights that could be used in the design of other interactive experiences for teaching (and learning) ethics. This section focuses on explaining and justifying the decisions made concerning the most relevant features for the design of the narrative experience—that is, what wants to be conveyed to the players through the way they experience the narrative.

Regarding the setting, the prototype is set in the context of computer science professionals—as computer science studies is where it might be easier to test it with real students of a virtual campus. As such, the ethical principles and scenarios featured in the narrative are relevant to that professional field—although not exclusively.

The main goal of using an interactive narrative to teach ethics is to reach the student at a level of subjective experience that other teaching methods, based on descriptive approaches, or non-interactive forms of media, would not have access to. There are three main features, each of them including different elements, that are key in order to design an interactive narrative that can potentially involve its players at a subjective level: *attachments*, *meaningful choices* and the creation of *spaces for reflection*.

2.1   Designing Towards Attachments

Characters are a key element in stories, but designing branching stories involving multiple characters that show enough depth can quickly transform a simple, straightforward story into a titanic enterprise that needs dozens (if not hundreds) of alternative lines for each character. In order to keep the prototype within realistic margins of development, it includes 4 characters (called Peter, Alex, Claire and Mark), of which 2 (Claire and Mark) are considered *close characters*[1] and are significantly affected by the choices made by the player.

Approval system. In the case of the two close characters, an additional layer is added. Although constrained by the length and scope of the story, the relationship that these characters have with the player evolves independently throughout the story using an *approval system* [4] that reflects whether the players' choices are in accordance, or in contradiction to the characters' beliefs and goals. The inclusion of this approval system follows three objectives: 1) to increase the depth of the relationship that the player can establish with those characters, thus favoring the chance to form attachments; 2) to use those characters as props to expose conflicting views about choices that the player will have to make; and 3) to introduce the tension of having to make a decision that someone might approve, or disapprove of.

Character-driven narration. The prototype seeks to avoid using large chunks of passive, narrator-driven text: instead, the majority of important passages in the narrative are presented through the voice of the fictional characters. Aside from making the characters more relevant to the story, this also presents certain advantages and opportunities: it avoids the effect of believing that the narrator voice represents objective facts; it allows to expose the player to potentially conflicting views on a matter in an organic way, as different characters might have different opinions about it; and it naturally allows to embed the approval system in decisions where the characters have already intervened beforehand.

2.2 Designing Towards Meaningful Choices

As it has been argued before in this paper, choices offered to the players need to appear meaningful in order for the players to feel invested in the fictional narrative. One obvious way of making choices be meaningful is to have their outcomes affect the story in significant, potentially long-lasting ways that are coherent with that choice.

Overall narrative structure. Some interactive stories feature a choice structure based on a *diamond* shape (also called a *foldback structure*) in which as the story move forwards, the choices branch out, but to converge later on at the same final state, thus rendering those as ultimately inconsequential. Nevertheless, avoiding diamond shapes is easier said than done, as the amount of writing involved in making all choices branch out towards a different path in the story grows dramatically as choices increase. One solution lies in *world states* and the use of *variables*. Instead of branching into completely different paths, some choices can affect the value of certain variables, take the player down the same path as other choices, but resolve certain events differently due to the values recorded in those variables. For example, the player finding a key at an earlier stage of a story might allow them to open a treasure chest at a later stage, or a higher

---

[1] The term "close character" is used in this prototype to refer to a fictional character whose relationship with the player can significantly change throughout the development of the fiction, and for whom this relationship is explicitly tracked through an "approval system".

number in the "fishing" skill might result, when the player throws a fishing rod into a pond, to catch an actual fish, instead of just an old boot.

By combining branching paths, world states, variables and sections of foldback structures, choices can still be made meaningful without having to explicitly create independent branches for each choice. Considering this, the structure of the interactive narrative created for the prototype has been designed as follows:

- Sessions are structured as blocks of foldback structures (see Fig. 1).
- Some choices, instead of branching out, affect variables that will be relevant in future sections of the narrative.

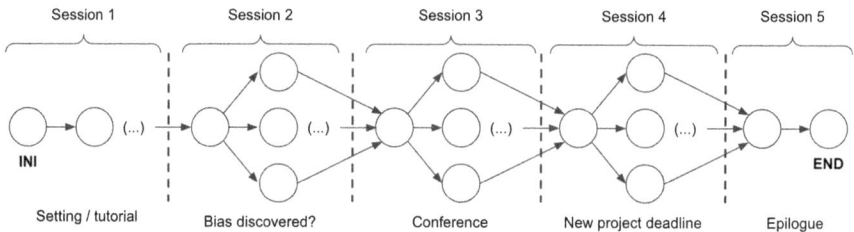

Fig. 1. A simplified schema representing the main structure of the interactive narrative (world variables are not represented for the sake of clarity).

Competing Interests. The underlying goal of this interactive narrative presented in this paper is to use it as a tool to teach ethics as a skill, in terms of exposing the students / players to the relevant ethical principles and presenting them with scenarios aimed to enhance their ethical awareness, sensitivity and reasoning capabilities.

Nevertheless, ethics is not something that happens in a vacuum: ethically relevant decisions are always affected by the context where they take place, and sometimes a choice that would be desirable can be in conflict with other interests that also need to be balanced out. In order to bring these potential conflicts into play, the narrative presents scenarios in where players may not need to just consider the ethically right choice, but where they may also be required to achieve certain goals, and in where it might not be possible to fulfill all these requirements at once. In the narrative, these different requirements are tracked using the following variables:

- A set of *ethical principles* that, through their choices, the player either supports, or disregards. These principles, which have been initially distilled from the IEEE Code of Conduct [11] and have then been complemented with other principles concerning user data rights, are: *honesty, cherry-picking, integrity, personal respect / companionship* (all from IEEE's code 1, 2, 3, 4, 7 and 10), and *privacy* and *informed consent* (from user data rights).
- The *resources* that the fictional company (where the player character works) allocates to the player's team, which affect the way the story ends.
- The *approval ratings* between the two significant characters, who might support, or disagree with certain choices made by the player.

The potential effects of each choice over the variables have been curated in a way that avoids offering a clearly better, non-conflicting option (see Sect. 2.3 for the reasons behind this decision). Figure 2 shows an abstracted schema of the relevant points, within the narrative, where each set of variables might be affected.

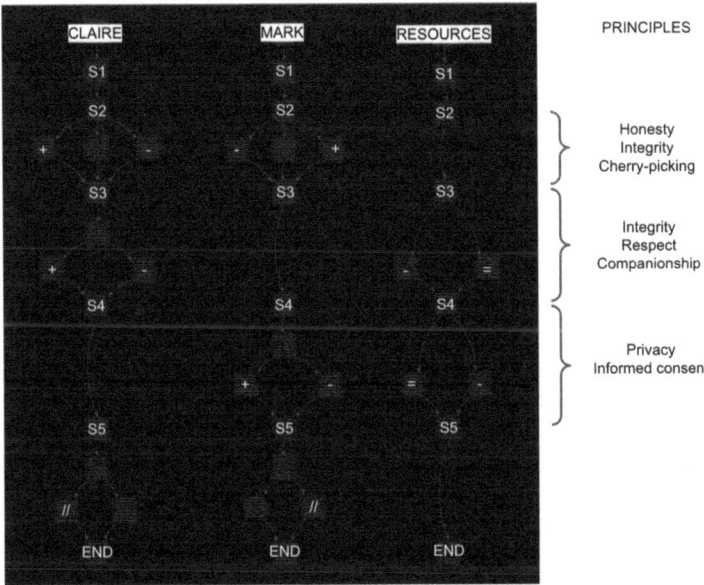

Fig. 2. Key points within the narrative that affect each set of variables. From left to right: approval (Claire), approval (Mark), resources, and ethical principles involved.

2.3 Designing Towards Spaces for Reflection

The main goal of this prototype is to foster reflection over certain relevant ethical principles, as well as to enhance awareness, sensitivity and reasoning skills on ethically relevant matters. It is as important to highlight what this prototype is about, as it is to understand what it is *not* about: in short, this interactive narrative is not about the (naive) dichotomy of a good, or a bad ending. Ethics is seldom as simple as that, and at the core of this narrative there is the explicit goal to highlight this fact as part of what defines this experience. As it has been argued before, ethical decisions do not happen in a vacuum, and thus contextual constrains, competing interests, and conflicting goals are bound to make acting right harder than it might seem when presented only via reductionistic, isolated examples. Nevertheless, the ethical dimension of our decisions should always be considered and taken into account when making a decision, but knowing that an optimal solution might not always be available; following virtue ethics, understanding when to act in which way is a matter of developing a practical wisdom that requires ethical awareness, sensitivity and reasoning skills.

According to this, the goal of this interactive narrative is to prompt and integrate ethical reflection as part of every scenario, every choice and every outcome, regardless of whether there exists a clearly better choice, or not, or whether the player decides to make the ethically correct choice in that case. This interactive narrative is, therefore, not a "test" in which the players will need to choose the correct options in order to get the good ending of the story: the value of this interactive experience is not in its end, but rather in its process. It is about training the habit of thinking about the ethical dimension of each and every decision, rather than about following the "correct" path to achieve a good ending. Even though this is not something that has an effect on the structure of the narrative, it does affect its content. The narrative has intentionally been shaped in such a way that there usually are no "absolutely correct" choices, and thus many choices require a compromise between following certain ethical principles, managing the in-game's resources, and following, or disregarding, the characters' advice.

Spaces for reflection. As it has been argued, this prototype aims to prompt ethical reflection in their players. But how can an interactive narrative create the appropriate spaces to allow for such reflection? This prototype tackles this question at the same time that tries to prevent tiredness and "attention overload" on the players, which could make them disengage from the narrative and make their choices without giving them too much thought. In order to address both considerations at once, the narrative has been designed as a series of separate, short sessions aimed to be played with a few days (at least one) of margin between them.

The expected outcome behind splitting the narrative into different chunks to be played at different times aims to provide the players with the space (a space of time, actually) to both reflect on the choices they have already make, as well as to try and foresee how these choices might affect further stages of the storyline. It should be noted, nevertheless, that the player is not given the option to change a decision once it has been made. These spaces of reflection are not, therefore, given in order to offer the player a chance to rethink and strategize the decisions they have already made in order to get the "right" one (if there is even such a thing), but rather to provide them with the space to reflect on how those choices might affect the storyline, the characters, or relate to ethical principles.

The prototype preliminarily features a story divided into 5 different sessions, designed so that they can be played in approximately 20 min, and allowing space of time between the sessions in order to allow the players to reflect on the events happening in the narrative, as well as potentially anticipating the outcomes of the last choices they would have made. In particular, the story have been divided as follows:

1. *Introductory session*: The setting, characters and plot are presented to the player. The session ends with the player being told that a piece of software developed by their team might be prone to showing bias towards some users. The session features mostly cosmetic choices that have minor effects on the narrative and that are used as a tutorial for the player.
2. *Potential bias discovered in software*: The player has to decide how to manage their team to address the bias and the current project they have been assigned. By the end of the session, the player has to decide what to do with regard to a conference presentation featuring the software.

3. *Conference presentation of the software*: The potential bias issue appears at the conference, and interests from the managerial team get in conflict with interests from developers in the player's team. The company's reputation might be severely affected, depending on how those events go down.

4. *Deadline of the new project*: The deadline for the new project is approaching and drastic measures might be needed, if the project is to be finished on time. In case the issue with the bias has not been already resolved, it reappears in this session.

5. *Epilogue*: The long-term consequences of the player's choices unfold. The player receives a final non-diegetic (i.e., happening outside of the fiction, or outside the game-space) report regarding the ethical principles that had been followed, or disregarded in their choices.

Hidden variables. If the player were to have the consequences and principles of their choices explicitly laid down to them, making those choices would be a matter of strategizing the numbers behind, but potentially without the need to engage in understanding the relevant parts of the choices and their consequences. As it has been introduced in Sect. 1.3, digital games that use implicit tracking systems can potentially prompt subjective reflection more easily. Due to this, all the relevant variables that appear in this interactive narrative are tracked implicitly. The player will never see the "numbers" behind those variables, or how their decisions might affect them, and the only way they might get to know about it is by reflecting on the way the story unfolds and by the way the characters relate to the player character.

3   Implementation of the Prototype

The final goal of the prototype is to be integrated as part of an online learning course, and thus it was implemented using a technology that would allow it to be easily integrated in a web-based context. In order to develop the interactive narrative, software tools like *Twine* [17] are tailored specifically to design these kind of narratives, and allow for an intuitive creation of branching storylines featuring variables, conditional structures, etc. Furthermore, Twine allows to export the resulting narratives in HTML format (also using JavaScript and CSS), in order to run them on a web browser quite effortlessly.

As it has been explained in Sect. 2.3, the story is divided into different sessions, where each session has its own branching structure starting and finishing at the same nodes, but potentially featuring variations due to the variables received from the previous sessions. Figure 3 shows a portion of a session in Twine (beware: spoilers ahead).

The content of each node within a session includes the text that will be shown to the player, as well as changes in variable values, conditional structures (which, in turn, affect what the player will see), inner monologue events, and potentially multimedia elements supported by web browsers. Figure 4 shows a reduced sample of a node, belonging to session 3, in the Twine editor.

There has not yet been the opportunity to integrate the prototype into an online classroom in order to test it with real students, but this is of course a natural next step in this research. In order to carry out this integration, the particularities of the technological requirements might vary depending on the online campus where the narrative should be

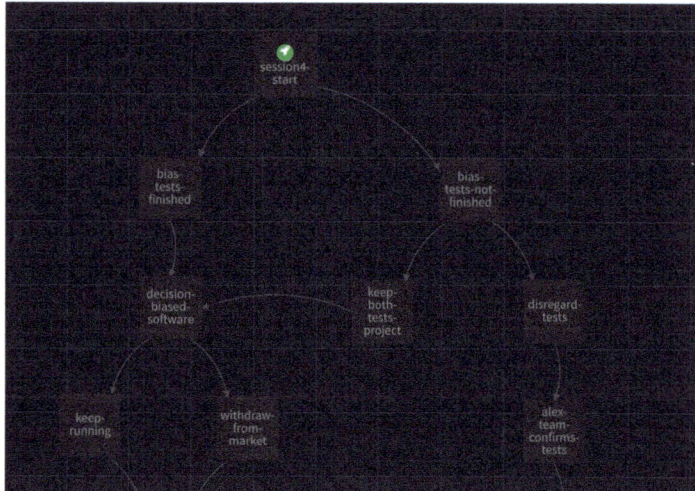

Fig. 3. Sessions start at the same node, regardless of previous decisions, but the content of that node is affected by those decisions. Sessions feature an internal branching structure.

```
• (if: $point_to_bias is false)[
• A hand rises at the back of the room. The lights are a bit dim, (...) it's
  [[Alex->question-alex]]. You can't help but feel a sudden thud in your
  chest as you wonder what in the world he could possibly ask you |A>
  [about]...
• ](else:)[
• A somewhat uncomfortable silence fills up the room. You can see some people
  in the audience whispering. (...)
• Just before the chair is about to speak, [[Peter->sales-speaks-up]] rises
  from his chair, faces the audience and starts speaking.
• ]
•
• \(click-append:?A)+(t8n:'dissolve')[ //(other than THAT, of course)//]
•
• \(click-append:?B)+(t8n:'dissolve')[ //(is this what doing the right thing
  feels like?)//]
•
• \(click-append:?chance)+(t8n:'dissolve')[ //(or maybe you just avoided
  it?)//]
```

Fig. 4. Sample of the (reduced) content of a node. The sample features conditional statements over values of a variable, links to further nodes within the session, and interactive parts of text.

integrated in, but it should be easy to incorporate the narrative as a link to an external resource. In that case, the narrative would need to be preceded by a user login screen to identify the students, and it would need to store each students' decisions and variables into a database in order to get them back in further parts of the story. As shown in Fig. 5, this could all be handled by storing the client side of the narrative (i.e., the files generated by Twine) in a server where a database is accessed via PHP files.

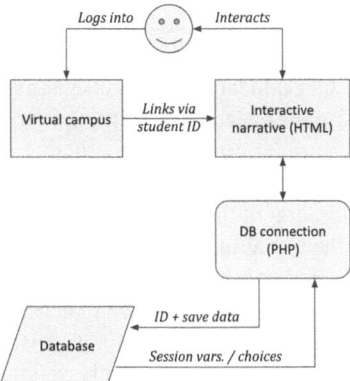

Fig. 5. A schema of the technologies planned to be used in the integration between the interactive narrative and a virtual campus.

4   Concluding Remarks and Future Work

In this paper, the motivation, the main design decisions and the implementation technologies of a prototype of an interactive narrative to complement the teaching of ethics as a skill and prompt subjective reflection about ethically relevant principles to the players have been presented. As it has been said, it is important to note that the goal of this narrative is not to teach the players about "right" and "wrong" (and thus "correct" and "incorrect") decisions, but rather about fostering the integration of ethical awareness and reflection throughout their choices. Even though the player will get a final summary of the choices made in the narrative (including the ethical principles they have either followed, or disregarded), this summary is not meant to classify, or evaluate players, but rather to provide them with a final, overarching layer of feedback to provide one further opportunity for their subjective reflection.

The main design decisions incorporate insights from the field of game design and focus on three main key features that can allow the appearance of subjective involvement in the player: attachments, meaningful choices and the creation of spaces for reflection. The insights around those decisions can be used as a series of guidelines towards important features to consider in the development of interactive experiences aimed at complementing the teaching (and learning) of ethics as a skill.

The following natural step in this research would be to test the prototype in a real setting—that is, by integrating it into a virtual campus within a relevant course. An evaluation, probably formed by a combination of questionnaires and semi-structured interviews, should be designed in order to check whether the interactive narrative can indeed be used to enhance subjective reflection in their players. This evaluation, therefore, should neither be designed in order to provide a snapshot of their comprehension and possession of ethically relevant skills, nor even a comparative change in such comprehension, but rather to assert whether their predisposition to incorporate ethical reasoning in their decisions has been enhanced. In this sense, the evaluation should aim to measure a change in tendency and predisposition (to think about the ethical dimension and ethical nuances of each and every decision), rather than an actual change in knowledge. Ideally, a first test could be an A/B where some students have access to the prototype, while some of them do not. The running hypothesis is that those students that use the prototype would likely enhance their ethical awareness, sensitivity and reasoning skills in a deeper way that those who do not—even if both groups of students have been exposed to the same non-interactive learning materials on ethics.

Once this first round of tests has been carried out, the outcomes should help shape improvements over a next iteration of the prototype. A feature that will most likely need to be incorporated refers to the use of multimedia content: even though the technologies used to implement this prototype allow the use of static images, audio files, and even video files, the amount of resources needed to create such material has not been available in the first stage of this project. Nevertheless, providing audiovisual materials to complement the narrative experience would likely make it not only easier to follow, but also easier for the students to feel immersed into, as well as to develop meaningful attachments with the setting and the characters in it. This, therefore, is something that is already expected to be included into a next iteration of the prototype.

Acknowledgments. This work has been supported by a UOC postdoctoral stay.

References

1. AMA Code of Medical Ethics: https://www.ama-assn.org/sites/ama-assn.org/files/corp/media-browser/principles-of-medical-ethics.pdf. Accessed 07 March 2022
2. Biedenweg, K., Monroe, M.C., Oxarart, A.: The importance of teaching ethics of sustainability. Int. J. Sustain. High. Educ. (2013)
3. Carstensdottir, E., Kleinman, E., El-Nasr, M S.: Player interaction in narrative games: structure and narrative progression mechanics. In: Proceedings of the 14th International Conference on the Foundations of Digital Games (2019)
4. Casas-Roma, J., Arnedo-Moreno, J.: Categorizing morality systems through the lens of fallout. In: DiGRA'19-Proceedings of the 2019 DiGRA International Conference: Game, Play and the Emerging Ludo-Mix, vol. 2019, pp. 1–16 (2019)
5. Davis, N.: Contemporary deontology. In: Singer, P. (ed.) A Companion to Ethics, Chapter 17, pp. 205–218. John Wiley & Sons, Hoboken, NJ (1993)
6. DC Bar's Rules of Professional Conduct: https://www.dcbar.org/for-lawyers/legal-ethics/rules-of-professional-conduct. Accessed 08 March 2022
7. EU Parliament.: Proposal for a regulation of the European parliament and of the council laying down harmonized rules on AI and amending certain union legislative acts (2021). https://eur-lex.europa.eu/legal-content/EN/TXT/PDF/?uri=CELEX:52021PC0206&from=EN. Accessed 03 Nov 2021
8. Hayse, M.: Ultima IV: simulating the religious quest. In Halos and avatars: playing video games with god, pp. 34–46 (2010)
9. Heron, M., Belford, P.: It's only a game—ethics, empathy and identification in game morality systems. Comput. Games J. 3(1), 34–53 (2014)
10. HLEG on AI.: High-level expert group on artificial intelligence: ethics guidelines for trustworthy ai (2019). https://www.aepd.es/sites/default/files/2019-12/ai-ethics-guidelines.pdf. Accessed 03 Nov 2021
11. IEEE Code of Ethics: https://www.ieee.org/about/corporate/governance/p7-8.html. Accessed 10 Jan 2021

12. IEEE.: Ethically aligned design: Prioritizing human wellbeing with autonomous and intelligent systems (2016). https://standards.ieee.org/content/dam/ieee-standards/standards/web/documents/other/ead1e.pdf. Accessed 15 Nov 2021
13. Mill, J.S.: Utilitarianism and other essays. Penguin Classics, New York, NY (1987)
14. Sicart, M.: Moral dilemmas in computer games. Des. Issues **29**(3), 28–37
15. Saw, P.S., Chuah, L.H., Lee, S.W.H.: A practical approach toward teaching ethics to community pharmacists. Int. J. Clin. Pharm. **40**(5), 1131–1136 (2018). https://doi.org/10.1007/s11096-018-0707-8
16. Švelch, J.: The good, the bad, and the player: The challenges to moral engagement in single-player avatar-based video games. Ethics and game design: teaching values through play. IGI Global 52–68 (2010)
17. Twine homepage: https://twinery.org/. Accessed 20 Dec 2021
18. Vallor, S. (2016). Technology and the virtues: A philosophical guide to a future worth wanting. Oxford University Press
19. Zagal, J.P.: Ethically notable videogames: moral dilemmas and gameplay. DiGRA conference (2009)

Influence of Immersive Virtual Reality on Cognitive and Affective Learning Goals

Janika Finken[1(✉)] [iD] and Matthias Wölfel[1,2] [iD]

[1] Karlsruhe University of Applied Sciences, Karlsruhe, Germany
{fija1013,woma0005}@h-ka.de
[2] Faculty of Business, Economics and Social Sciences, Stuttgart, Germany

Abstract. With the increasing prevalence of immersive virtual reality (VR) systems in educational contexts, it is important to understand how the use of VR impacts cognitive, affective, and psychomotor learning processes and learning outcomes. Proponents argue that VR can improve the quality of learning by stimulating individual learning processes to generate interest and motivate them to learn. Others claim that immersive media such as VR disrupts learning by increasing distraction, which leads to a decrease in cognitive performance and a reduction in acquired knowledge. To collect evidence on how VR influences learning, we performed a quantitative study. One group received information (about environment protection) before and after being immersed in two different VR applications, which incorporate content-related informational context. The second group was presented the same information but without the VR experience. We found that cognitive learning was better for the participants without the complementary VR experience, as they remembered information better. However, affective learning was better for the participants with the complementary VR experience. One week after the presentation of the content, participants with the complementary VR experiences had an increased awareness of threats to nature as well as the commitment to act environmentally conscious. Our findings, thus, support voices stating that VR can support learning, as well as those arguing against it. Whether the use of VR is supportive depends on the learning goals.

Keywords: Virtual Reality · Affective Learning · Empathy

1 Introduction

Already in the 1990s, the influence of immersive media on the learning process, as well as the learning effects and the achievement of different learning goals, was investigated. Recently, the tremendous technological progress in *virtual reality* (VR) technology and the availability and affordability of *head-mounted displays* (HMDs) have raised a growing interest in applying immersive media to learning [26]. In [18], it was found that VR is particularly suitable for learning cognitive, psychomotor, and affective skills. Regarding the transmission of procedural-practical and declarative, as well as analytical and problem-solving knowledge, some studies have already been conducted [15, 26]. Through many favorable and effective training scenarios with reduced risk, VR has been able to assert itself in teaching procedural, practical knowledge professionally in education [8, 27, 33]. VR has also proven to be very helpful in the acquisition of factual knowledge such as theoretical concepts and scientific principles, as well as analysis and problem solving [24, 29, 32]. However, little evidence could be found about the influence of VR on empathy-related skills. Both literature reviews [15, 26] identify VR-induced growth and change in emotions and attitudes as one of the least studied learning outcomes of VR. Thus, research on the effects of VR on affective and cognitive learning should complement the current focus on cognitive processes with a deeper understanding of relevant emotional impact. Learning experiences should evoke certain positive and negative emotions so that these emotions interact with cognitions to influence behavior and facilitate desired outcomes [16]. Based on this premise, this paper attempts to develop an integrative perspective for incorporating VR into learning processes that integrate emotions and cognitions in the context of desired outcomes. Due to the currently growing importance of nature conservation, the protection of the rainforest was chosen as the pivotal point for the investigation of empathic capacity towards nature. In this context, we investigate the question to what extent immersive VR influences empathy-related emotions towards nature.

2 Literature Review

In terms of affective learning and the acquisition of declarative knowledge with the help of VR, this literature review focuses on two different foci: the learning of empathy and the influence of VR on the learning process.

In the case of the former, a connection can be made to Bloom's affective learning goal taxonomy. This involves integrating new values into one's personality through observation and acting accordingly [6]. VR seems to be particularly well suited for this because of the possibility to take other perspectives; see, for example, the *body ownership illusion* or the *Proteus Effect*.

Whereas several studies [15, 24, 26] are dedicated to knowledge transfer within VR, it remains questionable to what extent virtual worlds influence what is learned outside VR. Due to this gap in previous research, we examine—regarding our second focus—whether the acquisition of factual knowledge is positively or negatively influenced by VR in general, as well as before or after being exposed to immersive media.

2.1 Learning Empathy with Immersive VR

Although many different learning VR applications have already been developed for different topics, only a few studies on affective learning can be found. Both examined literature reviews [15, 26] identify VR-induced growth and change in emotions and attitudes as one of the least studied learning content of VR. These behavioral influences can be classified under the affective learning goal taxonomy of Bloom's. According to Bloom, the affective learning process describes changes in interest, attitudes, and values, as well as the development of appreciation and appropriate adaptive skills [6]. In this process, an individual value system is built upon the basis of personal convictions until this finally leads to the growth of one's personality and character [21, p. 164–173]. Virtual body ownership can support this integration of newly acquired values, the illusion that the person has become the virtual body [2, 31]. A related effect is the Proteus Effect, a behavioral modulation caused by digital self-representation in virtual environments [5]. This effect can influence users to conform their behavior to the avatar. The appearance of the embodied avatar can change attitudes, beliefs, and actions. A large virtual body leads to more self-confident behavior in discussions than a small body, an avatar belonging to the Ku Klux Klan activates more negative thoughts and leads to more aggressive behavior such as revenge and murder. When embodying a Black avatar in computer games, more aggressive behavior can be seen and prejudices against other groups can be reinforced. Because the Proteus Effect is based on self-perception principles, positive attitudes and actions can also be triggered by observing external cues. For example, the embodiment of an inventor during a brainstorming session in VR leads to greater creativity and originality in ideas and the embodiment of a Sigmund Freud-like avatar leads to a better mood through self-counseling than talking to a self-presenting avatar [5].

As shown the interrelationship of presence in immersive VR can induce behavior change through body ownership. When searching for appropriate studies on changed behavior through the influence on empathy, one comes across some studies—as mentioned by [26]—that rather focus on the intended change of student behavior. This does not happen directly by acting according to newly recognized values but, for example, by motivating better learning habits or improved compliance with rules [7, 17, 23]. For example, raising awareness about environmental protection by picking up virtual litter left behind and learning to sort it properly in VR [17]. They intend to change perceptions, attitudes, and actions related to specific topics, but they do so without considering the influence of empathy on changed behavior [26]. Thereby, the development of affective learning goals, as described by Bloom, is disregarded.

However, it is equally important to develop the ability to see the world from another person's perspective. Tied to an emotional response, this creates a connection that inspires compassion and a need to help others [10]. An example is the VR application Clouds Over Sidra [22]. In this study, participants were randomly assigned to watch the documentary on a 2D screen or an HMD. Results illustrate that compared to the 2D video, the VR experience resulted in greater engagement and elicited higher levels of empathy for the refugee girl. This, in turn, suggests that VR-facilitated engagement and a greater sense of presence enhance a variety of emotional and cognitive responses such as empathy. Another study in this area illustrates that embodying a superhero in a virtual world also leads to prosocial behavior. A group of participants who were given superhero abilities in the virtual world to actively help people subsequently showed greater helpfulness in reality [28]. Interactive VR, thus, offers the potential to not only strengthen prosocial behavior in this virtual world, but also to transfer it to reality. The embodiment as a superhero could enhance the user's self-perception and identity such that they perceive themselves as someone who helps others. This idea of using body ownership to enhance empathy towards nature is reflected in another study [2]. Based on the compelling results of previous studies on increased engagement, caring, and supportive action through perspective-taking with VR, this study intends to build on these findings by using immersive virtual experiences to promote a sense of nature as part of people's self-identity. Embodying animals with multimodal sensory input led to a stronger sense of presence and connection with nature than watching the experience on video. This connection led further to a stronger perception of immediate threats to the environment and a deeper relationship with nature. Overall, the results suggest that VR experiences with a strong sense of presence and body transfer can be an effective tool to promote involvement in environmental issues.

Although the influence of VR on affective learning goals—including empathy-related emotions—has hardly been investigated so far, the presented initial results look promising. However, some questions remain regarding the influence on sustainable integration of the new values remain open. Likewise, the actual influence on changed behavior has not yet been clarified.

2.2 Influence of VR on the Learning Process

In general, there is an assumption that media consumption such as television or computer games after learning or during learning breaks, disrupts the learning process. Films and games often have a strong emotional impact and release adrenaline. This leads to the information conveyed being interpreted by the brain as more important than what was previously learned, thus devaluing it. Overall, regular television viewing, in particular, is associated with a decrease in cognitive performance and reduction of acquired knowledge, not only in children but also in adults [14]. Some, on the other hand, understand a disorder as something positive, as they are a necessary part of human development. The imbalance created by a disturbance initiates action to restore balance, a process that can be understood as learning. According to the learning theory of constructivism, disturbances trigger individual learning processes and are, thus, motivating for learning [20].

These findings complement a study, revealing that the learning experience with VR is evaluated more positively than some established teaching methods [24, 26]. Furthermore, this form of learning sparks interest and increases motivation, as well as enthusiasm, among students. According to [11], these effects and emotions have a significant positive impact on learning success. The arousal of interest and the strengthening of confidence in one's abilities favor the quality of learning, conceptual understanding as well as personal growth. However, in [24], those students who were taught the learning content via a PowerPoint rather than in VR scored better on the posttest. It was only when the VR group was asked to summarize elements of their virtual experience that they achieved significantly better learning outcomes without diminishing interest or motivation. In addition, several other studies have already demonstrated the potential of VR as a medium compared to traditional practices in an educational context [12, 30].

Compared to conventional teaching methods, VR offers clear advantages not only in the learning of practical skills—as has already been shown in many studies—but also the transfer of theoretical knowledge. In addition to the virtual experience, repetition of the learned content is useful to achieve the best possible learning results. However, it remains inconclusive how immersive VR applications affect the learning of information that is thematically complementary and supportive.

3 Description of the VR Applications Used

Various media have already failed to address the issue of environmental degradation adequately enough to evoke nature-conscious action among people. "It is not enough simply to know things or simply to teach things. Knowledge, in itself, is not enough" [13]. In addition to political decisions, an appeal to emotions and reason is also needed, as well as a broad-based change in ethics. This finding is consistent with social science research, which has shown, information alone cannot change behavior; for that, engagement with the content is necessary [4]. This is where the immersive nature of VR can bridge the spatial and psychological distance between environmental problems and people. Since experiences are the best teachers [9], this way, critical changes can be experienced and perceived firsthand. Through a personal event, people are more inclined to perceive the effects of environmental protection and destruction. Hence, they are more prone to perceive environmentally conscious action as important and to commit themselves to it [3].

Based on these findings, this paper intends to build on previous knowledge to investigate the effect of immersive and perspective-changing VR on the relationship between humans and nature. We pursue this goal by using two different VR applications about the rainforest in combination with a presentation about the rainforest to increase awareness of the risks to nature and the commitment to engage environmentally conscious.

The first VR application is *Inside Tumucumaque* by Interactive Media Foundation. This experience allows users to immerse themselves in the world of five different tropical forest inhabitants and view the tropical rainforest through their eyes. As a black caiman, harpy eagle, vampire bat, poison dart frog, and Goliath bird eater, users explore Tumucumaque, the largest rainforest reserve in northeastern Brazil. Perception from the perspective of wildlife is interpreted as a sensory experience comprehensible to the human perceptual system. Utilized for this purpose are ultraviolet color spectra, super slow-motion movement, visualizations of echolocation tracking and color night vision, and spatial 3D sound. This provides an insight into the fragile ecosystem of the Tumucumaque Reserve in the Amazon rainforest while developing a sense of the endangered species.

The second VR application used is *Tree VR* by New Reality. This experience lets the user slip into the skin of a rainforest tree. From a seed in the ground, the user grows into a large tree with their body as the trunk and their arms as the branches. Slowly seeing daylight, the user grows taller and taller as various animals climb or fly by and even land on his branch. Growing to full size, the user embodied as a tree proudly overlooks the other plants and trees when the scene turns into night. The calm atmosphere changes when suddenly birds fly away in alarm and a fire becomes visible in the distance. Unable to escape the forest fire or help other plants and animals, the observer slowly leaves the tree and watches the rainforest abandoned to its terrible fate.

The presentation intends to make participants aware of the importance of the rainforest to the global climate, the environment, as well as human life, while the VR applications bring them closer to the rainforest. The chosen VR applications Inside Tumucumaque and Tree VR complement each other, as well as the knowledge about the rainforest conveyed in the presentation.

A PowerPoint is used to educate participants about the location and spread of rainforests, their classification into different levels, biodiversity, benefits to human life and the environment, and the causes and consequences of their destruction.

4 Test Setup

To investigate the effects of using VR applications to extend a presentation, we divided the participants into two user groups (between-group design): the "VR group" was given a PowerPoint presentation and immersed in the two VR applications described in Sect. 3, while the "PT group" only saw the presentation.

The VR group was first presented with one part of the presentation, then they experienced the two VR applications using an Oculus Quest, and finally, they got presented the second part of the presentation. To investigate if the use of VR before or after the presentation of knowledge using slides has an effect, the participants in the VR group were again randomly divided into two sub-groups: half were presented Information Part I before the VR applications and then Information Part II. For the other half, these two parts of information were reversed.

To see what influence VR has on the acquisition of knowledge and the ability to retain it in the long term, the participants filled out the same questions at three different time points: before VR/PT, after VR/PT, and one week later. To investigate the development of environmental awareness and commitment to environmental responsibility the questions presented by [19] as first presented, supplemented, and applied by [22] were used as orientation and personally extended by additional questions (1 = strongly disagree, 5 = strongly agree; Cronbach's α = 0.769). In addition, demographic data were collected, and in the second questionnaire, the questions were followed by more questions about the feeling of the embodiment of an animal or tree and empathy towards the habitat of many endangered animals exposed to a forest fire (1 = strongly disagree, 5 = strongly agree; Cronbach's α = 0.834). Finally, more empathy-related questions were asked in the context of the rainforest as a habitat and environmental protector (1 = strongly disagree, 5 = strongly agree; Cronbach's α = 0.794). The questionnaire a week later asked application-related questions about the actions taken, such as "In the last 7 days, did you take action yourself to help protect the rainforest?" (1 = strongly disagree, 5 = strongly agree; Cronbach's α = 0.743). Finally, a quiz followed with 16 knowledge questions, 8 for each section of information about the rainforest. These were multiple-choice, single-choice, and in the case of numerical values, free-text questions.

For the PT group, the procedure was the same as for the VR group. The difference for these participants, however, was that the information did not have to be divided into two parts because they were not separated by a VR application. For the questionnaires, only the questions about the feeling of the embodiment of an animal or tree and empathy towards the habitat of many endangered animals exposed to a forest fire were omitted.

A total of 124 participants (79 male, 44 female, and 1 diverse) were recruited for the study. The majority are between 18 and 34 years old, with very few older or younger. With 86% of the participants, the majority are students, followed by professionals, and very few schoolchildren. A total of 78 participants were in the user groups with the complementary VR experience. The control group consisted of 46 participants.

5 Results

This section presents the results aggregated from the three questionnaires completed at three different times—each time once before the VR experience and/or presentation, afterward, and one-week following—as well as the quiz.

5.1 Learning Empathy with Immersive VR

In this section, we investigate the influence of awareness of environmental threats and commitment to environmental responsibility concerning represented information. The values in the tables refer to the Likert scales as described in Sect. 4. The tables present the mean values and standard deviations (in brackets). To test for significant differences (p-values ≤ 0.050) between the groups in the awareness of environmental threats and the commitment to environmental responsibility non-parametric repeated-measures ANOVA was conducted. Significant results are marked with an asterisk.

Awareness of Environmental Threats. The mean values in Table 1 do not show large differences between the VR and PT group; however, a pairwise comparison in Table 2 reveals some differences. For participants with the VR applications, the average awareness of environmental threats before and after the VR applications differ significantly ($p < 0.001$). This is also the case for the group with only the presentation ($p = 0.012$). However, the mean value for awareness after one week compared to before the VR experiences or presentation differ only for the VR group ($p < 0.001$). The difference is no longer significant for the PT group ($p = 0.384$). In addition, it is noticeable that, one week after the VR experience, the mean value for environmental awareness very clearly does not change significantly compared to immediately after the VR experience ($p = 1.000$). After only the presentation, however, there is a significant difference between these two mean values ($p = 0.012$). Thus, a significant increase in awareness takes place in both groups. However, one week later, the awareness of the participants who only received a presentation no longer differs significantly from the awareness before the presentation.

Since about 25% in the VR group and about 30% in the PT group have the highest possible awareness, an additional raise is not measurable. Therefore, we removed participants with high a-priori awareness (VR/PT $M < 4.5$) and repeated the significant tests. As shown in Table 2 for the VR group the significant difference between awareness before and after the VR applications ($p < 0.001$) as well as before the VR applications and one week later remains ($p < 0.001$). For the PT group, however, the difference between awareness before the presentation and after is now no longer significant.

Accordingly, the influence of VR on environmental awareness is more pronounced than the influence of the presentation. Thus, awareness of environmental threats can be better addressed by a VR experience in which one experiences life in the rainforest through the eyes of various animals or plants than with a PowerPoint presentation.

Commitment to environmental responsibility. As shown in Tables 1 and 3, the commitment to act environmentally aware is lower overall than the awareness of threats to nature. A pairwise comparison shows that the commitment to environmental responsibility before VR/PT and afterward is significant for both groups; VR ($p = 0.003$) and PT ($p < 0.001$). In contrast to awareness, there is no significant difference in commitment

Table 1. Unfiltered and filtered mean values and standard deviations of awareness of environmental threats for the VR as well as the PT group.

Group	VR	VR I $M < 4.5$	PT	PT I $M < 4.5$
Before VR/PT	4.55 (0.43)	4.07 (0.28)	4.66 (0.33)	4.22 (0.19)
After VR/PT	4.70 (0.35)	4.43 (0.35)	4.71 (0.37)	4.32 (0.43)
Week Later	4.69 (0.35)	4.40 (0.32)	4.64 (0.48)	4.32 (0.30)

Table 2. Significance of differences between mean values of environmental awareness.

Group	VR	VR I $M < 4.5$	PT	PT I $M < 4.5$
Before VR/PT - After VR/PT	<0.001*	<0.001*	<0.012*	0.095
Before VR/PT - Week Later	<0.001*	<0.001*	0.384	0.393
After VR/PT - Week Later	1.000	0.707	0.096	0.393

before the VR applications and one week later ($p = 0.263$). However, the difference stays significant for the PT group before the presentation and one week after ($p = 0.008$). The difference in commitment after VR/PT and one week following is not significant for both the VR ($p = 0.055$) and PT group ($p = 0.344$). Considering the individual measures taken to protect the environment in the week after VR/PT, it is apparent that the VR group took more measures. However, the difference to the PT group is not significant.

In line with awareness of environmental threats, commitment to environmental responsibility was filtered by initial perception. Due to lower mean values, the filter here is set to $M < 4.0$. The difference between the commitment before VR/PT and afterward remains significant for the VR ($p = 0.004$) as well as for the PT group ($p < 0.001$). However, the difference between initial engagement and engagement one week later also remains significant not only for the PT group ($p < 0.001$), but also for the VR group ($p = 0.021$). For both groups, the commitment does not decrease significantly after one week.

Consequently, a significant increase in commitment to environmental responsibility takes place for both groups after VR/PT. This increase does not decrease significantly even after one week. Both procedures leave a lasting impact on environmentally conscious behavior (Table 4).

Impact of cybersickness on affective learning through virtual reality. As demonstrated in Sects. 1 and 2, VR offers many different benefits in learning. However, various symptoms such as cybersickness occur frequently when using HMDs. Since it is important to consider their impact on humans, especially concerning the process of learning, this study also analyzes the influence of cybersickness on the learning process. A Mann-Whitney U test demonstrates a significant difference only for the commitment to environmental behavior after the VR experience compared to before. Commitment is even greater for participants who experienced cybersickness ($M = 4.42$) than those who had no problems with the VR ($M = 4.12$). However, performing a Mann-Whitney-U test on

Table 3. Unfiltered and filtered mean values and standard deviations of the commitment to environmentally conscious behavior for the VR as well as the PT group.

Group	VR	VR ∣ $M < 4.0$	PT	PT ∣ $M < 4.0$
Before VR/PT	4.08 (0.45)	3.63 (0.24)	3.84 (0.54)	3.43 (0.44)
After VR/PT	4.18 (0.48)	3.80 (0.35)	4.02 (0.48)	3.69 (0.38)
Week Later	4.14 (0.49)	3.79 (0.38)	3.97 (0.52)	3.71 (0.44)

Table 4. Significance of differences between mean values of the commitment to environmentally conscious behavior.

Group	VR	VR ∣ $M < 4.0$	PT	PT ∣ $M < 4.0$
Before VR/PT - After VR/PT	0.003*	0.004*	<0.001*	<0.001*
Before VR/PT - Week Later	0.263	0.021*	0.008 *	<0.001*
After VR/PT - Week Later	0.055	0.521	<0.344	1.000

the increase in awareness and commitment reveals no significant differences between participants with and without experienced cybersickness. Based on these results, it can be concluded that the HMD-induced symptom cybersickness does not have a negative effect on the learning of affective skills.

Difference between the sexes. During the study, it was found that a-priori environmental awareness (Mann-Whitney U $p = 0.021$) and commitment (Mann-Whitney U $p < 0.001$) are significantly higher for females. However, the increase in awareness and commitment did not show any significant differences. It can be concluded that, although women have a stronger environmental awareness and are also more engaged in this area, neither a PowerPoint presentation nor an immersive VR experience influences them more than men.

5.2 Influence of VR on the Learning Process

To investigate the influence of knowledge acquisition of immersive VR before or after the representation of information, a quiz was performed and analyzed. The PT group with an error rate of 23% achieved better results than the VR group with an error rate of 31%. However, according to a Mann-Whitney U test, only three questions show significant differences. No significant difference could be found if the knowledge was presented before or after the two VR experiences. Therefore, it can be concluded that information is better remembered solely on a PowerPoint presentation instead of the combination of a PowerPoint in conjunction with an immersive VR experience covering the same topic. Furthermore, since the results are similar to those of the study by Parong and Mayer [24], it can be assumed that a repetition of the content or a reprocessing of what was experienced in the virtual world could lead to better results.

6   Conclusion, Limitations, and Outlook

Emotion-evoked empathy represents significant new territory in the field of VR learning, which is somewhat surprising given the well-known interplay between learning and emotion [16]. We took a novel approach—integrating emotion into the learning process as well as behavioral changes elicited by empathy—to better understand the alignment of desired cognitions, emotions, and the effect of empathy on environmental awareness. Our results show that immersive VR experiences, in combination with additional information delivered via a PowerPoint presentation, positively affect environmental awareness not only in the short term, but also in the long term, compared to a presentation alone. However, no significant difference was found between the two groups in terms of environmental engagement. However, the VR group took more environmentally conscious actions than the PT group in the week following the experiment. In contrast to the positive impact of VR on affective learning, we found that cognitive learning was negatively influenced, as participants without the supplementary VR experience were better able to remember the presented information.

When interpreting the results, it should be considered that the majority of participants are students who are known to be more sensitive to environmental issues than other groups [1]. Consequently, participants were very environmentally aware and showed a high level of engagement, which made it difficult to show additional gains. About half of the participants in the VR group have never used an HMD before and are, thus, inexperienced with using VR. Students, however, spend a lot of time with digital media, which potentially makes the impact of immersive VR different for this population [25]. Another limitation of our study is that we were unable to measure long-term effects because we were only able to survey participants again one week after conducting the initial study.

To better generalize the results, future studies should examine the effects of VR experiences on empathic capacity towards nature in a broader range of populations and investigate the change in behavior and knowledge over several months.

References

1. Abbas, M.Y., Singh, R.: A survey of environmental awareness, attitude, and participation amongst university students: a case study. Int. J. Sci. Res. (IJSR) **3**, 1755–1760 (2014)
2. Ahn, S.J.G., Bostick, J., Ogle, E., Nowak, K.L., McGillicuddy, K.T., Bailenson, J.N.: Experiencing nature: embodying animals in immersive virtual environments increases inclusion of nature in self and involvement with nature. J. Comput.-Mediat. Commun. **21**, 399–419 (2016)
3. Akerlof, K., Maibach, E.W., Fitzgerald, D., Cedeno, A.Y., Neuman, A.: Do people "Personally Experience" global warming, and if so how, and does it matter? Glob. Environ. Chang. **23**, 81–91 (2013)
4. Bandura, A.: Self-efficacy: toward a unifying theory of behavioral change. Psychol. Rev. **84**, 191–215 (1977)
5. Bertrand, P., Guegan, J., Robieux, L., McCall, C.A., Zenasni, F.: Learning empathy through virtual reality: multiple strategies for training empathy-related abilities using body ownership illusions in embodied virtual reality. Front. Robot. AI **5**, 26 (2018)
6. Bloom BS (1956) Taxonomy of educational objectives: the classification of educational goals handbook I, Handbook I. McKay, Longman, New York, London

7. Carruth DW (2017) Virtual reality for education and workforce training. In: 2017 15th international conference on emerging elearning technologies and applications (ICETA), pp. 1–6. IEEE, Stary Smokovec

8. dela Cruz DR, Mendoza DMM (2018) Design and development of virtual laboratory: a solution to the problem of laboratory setup and management of pneumatic courses in Bulacan state university college of engineering. In: 2018 IEEE Games, Entertainment, Media Conference (GEM), pp. 1–23. IEEE, Galway

9. Dale, E.: Audiovisual Methods in Teaching, 3rd edn. Dryden, New York (1969)

10. Davis, M.H.: Measuring individual differences in empathy: evidence for a multidimensional approach. J. Pers. Soc. Psychol. **44**, 113–126 (1983)

11. Deci, E.L., Vallerand, R.J., Pelletier, L.G., Ryan, R.M.: Motivation and education: the self-determination perspective. Educ. Psychol. **26**, 325–346 (1991)

12. Deeks, H.M., Walters, R.K., Barnoud, J., Glowacki, D.R., Mulholland, A.J.: Interactive molecular dynamics in virtual reality is an effective tool for flexible substrate and inhibitor docking to the SARS-CoV-2 main protease. J. Chem. Inf. Model. **60**, 5803–5814 (2020)

13. Estok, S.C.: Ecomedia and ecophobia. Neohelicon **43**, 127–145 (2016)

14. Fancourt, D., Steptoe, A.: Television viewing and cognitive decline in older age: findings from the English longitudinal study of ageing. Sci. Rep. **9**, 2851 (2019)

15. Hamilton, D., McKechnie, J., Edgerton, E., Wilson, C.: Immersive virtual reality as a pedagogical tool in education: a systematic literature review of quantitative learning outcomes and experimental design. J. Comput. Educ. **8**, 1–32 (2021)

16. Hascher, T.: Learning and emotion: perspectives for theory and research. Eur. Educ. Res. J. **9**, 13–28 (2010)

17. Hu. X., Su, R., He, L.: The design and implementation of the 3D educational game based on VR headsets. In: 2016 International Symposium on Educational Technology (ISET), pp. 53–56. IEEE, Beijing, China (2016)

18. Jensen, L., Konradsen, F.: A review of the use of virtual reality head-mounted displays in education and training. Educ. Inf. Technol. **23**, 1515–1529 (2018)

19. Kals, E., Schumacher, D., Montada, L.: Emotional affinity toward nature as a motivational basis to protect nature. Environ. Behav. **31**, 178–202 (1999)

20. Knaus, T.: Technik Stört! Lernen Mit Digitalen Medien in Interaktionistisch-Konstruktivistischer Perspektive. fraMediale, München (2013)

21. Krathwohl, D.R., Bloom, B.S., Masia, B.B., Dreesmann, H.: Taxonomie von Lernzielen im affektiven Bereich, 2nd edn. Beltz, Weinheim (1978)

22. Müller, M., Kals, E., Pansa, R.: Adolescents emotional affinity towards nature: a cross-societal study. J. Develop. Process. **4** (2009)

23. Parmar, D., Isaac, J., Babu, S.V., D'Souza, N., Leonard, A.E., Jorg, S., Gundersen, K., Daily, S.B.: Programming moves: design and evaluation of applying embodied interaction in virtual environments to enhance computational thinking in middle school students. In: 2016 IEEE Virtual Reality (VR), pp. 131–140. IEEE, Greenville, SC, USA (2016)

24. Parong, J., Mayer, R.E.: Learning science in immersive virtual reality. J. Educ. Psychol. **110**, 785–797 (2018)

25. Prensky, M.: Digital natives, digital immigrants. Gifted 29–31 (2005)

26. Radianti, J., Majchrzak, T.A., Fromm, J., Wohlgenannt, I.: A systematic review of immersive virtual reality applications for higher education: design elements, lessons learned, and research agenda. Comput. Educ. **147**, 103778 (2020)

27. Rose, F.D., Attree, E.A., Brooks, B.M., Parslow, D.M., Penn, P.R.: Training in virtual environments: transfer to real world tasks and equivalence to real task training. Ergonomics **43**, 494–511 (2000)

28. Rosenberg, R.S., Baughman, S.L., Bailenson, J.N.: Virtual Superheroes: using superpowers in virtual reality to encourage prosocial behavior. PLoS ONE **8**, e55003 (2013)

29. Rosenfield, P., et al.: AAS WorldWide telescope: a seamless, cross-platform data visualization engine for astronomy research, education, and democratizing data. Astrophys. J. Suppl. Ser. **236**, 22 (2018)
30. Salzman, M.C., Dede, C., Loftin, R.B., Chen, J.: A model for understanding how virtual reality aids complex conceptual learning. Presence: Teleoper. Virtual Environ. **8**, 293–316 (1999)
31. Slater, M.: Implicit learning through embodiment in immersive virtual reality. In: Liu, D., Dede, C., Huang, R., Richards, J. (eds) Virtual, Augmented, and Mixed Realities in Education, pp. 19–33. Springer Singapore, Singapore (2017)
32. Ye, Q., Hu, W., Zhou, H., Lei, Z., Guan, S.: VR interactive feature of HTML5-based WebVR control laboratory by using head-mounted display. Int. J. Online Eng. **14**, 20–33 (2018)
33. Zhang, K., Suo, J., Chen, J., Liu, X., Gao, L.: Design and implementation of fire safety education system on campus based on virtual reality technology. In: 2017 Federated Conference on Computer Science and Information Systems pp. 1297–1300 (2017)

Integrating Video Timeline-Anchored Comments in Asynchronous Online Video-Based Presentation Lectures: Using Canvas Studio as an Example

Xi Lin[1](✉) ⓘ, Qi Sun[2] ⓘ, and Xiaoqiao Zhang[3] ⓘ

[1] East Carolina University, Greenville, NC, USA
linxi18@ecu.edu
[2] University of Tennessee, Knoxville, TN, USA
qsun8@utk.edu
[3] Shanghai Jiao Tong University, Shanghai, China
xiaoqiao.zhang@sjtu.edu.cn

Abstract. In asynchronous online learning, watching asynchronous video-based lectures while using a video timeline-anchored comment (VTC) technology tool for discussion could encourage students to actively interact with their instructor, learning content, and peers. This study introduces the VTC function and presents how this tool hosted by Canvas Studio was integrated into two graduate-level online courses, aiming to show its potential for motivating active learning in the asynchronous online learning context. Preliminary findings from 23 students' text-based feedback reveal that the VTC tool can motivate students' interactions with the instructor, peers, and learning content. This tool also creates a real-time connection between online lectures and offline social networks in asynchronous learning environments. Therefore, this study shows promising results that the VTC tool could motivate learners' online interactions and active learning in asynchronous settings when used in video-based lectures.

Keywords: Video Timeline-Anchored Comment · Asynchronous Online Learning · Online Interaction · Canvas Studio

1 Introduction

Asynchronous learning is a learning format. The delivery of learning content and instruction does not happen in real-time, and instructors use online discussion boards and emails to conduct interactions [20]. Students can engage in course content and learn anytime, anywhere with internet access through this modality. Second, students can learn at their own pace to thoroughly comprehend learning content [19]. Third, students can take time to reflect and develop thoughts for online discussion, which would lead to deep learning [5]. However, one major disadvantage of the asynchronous learning format is the unavailability of real-time interactions with peers and instructors. Students cannot receive an

© The Author(s), under exclusive license to Springer Nature Switzerland AG 2023
D. Guralnick et al. (Eds.): TLIC 2022, LNNS 581, pp. 522–531, 2023.
https://doi.org/10.1007/978-3-031-21569-8_49

immediate response from the instructor nor timely support from their peers [7]. The lack of real-time interaction would result in undesired learning experiences and outcomes, including a sense of isolation and passive and surface learning. The lack of interactions may also lead to a perceived disconnection from the online learning community [3].

Student engagement is defined as "the student's psychological investment in an effort directed toward learning, understanding, or mastering the knowledge, skills, or crafts that academic work is intended to promote" [18]. Student engagement is significant to online learning as it can be developed through interactions with learning content, peers, and instructors that further drive them to be active and engaged in online learning [9]. Student engagement also increases students' satisfaction, enhances their online learning motivation, and boosts their performance and academic achievement [14]. Student engagement additionally remediates student isolation, dropout, retention, and graduation rates in the online learning context [2]. In short, student engagement plays an essential role in stimulating online learning, while students' interactions through sustained communication in online learning environments impact their engagement and further influence their online learning experience.

Despite research that advocates for student engagement through interactive learning, limited studies look at the video timeline-anchored comment (VTC) to enhance the interactions in an asynchronous online learning environment. VTC is a newly developed computer-mediated communication tool with a screen commenting function that has been widely used for animation, comics, and game videos in East Asian countries. This technology tool has effectively increased viewers' interactions while watching videos or playing games. The VTC tool has also been preliminarily applied to online learning in some Asian countries and positively enhanced students' interaction with peers and instructors [23, 25]. With the increasing use of asynchronous teaching approaches in higher education institutions worldwide, mainly due to the COVID-19 pandemic, there is a need to motivate interactive learning in the asynchronous online context. However, using the VTC function for video-based lectures, an existing available technology from the East to expand it in the West may provide new insights to understand this tool for asynchronous online teaching and learning. This study introduced how the VTC function hosted by Canvas Studio was integrated into two graduate-level online courses in a US institution to present its potential for motivating active learning in the asynchronous online learning context.

2 Literature Review & Theoretical Framework

2.1 Using Asynchronous Videos for Online Learning

The video technology conveys immediacy behaviors, which develop students' perceptions of the instructor and peers' social presence. Therefore, instructors often use asynchronous videos to engage students in online learning [14]. Using asynchronous videos for interactions is essential to support students' critical thinking and communication [12]. However, commenting in asynchronous videos may reduce communication's immediacy, making it challenging to reference momentary video content [13]. Specifically, in many Learning Management Systems (LMSs), the instructor posts a video on one page and assigns the discussion task on another page. Students must leave the video page and share their comments on the other page.

Alternatively, a video can be posted on the discussion page, but students must finish watching the video before sharing thoughts. These approaches for comments may be distracting, time-consuming, and difficult to tie back to specific points of the video content. These ways of commenting—linking with video-based lectures to thinking and reflection—may limit student interactions. In short, while asynchronous videos have been widely used for teaching, limitations remain in motivating effective online interactions, leading to the next section introducing the VTC tool that could encourage online learning interactions.

2.2 Video Timeline-Anchored Comment (VTC)

The function of VTC was first used by the entertainment videos in East Asian countries, especially in Japan and China [22]. It has become popular because its interactive feature allows the audience to engage in conversations as if they are happening in real-time while watching. This commenting function is called danmaku (弾幕) in Japanese or danmu (弹幕) in Chinese. Danmaku is defined as a "real-time, horizontal, text-based display, where commentaries appear in the form of subtitles at the top of the video frame" [10] (see Fig. 1).

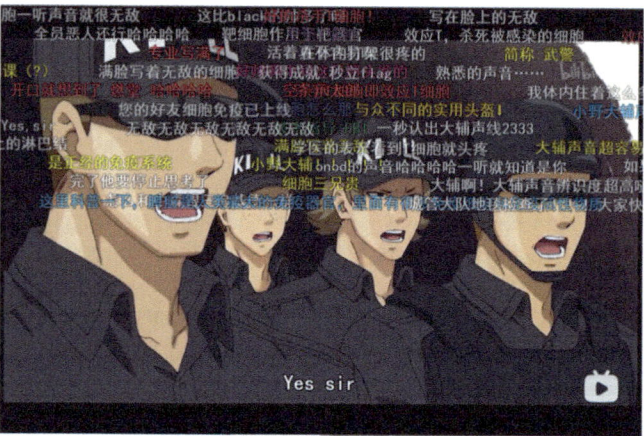

Fig. 1. Screenshot of a danmaku-commented episode of the Japanese anime series Cells At Work! on Bilibili.com.

This function is similar to live-streamed videos provided by social media platforms such as YouTube Live or Facebook Live, allowing viewers to watch while posting comments and reading real-time responses. These live-streaming videos often have "an embedded or adjacent chat channel, where the interaction develops as the streamer's activity proceeds" [24]. However, unlike live-streaming videos, viewers can still read the comments even when the video streamers and commentators are offline because those comments on danmaku video-sharing websites (e.g., Bilibili. tv based in China) are stored on the website server. Therefore, viewers who watch the video can still see previously posted comments and respond (see Fig. 2).

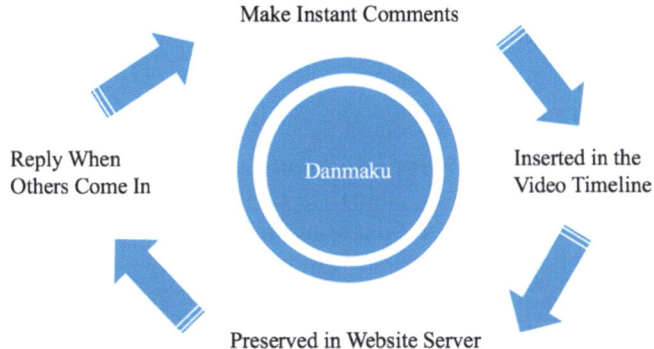

Fig. 2. Danmaku's primary operating mechanism [10].

Scholars defined the effect created by the VTC function when viewing videos as "pseudo-synchronicity" [8]. Those comments shown to the videos when watching seem to be a concurrent communicative act of viewers, while they are the written comments by the past viewers at a specific time during their views. Therefore, those comments are asynchronous [11]. This tool facilitates real-time interactions and becomes online and offline social networks for online viewers [25]. One primary motivator to use this function among viewers is to search for information through the subtitles' commentary because the video timeline-anchored feature makes it easy to find and dig for further information [6]. Viewers are also often amused by the timed comments they can identify with while watching it. Therefore, the interaction with others produces an entertaining effect that would reduce viewers' loneliness and isolation and increase a sense of belonging to the online learning community through discussing and sharing commonalities.

2.3   Canvas Studio

Canvas is a popular LMS platform that hosts online video-based lectures for asynchronous learning. Canvas Studio[1] was launched in 2016. It was designed to deliver asynchronous video content with video quizzing, editing, and archiving functionalities. Canvas Studio also provides features, such as adding captions or tagging keywords in the media files. One unique feature that Canvas Studio provides is the VTC function, which allows online students to make comments and ask questions at specific points while watching course videos. Despite the rapid increase of online learning in the US, including the COVID-led shift to online learning, six years had passed since 2016 when Canvas Studio was launched. This unique feature has not been actively applied to nor researched by the online learning community. Among the limited studies on danmaku, most investigated entertainment videos or focused on students' motivations, primarily

---

[1] https://community.canvaslms.com/t5/Studio/tkb-p/studio.

studies of East Asian countries. Therefore, by introducing how Canvas Studio was used in two graduate-level online courses offered by one research university in the United States, this study presents the incorporation of the VTC tool hosted in Canvas Studio into instructional activity and indicates its potential to stimulate active learning in the asynchronous online courses.

2.4 Theoretical Framework

This study applies Moore's [15, 16] interaction model as the theoretical framework. Moore's interaction model consists of three types of interaction for effective online learning to conceptualize students' interactions using the VTC function when watching video-based lectures for online learning and interactions: student-content interaction, student-instructor interactions, and student-student interactions (see Fig. 3). Student-content interaction indicates how students engage with the learning resources, and this interaction with course content facilitates students' understanding [15]. Student-instructor interaction refers to communication between students and the instructor [17] throughout the teaching and learning transaction. This interaction helps maintain students' interest and learning motivation. Student-student interaction refers to "the exchange of information and ideas among students about the course in the presence or absence of the instructor" [21]. This interaction consists of peer communication, including project collaboration and knowledge sharing that would enhance their learning experiences, engagement, and learning motivation [15, 17].

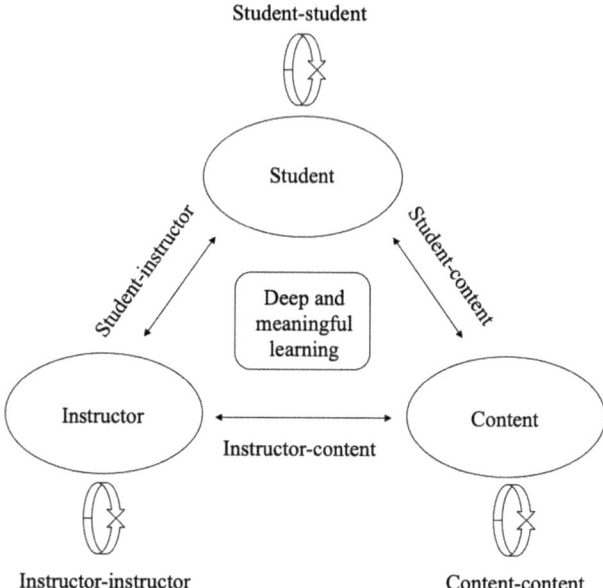

Fig. 3. Three types of interaction model [1].

3   Research Design & Method

3.1   Activity Design

Two sections of a 16-week graduate-level online course offered by the same professor were recruited. Students were required to develop a video presentation as an instructional lecture. Then they use the VTC tool hosted by Canvas Studio for the asynchronous online discussion on the video-based instructional lectures they develop. Students, either work alone or in a group of two members, select one chapter and decide a week to lead the class learning with the instructor's facilitation. There are four steps to complete this task: 1) the weekly presenters prepare a video-based instructional lecture based on the chapter; 2) the instructor reviews the video and offers suggestions for revisions; 3) after the student-developed instructional video is approved, the instructor publishes it through Canvas Studio, and the VTC function inserted in Canvas Studio is then available for class discussion; and 4) the presenters are required to regularly check and facilitate video discussion by responding to questions and commenting on posts by peers and the instructor. For this assignment, presenters must prepare two or more questions typed in their instructional videos for peers to discuss. To facilitate learning, the instructor also monitors and raises questions for the video discussion, especially when the video discussion needs further direction or redirection for content interactions. Lastly, the instructor replies to students' comments and questions in the video to ensure that the class learning is meaningful and towards the learning goals. Based on Moore's interaction model, this activity first motivates each presenter to interact with the learning content actively to comprehend the learning content they choose to present fully; second, students interact with their peers through the VTC discussion, viewing and answering questions, replying to peers' comments and insights for cognitive development and social and emotional support; and lastly, students interact with the instructor via feedback on video development and the discussion in the video.

When watching the videos, students can make comments and raise and answer questions at any time. The instructional videos will pause when students type comments. Students can press "Play" to continue watching the video afterward. The VTC function tool makes postings jumped out as a bubble with the words at the right corner. Depending on the comments' length, students can only read partially in the bubble. Once clicking the bubble, Canvas Studio directs students to the specific full remarks in the comment section below the video, where students can respond. Finally, Canvas Studio notifies the instructor and students whenever someone adds or replies to comments. Figure 4 shows an example of using VTC for a discussion hosted by Canvas Studio. The instructor made some notes with a digital pen in the slides including text and numbers while recording the video-based lecture.

A.   Comments made at 6:35 of the video jumped out at the right corner of the screen
B.   When clicking the bubble, Canvas Studio directs the viewers to the full remarks
C.   When clicking the time (i.e., 6:35), Canvas Studio directs the viewers to the video where the specific comment was made

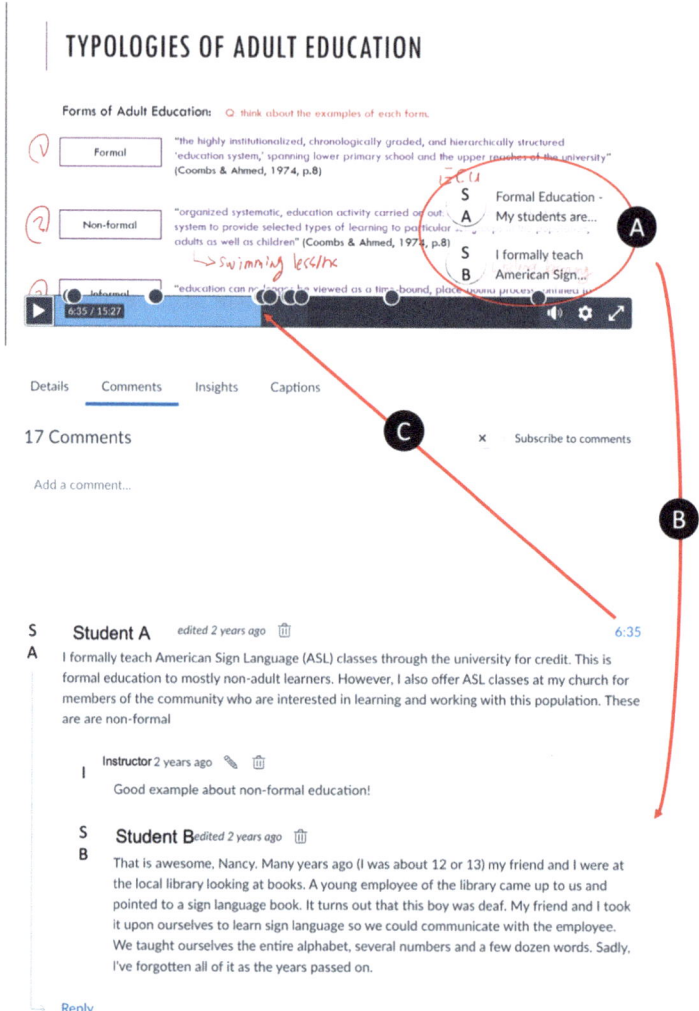

Fig. 4. Using the VTC in Canvas Studio (click it to watch a short instructional video using the VTC in Canvas Studio or use this link https://youtu.be/5w3zh-vbV1A).

3.2 Method

A multiple-case study method is applied to this study. A total of 24 instructional videos were presented through Canvas Studio. For most of the instructional videos, over 40 comments were posted by students in each video, from each course section (18 students for section 1 and 12 students for section 2). Data were collected from an anonymous survey to investigate students' perspectives on using the VTC function for interaction. Among 30 enrolled students, 23 of them (13 from section 1 and 10 from section 2) participated the study. Data were imported to *NVico 12* software. Braun and Clark's [4] thematic analysis approach was employed to investigate the "how" and "why" questions from students' feedback of an end-of-semester anonymous survey in this two-section course.

4   Findings & Discussions

Our preliminary findings generated four major themes of students using the VTC tool. 1) Developing a sense of real-time interaction. Students expressed that "online asynchronous instruction can be lonely at times. Having the instructor commenting in the video helps you feel more like a participant in a class and not the sole learner." They also believed that "Interacting with class peers via the commenting in Canvas video can at times keep the interaction in real-time." 2) Increasing interaction and engagement. Students reported, "Canvas video commenting makes interaction better with online courses," and "Canvas video commenting allows the students an opportunity to interact and engage while learning." Some of them concluded that the VTC function "was a very positive interaction tool" and "it works well for online discussion." 3) Preventing monotony and becoming more attentive and concentrated. Many students noted that using the VTC function prevents them from feeling monotonous or boring and "keeps presentations from being one-sided." They believed that this tool "provides variety instead of just reading and looking at words on a screen" and "the benefits are interactions with fellow students while engaging in the presentation." 4) Improving understanding of content. Students reported, "it [VTC function] may help students understand the topic better when discussing things while commenting on the other end." Additionally, some commented that this function "allows others to see what presentation information you are commenting on, enhancing their understanding of the information."

Briefly, incorporating the VTC function could engage students in active learning. Specifically, watching and discussing course materials may change the learning dynamic if watched together and with a real-time (i.e., pseudo-synchronicity) discussion. The VTC tool also allows students and the instructor to ask questions, make comments and respond to each other's thoughts, enabling real-time and togetherness for online learning. This tool also encourages students to learn and understand learning content because they can ask questions for clarification in the video towards specific lecture sections. Finally, this tool makes learners concentrate on the lectures and enjoy the entertaining effects.

5   Implications & Conclusions

Our study offers implications for online teaching and learning practice and future research. For instance, instructors may consider integrating the VTC tool for video-based learning materials. They could also guide students to develop questions in the video and encourage conversations. Instructors may also provide guidelines for students to use the VTC tool in Canvas Studio or other LMSs that provide similar video commenting tools as online learning skills. Lastly, they could regularly remind students to check comments and responses in the videos.

Due to the study's limitations, such as those students in this study were recruited from small online graduate-level courses in the field of education. We recommend that future studies may be conducted to apply the VTC function in large classes in other subjects (e.g., engineering). Moreover, students in this study were graduate students, thus future studies could be targeted at undergraduate-level online courses to investigate the influence of the VTC function on interactions among different student populations. Lastly, a mixed research method could be used to further explore the impact of the VTC function on students' interactions to better support asynchronous online learning.

In conclusion, this study introduces a course activity that incorporates the VTC tool hosted by Canvas Studio for asynchronous video-based lectures. Our preliminary findings indicate that the VTC tool could positively influence students' online interactions with their instructors, peers, and learning content by developing "real-time" "on-site" interactions. This tool could also link online and offline social networks and provides social presence and emotional support necessary for active learning in the asynchronous online context.

References

1. Anderson, T., Garrison, D.R.: Learning in a networked world: new roles and responsibilities. In: Gibson, C. (ed.) Distance Learners in Higher Education, pp. 97–112. Atwood Publishing (1998)
2. Banna, J., Lin, M.-F.G., Stewart, M., Fialkowski, M.K.: Interaction matters: strategies to promote engaged learning in an online introductory nutrition course. J. Online Learn. Teach. **11**(2), 249–261 (2015)
3. Bowers, J., Kumar, P.: Students' perceptions of teaching and social presence: a comparative analysis of face-to-face and online learning environments. Int. J. Web-Based Learn. Teach. Technol. (IJWLTT) **10**(1), 27–44 (2015)
4. Braun, V., Clarke, V.: Using thematic analysis in psychology. Qual. Res. Psychol. **3**, 77–101 (2006)
5. Brierton, S., Wilson, E., Kistler, M., Flowers, J., Jones, D.: A comparison of higher-order thinking skills demonstrated in synchronous and asynchronous online college discussion posts. NACTA J. **60**(1), 14–21 (2016)
6. Chen, Y., Gao, Q., Gao, G.: Timeline-anchored comments in video-based learning: the impact of visual layout and content depth. Int. J. Hum. Comput. Inter. 1–16 (2021)
7. Frimming, R.E., Bordelon, T.D.: Physical education students' perceptions of the effectiveness of their distance education courses. Phys. Educ. **73**(2), 340–351 (2016)
8. Johnson, D.: Polyphonic/pseudo-synchronic: animated writing in the comment feed of nicovideo. Japan. Stud. **33**(3), 297–313 (2013)
9. Lear, J.L., Ansorge, C., Steckelberg, A.: Interactivity/community process model for the online education environment. J. Online Learn. Teach. **6**(1), 71–77 (2010)
10. Lin, X., Huang, M., Cordie, L.: An exploratory study: using Danmaku in online video-based lectures. Educ. Media Int. **55**(3), 273–286 (2018)
11. Locher, M.A., Messerli, T.C.: Translating the other: communal T.V. watching of Korean T.V. drama. J. Pragmatics **170**, 20–36 (2020)
12. Loncar, M., Barrett, N.E., Liu, G.-Z.: Towards the refinement of forum and asynchronous online discussion in educational contexts worldwide: trends and investigative approaches within a dominant research paradigm. Comput. Educ. **73**, 93–110 (2014)
13. Ma, X., Cao, N.: Video-based evanescent, anonymous, asynchronous social interaction: motivation and adaption to medium. In: Proceedings of the 2017 ACM Conference on Computer Supported Cooperative Work and Social Computing, pp. 770–782. Association for Computing Machinery, New York (2017)

14. Martin, F., Bolliger, D.U.: Engagement matters: Student perceptions on the importance of engagement strategies in the online learning environment. Online Learn. **22**(1), 205–222 (2018)
15. Moore, M.G.: Editorial: three types of interaction. Am. J. Distance Educ. **3**(2), 1–7 (1989)
16. Moore, M.G.: Three types of interaction. In: Harry, K., John, M., Keegan, D. (eds.) Distance Education Theory, pp. 19–24. Routledge (1993)
17. Moore, M.G., Kearsley, G.: Distance education: a systems view of online learning. Cengage Learn. (2011)
18. Newmann, F.M.: Student Engagement and Achievement in American Secondary Schools. Teachers College Press (1992)
19. Pang, L., Jen, C.C.: Inclusive dyslexia-friendly collaborative online learning environment: Malaysia case study. Educ. Inf. Technol. **23**(3), 1023–1042 (2017). https://doi.org/10.1007/s10639-017-9652-8
20. Ruiz, J.G., Mintzer, M.J., Leipzig, R.M.: The impact of e-learning in medical education. Acad. Med. **81**(3), 207–212 (2006)
21. Sher, A.: Assessing the relationship of student-instructor and student-student interaction to student learning and satisfaction in web-based online learning environment. J. Interact. Online Learn. **8**(2), 102–120 (2009)
22. Wu, Z., Ito, E.: Correlation analysis between user's emotional comments and popularity measures. In Proceedings of 3rd International Conference on Advanced Applied Informatics (IIAIAAI), p. 31 (2014)
23. Yao, Y., Bort, J., Huang, Y.: Understanding Danmaku's potential in online video learning. In: CHI Conference Extended Abstracts on Human Factors in Computing Systems, pp. 3034–3040 (2017)
24. Zang, L. T., Cassany, D.: Making sense of Danmu: Coherence in massive anonymous chats on Bilibili.com. Discourse Stud. **22**(4), 483–502 (2020)
25. Zhang, Y., Qian, A., Pi, Z., Yang, J.: Danmaku related to video content facilitates learning. J. Educ. Technol. Syst. **47**(3), 359–372 (2019)

Nudging Lifelong Learning and Metacognition Tendencies in Engineering Management Undergraduates Utilizing the LinkedIn Learning Platform

Allan MacKenzie[✉]

McMaster University, Hamilton, ON, Canada
mackenza@mcmaster.ca

Abstract. Most engineering and technology-focused program curricula are firmly fixated on the required technical skills to meet the profession's needs. However, in today's rapidly changing, globalized world, engineers and technologists need more than technical competencies to meet the requirements of their professional work. This work illustrates how the LinkedIn Learning (LiL) platform was used as a "learning partner" to complement undergraduate engineering technology management courses to enrich reflective thinking and nudge lifelong learning tendencies. The rationale for integrating LiL into the course framework is examined, including study design and survey results. Summary research indicates that students appreciated the LiL coursework assignments. Most respondents perceived that the LiL courses increased their knowledge and skills in the subject matter presented. The study illustrated movement towards self-determined learning behaviour and improved reflective capabilities.

Keywords: Engineering education · Reflective thinking · Digital learning tools · Summarization · Asynchronous learning · Engineering management

1 Introduction

This paper illustrates how the LinkedIn Learning (LiL) platform was used as a "learning partner" to complement two engineering management courses' content to enrich metacognition reflection outcomes and nudge students toward lifelong learning tendencies. Research on the contemplative dimension of learning and the importance of lifelong learning for engineering students is surveyed. The rationale behind integrating LiL as a learning partner and how the curated third-party learning content was interwoven asynchronously into the course framework are discussed. Lastly, the research of the student perspective on using LiL as a complementary learning asset is described, including limitations.

Let's begin with an explanation of the LiL platform and its use in higher education. LiL is a self-service curated digital learning platform owned and operated by Microsoft Corp. With over 16,000 video tutorials in multiple languages within the topic categories of business, creativity, and technology [1]. Microsoft promotes LiL enterprise licenses to teams, companies, and organizations who wish to access the learning platform. As part of its Career KickStart strategy, the Ontario government funded access to the LiL platform on behalf of all higher education institutions in the province from 2017 to 2020 [2]. Building on this opportunity, McMaster University negotiated and secured a multi-year institutional enterprise license more recently, allowing free access to LiL for all active students, faculty, and staff. This institutional access enabled the author to integrate LiL video tutorial assets into two engineering management courses.

2 Learning Perspectives

2.1 Metacognitive Learning

Metacognition is thinking about one's thinking. More precisely, it refers to the processes used to plan, monitor, and assess one's understanding and performance. Metacognition includes a critical awareness of a) one's thinking and learning and b) oneself as a thinker and learner [3]. When learners engage in metacognitive reflection, it contributes to helping them understand what they have learned and transfer new knowledge into other contextual situations.

Most engineering and technology-focused program curricula are firmly fixated on the required technical skills to meet the profession's needs. However, in today's rapidly changing, globalized world, engineers and technologists need more than technical competencies to meet the requirements of their professional work. Reflection or the contemplative dimension of personal learning has not historically received much attention in engineering education, despite calls for more significant consideration of the use of reflection. For example, in a National Academies piece calling for curricular change in undergraduate engineering, Ambrose [4] suggests that learning happens with reflection and instructors should "provide structured opportunities to ensure that reflection occurs." Indeed, published evidence indicates that students reflecting on their learning enhance metacognition and learner agency [5–7].

Reflective practice is not new in engineering education, although it is by no means mainstream. Many have drawn on Schon's [8] work on the "reflective practitioner" and how "reflection-in-action" and "reflection-on-action" can influence professional education [9–13]. Other researchers have emphasized the value of reflective thinking and underlined that students do not automatically learn from experience [14]. Instead, reflection as an intentional and dialectical way of thinking about an experience to inform future actions should be encouraged in engineering education [15].

Both technical skills and metacognitive development are essential for achieving the goals of a "whole" engineer education, but the latter is often shortchanged or not deliberately explored. It's usually only implicitly hinted at in teaching, if mentioned at all. One reason is that facts, technical knowledge, and skills are easier to measure, but the reflection on one's learning is much harder to assess. However, if you ask employers what they are looking for in an engineering graduate, they often state elements related to the candidate's learning character. They are not looking for applicants solely focused on technical abilities but individuals who are more metacognitively aware and reflect on their process for achieving specific results within organizational parameters.

2.2 Lifelong Learning

The research literature on lifelong learning has grown exponentially in the past few decades [16]. The emergence of governmental and economic policies promoting lifelong learning and the proliferation of curated digital learning platforms has ushered in a new era in which education is ongoing. Changes in technologies, increasing demands of the new economy, fierce global competition, and the growth of increasingly well-informed and well-educated consumers create new markets for the education sector [17]. Lifelong learning is, thus, becoming a sector of mass participation, particularly as people in developing countries realize that their financial survival depends on it.

For this reason, the Government of Ontario's Career KickStart strategy has placed increasing emphasis on the issue of lifelong learning [2]. According to Knapper and Cropley [18], lifelong learners are active learners who plan and assess knowledge rather than waiting for others to prepare it for them. They can learn in formal and informal settings from their peers, teachers, and mentors. They can apply their knowledge to different contexts and are astute users of different learning strategies for unique situations. This self-directed learning mindset is imperative in this era of unprecedented rapid and fundamental change, in which some graduates will never directly use the disciplinary knowledge they acquired in university [19].

Today's engineering technology professionals work in a continual change and innovation ecosystem. To meet this challenge head-on and remain competitive in the workplace, technical professionals need to be content experts, highly skilled problem solvers, team players, and lifelong learners [20]. So one of the critical issues for higher education should be whether students are developing a belief and commitment to lifelong learning. Nudging students to adopt early habits and tools for lifelong learning is something we need to help learners embrace before they leave our institutions. One way to enhance this awareness is to interact with curated learning platforms, such as LiL, typically outside university parameters. Indeed, as educators, we should encourage metacognitive reflection and endeavour to nudge students towards lifelong learning tendencies to achieve the ambitions of a "whole" engineer education.

3 The Study and Results

3.1 The Coursework

The undergraduate engineering technology programs within McMaster's University W Booth School of Engineering Practice and Technology integrate technical comprehension with cross-boundary skills in business and management. The author integrated LiL into two engineering management courses, a fourth-year Entrepreneurial Thinking and Innovation course and a second-year Management Principles course. Both courses had students enrolled across the program streams of Automotive and Vehicle Engineering Technology, Biotechnology, and Automation Engineering Technology.

The Entrepreneurial Thinking and Innovation course introduces students to the interrelationship of entrepreneurial thinking and innovation at industrial and individual levels. It is project-based learning (PBL) course focused on developing an enterprise-level business case for a real organizational opportunity. The Management Principles course, on the other hand, is a fundamentals course examining the management principles of planning, organizing, leading, and controlling in technology organizations.

In both courses, students were assigned to watch three separate LiL video courses throughout the term and complete an individual written Video Tutorial Report assignment for each. The entire report was limited to 1500 words (3 pages single-spaced) with two parts. In Part A, the students were required to summarize what they considered the most important ideas/concepts from the video tutorial, written in a straightforward narrative that assumed the "reader" had not watched the LiL video tutorial course material. The reflection component of the assignment was Part B. Learners were required to explain and articulate multiple connections between what they comprehended from watching the video tutorial and connect it to prior learning in other courses or life/work experiences and future goals. Students were provided but were not limited to the following questions to help guide and facilitate their reflection process:

- The most important part of this video tutorial for me was? Why?
- What new skill or "piece" of knowledge did I come away with after the video tutorial? Why?
- I could see myself using this knowledge in my course or a future (or previous) workplace role. Why?
- After the video tutorial I will change_____. Why?
- Now I understand _____ after watching the video tutorial. How will this new understanding be helpful for you?

The video tutorial reports were worth 15% of the final course grade. However, the worth of each assignment is scaffolded, starting at 3% for the first report, 5% for the second and 7% for the third. Having a lower percentage assigned to initial reports enabled students to practice and learn from their shortcomings. Each student was provided extensive written feedback from the Teaching Assistants and allocated a standardized rubric score. Grading was completed promptly, so students could incorporate the feedback to enhance their performance before submitting the subsequent video tutorial assignment.

3.2 The Assignment Rationale

The author found LiL to be an efficient way to reinforce industry-specific approaches and bring complementary skill attainment into the course learning environment. It also allowed students to experience other voices through the LiL course instructors. There is a clear pedagogical advantage when students can access experts through platforms that ensure a rigorous talent selection process, such as LiL, versus the sometimes-dubious origins of many open-source videos.

Being able to summarize has become a skill that is more important than ever in today's information overflow. Learning how to summarize helps learners understand novel and challenging subject matter, which they can then apply to solving problems or developing a project. According to Kintsch, Eileen et al. [21], summarization has several advantages: promoting deeper thinking and analysis to select the relevant information; teaching essential study skills, such as identifying important content and separating main ideas from details. Summarizing is a way to develop a solid understanding of complex material and articulate one's understanding to be shared with others. Being able to convey the most important information concisely and accurately, without wasting time or causing misunderstandings, is a skill that many engineering managers prize in their employees and engineers appreciate in their supervisors [22].

However, the fundamental pedagogical rationale for incorporating coursework requiring students to interact, specifically with the LiL platform, was to encourage the development of contemplative learning and nudge lifelong learning tendencies. The work presents an innovative undergraduate training experience using LiL as a "learning partner" in two undergraduate engineering technology management courses. The analysis of students' perceptions and the impact on knowledge and skills allows for an understanding of the real effects of self-reflection and self-determined learning in the short term. The work is positioned as a forerunner concerning improving the university engineering education models to prepare students for today's dynamic workplaces.

3.3 Student Perceptions

At the end of the term, students were invited to complete a short online questionnaire to explore their perceptions about the Video Tutorial Report assignments and their experience using the LiL platform. The survey was entirely anonymous, and participation was optional. The questionnaire consisted of eight closed-ended question items. The first two dichotomous questions explored their use of the LiL platform for academic credit and usability. The following four questions surveyed their perception of the knowledge and skills gained in the subject matter from each of the three LiL courses. These questions used a five-point Likert scale, ranging from (1) strongly agree, (2) agree, (3) neutral, (4) disagree, and (5) strongly disagree, along with not applicable option. The final two questions probed the likelihood of the students using the LiL platform in the future and sharing their digital certificates of achievement on their social media platforms. These final two questions used a five-point Likert scale for likelihood, ranging from (1) extremely likely, (2) very likely, (3) moderately likely, (4) slightly likely, and (5) not at all likely.

The surveys were conducted across different years in two separate courses—the first measured students in a fourth-year Entrepreneurial Thinking and Innovation course in the fall 2019 semester. There were 85 students enrolled across two sections that the author taught. The overall participation rate was 44.7%, with 38 (n) students completing the survey. The other course was a second-year Management Principles course delivered in the winter semester of 2021. There were 250 students enrolled across four sections that the author taught. The overall participation rate was 34.4%, with 86 (n) students completing the survey. The questionnaire results indicated that students overwhelmingly felt the LiL platform was easy to use. The assigned LiL video tutorials were well received, and between the two surveys, there was an increased interest in using LiL for self-directed learning in the future.

Table 1 is the results from the first survey question, "Was this the first time you have used the LiL platform as part of a graded assignment in a university course?" For the most part, this was the first-time students had used LiL for a graded assignment in their courses. In 2019, a small cohort of students used LiL in another technical class, which would account for the 3% answering no to the question. The no response increased slightly in 2021, as more instructors within the school incorporated LiL into their courses during the switch to online instruction during the COVID-19 pandemic.

Table 1. LinkedIn Learning usage within academic courses.

	Entrepreneurial thinking and innovation (F2019) (%)	Management principles (W2021) (%)
Yes	97	91
No	3	9

Table 2 illustrates the results from the second question, "Was the LinkedIn Learning platform easy to use?" Again, the majority of students indicated the LiL was easy to navigate. One of the contributing factors to the higher percentage in the "somewhat" category in the 2019 survey was that some students had challenges with the export functionality of the notebook feature within LiL that allowed users to take notes within a course while watching. The problem was detected after the first video tutorial report assignment. Subsequently, the instructor encouraged the students not to use the notebook feature within LiL and to create their summary notes outside of the platform to reduce difficulties. The most likely reason for fewer problems with the LiL platform from the 2021 survey was that students were exploring LiL for personal use and becoming more adept at navigating online technology platforms due to the mandatory virtual classes during the pandemic.

Table 2. Ease of use of the LinkedIn Learning platform.

	Entrepreneurial thinking and innovation (F2019) (%)	Management principles (W2021) (%)
Yes	87	95
Somewhat	13	5
No	0	0

Figure 1 graphically depicts the four items that dealt with the educational value of the LiL Video Tutorials in the F2019 Entrepreneurial Thinking and Innovation course. The students generally perceived the grading criteria positively for the Video Tutorial Report assignments. The majority of students concurred the first two LiL Video Tutorials offered educational value. For both these LiL courses, 79% agreed or strongly agreed that the LiL course increased their knowledge and skills in the subject matter presented. In total, 68% perceived the third LiL Video Tutorial, "Presenting as a Team," was not as valuable in enhancing their knowledge and skills. Anecdotally, students believed they already had sufficient experience presenting, given they were seniors, so this LiL course content was allegedly less valuable from their perspective.

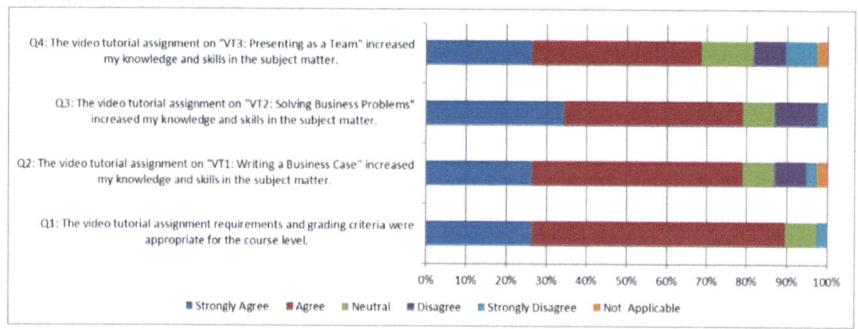

Fig. 1. Descriptive F2019 survey results for the LiL educational value in the entrepreneurial thinking and innovation course

Figure 2 graphically depicts the four items that dealt with the educational value of the LiL Video Tutorials in the W2021 Management Principles course. Again, the students generally perceived the grading criteria positively for the Video Tutorial Report assignments. The majority of the students surveyed indicated all three LiL Video Tutorials offered educational value. For the video tutorial on "Being an Effective Team Member," 88% agreed or strongly agreed that the LiL course increased their knowledge and skills in the subject matter presented. Just over 89% perceived the "Giving and Receiving Feedback" as valuable, and regarding the third LiL Video Tutorial, "Management Foundations," just shy of 92% agreed or strongly agreed that it enhanced their knowledge and skills in the subject matter.

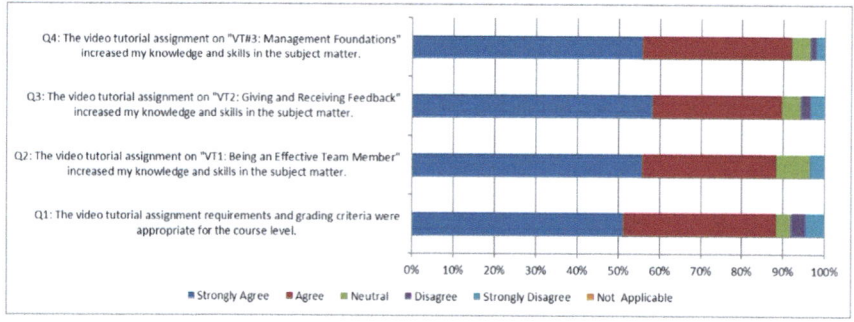

Fig. 2. Descriptive W2021 survey results for the LiL educational value in the management principles course

Table 3 displays the results from the seventh survey item focused on the tendency to use the LiL platform in the future. This question's objective was to validate whether the nudge toward lifelong learning predilection was beginning to take hold. In the 2019 survey, 31.6% indicated they were extremely likely or very likely to continue using LiL independently. Conversely, this measure rose to 47.7% in 2021 by survey respondents. The 2021 students exemplified a positive shift towards self-determined learning behaviour in the short term compared to the 2019 respondents.

Table 3. Future use of the linkedin learning platform.

	Entrepreneurial thinking & innovation (F2019) (%)	Management principles (W2021) (%)
Extremely likely	10.6	18.6
Very likely	21.1	29.1
Moderately likely	36.8	34.9
Slightly likely	18.4	12.7
Not at all likely	13.2	4.7

The Video Tutorial Report assignment's written deliverables clarified that a digital certificate of achievement would be awarded to students when they completed each LiL course. They had the option to publish this digital certificate on their LinkedIn professional profile to display to potential employers and other career influencers. Table 4 indicates the respondent's likelihood of exhibiting their digital certificates. In the 2019 survey, 32% indicated they were extremely or very likely to display their earned LiL course digital certificate. The 2021 survey revealed that 55% of respondents were extremely or very likely to exhibit their digital certificates. Given the 2021 students were sophomores, this positive difference could be rationalized because they wanted to enhance their professional profile to help secure a future paid workplace internship, a requirement for all students in the W Booth School.

Table 4. Likelihood of displaying linkedin learning digital certificate of achievement.

	Entrepreneurial thinking and innovation (F2019) (%)	Management principles (W2021) (%)
Extremely likely	13	25
Very likely	19	30
Moderately likely	24	29
Slightly likely	21	7
Not at all likely	18	7
Did not know there was a certificate of completion	5	2

3.4   Student Reflection Performance

As stated in Sect. 3.1, students were required to explain and articulate multiple connections between what they comprehended from watching the LiL video tutorial course and connect it to prior learning in other courses or life/work experiences and future goals. This reflective exercise aimed to help students enhance metacognition and learner agency.

The Video Tutorial Report (VT) assignment's reflective component was weighted at 35% of the overall assignment's worth. It was assessed using a rubric that assigned points ranging from 17.5 to 0 based on performance identified as (1) target, (2) acceptable, (3) developing, (4) unacceptable, and (5) incomplete. A student achieving target performance exhibited an in-depth analysis that demonstrated the value of the derived learning to self and the enhancement of the learner's appreciation of the concepts. This involved articulating multiple connections between prior learning in other courses or life/work experiences and future goals. On the other hand, if a learner's reflection only described the video tutorial learning experience and did not articulate any connection to prior learning or life/work experience, they would earn an unacceptable performance score. Scores landing between the target and unacceptable performance levels were more descriptive than reflective. Generally, they lacked a personal connection to the learning, or the linkages were vague or unclear.

Table 5 displays the change in the overall average student scores for the reflective component from each Video Tutorial Report (VT) assignment to measure the difference in reflective performance. The 2019 cohort of students enrolled in the Entrepreneurial Thinking and Innovation course exhibited positive change, with just over 5% in their reflective performance from the first to second VT Report. However, there was a slight decline in performance from the second to third VT Report. However, the reflective performance improved slightly from the first to the final (third) Video Tutorial Report.

In comparison, the student cohort enrolled in the 2021 Management Principles course exhibited a positive change in reflective performance across all the VT Reports. From the first to the second, just over 7% improvement and from the second to third, VT Reports indicated a 6.5% positive difference. The performance change from the first to the third VT Report exhibited slightly over a 14% improvement.

Overall, the students enhanced their reflective capabilities as they completed the VT Reports. This would suggest they benefited from the repetition of reflective thinking and receiving guided feedback to improve their learner agency.

Table 5. Changes reflective performance across video tutorial (VT) assignments.

	Entrepreneurial thinking & innovation (F2019) (%)	Management principles (W2021) (%)
VT1 to VT2	+5.41	+7.11
VT2 to VT3	−0.64	+6.52
VT1 to VT3	+4.73	+14.10

3.5 Study Limitations

The study, as described, had several limitations. First is its small scope, with only 123 students surveyed across two courses with the same instructor. The small sample limits the study's transferability, and the positive impact could be linked to the instructor's familiarity with the students and unconsciously advocating for the LiL platform. Another limitation of the research was reliance on only eight closed-ended participant-reported questionnaire statements that primarily focused on students' perception of the video tutorials and the LiL platform. Finally, students lacking English proficiency could have had difficulties understanding and summarizing the LiL video courses. This notable lack of mastery and confidence in language skills has been identified by other researchers regarding writing tasks, like summarizing, which require articulating ideas, not their own [23].

The shift between the 2019 and 2021 student cohorts may have been influenced by the changes that educational institutions underwent through the 2020 pandemic, which involved much more intensive use of technology and remote learning platforms, such as LiL. The 2021 cohort's positive shift towards self-determined learning could be a collateral outcome of comfort and familiarity with online asynchronous digital learning platforms. This cohort may also have a different predisposition toward diverse ways of learning or a more active interest in honing other skills, given they were sophomores.

Future studies would benefit from a more deliberate research design incorporating longitudinal pre and post surveys from two different survey instruments: one instrument measuring self-efficacy and the other lifelong learning tendencies. Self-efficacy is a construct that has been studied in many different contexts, including learning, individual entrepreneurship, technology solutions, innovativeness, change, and task completion. Studies have found that self-efficacy is significantly related to people's engagement in change and personal development [25].

The challenge is to provide students with educational experiences that enhance their aptitude for continued self-directed learning and help them gain enough confidence to initiate, maintain, and finish any endeavour they like. So, one of the most critical issues for engineering education should be whether students are developing a belief in and commitment to lifelong learning [25]. Investigating the factors contributing to reflection and lifelong learning is critical to encouraging dynamic engineering professionals.

4   Conclusion

Specifics were shared about the experience of using LiL as a "learning partner" in two undergraduate engineering technology management courses. The study revealed students valued having the LiL assignments integrated into the coursework. The work also demonstrated the majority of the respondents strengthened their skills in summarizing industry-related best practices and their self-reflection capabilities. Respondents showed a shift towards self-determined learning behaviour in the short term, indicating a nudge toward lifelong learning behaviours, an essential 21st-century attribute graduates need to succeed in their careers, given the lightning pace of change in today's technology organizations. Lastly, the study discussed several limitations that impacted its potential for replicability and recommended a more deliberate longitudinal research design for future exploration of metacognition and lifelong learning enrichment.

References

1. LinkedIn Learning Homepage. https://learning.linkedin.com/product-overview. Accessed 05 March 2022
2. Ministry of Finance, Ontario Boosting On-the-Job Learning Opportunities for Students. https://news.ontario.ca/mof/en/2017/04/ontario-boosting-on-the-job-learning-opportunities-for-students.html. Accessed 05 March 2022
3. Chick, N.: Metacognition. https://cft.vanderbilt.edu/guides-sub-pages/metacognition/. Accessed 06 March 2022
4. Ambrose, S.A.: Undergraduate engineering curriculum: the ultimate design challenge. Bridg.: Link. Eng. Soc. **43**(2), 16–23 (2013)
5. Burrows, V.A., McNeil, B., Hubele, N.F., Bellamy, L.: Statistical evidence for enhanced learning of content through reflective journal writing. J. Eng. Educ. **90**, 661–667 (2001)
6. Cowan, J.: On Becoming an Innovative University Teacher, Reflection in Action. Open University Press (2006)
7. Moon, J.A.: Learning Journals. A Handbook for Reflective Practice and Professional Development, 2nd edn. Routledge (2006)
8. Schon, D.A.: Educating the reflective practitioner. Jossey-Bass, San Francisco (1987)
9. Adams, R.S., Turns, J., Atman, C.J.: Educating effective engineering designers: the role of reflective practice. Des. Stud. **24**(3), 275–294 (2003)
10. Hicks, N., Bumbaco, A.E., & Douglas, E.P.: Critical thinking, reflective practice and adaptive expertise in engineering. In: 121st ASEE Annual Conference, Indianapolis, IN, USA (2014)
11. Lindsley, L.L., & Burrows, V.A.: Instructor credibility: an analysis of engineering students' reflective writing for evidence of attitude shifts. In: 37th ASEE/IEEE Frontiers in Education Conference, Milwaukee, WI, USA (2007)
12. Nilsson, P.: Developing a scholarship of teaching in engineering: supporting reflective practice through the use of a critical friend. Reflective Pract. **14**(2), 196–208 (2013)
13. Ryan, M.: Improving reflective writing in higher education: a social semiotic perspective. Teach. High. Educ. **16**(1), 99–111 (2011)
14. Verdonschot, S.G.M.: Methods to enhance reflective behaviour in innovation processes. J. Eur. Ind. Train. **30**(9), 670–686 (2006)
15. Turns, J., Sattler, B., Yasuhara, K., Borgford-Parnell, J., & Atman, C.J.: Integrating reflection into engineering education. In: 121st ASEE Annual Conference, Indianapolis, IN, USA (2014)
16. Schuller, T., Desjardins, R.: Understanding the Social Outcomes of Learning. OECD, Paris (2007)
17. Green, A., Wolf, A., Leney, T.: Convergence and Divergence in European Education and Training Systems. Institute of Education, University of London, London (1999)
18. Knapper, C., Cropley, A.J.: Lifelong Learning in Higher Education, 3rd edn. Kogan Page, London (2000)
19. Kirby, J.R., Knapper, C., Lamon, P., Egnatoff, W.J.: Development of a scale to measure lifelong learning. Int. J. Lifelong Educ. **29**(3), 291–302 (2010)
20. Dunlap, J.C.: Problem-based learning and self-efficacy: how a capstone course prepares students for a profession. Educ. Tech. Res. Dev. **53**(1), 65–85 (2005)
21. Kintsch, E., et al.: Developing summarization skills through the use of LSA-based feedback. Interact. Learn. Environ. **8**(2), 87–109 (2000)
22. Fergusson, K.: How to summarize and paraphrase. https://owlcation.com/academia/Learn-to-summarize-and-paraphrase. Accessed 06 March 2022
23. Lin, O.P., Maarof, N.: Collaborative writing in summary writing: student perceptions. Soc. Behav. Sci. **90**, 599–606 (2013)

24. Smylie, M.: The enhancement function of staff development: organizational and psychological antecedents to individual teacher change. Am. Educ. Res. J. **25**(1), 1–30 (1988)
25. Bath, D.M., Smith, C.D.: The relationship between epistemological beliefs and the propensity for lifelong learning. Stud. Contin. Educ. **31**(2), 173–189 (2009)

A Pedagogical Conversational Agent for Tutoring in the Development of Educational Research Projects

Elvis Gerardo Ortega Ochoa$^{(\boxtimes)}$

Doctoral School, Universitat Oberta de Catalunya, 08035 Barcelona, Spain
egortega@uoc.edu

Abstract. Some countries struggle to respond in a personalized way to a growing student base, owing to the student-teacher ratio being 13:1 on average. Practice and theoretical review indicate that there is an analysis lack in e-learning of the Pedagogical Conversational Agent (PCA) for tutoring in the development of educational research projects. Therefore, the objective was to determine the algorithm of the PCA dialogic system for tutoring the Integrating Knowledge Project at the National University of Education. The population was students of the Basic Education Career, distance learning, academic period semester I—2021 ($N = 1,124$), and the sample, with a 95% confidence level and an estimated sampling error of 5%, was 287 participants. They were selected by cluster sampling technique. Paradigm was pragmatism, and method was a light mixed method of convergent design (questionnaire variant) of descriptive scope, cross-sectional and empirical. The research used descriptive statistics and content analysis. The main results were the algorithm's meta-inferences, which should have a teacher/tutor's role, offer emotional support, identify errors in the dialogue topics of the domain, and so forth. This research is significant because it determines the generalities and particularity of the PCA dialogue system from and for the students. Moreover, it initiates the linkage of the PCA dialogue system and the tutoring of educational research projects.

Keywords: E-learning · Dialogue system · Pedagogical Conversational Agent · Project design · Mixed method research

1 Introduction

The average student-teacher ratio in vocational training programs is 13:1. In this context, some countries struggle to provide a personalized response to a growing student base [10]. As a result, there is a need to adapt the educational process to the students' characteristics. However, in the last decade, emerging technologies and practices have been addressing this situation through adaptive learning. It is the case of Pedagogical Conversational Agents (PCAs), which guide interaction with the student. Therefore, the research topic is the algorithm of the PCA dialogic system.

© The Author(s), under exclusive license to Springer Nature Switzerland AG 2023
D. Guralnick et al. (Eds.): TLIC 2022, LNNS 581, pp. 544–556, 2023.
https://doi.org/10.1007/978-3-031-21569-8_51

The National University of Education (UNAE, by its acronym in Spanish) of Ecuador initiated at the end of 2017, among other careers, the Basic Education Career, distance learning, for teacher professionalization. Based on teaching practice, in each professional training cycle, there is an educational research project developed called Knowledge Integration Project (PIENSA, by its acronym in Spanish). In this project, tutoring is provided to help, support, and accompany the training process.

1.1 Research Problem

The situation has several approaches. On the one hand, although there has been advancement in adaptive learning—especially related to the algorithm of the PCA dialogic system linked to different domains, for instance, mathematics [12], programming [8], physics, and computer science [5], and so forth—there is a research lack between the connection of these systems with tutoring of educational research projects, as well as a comprehensive approach to the study object through the mixed method [16]. On the other hand, considering that tutorial accompaniment is a process directed and personalized to each PIENSA in theoretical, methodological, and practical aspects of the research process [15], the dialogue mediated by Information and Communication Technologies (ICT) with the learner must activate the teaching and learning process [6, 11]. As a result, there is a problem based on research and practice.

From deductive reasoning, the algorithm of the PCA dialogic system for tutoring the PIENSA is the variable, which corresponds to a categorical type because it allows characterization of the algorithm according to the participants. Thus, the research question is: what is the algorithm of the PCA dialogic system for tutoring the PIENSA based on the students' preferences? The question has a descriptive orientation and requires the consideration of the particularities, characteristics, and needs of university students.

1.2 Review of Relevant Scholarship

Project tutoring is a guiding activity that must be highly personalized. However, the student-teacher ratio does not facilitate this activity. Nowadays, educational technology, specifically the PCAs, allows attention to this situation through adaptive learning. Conversations between the PCA and students are composed of intentions and entities based on the learning content. The first word refers to a dialogue topic, and the second is terms relevant to a specific and personalized context for the intention [2]. The PCA algorithm refers to the set of instructions established, organized, and circumscribed, in this case, for tutoring the educational research project. Consequently, some studies have determined the characteristics of the algorithm of the PCA dialogic system.

Tamayo's doctoral thesis [12], *Methodology Proposal for the Design and Integration in the Classroom of a Pedagogical Conversational Agent from Secondary Education to Early Childhood Education*, proposes a type of Design, Integration, and Evaluation Methodology (MEDIE, by its acronym in Spanish) of the PCA that considers students' preferences so that any teacher can use it in his or her classes. The methodology has phases that emphasize the algorithm's characteristics (role, general and specific functions) through quantitative responses, designed from the perspectives of the educational actors (User-Centered Design). In the practical sessions, one of the main results was to achieve the PCA's use in the classroom; in addition, the teaching staff has been able to adapt to the courses' characteristics with relevant content for the students.

Graesser's research [5], *Conversations with AutoTutor Help Students Learn*, adapted the PCA to the student and their emotions, trying to have a natural dialogue. The PCA mimicked the conversational movements of human tutor teachers, as well as pedagogical methods for physics and computer science students. Mainly, the researcher constructed an expectation and misconception-tailored dialogue (EMT dialogue), which is based on the conjunction of semantic patterns, and attempted to finalize the pattern using recommendations and announcements. Furthermore, EMT created reasonably fluent dialogues in AutoTutor and prompted the learner to learn.

Ocaña et al.'s research [8], *Pedagogic Conversational Agent dialogue management to learn how to program*, established PCA conversation management guidelines for teaching programming based on the conversations' particularities. The MEDIE [12] was used in the drafting, progress, and assessment. The study analyzed 66 children's conversations (dialogic simulation) about what they would like to learn in programming, which allowed them to know the dialogue system. Content analysis categories were course, text type, dialogue topic, number of questions and exclamations, dialogue length, knowledge level, and friendship intention. Thus, the analysis facilitated the suggestions generation for the PCA's design linked to the programming's teaching.

All in all, the review demonstrates that there is no research on the algorithm of the PCA dialogic system whose domain is specifically tutoring in the development of educational research projects. Despite this, the studies presented are the research base because they consider transversal elements (e.g., theoretical, methodological), these are processes for the PCA co-design and its link with the domains, dimensions of the dialogic system, and instruments to determine the students' preferences (quantitative responses and dialogic simulation).

1.3 Study Objective

The purpose is to determine the algorithm of the PCA dialogic system for tutoring the PIENSA in the Basic Education Career, distance learning, at the UNAE, academic period semester I (SI)—2021. Considering that purpose exposes two realities, objective and subjective (established by the interpretation of each student), the study has a mixed method, which uses the survey to collect data from a pragmatic view because the dimensions of the algorithm of the PCA dialogic system require it.

2 Method

The research applied pragmatism. Creswell and Plano-Clark [3] indicate that this paradigm highlights the application's value, what works, what solves, and provides responses to questions from a mixed method. That is, pragmatism promotes an integrative approach to resolve traditional philosophical dualisms, as well as to make methodological decisions. The reason for selecting this paradigm was that the study required in-depth research, which required a quantitative and qualitative approach.

Delimitation of a quantitative or qualitative research approach is not relevant in the paradigm since it will be pertinent to integrate all paths to respond to the study purpose. The reason for using mixed methods research (MMR) was the need to obtain complete and corroborated results [3, 13, 14]. The study was descriptive scope, cross-sectional, and empirical. The author applied a convergent design and used the questionnaire variant (survey) called mixed method light [3] to respond to the research purpose. Design implied collecting and analyzing quantitative (QUAN, mostly) and qualitative (qual) data simultaneously. In both approaches, the algorithm of the PCA dialogic system was determined (see Fig. 1). The population was the students of the Professional Training Unit of the Basic Education Career, distance learning, at the UNAE, academic period SI—2021 ($N = 1,124$). The students fulfill the role of apprentices, and they are teachers in each of their schools; furthermore, their knowledge and skills are very enriching since they have had different formative contexts.

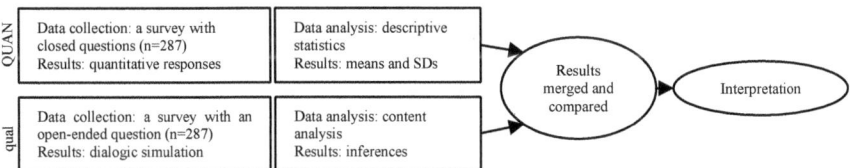

Fig. 1. Procedural diagram of the convergent design, questionnaire variant.

The author's background in applying the mixed method is varied. At the date of study application, the author had seven years in the field of educational research using the socio-critical and pragmatic paradigm to respond to the research problems. Therefore, the author's experiences with qualitative and quantitative research helped him to conduct the mixed methods research and report it transparently and completely.

2.1 Participant Recruitment

The population sample was identical for the two approaches; it was selected with the one-stage cluster sampling technique since units of analysis were grouped in courses (clusters = 36). In this regard, the risk assumed was that the students in the different clusters would also have different characteristics. The sample (n_h), with a 95% confidence level and an estimated sampling error of 5%, was 287 students (Men = 122, Women = 194). It is worth mentioning that the sample was increased by 10% to allow freedom in particular cases. They were mostly in the age range of 30–39 years.

Simple Random Sampling (SRS) was applied to select clusters and participants. Since each cluster had at least 20 students, the selected courses were 16 and their students' number was 511 (N_h). Equation 1 displays the Kish stratification formula [7]. In this case, the sampling fraction (f_h) of the stratum is 0.6183. Then, students' numbers from each selected cluster had to be multiplied by the proportionality constant (kS_h) to determine their cluster sample. Moreover, a pilot test was applied to 100 [4] of the 316 participants, resulting in a f_h value of 0.3164. After determining the sample size for each cluster, the participants also were selected using SRS.

$$f_h = \frac{n_h}{N_h} = kS_h \tag{1}$$

2.2 Data Collection

The instrument was the questionnaire, which allowed data collection [4]. The study adopted and unified the questionnaires proposed by Tamayo [12] and Ocaña et al. [8]. Accordingly, tool consolidation had three stages: adjustment, pilot test, and final version. The final version has three sections: role, general and specific functions (see Appendix). Each of the first two sections has one closed question, and the third has two closed questions and one open-ended question (dialogic simulation). The scale of the closed responses is nominal or interval (5-point Likert scale). On the other hand, the open-ended question requires further categorization for content analysis. In addition, each of the questions and responses is defined by a data coding scheme, identifiers of the different response alternatives, and treatment of missing values.

The research procedure had the first MEDIE phase [12]; the author conducted the student survey. The survey was voluntary, and the questionnaire was administered electronically in Microsoft Forms, open from November 16 to November 23, 2021. Participants received information in the emails of the UNAE institutional server to ensure correct reception in the inbox. Mailing lists corresponded to the institution's databases; therefore, quality was optimal. The students received specific and homogeneous instructions in the pilot test and final administration. The process began with a first email containing the survey information. Subsequently, around halfway through the stipulated deadline, a follow-up email was sent to each participant who had not completed it, accompanied by a message and a telephone call. Finally, 12 h before completion, a reminder was sent by mail. Regarding the monitoring strategy, the participants were accompanied synchronously and asynchronously during the process through the mail, phone call, and chat group to determine the progress, give any help or clarification and achieve a high response rate. In this context, communication is not considered intrusive because they are used to constant dialogue with their teacher/tutor.

2.3 Data Analysis

Meta-inferences were obtained through the horizontal mixed analysis of the quantitative and qualitative data collected [9]. It involved a mono-analysis of quantitative data and a mono-analysis of qualitative data; nevertheless, these results did not interact with each other in any way until the data interpretation phase (see Fig. 1). Data analysis and interpretation were developed using descriptive statistics and content analysis. The first technique allowed the data systematization in indicators and charts of the nominal scale and interval questions, and the second favored the text inference of the open-ended question. The quantitative and qualitative analyses were performed using IBM SPSS Statistics (version 27) and NVivo (version 12), respectively.

2.4 Validity, Reliability, and Methodological Integrity

As for the instrument's evaluation, the author explained the validity and determined its reliability. Firstly, instrument validity [1] was based on positive consequences obtained in its application [8, 12]. Secondly, instrument reliability was determined in the pilot test applied from November 12 to November 15, 2021, to identify unclear or confusing instructions and questions [4]. Results indicate that the participants agreed with the instructions' clarity in the process ($M = 3.44$, $SD = 0.539$), the questions' clarity in the questionnaire ($M = 3.44$, $SD = 0.539$), and agreed that there were no problems in understanding the questions' type or in responding them as they have been approached ($M = 3.35$, $SD = 0.683$). In addition, the students' comments are positive (e.g., all the questions are clear and understandable). Moreover, a financial reward was given to encourage the students' participation in the final administration because the response rate of the pilot test was improvable (53.06%). The financial reward was drawn among the surveys received using Excel (version 2203). The winner was the one with the highest random number, this information was sent by email. Therefore, the data indicate a high relevance of the implementation procedure and the instrument content.

The ethical decision-making approach was to establish rules and assume the risk that a detailed explanation might distort the participants' responses [4]. Then, academic management and students were informed of the objective, phases, and results of the research. Likewise, the ethical guidelines were followed, obtaining informed consent during data collection. The identifying information in the data was separated from the participants' responses to ensure that privacy was maintained. As a result, strict ethical criteria were kept certifying free and responsible collaboration in the design, implementation, data collection, and data processing.

3 Results

The final administration of the questionnaire achieved a slightly higher response rate than the pilot test. The response rate was calculated using Eq. 2. After the cleaning of duplicates, the number returned was 149 surveys. However, although the survey was sent to the total sample, six of them reported that they had withdrawn from the training process and one email address bounced due to erroneous registration in the database. That is, this number was 280. Thus, the response rate was 53.21%.

$$\frac{Number\ returned}{N\ in\ sample - (inelegible + unreachable)} \times 100\% \tag{2}$$

3.1 Quantitative Analysis

The measures of central tendency and dispersion with frequencies and percentages of the students' preferences allowed the author to obtain generalizations about the general functions of the PCA dialogic system. Although there are no major differences in the responses, it can be observed that in the data distribution there are trends in the top two levels: agree and strongly agree (see Fig. 2). Participants agreed that the PCA should encourage them to keep working ($M = 3.34$, $SD = 0.74$), tell them what they do wrong ($M = 3.28$, $SD = 0.69$), remember what they say ($M = 3.15$, $SD = 0.77$), and give them good advice ($M = 3.15$, $SD = 0.89$). That is, the PCA should offer motivation, identify errors, remember the students' dialogues, and give recommendations.

Results also allowed the author to obtain generalizations about the role and specific functions of the PCA. Students indicated that the PCA should have the role of teacher/tutor (65.8%), given that the other options were below (classmate = 22.1%, student = 11.4%, and missing values = 0.7%). In addition, when they are not attentive to the development orientations of the PIENSA, the PCA should tell them that it is going to contact the teaching staff (45%), advise them to study more (45%), and indicate that if they do not study, they will fail (40.3%), instead of doing nothing (0.7%). Likewise, when they do not understand the activity indications, the PCA should mainly explain in detail all the steps to be developed (61.7%) because it was superior to the option of explaining them with one or more similar tasks (48.3%), giving them hints, tips, and recommendations (44.3%), repeating orientations as many times as necessary (38.9%) and receiving questions about the activity (26.2%). Thus, the results show a framework of specific functions for the PCA dialogic system.

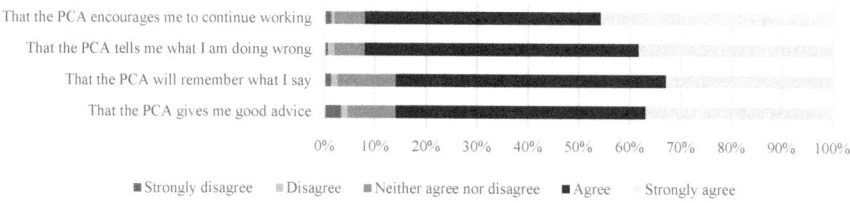

Fig. 2. Students' preferences about the general functions of the PCA dialogic system.

3.2 Qualitative Analysis

Content analysis of dialogic simulation was applied by transforming qualitative data into quantitative data. Dialogic simulation (chat log) has 313 interventions, the students 159 (50.8%) and the PCA 154 (49.2%) times. The categories established were six, which encompass the student and PCA dialogues. These are shown in Table 1 together with the quantitative results. The purpose was to evaluate them on an organized and systematized basis and make in-depth evaluations of their particularities.

Table 1. Categories and subcategories of the chat log between students and the PCA.

Categories and subcategories	f (%)	
	Student	PCA
Dialogue topic (intention)		
1. Activities' methodological orientations	41 (25.8)	28 (18.2)
2. Appreciation/Motivation	33 (20.8)	22 (14.3)
3. General course guidelines	28 (17.6)	30 (19.5)
4. Activity resources	25 (15.7)	58 (37.7)
5. Educational research	19 (11.9)	7 (4.5)
6. Learning management system technical support	9 (5.7)	8 (5.2)
7. Connectivity problems	1 (0.6)	–
8. Technological applications	1 (0.6)	–
9. Standards of the American Psychological Association	1 (0.6)	–
10. Activity evaluations	1 (0.6)	1 (.6)
Knowledge level		
Low (less than 5 words)	151 (95)	145 (94.2)
Medium (between 5 and 15 words)	8 (5)	9 (5.8)
Dialogue length		
Short (less than 10 words)	104 (65.4)	90 (58.4)
Medium (between 10 and 20 words)	54 (34)	60 (39)
Long (more than 20 words)	1 (0.6)	4 (2.6)
Questions, spelling, and grammar		
No questions	119 (74.8)	–
One question	40 (25.2)	–
Low (less than 2 spelling errors)	138 (86.8)	–
Medium (between 2 and 5 spelling errors)	21 (13.2)	–
Low (less than 2 grammar errors)	150 (94.3)	–
Medium (between 2 and 5 grammar errors)	9 (5.7)	–

Results presented some particularities. Some students' dialogue topics did not correspond to the same dialogue topic in the PCA's responses; it was since the entities modified the intention (dialogue topic). An example can be seen in Fig. 3. Additionally, some students demonstrated great knowledge. Hence, the PCA should be able to favor higher cognitive processes. Likewise, it was characteristic that dialogue length was mostly short both from the students and PCA, which may be for many reasons. In addition, repeatedly, it was possible to recognize implicit questions. However, these were not counted because they would create ambiguities. Finally, frequent errors were found in the writing; this situation increased when the dialogue length was greater. These particularities made it possible to analyze in-depth the PCA dialogic system.

PCA: remember that we are about to deliver the first draft of PIENSA. Intention (I) = 3
S: I have difficulties in the development of activity 4.6 that allows me to develop my theoretical framework. I = 5
PCA: you can find relevant information on your topic at the following links. I = 4
S: excellent, the links are adequate and will help me to improve and complete my project. I = 2

Fig. 3. Excerpt of the chat log between the students and PCA.

3.3 Meta-inferences: Interpretation of Information

Results integration allowed to obtain generalizations and particularities of the algorithm of the PCA dialogic system, that is, response to the research question (see Table 2). A certain point of convergence is the PCA's role as teacher/tutor. It should offer motivation and error identification, including writing errors. Likewise, mono-analyses complement each other in several respects. Requested dialogic topics should be explained in detail, and the PCA should recognize explicit and implicit questions. At the same time, it should remember dialogues and give recommendations. Furthermore, it will have to identify knowledge level, if it is the case, favor higher cognitive processes. It should correspond in dialogue length with that of the student, and if the student is not attentive, apply respective functions. It is worth mentioning that no divergence data were found. Finally, Fig. 4 illustrates the initial algorithm of the PCA dialogic system, which is based on the exposed meta-inferences.

Table 2. Representation of integration results through joint display.

Quantitative	Qualitative	MMR meta-inferences
Offer motivation ($M = 3.34$), identify errors ($M = 3.28$), remember dialogues ($M = 3.15$), and give recommendations ($M = 3.15$) Teacher/tutor's role (65.8%) When they are not attentive to orientations, tell that it is going to contact the teaching staff (45%); advise to study more (45%); and indicate that if they do not study, they will fail (40.3%) When they do not understand orientations, explain in detail all the steps to be developed (61.7%)	Recognize the dialogue, intention, and entity. Mainly, respond to requests for activities' methodological orientations and general course guidelines, and offer activity resources and motivation Correspond to the knowledge level and dialogue length Train to promote higher cognitive processes. Identify implicit questions. Distinguish errors in writing	Convergence: the PCA is assumed with the role of teacher/tutor, characterized by offering motivation and identifying errors, including writing Complementarity: the topic should be explained in detail, remember dialogues, and give recommendations. Correspond with the knowledge level, if applicable, to promote higher cognitive processes, correspond in dialogue length with that of the student, and in case of inattention, apply the respective functions

4   Conclusions

The most noteworthy aspect of the algorithm determination was to analyze the PCA dialog flow for tutoring in the development of educational research projects since it allowed for the interpretation of the population's perspectives, both generalities, and particularities. The results indicated general and specific functions, and the role, that is, pedagogical and social criteria of PCA [11]. The research and co-design resulted in that the algorithm of the PCA dialogic system should have the teacher/tutor's role, offer emotional support, identify errors in the dialogue topics, and so forth (see Fig. 4). What most supported the research was to unify, adapt, and validate the questionnaire [8, 12] because it addressed dimensions that make up the variable.

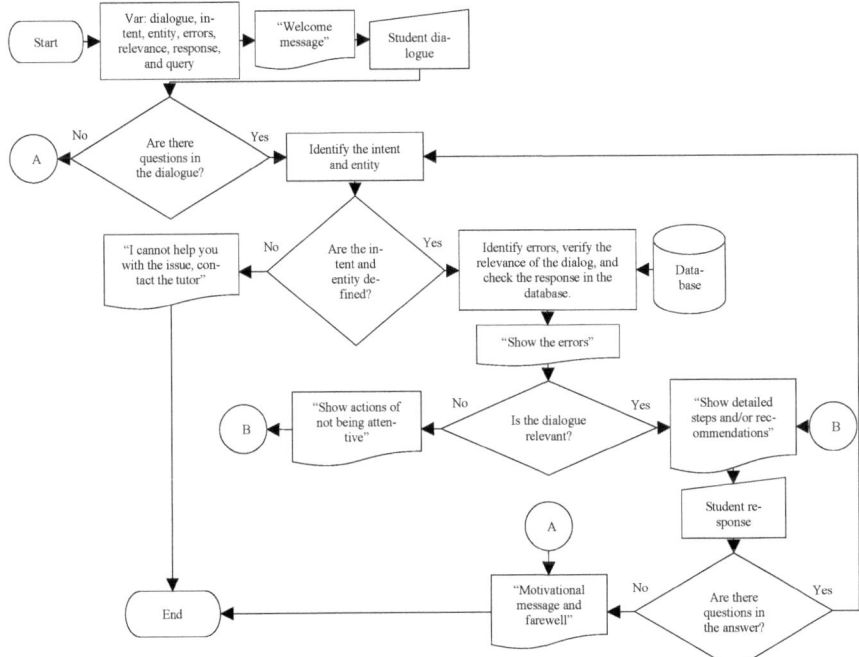

Fig. 4. Flowchart of the algorithm of the PCA dialogic system for tutoring the PIENSA.

There are several aspects of relevance and usefulness of the research results. The research linked the PCA dialogic system with the PIENSA tutoring, and the results were generated from an integrative and pragmatic approach with the mixed method. In addition, the study focused on the needs of the students and had a proposal of the algorithm for tutoring personalization (help, support, and accompany). Therefore, the study will allow teachers and engineers to have a solid context in the decision-making for the next phases of the design of the PCA dialogic system (pedagogical sequence). Overall, the inquiry helped in the design of ICT-mediated dialogic learning applied in tutoring and contributed to adaptive learning in higher education.

Acknowledging difficulties encountered and tasks partially accomplished is scientifically ethical to provide solutions for future research. In the population and with the research procedure, the electronic survey is not the best alternative to have a higher response rate. Consequently, other administration ways need to be evaluated. In addition, data from the dialogic simulation are few to determine the students' behavior. Hence, a broad database with diverse interactions is necessary to validate categories. Furthermore, one of the gaps in the algorithm validation with the students is because they are part of the techno-pedagogical co-design. Future studies should consider this task. The participation mechanism that guarantees the voice and vote of all, the big data, as well as the algorithm validation are the weak points and shortcomings that allow defining the possible lines of research continuity.

Data Availability Statement.   Data are available upon request to the author; these data are not publicly available for confidentiality reasons.

Funding.   The author has no support or funding to report.

Appendix

Questionnaire

Role. During the tutoring of the development of the PIENSA, what role would you like the program to play?
☐ Teacher/tutor
☐ Classmate
☐ Student
Others Click or tap here to enter text.

General Functions. If you could have a virtual program, in addition to your tutors and classmates, to help you with the development of the PIENSA, what would you like its role to be?

Affirmations	Strongly disagree	Disagree	Neither agree nor disagree	Agree	Strongly agree
That the PCA gives me good advice	☐	☐	☐	☐	☐
That the PCA encourages me to continue working	☐	☐	☐	☐	☐
That the PCA tells me what I am doing wrong	☐	☐	☐	☐	☐
That the PCA will remember what I say	☐	☐	☐	☐	☐

Specific Functions

1. If you are not attentive to the development orientations of the PIENSA, the program should ... *Multiple answers are possible*
☐ To do nothing
☐ Tell me that I must study more
☐ Show me that if I do not study, I am going to fail

☐ Tell me that you are going to communicate with my tutors
☐ Others Click or tap here to enter text.

2. If you do not understand the orientations of a praxis or experimentation activity of the PIENSA, what would you like the program to do? *Multiple answers are possible*
☐ Repeat orientations as many times as necessary
☐ Explain in detail all the steps to be developed
☐ Explain with one or more similar activities
☐ Giving hints, tips, and recommendations
☐ Receive questions about the activity
Others Click or tap here to enter text.

3. Imagine the interaction with the virtual program during the realization of the activity of the PIENSA and write at least two interventions of each character in the following lines. For example, Student: I cannot find the activity resources 2.3; Program: You can find the activity resources 2.3 in this link ...; Student: Thank you very much, I have read the articles, but I still do not know what innovation is; Program: Good job, this video can help you. Check the spelling and grammar.
Click or tap here to enter text.

References

1. American Educational Research Association, American Psychological Association, National Council on Measurement in Education: The Standards for Educational and Psychological Testing. American Educational Research Association, Washington (2014)
2. Clark, A., Fox, C., Lappin, S. (eds.): The Handbook of Computational Linguistics and Natural Language Processing. Wiley, Chichester (2010). https://doi.org/10.1002/9781444324044
3. Creswell, J., Plano-Clark, V.: Designing and Conducting Mixed Methods Research, 3rd edn. SAGE Publications Inc., Los Ángeles (2018)
4. De Vaus, D.: Surveys in Social Research, 6th edn. Routledge, London (2013). https://doi.org/10.4324/9780203519196
5. Graesser, A.C.: Conversations with AutoTutor help students learn. Int. J. Artif. Intell. Educ. **26**(1), 124–132 (2016). https://doi.org/10.1007/s40593-015-0086-4
6. Graesser, A., Li, H., Forsyth, C.: Learning by communicating in natural language with conversational agents. Curr. Dir. Psychol. Sci. **23**(5), 374–380 (2014). https://doi.org/10.1177/0963721414540680
7. Kish, L.: Survey Sampling. Wiley, New York (1965)
8. Ocaña, J., Morales-Urrutia, E., Pérez-Marín, D., Tamayo, S.: Pedagogic Conversational Agent dialogue management to learn how to program (in Spanish). Iberian J. Inf. Syst. Technol. (E19) 239–251 (2019). http://www.risti.xyz/issues/ristie19.pdf
9. Onwuegbuzie, A., Johnson, B. (eds.): The Routledge Reviewer's Guide to Mixed Methods Analysis. Routledge, New York (2021). https://doi.org/10.4324/9780203729434
10. Organisation for Economic Co-operation and Development: Education at a Glance 2020: OECD Indicators. OECD Publishing, Paris (2020). https://doi.org/10.1787/69096873-en
11. Pérez-Marín, D.: Review of the practical applications of pedagogic conversational agents to be used in school and university classrooms. Digital **2021**(1), 18–33 (2021). https://doi.org/10.3390/digital1010002

12. Tamayo, S.: Methodology Proposal for the Design and Integration in the Classroom of a Pedagogical Conversational Agent from Secondary Education to Early Childhood Education (in Spanish) [Doctoral thesis, Rey Juan Carlos University]. Rey Juan Carlos University Institutional Repository, Móstoles (2017). http://hdl.handle.net/10115/14691
13. Tashakkori, A., Johnson, B., Teddlie, C.: Foundations of Mixed Methods Research: Integrating Quantitative and Qualitative Approaches in the Social and Behavioral Sciences, 2nd edn. SAGE Publications Inc., Thousand Oaks (2021)
14. Tashakkori, A., Teddlie, C. (eds.): SAGE Handbook of Mixed Methods in Social & Behavioral Research, 2nd edn. SAGE Publications, Inc., Thousand Oaks (2010). https://doi.org/10.4135/9781506335193
15. Technical University of Manabí: Methodological Instructions on Knowledge Integration Projects (in Spanish). UTM, Portoviejo (2017)
16. Winkler, R., Söllner, M.: Unleashing the potential of chatbots in education: a state-of-the-art analysis. In: Taneja, S. (ed.) Academy of Management Annual Meeting, vol. 2018, p. 15903. AOM, Chicago (2018). https://doi.org/ghptjr

Improving a Gamified Language Learning Chatbot Through AI and UX Boosting

Polina Tsvilodub[(✉)], Esther Chevalier, Vera Klütz, Tobias Oberbeck, Kristina Sigetova, and Frederik Wollatz

Institute of Cognitive Science, Wachsbleiche 27, 49090 Osnabrück, Germany
{ptsvilodub,echevalier,vkluetz,toberbeck,ksigetova,
fwollatz}@uos.de

Abstract. In this paper, we show how the chatbot-based collaborative English learning app *Escapeling* can be enhanced through state-of-the-art artificial intelligence (AI) techniques and user experience improvements. In this app, intermediate learners can practice different skills in the available tasks, which are integrated in a gamified escape room scenario. To enhance the learning experience, we implement AI-powered grammar-learning support in a free writing task and we improve adaptive feedback for other tasks. Furthermore, we address affective aspects of learning and improve the user experience by integrating different types of situation-dependent visualizations and an improved contextualization in the storyflow into our app. Finally, students can now also practice listening comprehension within the app. Preliminary results gathered in three qualitative evaluation sessions indicate a noteworthy boost of the didactic value, user satisfaction, and enjoyment of Escapeling due to reported updates.

Keywords: Language learning · Artificial intelligence · User experience

1 Introduction

This article describes recent developments of the *Telegram bot Escapeling*[1]. This mobile language-learning application was presented previously [11]. Its goal is to provide a cooperative gamified setting for intermediate ESL learners, embedded in a science-fictional narrative. Escapeling is designed as an escape room game where users must solve different language-practice tasks to escape a spaceship where they are held as captives. Similarly to the offline escape room design, the participants have to rely on active group collaboration to be able to complete the game successfully.

   This mobile game setting provides a set of advantages with respect to learners' engagement and learning progress. First, the application relies on conversational language-learning scenarios and a narrative-based contextualization of the target language. Furthermore, Escapeling is played in groups of 3–4 players, which fosters cooperative interdependence and is beneficial for social and cognitive development, while

---

[1] https://t.me/Escapeling_Bot.

D. Guralnick et al. (Eds.): TLIC 2022, LNNS 581, pp. 557–569, 2023.
https://doi.org/10.1007/978-3-031-21569-8_52

closely matching natural social context of language use [10]. Finally, its mobile format makes the access to Escapeling easier for a large range of different students [3, 24]. Thus, Escapeling makes use of engaging remote learning methods, which have proven especially critical during the pandemic of the last two years.

Three different tasks have been introduced previously [11], specializing in vocabulary, written language production, and grammar practice. Presented results of a successful concept validation show high satisfaction among test users. These results indicate the potential of Escapeling and deliver an incentive to continue the project.

In this paper, we first present a new listening comprehension task that has been added to the app (Sect. 2.1). We then show how the learning experience within the app can be improved through the integration of AI-supported or adaptive feedback in the tasks (Sects. 2.2). In Sect. 3, we discuss how the user experience has been enhanced by adding visualizations and by introducing a more complex storyline, while keeping the overall app structure from a previous source [11]. Finally, we present initial user and instructor feedback in Sect. 4.

2 Learning Experience

The version of Escapeling described previously [11] presents three tasks covering four areas of language learning.

The *sentence correction task* is used to practice English grammar. The users are asked to spot and correct grammatically erroneous words in a sentence, if there are any. Every group member has to correct a sentence at least once to complete a round.

The *vocabulary guessing task*'s goal is to actively use and interpret English vocabulary. In each round, one player has to describe an English word, without using it. If another user finds the correct word within 90 s, the round is successful.

The *discussion task* acts as practice for text comprehension and written expression. After reading a short text about a specific topic, the users are prompted to discuss their personal opinions. This is encouraged by three questions, to be discussed in three minutes each. The group's success depends on the number of questions each user contributed to and the number of words they used to do so. Further comments and details about the aforementioned tasks can be found elsewhere [11].

2.1 Listening Task

The first three tasks have been designed to provide practice with the ability to read, write, explain words and use grammar. However, complete language proficiency crucially depends on another skill, namely listening [21]. Being able to comprehend spoken words is necessary for face-to-face communication, which is the most common end goal of studying a foreign language [15]. It also helps students to acquire the correct pronunciation, word stress, and a more confident use of vocabulary and syntax, which all improve the students' speaking competences. For this reason, we created the *listening task* aiming to train the user's listening comprehension skills.

In total, there are three audio tracks (i.e., three rounds) to be completed within the task. The task flow is presented in Fig. 1.

At the beginning of each round, an audio file is sent to the group chat (Fig. 2A). Each player is able to listen to the audio track at their own pace. All audio files last for about one minute. Then, a multiple-choice question related to the audio file's content is displayed with three corresponding answers (Fig. 2B).

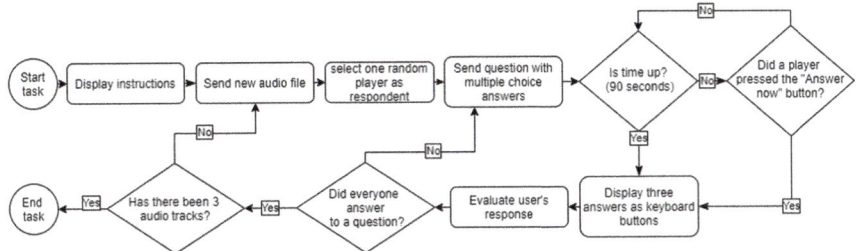

Fig. 1. Listening task flow.

The audio database contains 45 recordings, each centered around a certain topic and labeled according to a difficulty level. In order to ensure an optimal learning experience, we aim to provide audios with a good variety of speaker voices and accents, similar speed and length, as well as a wide range of everyday topics for which intermediate learners are expected to have the required core vocabulary. Some example topics include *Health*, *Media* and *Job Duties*.

Since the intermediate proficiency group tends to be quite heterogeneous in terms of the individuals' strengths and weaknesses, we also include slightly simpler audios targeting the A2 level. Each recording was carefully reviewed by at least three authors, checking for the audio and content quality of the recording. All files are publicly available for educational purposes from different online resources [5, 6, 14]. In sum, we hypothesize that this new task extends pedagogical advantages of the app.

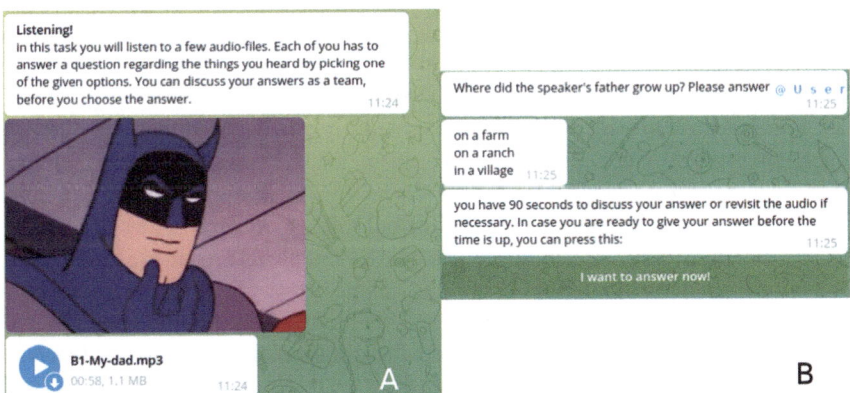

Fig. 2. A: Example GIF visualization and task presentation. B: Example question and answer options in the listening task.

2.2 Pedagogical Feedback

Building upon the developed task flow of Escapeling, we identify paths for enhancing the *learning experience* within the tasks, so as to further increase the pedagogical value and user satisfaction with the app [8, 23].

One critical aspect of learning experience is the feedback that learners receive (see [22] for an overview). During the vocabulary guessing and sentence correction tasks, users previously only received binary feedback on their response—i.e., whether the described word was guessed successfully, and whether the grammaticality of a sentence and potential mistakes were identified correctly, respectively. Due to the free production nature of the users' responses in the discussion task, no feedback was provided in this task in the previous version of the app at all.

We focus on using innovative tools for improving the learning experience and provide a first step towards integrating artificial intelligence (AI) techniques and adaptive mechanisms for improving task feedback within the application, which is still a challenging aspect of mobile language learning app design [9].

Grammatical Error Correction in the Discussion Task. As a first step towards improving the learning experience provided by Escapeling, we include a so-called *grammatical error correction* (GEC) feature in the discussion task. In order to aid the challenging acquisition of correct grammatical structures, we provide diverse feedback on grammatical errors made by the users during the discussion task. Improving the users' learning experience by presenting targeted feedback is in line with literature showing that it is pedagogically beneficial to provide more detailed explanations of the concepts practiced in the task [22].

As argued in prior research, it is important not to demotivate and not to confuse the learner whether the feedback is corrective or provides positive alternatives [22]. Therefore, in the case of such a GEC tool, we deem it very important to keep the rate of false corrections at a minimum (i.e., the rate of alternations of originally correct sentences, or, false positives, determined via the False Positive Rate FPR).

To achieve this, we use state-of-the-art AI-based tools for GEC (see [17] for an overview). More specifically, we combine two open-source models in a single grammatical error correction pipeline—a Python implementation of *LanguageTool* (LT) [16] and a neural network-based model called *GECToR* [19].

LT operates on a pre-specified set of errors and suggested corrections in a rule-based manner; given an erroneous input sentence, it outputs corrections and the underlying grammatical rules that were violated. In order to construct the precise GEC pipeline, we focused on the evaluation of the FPR for our GEC tools (see below for details of the evaluation procedure). Qualitative evaluation of sample test results revealed that LT may overcorrect intermediate English learners. Thus, in order to avoid false corrections, we use LT to track certain grammatical errors, but not to provide corrections across the board. Based on our evaluation, we identified 20 error types that are common to learners at the target proficiency level, and are reliably detected by LT with high accuracy. Among the reliably detected error types, as a proof of concept, we choose six for which users receive instructive feedback at the end of the task, if that error type was detected at least three times throughout the task completion. For instance, if users repeatedly use incorrect third person singular verbs they might read "In present tense, an '-s' ending is added to the verb when the pronouns 'he/she/it' appear in the subject of the sentence. Exception: The '-s' is not added to modal verbs." at the end of the task.

In order to further increase the accuracy of detecting errors, we use the second GEC model—GECToR—to check if it also detects errors in a given sentence. GEC-ToR is a pretrained open-source neural model [19]. At its core, the model consists of a transformer-based sequence tagging encoder with two linear layers on top [26]. The encoder applies error corrections by tagging the tokenized input sentence with transformations that need to be applied in order to transform the erroneous input text into a correct sequence [19]. In practical terms, when receiving an erroneous input sentence, the model outputs a suggested correct sentence but does not output grammatical rules that were violated. Therefore, we choose LT for tracking particular errors, while using GECToR for double-checking the performance of LT.

The choice of these particular models among other alternatives was motivated by the accuracy and the FPR of LT and GECToR, which were both evaluated on several datasets. The first evaluation step was performed using a corpus constructed from the first test split of the WI + LOCNESS 2019 corpus [2] (N = 1300 sentences) from which sentences containing the 20 target grammatical structures defined as prone to errors were extracted, combined with hand-crafted sentences.

Next to the overcorrection tendency of LT mentioned above, the evaluation of GEC-ToR on this corpus also indicated a rather high overcorrection rate (FPR = 0.15). To counteract this issue, we fine-tuned the last two layers of the GECToR neural network on an augmented dataset of erroneous sentences (N = 800), containing instances of the 20 common error types and their respective corrections. That is, the pretrained neural network provided previously [19] was trained further to better detect errors common to our target learners. The fine-tuning dataset was augmented using part-of-speech tagging in a rule-based fashion, based on the corpus used above.

For a final performance evaluation, we used the dataset created for the sentence correction task of the app (N = 1500 sentences). The fine-tuning of GECToR allowed for a remarkable decrease of the overcorrection rate (see Table 1).

Table 1. Evaluation results of the final GEC models on the sentence correction dataset. The absolute numbers of sentences classified for each category are shown for the first four columns. Relative proportions of the entire dataset are shown for accuracy and FPR.

Model	True positives	True negatives	False positives	False negatives	Accuracy	False positive rate
LT	767	1467	33	733	0.745	0.022
GECToR	1056	1405	95	444	0.820	**0.063**

On top of increasing the learning efficacy of the app, using two GEC tools allows to introduce a gradual scale of confidence with respect to the grammaticality of the input. Below, we suggest a more fine-grained free writing task evaluation pipeline integrating GEC, using the discussion task for an example operationalization.

The evaluation pipeline for the discussion task consists of four steps: input pre-processing, error classification via our GEC tools, correctness score calculation, and user participation scoring. We first perform basic message preprocessing and remove non-encodable symbols, images or GIFs. The preprocessed messages are passed to the pipeline which chains the LT and GECToR. The pipeline calculates a *correctness score* ranging from 0 to 1, encoding how confident it is that a given sentence is incorrect and provides a list of error types that were found in that sentence. The computation of the score is presented in Fig. 3. Furthermore, the pipeline increments the counts of the six common errors if they are detected, for which feedback is provided at the end of the task. If LT identifies one of these errors in a sentence, the correctness score is set to 0, ignoring the GECToR output.

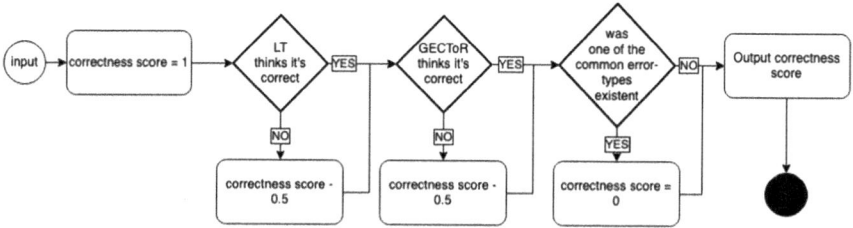

Fig. 3. Correctness score computation in the GEC pipeline in the discussion task.

Based on the correctness score, an individual user score is calculated for each user's participation in each discussion question. This score is the product of total sentence length and the correctness score. Only sentences longer than two words are considered for this. A group score is also calculated as the mean of all users' scores.

Within the game, intermediate feedback is sent by the bot after the first and second discussion questions have been debated. If not every user has written at least 15 words, in order to provide quantitative feedback, the bot sends a message encouraging the group to write more in the next round. For the qualitative aspect of the feedback, the pipeline checks if the ratio of grammatically correct sentences in the user's contribution is at least 0.5. If not all users have passed this threshold, they are advised to pay more attention to writing correct sentences in the next round.

At the end of the discussion, final feedback is sent to the group. The final feedback informs the users which error among the six tracked by LT was most often identified during the discussion, if at all. The bot then provides the violated grammatical rule and suggests how to avoid this error in the future. The group passes the task if each user has reached a minimum score in at least two of the three discussion rounds and if the group reaches a minimum group threshold. In future versions of the app, the passing thresholds within the GEC evaluation might be adaptive based on anticipated user proficiency.

In sum, given our corpus evaluation and initial user feedback (see Sect. 4), we hypothesize that the proposed AI-powered grammar evaluation pipeline is suitable for enhancing the grammar learning and thus learning experience in a free writing task.

Adaptive Feedback. As a second step, we improve the learning experience within the sentence correction and vocabulary guessing tasks by improving the *adaptive module* of

Fig. 4. A: Example adaptive feedback in the sentence correction task. Incorrect word suggested by user is not the target word (respective message omitted), yielding instructive feedback. B: Example adaptive feedback in the vocabulary guessing task. The shown feedback is rendered when guessing users are not able to guess the intended word.

Escapeling. The adaptability of the bot is an important feature justifying the development of an interactive system in contrast to a classical learning setting. The chatbot adapts its output and prompts depending on the user's input, ultimately addressing individual strengths and weaknesses. This feature ensures that users get the assistance they need to complete a task, but not more, to protect the pedagogical value of the exercise at hand. We expect the users to feel more addressed during the interaction with the Escapeling bot, leading to a higher immersion into the game.

In the presented version of the Escapeling bot, the adaptive module gives specific and targeted hints or feedback if the user made a mistake. In the sentence correction task, the hint's purpose is to call attention towards the mistake and to give the user a second chance at solving the current task. If the user still makes a mistake, the bot assumes the user lacks the formal knowledge needed to solve this task, and a small targeted lesson is presented in the chat, covering only the formal rules required for solving the task (Fig. 4A). In the vocabulary guessing task, the bot iteratively provides hints regarding the category, and then letters constituting the target word. Importantly, the bot tracks users' attempts to solve the task and only provides these adaptive hints after a predefined number of wrong guesses (Fig. 4B). We hypothesize that this personalization enhances the efficacy of learning with the chatbot.

Finally, spelling feedback was also added to the vocabulary guessing task of the bot. We employ Levenshtein distance computation between the user's response and the correct answer in order to check if the responding user made a typo [13]. If so, the user receives a hint to check their spelling (Fig. 4B). We hypothesize that, pedagogically seen, this gives users a chance to correct themselves and learn from their own mistakes.

3   User Experience

One further strength of Escapeling is the contextualization of learning the target language into an engaging storyline. In this fictional story the users play the main characters. They are assisted by the help of two different in-game characters—the friendly alien Elias and the human hacker Harriet who needs to pose the tasks to the players in order to help them escape. The storytelling is accompanied by a narrator. In order to enhance this gamification aspect and the learners' motivation to solve the tasks, we identify the following paths to improve the *user experience* in the game.

3.1   Storyline

First, in order to facilitate Telegram group creation for the game, users can meet on our newly implemented lobby platform.[2] Upon entering the main Escapeling game, the storyline elements are embedded between three mandatory tasks (see Fig. 5). To improve the outcome of the story the users can additionally complete the listening task. In order to increase the user satisfaction and replayability of the game, the players now get a chance to redo a task or to do another task first if they fail one. The order of the tasks is not fixed so that, each time the users get to choose the next task, all failed or not attempted tasks are presented as an option. The storyline is presented to the users via different chat messages they receive from one of the in-game characters or from the narrator.

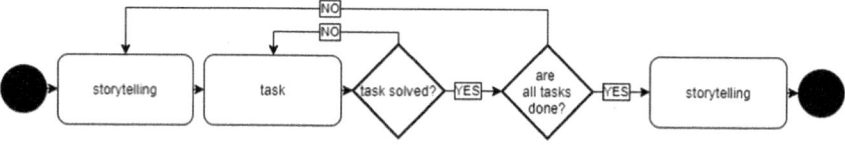

Fig. 5. A simplified diagram of the game and storytelling flow.

3.2   Visualizations

Secondly, to enhance the effects of the storyline, different forms of visualizations were added to the game. It has been shown that visualizations "displayed in online interactive e-learning (…) [represent] a critical aspect in teaching and learning" ([12] p. 4041) as they increase the interest in the subject matter [12]. Furthermore, visual stimuli can help

---

[2] https://escapeling-lobby.herokuapp.com/.

to focus on the learning content as well as to remember what has been learned [12, 27]. The selection of the type of visualization elements is based on both the use of these elements by our target audience (i.e., students at different levels of education) and the functionalities of our app. These two points led us to employing GIFs, emojis, stickers and adaptive images.

GIFs are short, constantly repeating videos. According to a previous study [27], "it is easier for students to better understand certain topics after showing emoticons or GIFs. (…). This is an element that highly stimulates the attention and by levering on emotions, students are more likely to focus their attention after the given emotional stimul[us]" (p. 3). Literature further underlines the importance of emotions in e-learning, as the user's emotions are also part of the user experience [4]. In a previous study [27] that examined the use of emojis and GIFs to promote student engagement, these visualizations were successfully used "in order to stimulate the emotions such as hilarious and ironic feelings among young students." Such emotional associations have been shown to lead to better long-term retention of the respective content [27].

For this reason, each task is introduced with a GIF (Fig. 6A). This marks the intersection between the previous storytelling part and the subsequent task demanding the user's attention. Another function of GIFs in our app is to provide feedback and to motivate. For example, a GIF can be sent in response to a correct answer.

An additional type of visualizations used in our application are emojis. For example, emojis were inserted in a selection menu for quicker orientation. This also made the previously completely text-based selection more colorful, following various studies that have proven that colors can stimulate the brain during learning [1, 7].

In addition to GIFs and emojis, adaptive images have been inserted to create a custom timer for the discussion task. Since Telegram's API does not allow the implementation of a time progress bar, this challenge is solved by sending an adaptive image showing the remaining time to complete the task (see Fig. 6A).

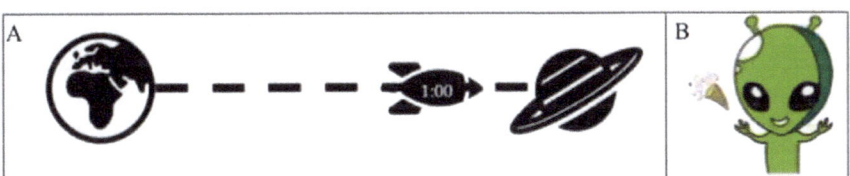

Fig. 6. A: Image showing the time remaining for discussing a question in the discussion task. B: Sticker of happy Elias.

Lastly, stickers are introduced as a visualization type. These are images that have no background and are displayed much larger than emojis in Telegram [25]. To enhance the effects of storytelling, a standalone sticker pack was developed that incorporates various components of the storyline. Since emotions can provide a better immersion into the story, different emotional states of the main characters, Harriet and Elias, such as joy, mistrust or sadness are visualized (see Fig. 6B).

However, a previous study [27] notes that too many graphics could distract learners more than focus them. In addition, especially GIFs and emojis could lower the seriousness of the application [27]. While it is desired to break away from what is often perceived as a strict school learning environment and also to convey the joy of learning, Escapeling should also not be perceived as a chat game without a serious English learning intention. In order to avoid these effects, we carefully chose the placement of the visualizations to only mark important storyline and task milestones or to highlight feedback, such that the learners aren't distracted from main learning contents.

In addition to boosting emotional engagement in the storyline, visualizations also contribute to an appealing design of the learning application. According to a previous study [18], attractively designed applications are evaluated as more usable by the user than unattractive ones. To a large extent, this is due to the influence of emotions [18]. Thus, we hypothesize that our novel additions improve the users' evaluation of the enjoyment and usability of the Escapeling app.

4 Evaluation

In this section, we describe initial experimental evaluation results for the presented improvements of the user and learning experience. The described features were tested in three qualitative evaluation sessions [20]. All subjects participated voluntarily and were not reimbursed for their participation. Due to the ongoing pandemic, the second and third evaluation rounds are still in progress.

First, a subset of the users recruited by a previous study [11] was recruited again. Given these subjects' prior experience with the old version of the game, the qualitative evaluation focused on outlining the user experience updates in this current version of the bot and presented the participants with visualizations or descriptions of respective changes.

Five testers received a PDF file with examples of modified application functions and descriptive information about the listening task collected in presentation style and a feedback questionnaire. In the questionnaire, participants were asked how much they liked an updated feature (i.e., for their satisfaction), and how well they thought each feature improved the overall experience with the bot. Both questions were answered on a Likert-scale ranging from 1 ("completely dissatisfied with the update"/"doesn't improve anything") to 10 ("absolutely happy with the update"/"the update is very helpful"). They also had the opportunity to provide free feedback.

The first update concerning the addition of the new stickers was rated by the users with an average of 9.6 points in terms of satisfaction with the idea and 8.6 points for the game improvement, while some free feedback raised concerns about device memory usage by the media. The addition of GIFs to the various sections of the bot was rated with 8.8 and with 8.2 average points in terms of satisfaction and in favor of the game improvement, respectively. The participants rated the addition of adaptive graphics for the remaining time in the discussion task with average scores of 9.6 and 10 for the satisfaction and the app improvement, accordingly. Almost every tester indicated that they had missed it earlier. The first survey ended with an assessment of the new listening task. Subjects gave average scores of 9.8 for the anticipated satisfaction and 10 for the app improvement, respectively. Overall, user reviews were highly satisfied with respect to all user and learning experience improvements.

In the second evaluation round, which is still in progress, we recruit students of Cognitive Science and related fields at the Osnabrück University. The goal is to further evaluate the end user experience with our bot. To this end, participants are invited to play the game and fill out a short online questionnaire. They answer the questions via free typing, multiple choice or 5-point Likert-scale ratings. Participants answer questions about the pedagogical value of the different tasks, their enjoyment of the single tasks and the game overall, and the UX value of the visualizations. Data from 14 participants has already been collected (71% Osnabrück University Cognitive Science students, average English proficiency: B2-C1). In total, 78% of participants reported that they would recommend the app to other English learners, and more than 90% would likely play the game again. On average, they indicated that they could best improve their general communication, reading, and listening skills with Escapeling. The new lobby platform was rated with 3.3 average points. Furthermore, participants stated that the improved visualizations were very important for their enjoyment of the game, providing an average score of 4.3. Finally, they reported high satisfaction with Escapeling compared to other educational apps they used, providing an average score of 4.3. These preliminary results indicate the success of the UX and task improvements presented in this paper among end users.

Finally, our third on-going evaluation focuses on assessing the pedagogical value of the game, next to user experience. To this end, we already conducted in-depth interviews with three ESL students for the German secondary and high school system. The interviewees first played the game and then completed interviews about it. The interview questions addressed the following aspects of the bot: target audience appropriateness, possible contexts of application, gamification, visualizations, learning experience and potential for improvements. Overall, all interviewees indicated high usability and enjoyability of the game and stated that they would definitely use it as a supplementary teaching tool. They indicated that the difficulty of the tasks in Escapeling is suitable for students in their 10th grade or above but noted that they would consider employing it from 5th grade onwards. Furthermore, two interviewees stated that, due to the fun nature of the game, students might find more motivation to learn English through using it informally in Escapeling. Crucially, all interviewees stated that the visualizations were a critical motivational force and would be very appealing to learners, especially in school age. This confirms the efficacy of our user experience improvements. The interviewees also stated that the cues and feedback given throughout the tasks are crucial from a pedagogical and didactic point of view. Specifically, they stated that it is preferable that students first receive hints and explanations and have the chance to correct their mistakes themselves. We take this feedback as strong confirmation of the value of our learning experience improvements suggested in this article.

These preliminary results indicate that the described improvements are highly successful and might largely improve the user and learning experience for future bot users. Results from the ongoing evaluations, as well as our planned field classroom studies, might provide further insights.

4.1 Conclusion

This paper proposes several ways to boost the user and learning experience within the conversational agent based language learning app *Escapeling*. We propose improving the learning experience by providing a solution for including an AI-powered grammar acquisition support tool in the bot, which goes to show how free text tasks can be embedded in a gamified collaborative setting. Furthermore, we suggest ways to embed adaptive pedagogical feedback within other tasks and propose a new task covering the listening skill set. Finally, we suggest improving the user experience by including various visualizations and motivational graphics. Given initial user and instructor feedback, these elements clearly contribute towards the enjoyment, optimal user experience, satisfaction, and a didactically valuable learning experience of the game.

Acknowledgments. The authors gratefully acknowledge the following students at the Osnabrück University for their invaluable collaboration in this project (listed alphabetically): Janosch Bajorath, Eliasz Ganning, Malte Heyen, Ieshia Hickey-Williams, Paribartan Humagain, Karina Khokhlova, Elena Korovina, Hanna Linder, Ivan Polivanov, Jannik Schmitt, Polina Shamraeva, Liling Wu, Hin Sing Yuen, amd Emma Zanoli. Additionally, the authors would like to thank Dr. Tobias Thelen for his help and for making this study project possible.

References

1. Althouse, R., Johnson, M.H., Mitchell, S.T.: The Colors of Learning: Integrating the Visual Arts into the Early Childhood Curriculum, vol. 85. Teachers College Press, New York (2003)
2. Bryant, C., Felice, M., Andersen, Ø. E., Briscoe, T.: The BEA-2019 shared task on grammatical error correction. In: Proceedings of the Fourteenth Workshop on Innovative Use of NLP for Building Educational Applications, pp. 52–75. Association for Computational Linguistics, Florence, Italy (2019)
3. Chen, Z.-H., Chen, H.H.-J., Dai, W.-J.: Using narrative-based contextual games to enhance language learning: a case study. Educ. Technol. Soc. **21**, 186–198 (2018)
4. DIN EN ISO 9241-210:2020-03: Ergonomics of Human-System Interaction—Part 210: Human-Centred Design for Interactive Systems (ISO 9241-210:2019)
5. English Listening Lesson Library Online. www.elllo.org. Last accessed 08 March 2022
6. English as Second Language lounge. www.esl-lounge.com. Last accessed 08 March 2022
7. Ferrari, V., Zisserman, A.: Learning Visual Attributes. Advances in Neural Information Processing Systems, pp. 433–440 (2008)
8. García-Martínez, I., Fernández-Batanero, J.M., Cobos Sanchiz, D., Luque de La Rosa, A.: Using mobile devices for improving learning outcomes and teachers' professionalization. Sustainability **11**(24), 6917. MDPI, Basel, Switzerland (2019)
9. Heil, C.R., Wu, J.S., Lee, J.J., Schmidt, T.: A review of mobile language learning applications: trends, challenges, and opportunities. EuroCALL Rev. **24**(2), 32–50 (2016)
10. Johnson, D.W.: Cooperative Learning: Increasing College Faculty Instructional Productivity. ASHE-ERIC Higher Education Report No. 4. ASHE-ERIC Higher Education Reports, George Washington University, Washington, DC 20036-1183 (1991)
11. Johnson, C., Urazov, M., Zanoli, E.: Escapeling: a gamified, AI-supported chatbot for collaborative language practice. In: Guralnick, D., Auer, M.E., Poce, A. (eds.) TLIC 2021. LNNS, vol. 349, pp. 141–148. Springer, Cham (2022). https://doi.org/10.1007/978-3-030-90677-1_14

12. Jusoh, S., Almajali, S., Abualbasal, A.M.: A study of user experience for e-learning using interactive online technologies. J. Theor. Appl. Inf. Technol. **97**(15), 4036–4047 (2019)
13. Levenshtein, V.I.: Binary codes capable of correcting deletions, insertions, and reversals. Sov. Phys. Doklady **10**(8), 707–710 (1966)
14. Listen a minute website. www.listenaminute.com. Last accessed 08 March 2022
15. Morley, J.: Aural comprehension instruction: principles and practices. In: Celce-Murcia, M. (ed.) Teaching English as a Second or Foreign Language, pp. 69–85. Heinle and Heinle, Boston (2001)
16. Naber, D.: A rule-based style and grammar checker (2003)
17. Ng, H.T., Wu, S.M., Briscoe, T., Hadiwinoto, C., Susanto, R.H., Bryant, C.: The CoNLL-2014 shared task on grammatical error correction. In: Proceedings of the Eighteenth Conference on Computational Natural Language Learning: Shared Task, pp. 1–14 (2014)
18. Norman, D.A.: Emotional Design: Why We Love (or Hate) Everyday Things. Basic Books (2005)
19. Omelianchuk, K., Atrasevych, V., Chernodub, A., Skurzhanskyi, O.: GECToR—Grammatical Error Correction: Tag, Not Rewrite. In: Proceedings of the Fifteenth Workshop on Innovative Use of NLP for Building Educational Applications, pp. 163–170. Association for Computational Linguistics, Seattle, USA (2020)
20. Patton, M.Q.: Qualitative Research & Evaluation Methods: Integrating Theory and Practice. Sage Publications (2014)
21. Renukadevi, D.: The role of listening in language acquisition; the challenges & strategies in teaching listening. Int. J. Educ. Inf. Stud. **4**(1), 59–63 (2014)
22. Russell, J., Spada, N.: The effectiveness of corrective feedback for the acquisition of L2 grammar. Synth. Res. Lang. Learn. Teach. **13**, 133–164 (2006)
23. Schiefelbein, J., Chounta, I.-A., Bardone, E.: To gamify or not to gamify: towards developing design guidelines for mobile language learning applications to support user experience. In: Scheffel, M., Broisin, J., Pammer-Schindler, V., Ioannou, A., Schneider, J. (eds.) EC-TEL 2019. LNCS, vol. 11722, pp. 626–630. Springer, Cham (2019). https://doi.org/10.1007/978-3-030-29736-7_54
24. Sharp, H.: Interaction Design: Beyond Human Computer Interaction. Cornell. Encyclopedia of the Sciences of Learning. Springer, Boston, MA (2006)
25. Telegram stickers. https://core.telegram.org/stickers. Last Accessed 10 March 2022
26. Vaswani, A., Shazeer, N., Parmar, N., Uszkoreit, J., Jones, L., Gomez, A.N., Polosukhin, I.: Attention is all you need. In: Advances in Neural Information Processing Systems, vol. 30 (2017)
27. Zallio, M., Damon, B.: Computer aided drawing software delivered through emotional learning. The use of emoticons and GIFs as a tool for increasing student engagement. In: 32nd International BCS Human Computer Interaction Conference, pp. 1–4 (2018)

Inclusive Learning Special Track

Open-Access Learning as a Pathway to Equity During Health Emergencies

Melissa Attias[✉] ⓘ, Heini Utunen ⓘ, Ngouille Ndiaye ⓘ, and Lama Mattar ⓘ

World Health Organization, Avenue Appia 20, 1211 Geneva 27, Switzerland
`outbreak.training@who.int`

Abstract. Real-time learning in health emergencies such as the COVID-19 pandemic is a critical mechanism to provide frontline health workers, responders, decision-makers and the public with equitable access to the latest knowledge to save lives, reduce disease transmission and protect the vulnerable. The World Health Organization (WHO) established the OpenWHO.org learning platform to meet this need. Courses are free, self-paced, accessible in low-bandwidth and offline formats, and available in national and local languages. Enrolment data from OpenWHO's introductory COVID-19 course, which has more than 1 million enrolments across 45 language versions, were examined according to language and geographical reach to assess how multilingual availability contributes to equity in learning. The analysis found that most language versions had uptake clustered in key countries where native speakers are concentrated, while use of some translations was more broadly dispersed. In nearly three-fourths of the available language versions of the course, more than one-third of enrolments were found in the top country of use. The findings suggest that courses available in the United Nations languages, as well as national and local languages, served as entryways for learners who may not have otherwise been able to participate. A production policy that prioritizes translation of open online courses into diverse languages contributes to equity in access to public health knowledge at the global and country levels during health emergencies.

Keywords: Online learning · Multilingual learning · Equity · Pandemic

1 Introduction

The World Health Organization (WHO) launched the OpenWHO.org online learning platform in 2017 to facilitate the transfer of public health knowledge for emergencies on a massive scale in anticipation of the next pandemic. Grounded in the principles of open access and equity, courses are free, self-paced, accessible in low-bandwidth and offline formats, and available in national and local languages [1]. The platform builds on the massive open online course (MOOC) model, which aims to make education accessible for all learners, including vulnerable communities, learners with low education levels, non-English speakers and learners with disabilities [2]. Although the model has yet to achieve its full potential, open programs have also been found to meet their aims of enabling student equity and social inclusion based on program outcomes [2–4].

D. Guralnick et al. (Eds.): TLIC 2022, LNNS 581, pp. 573–580, 2023.
https://doi.org/10.1007/978-3-031-21569-8_53

After serving frontline responders in regionalized outbreaks from Ebola to plague, OpenWHO dramatically scaled up course production for the COVID-19 pandemic, making life-saving information from WHO experts available online at a time when lockdowns and social distancing limited the ability to physically come together to learn [1]. The platform offers courses on 44 different COVID-19 topics, as well as 100 courses on additional health topics. Courses are available across 64 languages, including the 15 most-spoken languages worldwide and the official languages of 44 out of 46 of the least-developed countries, in recognition that it is easier to learn and understand in one's native tongue [5].

Prioritizing access to learning has enabled OpenWHO to have tremendous reach in line with WHO's mission to promote health, keep the world safe and serve the vulnerable. Demand has surged during the COVID-19 emergency, with course enrolments increasing from 160 000 in January 2020 to 6.5 million in March 2022. OpenWHO has served as a growing source of knowledge for demographics that are typically underserved online, including women and people aged 70 and older as they actively sought information about COVID-19 [6]. Learning reach has also extended beyond the use of the online platform as countries and communities have adapted materials to local contexts and offline demands to bypass technological and connectivity barriers, creating a multiplier effect that further advances equity in learning [7, 8].

This paper analyzes enrolment trends in the introductory COVID-19 course, which has more than 1 million enrolments across 45 different language versions and is the most popular course on the OpenWHO platform. It presents findings on language and geographical reach to examine how multilingual availability contributes to equity and inclusion in learning.

2   Methods

Enrolment data were drawn from OpenWHO's built-in reporting system, which tracks learners' enrolments, completion rates, demographics and other key course-related metrics. Data were collected from the launch of OpenWHO in 2017 up until December 2021 and aggregated in the R environment. Descriptive statistics were calculated using the Microsoft$^{®}$ Power BI tool. Key outcome variables of interest were learners' locations and language of learning, as well as self-reported data on sex, age and affiliation. Data for the Introduction to COVID-19 course, from its launch in English on January 26, 2020, to December 31, 2021, were segregated to examine enrolment trends, including the distribution of learners by language version and country.

3   Results

Overall, the Introduction to COVID-19 course was slightly more popular among female learners (53.15%) than males (46.68%). Learners ages 20–29 were the most represented age bracket (42.14%), followed by learners ages 30–39 (23.41%), younger than 20 (13.08%), 40–49 (11.04%), 50–59 (4.76%), 70 and older (4.54%), and 60–69 (1.06%). Students were the most represented learners in the course (42.32%), followed by "other" affiliations (21.27%) and health care professionals (16.38%). The countries with the most course enrolments were India (26.70%), Mexico (10.70%), Ecuador (8.74%), the United States of America (6.68%) and Colombia (5.27%).

Of the 45 languages available for the introductory course, 6 represented the official languages of the United Nations (UN) (Arabic, Chinese, English, French, Russian and Spanish), and the remaining 39 were national and local languages spoken across the globe. Almost half (48.82%) of course enrolments were in English, 29.12% were in Spanish, 7.31% were in another UN language, and 14.75% were in another national or local language, for a total of 512,095 enrolments in language versions other than English.

An analysis of the UN language versions found multilingual availability attracted learners from key geographical demographics. For example, the Arabic language course was most used in Iraq, with 20.25% (5,269) of Arabic course enrolments located in that country. The Arabic course was also the most popular language version in Iraq overall, representing 55.09% of Iraq's course learners. In addition, Arabic was notably popular in Saudi Arabia (18.49%, 4,812) and Egypt (17.29%, 4,499). The Spanish language course was most used in Mexico, representing 28.65% (70,992) of Spanish course enrolments, and comprised 78.94% of course enrolments from learners in Mexico.

The same trend was observed in the analysis of national and local languages. The Macedonian language course was most used in North Macedonia, for example, with 80.85% (650) of Macedonian course learners located in the country. The Macedonian version of the course was also the most utilized language version in North Macedonia overall, accounting for 84.75% of North Macedonia's learners. Similarly, the Portuguese language course was most used in Brazil, representing 39.76% (5,932) of enrolments in the Portuguese course and comprising 74.01% of Brazil's learners.

In other contexts, multilingual availability attracted additional learners in countries where the English language version remained the most-utilized course. For example, the Amharic course was most used in Ethiopia, which provided 71.03% (255) of Amharic course learners, even as 13.04% of Ethiopia's learners used the Amharic course and 82.20% used the English version. Similarly, the Indonesian course was most used in Indonesia, which contributed 71.40% (2,007) of Indonesian course enrolments, while 32.04% of Indonesia's learners used the Indonesian course and 64.72% used the English version.

India also offered a snapshot of this trend at the country level. The most used language version of the course in India was the English course, serving 57.43% of India's learners. Even so, the following language versions also had the highest levels of popularity in India, making the course accessible to additional demographics within the country: Hindi (97.84% of enrolments came from India), Indian Sign Language (88.07%), Marathi (91.16%), Oriya (93.46%) and Punjabi (82.58%). India also emerged as the top country for many additional language versions of the course, as more than 1 in 5 OpenWHO learners are from India.

In terms of the overall geographical distribution, Fig. 1 shows that, in nearly 3/4 (32/45) of the available language versions of the introductory course, more than 1/3 of enrolments were found in the top country of use. In fact, 19 out of 45 languages had more than half of their enrolments in the top user country. In contrast, 4 of the 6 official UN languages were among the remaining language versions (13/45) with the least observed geographical concentration in the top country of use: Russian (18.47%, Ukraine), Arabic

(20.25%, Iraq), Spanish (28.65%, Mexico) and English (30.74%, India). French ranked in the 19th position with 39.06% of its users localized in India, while Chinese ranked 33rd out of 45, with 61.54% of its users localized in China. Although almost 2/5 of the French version's enrolments were collected from India, additional key countries of note were France, Congo (DRC) and Mexico, totaling 14.35% of French enrolments.

Table 1 shows the start date for each of the top 10 language versions and the total enrolments accumulated. Languages published later naturally have fewer entries, except for languages whose popularity outweighed the impact of the publication date. The Spanish language version published on February 10, 2020 (291,399 enrolments) has been 33 times more popular than the Chinese version (8,763 enrolments) released one day before, and has had about 10 times the uptake of the French version released 3 days before. Similarly, with the exception of the Spanish version, the Indian Sign Language translation (488,460 enrolments) has met greater popularity compared to the 9 other languages that preceded its publication after the original English course.

4   Discussion

The findings suggest that multilingual production of the introductory COVID-19 course has enabled additional learners to access and understand critical public health learning materials, advancing equity at the global and country levels. In some cases, like in Iraq and Mexico, a UN language course other than English was the most popular course in the country and brought thousands of learners to the course. In other cases, like North Macedonia and Brazil, a national language course was the most popular as it served as the entryway for learners. In other cases, still, the English language course was the most popular in the country overall, but UN and other language courses brought additional learners within the country that may not have otherwise been able to participate in the learning.

Overall, the geographical distribution of enrolments shows that nearly three-fourths of language versions had centralized uptake in key countries where native speakers are concentrated, while use of the remaining language versions was more widely dispersed. Such multilingual access is critical because the ability to learn in one's preferred language has been proven to increase uptake and understanding [5]. This can perhaps be most directly observed in courses targeting populations with special needs. For example, enrolments in the Indian Sign Language course comprise 18.10% of total introductory course enrolments from India, reaching a targeted demographic that would otherwise not be served. The focus on multilingualism supports inclusiveness so that learners feel like valued members of the learning community, alongside platform features like multi-format and low-bandwidth options designed to meet learners' diverse needs. Open-WHO's tagline of "open to all, anytime, from anywhere" embodies this commitment to inclusion.

A particularity of the OpenWHO production process is that the English language version of a course is generally launched first based on the relevant technical guidance and then translated and published in additional languages. English production is prioritized for courses of global interest because the language is shared by many native and non-native speakers around the world and is the most widely used working language of the

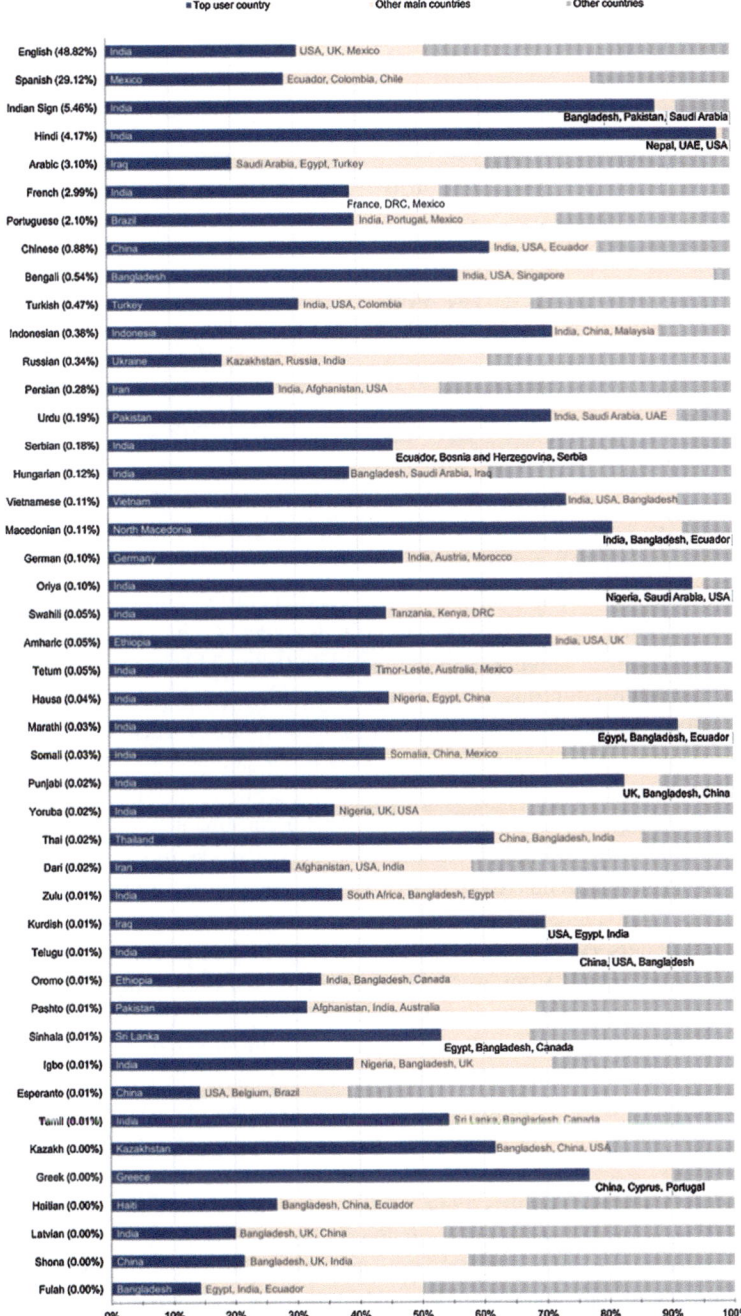

Fig. 1. Geographical distribution of the multilingual OpenWHO COVID-19 introductory course.

Table 1. Top 10 language versions of the multilingual OpenWHO COVID-19 introductory course as of December 31, 2021.

Language	Start date	Total enrolments
English	January 26, 2020	488,460
Spanish	February 13, 2020	291,399
Indian sign	March 23, 2020	54,623
Hindi	March 18, 2020	41,714
Arabic	February 28, 2020	30,968
French	February 10, 2020	29,892
Portuguese	February 27, 2020	21,051
Chinese	February 12, 2020	8,763
Bengali	April 24, 2020	5,443
Turkish	March 10, 2020	4,677

UN system. In fact, it is one of the language versions of the introductory course with the least geographical centralization. Similarly, Russian, Arabic and Spanish are also among the most-spoken languages in the world and showed less centralization due to their use across many countries.

As a result of the prioritization of English materials, some learners may have enrolled in the English version to have immediate access to the course knowledge even though they would have otherwise enrolled in another language if that version was concurrently available. This suggests that analysis of OpenWHO enrolment figures may underestimate the demand for multilingual courses. It also serves as a reminder of the challenges of relying on enrolment trends to define the relevance and need for multilingual production that caters to all audiences through a policy of equity, as production of courses in neglected and underserved languages may appear to be less effective when measured exclusively by enrolments. Empowering one community or even a single individual with knowledge in their language can also have a critical impact as these learners take their knowledge forward to support and educate others, yielding benefits for equity that are challenging to quantify. Additional research could systematically explore how multilingualism contributes to this multiplier effect, both in terms of offline and adapted use of learning materials, and how knowledge gained is implemented at the individual and local levels.

Beyond the language analysis, course data on sex, age and affiliation also illustrate OpenWHO's progress in promoting equitable access to knowledge. The course reaches a higher proportion of women (53.15%) than the overall platform average, which shifted from having 40.39% female learners before the pandemic to 51.80% at the end of 2021. This trend contributes to equity, as OpenWHO survey results have found that women are more likely than men to find time and cost to hinder their access to learning, echoing broader gender equality concerns [1, 9, 10]. The presence of both older and younger users in the introductory course also shows progress towards more inclusive learning, as learners aged 70 and older were statistically unrepresented on the platform prior to the pandemic compared to 4.54% in this course, and learners under age 20 were only 3.22% of platform enrolments compared to 13.11% here. Finally, the large numbers of learners who are students and whose affiliations do not align with one of the health, government or international organization categories that are traditional constituencies of the OpenWHO platform suggest that the course's introductory nature serves a general audience, increasing equity and access to critical knowledge during the pandemic.

5 Conclusion

Since 2020, WHO's OpenWHO learning platform has heightened its focus on multilingualism to support effective uptake of critical learning resources for COVID-19. The geographical distribution of learners in the 45 language versions of the introductory COVID-19 course suggests that language availability makes learning accessible to learners concentrated in key countries, as well as populations dispersed across many countries and regions. Overall, the analysis finds that prioritization of translation into national and local languages contributes to equity in access to public health knowledge at the global and country levels, attracting learners that may not have otherwise been able to participate in open online learning.

Acknowledgments. The authors would like to acknowledge the massive efforts of the Learning and Capacity Development Unit of the WHO Health Emergencies Programme, as well as WHO technical experts, course producers and leadership, without whom the OpenWHO.org learning model would not be possible.

References

1. George, R., Utunen, H., Ndiaye, N., Tokar, A., Mattar, L., Piroux, C., Gamhewage, G.: Ensuring equity in access to online courses: Perspectives from the WHO health emergency learning response. World Med. Health Policy 1–15 (2022)
2. Lambert, S.R.: Do MOOCs contribute to student equity and social inclusion? A systematic review 2014–18. Comput. Educ. **145** (2020)
3. Chandler, C.B., Quintana, R.M., Tan, Y., Aguinaga, J.M.: Realizing equity and inclusion goals in the Design of MOOCs. J Appl. Instruct. Des. **10**(4), (2021)
4. Barcena, E., Read, T., Vilhelm, M.. Open online courses and the democratization of knowledge for vulnerable groups. In: European Distance and E-Learning Network (EDEN) Proceedings, pp. 322–331. Madrid (2021)
5. Alidou, H., Glanz, C. (eds): Action Research to Improve Youth and Adult Literacy: Empowering Learners in a Multilingual World. UNESCO Institute for Lifelong Learning (UIL), Hamburg (2015)
6. Utunen, H., Ndiaye, N., Mattar, L., Christen, P., Stucke, O., Gamhewage, G.: Changes in users trends before and during the COVID-19 pandemic on WHO's online learning platform. Stud. Health Technol. Inform. **287**, 163–164 (2021)
7. Utunen, H., Attias, M., George, R., O'Connell, G., Tokar, A.: Learning multiplier effect of OpenWHO.org: use of online learning materials beyond the platform. Weekly Epidemiol. Record **96**(01–02), 1–8 (2022)

8. OECD: The Potential of Online Learning for Adults: Early Lessons from the COVID-19 Crisis, OECD Policy Responses to Coronavirus (COVID-19). OECD Publishing, Paris (2020)
9. Jena, A.B., Olenski, A.R., Blumenthal, D.M.: Sex differences in physician salary in US public medical schools. JAMA Intern Med. **176**(9), 1294–1304 (2016)
10. Del Boca, D., Oggero, N., Profeta, P., Rossi, M.: Women's work, housework, and childcare before and during COVID-19. Rev Econ Household **18**, 1001–1017 (2020)

Scenario-Based Training and On-The-Job Support for Equitable Mentoring

Danielle R. Chine[✉][ID], Pallavi Chhabra, Adetunji Adeniran, Joseph Kopko, Cindy Tipper, Shivang Gupta, and Kenneth R. Koedinger

Carnegie Mellon University, Pittsburgh PA, USA
dchine@andrew.cmu.com

Abstract. Personalized Learning^2 (PL^2) is a mentor professional development platform designed to improve efficiency and workplace training through scenario-based instruction and personalized support. Combining research-driven mentor training with artificial intelligence-powered (AI-powered) software, PL^2 connects mentors, often under-trained tutors, to personalized resources with the click of a button. These curated resources cover a range of topics from social-emotional learning and math content to culturally responsive teaching practices. PL^2 is addressing the opportunity gap among marginalized students by recommending specific instructional supports and social-emotional resources based on a student's individual math performance and effort. This work in progress showcases our recent development of the PL^2 approach to mentoring in three ways. First, we highlight the key functions of the PL^2 system. Second, we present recent research results determining the most effective competencies for successful mentorship. Last, in response to partner feedback on which mentor competencies are most needed, we detail the development of asynchronous, scenario-based lessons, housed within the PL^2 platform for on-demand training. The use of human-computer teaming for on-the-job support offers a lower cost option for deliberate practice using scenario-based training to increase the impact and learning capacity of mentors in the workplace.

Keywords: Tutor training · Mentor training · Scenario-based · Equitable mentoring · Adult learning · Workplace training

1 Introduction

The demand for experienced quality mentors or tutors in public education and afterschool organizations is at an all-time high [8]. Due to the repercussions of the COVID-19 pandemic on the educational system, schools and organizations are struggling to recruit not only licensed teachers, but also experienced and qualified mentors [8]. Historically marginalized students (e.g., Black, Hispanic, students experiencing poverty, first-generation college students) are at the greatest risk of not meeting annual growth gains compared to their peers [7, 10]. In fact, students in high-poverty schools or marginalized groups have been impacted the most by recent educational disruption, with the

D. Guralnick et al. (Eds.): TLIC 2022, LNNS 581, pp. 581–592, 2023.
https://doi.org/10.1007/978-3-031-21569-8_54

biggest declines in overall performance occurring in math [10]. A possible solution to attend to the opportunity gap experienced by these under-resourced student populations is individualized instruction through mentoring. Individualized instruction through mentoring has been repeatedly found to be one of the most impactful interventions on student achievement [9]. One challenge to using mentors to address the gaps of learning loss and provide additional instructional support is the lack of experienced and skilled mentors. For purposes of this work, a mentor is an unlicensed paraprofessional who provides academic support similar to a tutor, along with providing social-emotional support and relationship-building skills. In addition, a "mentor" is defined as a person, oftentimes with more experience, who provides support and guidance to somebody with lesser experience and is typically younger [6]. This work discusses a human-computer teaming platform that provides both scenario-based training and personalized resources to mentors—many of whom are unskilled or lack experience. We discuss the need for quality upfront training and continuous on-the-job professional development. Two competencies that surveyed organizations highlighted as critical for training were 'culturally responsive teaching practices' and 'awareness of biases.' We created both synchronous and asynchronous scenario-based workplace training attending to these mentor competencies within a human-computer tutoring platform. Included with the scenario-based training details are preliminary research results and feedback from mentors and our partner organizations.

2   The Personalized Learning^2 Approach

Personalized Learning^2 (PL^2) is software designed to improve K-12 student learning by combining the benefits of both human mentoring and computer-based learning with EdTech tools [12][1]. By syncing with students' math learning software, mentors can easily and effectively determine areas of student's need and access personalized resources. PL^2 ensures customized mentor support via continuous, on-the-job training and resource assistance in both math content skills and social-emotional and motivational teaching. Students, particularly marginalized students, benefit as they receive personalized resources based on their individual math learning performance, gain access to social-emotional supports, and receive resources attending to equity [4].

2.1   Key Features of the Personalized Learning^2 App

The PL^2 platform contains a mentor library of resources organized by competency. The PL^2 mentor resource library allows mentors to use the *Resource Assistant*, which assists mentors with finding the appropriate resource given a student's individual need (e.g., a student lacking motivation, a student struggling with a certain math skill). Mentors can view pinned resources for easy access and also create their own resources. Mentors can also add feedback and input into the AI-driven software system using the *Make a Reflection* feature and participate in asynchronous lessons by clicking the *Lessons* tab (See Fig. 1).

---

[1] http://personalizedlearning2.org/index.html.

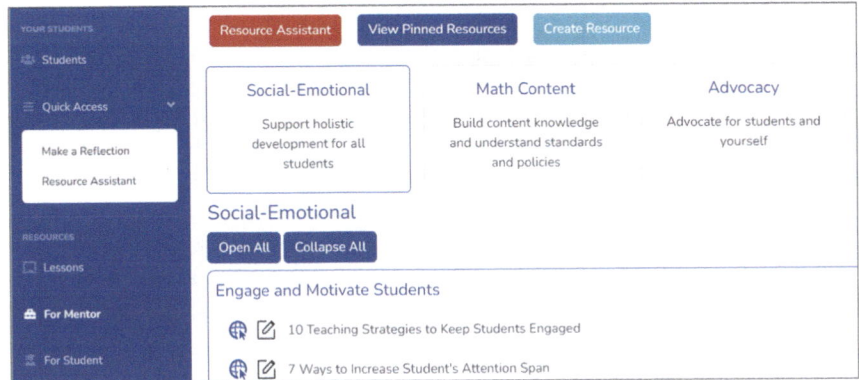

Fig. 1. The PL^2 resource library gives mentors access to the *Resource Assistant*, allows them to view pinned resources for easy access, and helps mentors to create their own resources. Mentors can also add feedback and input into the system using the *Make a Reflection* feature and participate in asynchronous lessons by clicking the *Lessons* tab.

Released three years ago, PL^2 contains both a web app for mentors and system administrators and another for students. Math learning software (e.g., ALEKS, MATHia, Dreambox) used by students is connected to the mentor platform to assist mentors with making decisions regarding recommended resources and goal setting. From the platform, mentors can see their student's performance and customize goals by adjusting student effort and progress goals based on math software performance and mentor observation. Methods of measuring student effort include the number of minutes spent using the EdTech, number of topics attempted, and similar measures unique to different EdTechs. Students are assigned two effort goals within PL^2 aligned to the amount of time and the number of lessons/topics attempted. Progress is determined by measuring student achievement via the number of problems correct or lessons completed demonstrating mastery development.

The status of each student (i.e., *Missed you*, *Ramp it up*, *Wow*, and *Keep it up*) is indicated by giving mentors information regarding a student's performance on both their effort and progress goals. Figure 2 displays the status of a group of students assigned to a particular mentor. A status of *Missed you* means a student did not meet their effort goals. A status of *Ramp it up* means they met their effort goals but did not meet their progress goal. The *Keep it up* status means the student met both their effort and progress goals, with students displaying a *Wow* status upon exceeding both goals.

Resources for both mentors and students are then recommended based on the student's performance toward reaching their goals and mentor feedback. Mentors give feedback by providing post-session reflections via a reflection form within PL^2. The form asks non-mandatory questions regarding the mentoring session (e.g., What's one thing that went well? What's one thing that could have gone better? What do you want to remember for your next session with the student?). The mentor is asked to choose an area where a student faces challenges (e.g., social-emotional health, math, relationship building). Finally, the reflection form requires mentors to rate the session based on a five-star rating system.

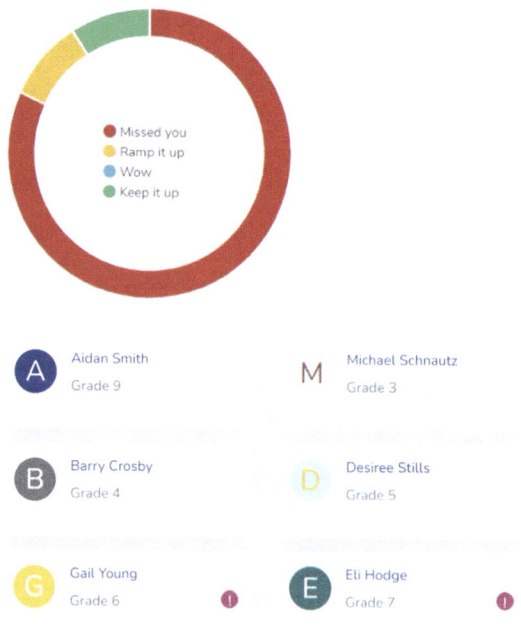

Fig. 2. The student status indicator shown on the PL^2 homepage detailing the status of students, giving mentors information regarding a student's performance on both their effort and progress goals. Although not all students' cards are shown due to space, from the circular graphic, it is apparent that the majority of students are in the *Missed you* status, meaning they have not met either of their effort goals.

Mentors can track their students' progress toward their goals and see individual and group progress in graphical form indicated by time spent using EdTech, number of problems correct, and percent accuracy with options varying based on EdTech used (See Fig. 3).

This information greatly assists mentors with seeing students' progress over longer periods of time (e.g., weeks, months). It also helps mentors identify trends in groups of students' progress, effort, and learning to assist mentors with personalizing the mentoring experience for students and recommending appropriate resources.

The PL^2 approach to mentoring provides support in two ways. First, through the use of human mentoring providing support in not only math content, but also social emotional learning, culturally responsive teaching practices, and relationship building. Students meet with mentors several days per week for one to two hours. This transcends the concept of tutoring into "high dosage" mentoring, an intervention found to produce significant growth gains among marginalized students [8]. Human mentoring occurs in conjunction with AI-driven software to recommend resources using student EdTech data and mentor input and feedback to provide AI-driven, personalized support for mentors. Second, the PL^2 approach is a platform for delivering and providing customized synchronous and asynchronous lessons to mentors based on individual need. The main

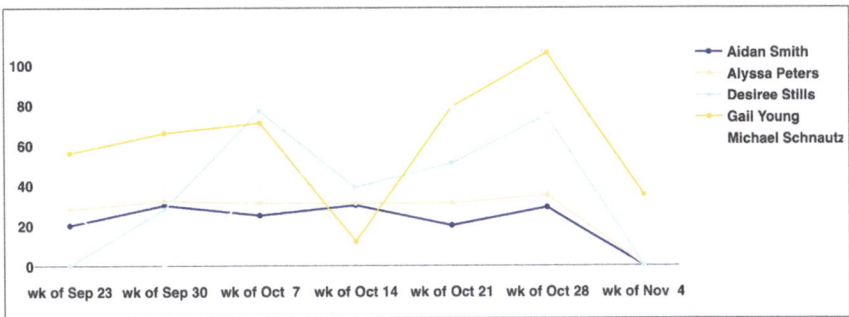

Fig. 3. A data report displaying the progress of five students assigned to a mentor within PL^2 showing the time spent using EdTech over several weeks. Mentors can use this data to identify trends and abrupt changes in student performance.

purpose of this present work is to showcase the topics and methods of creating these scenario-based lessons.

Preliminary work on the use of the PL^2 approach to improve student learning in math was conducted through the Ready to Learn (RTL) program [4]. RTL is an after-school math intervention program designed to reach marginalized students. In partnership with the University of Pittsburgh's Center for Urban Education, RTL focuses on educational equity by providing mentor on-the-job training in the areas of mentor bias awareness and providing culturally relevant teaching practices. Preliminary data found students (~80% Black) grew by an average of 6.8 scale points on the NWEA Measure of Academic Progress (MAP) test compared to a 3.6-point improvement by matched students not in the program. This impressive growth from 2019 to 2020 is higher than typical one-year growth in MAP scores (~5.5 points), despite pandemic-related challenges [4]. Further research is needed to determine the impact of using the PL^2 approach of using low-cost mentoring in synergy with computer-based tutoring. In addition, more research needs to be conducted on the influence of mentor on-the-job training and scenario-based lessons and the subsequent impact on student learning.

Many schools and out-of-school time organizations, such as the Boys and Girls Club and AmeriCorps, have partnered with PL^2, with over 150 mentors using the app. Due to the COVID-19 pandemic, both students and mentors have lower baseline knowledge and the gap for marginalized students has further increased [10]. Virtual mentoring is being used more often, with mentors needing support especially related to using technology. EdTech usage has increased, leading to the need for a singular data collection platform with one early PL^2 success of being able to pull in data from multiple types of software. Another challenge is the recent decrease in both quality and quantity of available mentors. With PL^2's mentoring organizations using a variety of implementation models (e.g., in-person, online, hybrid), particularly post-COVID, there is an increased need for development of asynchronous training that allows mentors to complete as-needed and on their own time [2].

3   Determining Competencies of Successful Mentorship

Upon discussing with our partners and feedback from mentors, there is an expressed need for mentor training in specific mentoring areas (e.g., How can mentors increase student engagement and motivation? How can mentors gain math content knowledge? How can mentors be more culturally responsive?). We have conducted preliminary research in examining the qualities of effective mentorship among our partners determining educational practitioners' attitudes toward different mentoring competencies in order to identify specific skills or competencies for which mentors need training most urgently [1].

3.1   SMART Supports Creation

Mentor resources in the PL^2 platform were organized by categories aligned with the appropriate mentor competencies. To determine appropriate mentor competencies, we synthesized from triangulation of competencies curated from educational wellness organizations, such as Danielson Group, CASEL, and KnowledgeWorks[2]. In addition, we received input and feedback from our PL^2 partners, which were all mentoring and tutoring organizations. We also surveyed 18 organization administrators to determine how important they felt each skill or competency was to successful mentorship [1]. A 4-point modified Likert scale was used with respondents rating each of the 14 competencies based on how important they felt the skill, ability, or behavior was for successful mentoring within their organization. The scale options for each randomized competency indicated with increasing importance were: (1) low priority, (2) medium priority, (3) high priority, (4) urgent. From these results, the competencies were organized by theme, or super-competences. We call these super-competencies SMART supports (i.e., Social-emotional, Math, Advocacy, Relationships, Technology). This preliminary work was submitted in a prior paper [1]; however, the results are an impetus for the creation of asynchronous and synchronous lessons which are the focus of this present work. Table 1 displays the 14 competencies organized into the schema of SMART supports— or super-competencies. *Engage and Motivate Students* received the highest rating (3.7) and *Understand Educational Policies and Norms* received the lowest (2.2). The average priority rating is shown in the first parenthesis, and the number of mentor-specific resources within the PL^2 app is shown in the second parenthesis indicated in italics. There is a lack of resources located in the following mentor competencies: *Demonstrate Awareness of Bias, Practice Self-Care, Manage Learning Environment, Stay Organized.*

This novel method of categorizing mentor competencies into SMART supports calls attention to the *Advocacy* super-competency with *Demonstrating Awareness of Bias* and *Practice Self-Care* containing zero resources. *Demonstrate Awareness of Bias* had one of the highest ratings (3.4) according to PL^2 partners, and there is a need for more resources related to *Use of Culturally Responsive Teaching Practices,* which had a high priority average rating of 3.5 and 3.4. The dedication of PL^2 to attending to equity and the expressed need from our PL^2 partners for training and resources related to these competencies sparked creation of lessons within the PL^2 app [1].

---

[2] https://danielsongroup.org/, https://casel.org/, https://knowledgeworks.org/.

Table 1. SMART supports illustrating the schema developed from the survey data organizing the 14 competencies into main themes. The first number in parentheses indicates the average priority rating. The second number in parentheses (indicated in italics) is the current number of mentor resources located within the PL^2 app. Notice the lack of resources within the *Advocacy* super-competency.

Social-Emotional	Math content	Advocacy	Relationships	Technology tools
Engage and motivate students (3.7)(32) Foster independent learning (3.5)(3) Apply Social-emotional learning Practices (3.3)(12) Manage learning environment (3.1)(0)	Demonstrate content Understanding (3.4)(19) Understand educational policies and norms (2.2)(5)	Demonstrate awareness of biases (3.5)(0) Use culturally responsive teaching practices (3.4)(9) Practice Self-Care (2.9)(0)	Build relationships with students (3.5)(7) Personalize learning (3.4)(4) Communicate with caregivers (2.5)(10)	Use technology effectively (3.0)(17) Stay organized (2.7)(0)

4   Scenario-Based Training

The PL^2 mentor platform contains mentor-specific resources, student EdTech data, and scenario-based training organized by mentor SMART supports. Deliberate practice of scenario-based learning mimicking workplace fundamentals (e.g., mentors working with students) are at the forefront of mentor support. Scenario-based training has shown to be an impactful method of achieving learning gains by providing simulated experience and deliberate practice in a safe environment [13]. Deliberate practice is difficult and requires mentors to persevere in the learning process. Fortunately, focused and deliberate practice can yield substantial rewards—large growth gains [5].

Within the PL^2 platform, we offer both synchronous and asynchronous lessons. Synchronous lessons are presented live to mentors as interactive presentations containing live links and videos housed within *Lessons*. Currently, upfront lessons provided include: *Strategies for Effective Tutoring & How PL^2 Supports Them, Student Engagement & Motivation Strategies Using PL^2, Introduction to Teaching Math and Gaining Math Support*, and *Using Culturally Responsive Teaching Practices*. From the feedback received from PL^2 partners, we are expanding our synchronous lessons offered and are currently creating on-the-job, in-person lessons related to demonstrating an awareness of bias and strategies for fostering independent learning. Asynchronous lessons are currently being added with a focus on areas indicated as high priority according to our partners surveyed and feedback showing expressed areas of need (e.g., *Demonstrate Awareness of Bias, Build Relationships with Students, Engage and Motivate Students*). This work focuses on the development and structure of mentor asynchronous training modules.

4.1   Asynchronous Scenario-Based Training

Asynchronous scenario-based lessons have been created for continuous, on-the-job support that mentors can complete on their own time and at their own pace. Some asynchronous lessons already created include: *Supporting a Growth Mindset, How to Present
the Lesson to Students: Framing Task Difficulty, Using Intrinsic and Extrinsic Motivation Strategies* [3]. Recent plans upon receiving survey feedback from partners include
creating more asynchronous lessons related to the following competencies: *Demonstrate
Awareness of Bias* and *Manage the Learning Environment*. A segment from part of an
asynchronous, scenario-based training titled *Reacting to Errors,* instructing a mentor
on how to best respond to a student making errors while being cognizant of student
self-efficacy, is shown in Figs. 4–6. This module uses an introductory training scenario,
walking a mentor through predicting and explaining the best mentor response. Then,
they observe the research-based recommendation and explain their thoughts of applying
the research-driven recommended strategy. Finally, as a post-assessment and to provide further formative instruction, the mentor completes the cycle again with a transfer
scenario to determine their ability to apply their learning from the initial training scenario. The constructed-response question in Fig. 4 asks mentors to predict or choose
how they would approach the student in recognizing his mistake while attending to his
self-efficacy. This lesson aligns with the competency of *Engage and Motivate Students*
within the *Social-Emotional* SMART support (Figs. 5, 6).

1. Imagine you are a mentor to a student, Aaron, who has a long history of struggling with math.
Aaron is not particularly motivated to learn math. He just finished a math problem adding a 3-digit
and 2-digit number and has made a common mistake (shown below).

Briefly explain how you would approach Aaron in recognizing his mistake and be sure to the
consider the impact of your comments on Aaron's self-efficacy, that is, his belief that he can learn
math.

Fig. 4. A segment of the *Reacting to Errors* asynchronous lesson showcasing the initial training
scenario asking mentors to predict or choose how they would respond to the given scenario while
attending to the student's self-efficacy.

2. With respect to Aaron's mistake, which of the approaches below do you think would be best to correct Aaron's mistake, and subsequently, improve his self-efficacy, or belief that he can learn?

○ [A] I would ask him to walk me through his thinking/reasoning, and continue asking questions to try to get him to see the error himself.

○ [B] I will first praise him for the effort he made, then will tell him to remember to carry the 'tens' place each time we add a column together. Also, I will explain his error, by stating, 'We also have to carry the 1 up to the tens-column, that way we have all our numbers ready to add in the tens-column.' I will ask him to try the problem again but this time remembering to carry.

○ [C] I will say to Aaron: 'What do you get when you add the ones column (if he doesn't know which is the ones I might point it out or say 8 plus 8)? Good, 16. You put 6 here, where do you put the 1? Yes, you have to carry it here to the tens column. So then what number belongs here (pointing to the 2)?'

○ [D] I will let Aaron know that the answer is incorrect, and show him the right approach.

3. Why do you think that the approach you selected in Question 2 will best promote Aaron's self-efficacy, or belief that he can learn math?

Fig. 5. A segment of the *Reacting to Errors* lesson within the training scenario displaying the selected-response question intended for the mentor to predict or choose the best approach, followed by a constructed-response question asking them to explain.

Studies have shown that the way tutors intervene when students make mistakes or show misconceptions during learning activities can contribute to strengthening or weakening the student's self-efficacy. According to experts the intervention/approaches in: According to these studies, option [C] (shown below) would be the best to support a low motivated student and boost their self-efficacy.

[C] I will say to Aaron: 'What do you get when you add the ones column (if he doesn't know which is the ones I might point it out or say 8 plus 8)? Good, 16. You put 6 here, where do you put the 1? Yes, you have to carry it here to the tens column. So then what number belongs here (pointing to the 2)?'

4. Why do you think experts recommend the approach in [C] to support a student with low self-efficacy, or belief that they can learn?

Fig. 6. A segment of the *Reacting to Errors* lesson within the training scenario revealing the research-recommended strategy for the formative training scenario. Mentors are then asked why they think the research-recommended response is suggested by experts as both a constructed-response question and then as a selected-response (not shown).

Mentors complete the training scenario, which concurrently involves deliberate practice and applying the research-recommended strategy to the initial training scenario. Then, as a post-assessment quantifying their learning gains, they participate in a similar transfer scenario (e.g., predict the best response, explain your response, observe the expert-recommended strategy, explain the expert-recommended strategy). The comparison of mentor performance from the initial scenario (training scenario) to the final scenario (transfer scenario) is used to determine mentor learning gains. The transfer scenario uses a similar scenario and follows the same predict-explain-observe-explain pattern of inquiry. The inquiry model (i.e., predict-explain-observe-explain) of asking mentors to respond to a given scenario and apply the research-recommended strategy for both the training and transfer scenarios has been developed to give unskilled or under-experienced mentors the situational experience needed to be successful. In addition, mentors receiving feedback, particularly corrective feedback, throughout the lesson is associated with high learning gains [15]. We are using preliminary data from the constructed-response questions to create more authentic selected-response question options. The use of pilot mentor response data from the constructed-response questions to optimize the selected-response question options is modeled from Wang et al. [14]. The creation of more authentic and "real life" mentor responses for selected-response questions from constructed-response data allows us to capture common misconceptions among mentors and use them as "less-desirable" or "incorrect" answer options. This increases not only the validity of the mentor training as an accurate assessment of mentor understanding, but also scalability through the use of selected-response question types.

5   Discussion & Conclusion

Currently, the PL^2 platform is transitioning from a primarily mentor-specific resource app to a holistic system. We are providing not only personalized resources, but also dual functionality as a training platform for mentors. Since initial submission of this article, we have repackaged the PL^2 platform into a system of three focused solutions. The first solution is PL^2 *Training,* which contains the resource library discussed previously and also houses asynchronous lessons for on-demand and as-needed access by mentors. At the time of press, we have created a dozen asynchronous lessons that were rated high-priority by competency and/or have few resources located within the mentor library (i.e., *Foster Independent Learning, Demonstrate Awareness of Biases, etc.*).

The second solution, PL^2 *Toolkit,* repackages student EdTech status features, goal setting, and recommended resources allowing mentors to personalize the learning experience for their students. Within the *Toolkit*, mentors can get a quick snapshot of a student's progress with data from both EdTech software and embedded tools for reflections and goal setting. Presently, we are seeking out ways of adding more AI-driven features to increase personalization of resources, such as using natural language processing to assess open-response questions for purposes of scaling with our nationwide partners.

The last focused solution in the PL^2 solution trifecta is the *Tutoring Corps.* The *Tutoring Corps* is a team of trained mentors ready to go on-demand for as-needed mentoring support for our partners. Recent changes due to ramifications of the COVID pandemic have caused a nationwide shortage of qualified teachers and mentors. PL^2 has approached this obstacle by developing a strategic plan in recruiting quality mentors. Based on a model proposed by Kraft and Falken [8] in *A Blueprint for Scaling Tutoring and Mentoring Across Public Schools*, we designed a mentor recruitment plan targeting undergraduate college students. Recruitment of college students is often easier because of their need for volunteer hours for graduate applications, internships, and resume building. As a benefit to organizations, college students are often unable to accept compensation. In addition, aside from volunteer work, college students can receive mentoring certifications, participate in training, gain valuable experience working with children, and contribute to the community. Necessary mentor qualifications include being English speaking; possessing quality reading, math, and writing skills; and having a willingness to work with children. Fortunately, many college students today come with math competence and an ability to tutor up through middle school, oftentimes matching in math skills to trained mentors.

We are also working on several functionality advances, such as a student-facing PL^2 version allowing students to directly access resources, monitor their own goals, and directly message mentors. The latter feature creates a method of asynchronous, as-needed communication between mentor and student to improve the efficiency and accessibility of supportive tutoring for students.

Future work consists of surveying more partners in differing mentoring roles (e.g., supervisors, mentors, coaches) to determine their perspectives regarding mentoring competencies. This will assist us with the creation of asynchronous lessons based on mentoring roles and organization types (e.g., public school, tutoring organization, support network). Moreover, research is being conducted on our asynchronous lessons by analyzing the pre- to post-instruction learning gains determined by the selected-response and constructed-response data by comparing mentor performance of training and transfer scenarios [3]. Some of the research questions we are investigating include: What differences exist among mentors' perceptions of learning and their demographics (e.g., race, gender, age) and also self-reported level of experience? Do mentors reporting a high level of experience perform better? Do mentors learn new mentoring strategies and skills from short scenario-based lessons? This last question is of particular importance with preliminary evidence supporting vast improvements in mentor learning as a function of our scenario-based mini lessons [3].

Lastly, the high cost of one-on-one coaching programs makes it hard to scale [8, 11]. The PL^2 platform is working towards enabling lower-cost mentoring without sacrificing learning quality. The use of human-computer teaming for on-the-job support offers a lower cost option for deliberate practice using scenario-based training to increase the impact and learning capacity of mentors in the workplace.

Acknowledgments. This work is supported with funding from the Bill and Melinda Gates Foundation, Chan Zuckerberg Initiative (Grant # 2018–193694) and the Richard King Mellon Foundation. Any opinions, findings, and conclusions or recommendations expressed in this material are those of the authors.

References

1. Authors. An evaluation of perceptions regarding mentor competencies for technology-based personalized learning. [Manuscript submitted for publication] (2022)
2. Authors. Can we pivot? a human-ai mentoring program's response to today's tutoring challenges. [Manuscript submitted for publication] (2022)
3. Chine, D.R., Chhabra, P., Adeniran, A., Gupta, S., Koedinger, K.R.: Development of Scenario-based mentor lessons: an iterative design process for training at scale. In Proceedings of the Ninth ACM Conference on Learning@Scale (2022)
4. Chine, D.R., Brentley, C., Thomas-Browne, C., Richey, J.E., Gul, A., Carvalho, P.F., Branstetter, L., Koedinger, K.R.: Educational Equity Through Combined Human-AI Personalization: A Propensity Matching Evaluation. In International Conference on Artificial Intelligence in Education. Springer, cham. (accepted) (2022)
5. Duckworth, A.L., Kirby, T.A., Tsukayama, E., Berstein, H., Ericsson, K.A.: Deliberate practice spells success: Why grittier competitors triumph at the National Spelling Bee. Soc. Psychol. Pers. Sci. **2**(2), 174–181 (2011)
6. Ellison, R.L., Cory, M., Horwath, J., Barnett, A., Huppert, E.: Can mentor organizations impact mentor outcomes? Assessing organizational norms on mentor intent to stay and willingness to "go the extra mile." J. Community Psychol. **48**(7), 2208–2220 (2020). https://doi.org/10.1002/jcop.22391
7. Guryan, J., Ludwig, J., Bhatt, M.P., Cook, P.J., Davis, J. M., Dodge, K., Steinberg, L.: Not Too late: improving academic outcomes among adolescents (No. w28531). Natl. Bur. Econ. Res. https://www.nber.org/system/files/working_papers/w28531/w28531.pdf (2021)
8. Kraft M.A., Falken G.T.: A Blueprint for scaling tutoring and mentoring across public schools. AERA Open. **7**(1). 1–21. https://doi.org/10.1177%2F23328584211042858 (2021)
9. Nickow, A., Oreopoulos, P., Quan, V. The impressive effects of tutoring on PreK-12 learning: a systematic review and meta-analysis of the experimental evidence. Working paper 27476. Natl. Bur. Econ. Res. (2020)
10. Lewis, K., Kuhfeld, M.: Learning during COVID-19: An update on student achievement and growth at the start of the 2021–22 school year. NWEA (2021)
11. Oreopoulos, P., Petronijevic, U.: Student coaching: How far can technology go? J. Hum. Resour. **53**(2), 299–329 (2018)
12. Schaldenbrand, P., Lobczowski, N.G., Richey, J.E., Gupta S., McLaughlin, E.A., Adeniran, A., Koedinger, K.R.: Computer-Supported Human Mentoring for Personalized and Equitable Math Learning. In International Conference on Artificial Intelligence in Education. Springer, Cham. (pp. 308–313) (2021, June)
13. Thompson, M., Owho-Ovuakporie, K., Robinson, K., Kim, Y.J., Slama, R., Reich, J.: Teacher Moments: A digital simulation for preservice teachers to approximate parent–teacher conversations. J. Digit. Learn. Teach. Educ. **35**(3), 144–164 (2019)
14. Wang, X., Rose, C., Koedinger, K. R. (2021, May). Seeing beyond expert blind spots: online learning design for scale and quality. In Proceedings of the 2021 CHI Conference on Human Factors in Computing Systems (pp.1–14). https://doi.org/10.1145/3411764.3445045
15. Wisniewski, B., Zierer, K., Hattie, J.: The power of feedback revisited: A meta-analysis of educational feedback research. Front. Psychol. **10**, 3087 (2020). https://doi.org/10.3389/fpsyg.2019.03087

Community Education and Diversity in Digital Contexts: Curricular and Empirical Perspectives

Christoph Knoblauch[(⊠)] [iD] and Anselm Böhmer[iD]

Ludwigsburg University of Education, Reuteallee 46, 71634 Ludwigsburg, Germany
`{christoph.knoblauch,boehmer}@ph-ludwigsburg.de`

Abstract. As many higher education institutions have begun to implement matters of community education within their missions, teachers and students try to connect universities, schools, and neighborhoods to develop civic capacity and foster democratic citizenship. Focusing on the collaboration of communities and educational institutions in the context of diversity and digitalization, this paper discusses (a) international curriculum modules and (b) findings from a digital project-based course in the higher education sector. The curriculum analysis discusses principles for community education from an international and interdisciplinary perspective. The qualitative empirical perspective focuses on students' attitudes and experiences towards community education in the context of digitalization and diversity. The interviewed students autonomously designed and implemented a project, connecting the Ludwigsburg University of Education (Germany) and social spaces, thus, linking their academic studies with social inquiry. Against this backdrop, this paper combines curriculum analysis and students' experiences in community education, discussing the potentials and challenges of digitalization and diversity in universities, schools, and neighborhoods.

Keywords: Community education · Digitalization · Diversity education · International curriculum analysis · Higher education sector · Empirical qualitative evaluation

1 Introduction

This paper analyzes and discusses curriculum perspectives[1] and students' learning experiences in community education in digital and diverse settings in the higher education sector. It reports on the development, implementation, and evaluation of a course in the Master's program "Teacher Education"[2] at Ludwigsburg University of Education (LUE) in Germany. Participants were students enrolled in teacher education programs at LUE in the summer term of 2021. The course discussed shows a mix of synchronous and asynchronous course sessions [1]. In addition to this, the participating students reserve

---

[1] In this context, the curricula for elementary and high schools in the state of Baden-Württemberg are discussed, as they serve as a basis for teacher education at the Ludwigsburg University of Education.

[2] https://www.ph-ludwigsburg.de/7684+M5054de7a952.html (access on 2021/4/8).

© The Author(s), under exclusive license to Springer Nature Switzerland AG 2023
D. Guralnick et al. (Eds.): TLIC 2022, LNNS 581, pp. 593–607, 2023.
https://doi.org/10.1007/978-3-031-21569-8_55

several weeks within the semester for a project-based, autonomous development and execution of a small-scale project in the field of community education. The focus of this paper is to discuss possible potentials for community education and diversity in digital environments to present results on the possible impacts of digital learning scenarios. To facilitate an insight into the learning experiences of students in the described digital settings, the paper presents a qualitative study focusing on the reflections of students towards personal learning experiences in digital contexts. By doing so, the study assesses current practice and analyzes the course mentioned in detail.

The discussion of fundamental pedagogical characteristics [2] and curriculum principles that influence learners' experiences, combined with the perspectives of a digital environment, leads to a theoretical framework for the development and analysis of the qualitative interviews [3]. The paper displays the theoretical framework of this study (2) with regard to education in general and community education in particular. Having presented the theoretical approach, this paper then gives insight into the discussed course at LUE (3). After this, the study conducted is presented with its design (4) and its core findings (5). The discussion of these findings (6) reveals the importance of community education for pre-service teachers as well as for the University itself.

2 Theoretical Aspects of Community Education

2.1 Education

As education is a wide field of theory, interpretation and practice, here we want to concentrate on its relevance for fostering democratic citizenship [classical: 4]. As recent research has shown, it is of increasing importance for many institutions of higher education to gain knowledge that helps address problems dealing with issues of everyday life. By this, universities are more successful to optimize their social and societal outcomes [5]. One core factor of higher education regarding societal impact is motivation and effectiveness in a learning environment that promotes students' autonomy. Here, specific and learner-adopted materials are important, they also need to be related to the given cultural situations and contexts [6].

As a special condition of learning, the power of economic and managerial think-ing is introduced—to society and university as well [7]. To widen this economic-only assessment in higher education, educational standpoints are required by integrating alternative approaches within universities. Regarding this mission, different forms of education are defined: building a learning community, co-creating knowledge also from subaltern positions, opening spaces for different intensity of participation, de-centering Western ideas and knowledges, focusing the cycle of critical thinking—reflection—action and connecting in virtual spaces [8]. Thus, education for emancipatory citizenship means deconstructing ongoing epistemologies and bringing their consequences in contact with organizations and actors of civil society. Education understood as promoting students' inclusion in the academic field as well as in the community's challenges, therefore, becomes a more tentative and experimental, but also empathetic and situated process of developing knowledge [9] with focus on ethnographic research [10].

As slightly mentioned before, higher education at universities is under the expectation of combining its knowledge outcomes with 'everyday world challenges'. To achieve this combination, problem-oriented research and teaching are common educational practices. The goal then is to connect education with community or with industry and economy [11]. Here, challenges of interdisciplinarity and divergent interests within the common projects may occur.

A crucial role in connecting higher education with its social or economic environment play professionals who can combine knowledge work at university with challenges of the surrounding communities and companies [12]. These professionals also lead students' learning activities in direction of the surroundings and support organizations of higher education in executing their third mission for society and further stakeholders.

By realizing, that the COVID-19 pandemic urged organizations of higher education to bring new solutions to the table, digital learning became of increasing relevance within a very short time [13]. But not only new opportunities arose by the ongoing digitalization of higher education. Also new or just more intense challenges occurred such as issues of connectivity, access to infrastructure in general or didactical competencies. Hence, inequalities in education were rather increased with regard to learners, teachers, organizations, and national education systems [14].

As a consequence, education not only is a question of competence, access, or individual giftedness. It also is to be addressed as a structural, infrastructural, and methodological challenge for all participants in their diverse social positions and labeling.

These theoretical approaches reveal an intense connection of the third mission in organizations of higher education and their community-related engagement. Thus, reflective pedagogy and didactics of higher education find their consequences in an inclusive and dialogical connection with its spatial and social environment, its organizations, and core actors and, thus, leads to a form of caring pedagogy and cognitive compassion [15]. This means, that special forms of education and collaboration result from such an educational concept as they highlight social spaces as physical and/or digital networks of individuals, groups, or organizations. What this means in detail, especially for bridging the gap between different social spaces, will be shown in the following paragraph.

2.2 Community Education

After displaying higher education concerning universities' third mission and challenges of digital learning, a further question of this paper is the relevance of community education for academic learning.

The first aspect of community education is the issue of "models" and definitions [16, regarding international collaborations]. Care should be taken using those general approaches and explanations, as community education is a particular form of the aforementioned tentative and experimental, empathetic, and situated processes of developing knowledge and evaluating it. Thus, reflexivity is needed in the theoretical and empirical work of community education [16].

From a perspective of human capital, collaboration in the spatial surrounding of universities leads to gaining more of this capital, improves knowledge and its production, links research and business, and supports academic entrepreneurship [17]. Although these arguments are prominent in the governance of higher education, it should be noted that this economic perspective needs to be supplemented by a social approach that demonstrates the social and societal impact of higher education [18].

A unique form of interacting with the local communities is service-learning. Here, actors from the university's spatial environment bring in their own experiences, knowledges, and perspectives. At the same time, academics and their students learn to interact and to apply their research findings to local challenges [19]. Thus, "the university students have an opportunity to practice what they learn in the classroom while the adult learners [in the community] benefit from their knowledge and skills." [20, p. 208].

Regarding this local connection, special attention needs to be given to students' experiences before they meet their local partners. As in many cases, university students are middle-class members, they need to foster further dispositions that connect to everyday experiences within the social spaces of local residents [20, p. 211], as well as those residents, should be involved in research projects to better understand and use the outcomes of those collaborations [21].

One main concern here is to bridge the gap between education and work [22], especially pre-service teachers who might benefit a lot from service-learning projects within the local communities. Thus, students might better understand their future pupils' everyday challenges and tasks but also adopt questions and experiences from here into their further studies. One important aspect of this approach into the local community is peer-learning as a learner-centered form of interacting with students' own experiences, their involvement in the community, and its challenges. But also, common questions within the community are found so that another form of peer learning is realized [23].

Another aspect needs to be mentioned: as digital learning becomes more important in post-pandemic times, community education by academics—scholars and students—needs to be aware of ways creating relationships, engagement, and identities in digital environments of their organizations as well as their local communities [24]. Thus again, local and situated aspects need to be considered in practicing this form of community education to reach the locals, their knowledge, and also their challenges [regarding indigenous education, 25].

For collaboration of university and community, this means that scholars and students have to face "multiple roles and responsibilities housed within one position that spanned two contexts" [26, p. 46]. Academics then need to establish their role in both contexts and manage the intersections of both.

A specific view on this doubled engagement of academics in community education was offered by Furco [27] and his concept of the *Engaged Campus*: "No longer is community-engaged work seen as something that fulfills only the public service and outreach component of higher education's overarching mission. Rather, public engagement serves all parts of the tripartite mission, including facilitating institutions' achievement of their research/discovery and teaching/education goals" [27, p. 381]. Under this perspective, community education of universities' staff and students is not only a "nice to have" or a necessary "add on," but a concise operationalization of higher education's missions in general.

Working like this might lead to academics as community workers who find new ways to realize informal learning processes [28]. This not only means another structural widening of higher education and its fields, but also a change of education concepts – from just individual learning to community knowledge and its individual and structural impact [29].

   To summarize these suggestions on education, community education of universities in general and pre-service teachers in particular means collaborating with local groups and actors not only for reasons of knowledge transfer but also for establishing wider and—at the same time—more specific competencies for academics and residents. Important are peers, structures, and opportunities to build collaboration networks. If this is possible, the university is not only engaged in its own fields of producing and distributing its knowledge but building collaborative knowledge in joint engagements within local communities. This is of particular relevance for pre-service teachers, as they need to bridge possible gaps between their class of origin and other social classes in way of their own (digital) learning and future teaching.

2.3 Community Education and Diversity—Curriculum Perspectives

The described collaboration with local partners, and the mutual benefits of it, are widely discussed in the Canadian school and higher education sector [30]. Canada's schools and universities have a long history of community education and emphasize the importance of partnerships between education organizations and communities in various contexts [31]. Therefore, Canadian Curricula serve as an international perspective in this discussion. Ontario's schools for example describe themselves as "community hubs where all people can gather to learn and participate in a range of activities offered by community organizations" [32]. Against this backdrop various partners within the community are discussed as enrichment for learning, offering valuable support for teachers and learners. It is noteworthy that the Canadian understanding of community education involves many different partners on different levels in different social spaces. Diversity plays a major role in these collaborations: "These partners may include conservation authorities; provincial and national parks; service providers such as fire departments and social service agencies; non-governmental organizations; museums and historical societies; First Nation, Métis, and Inuit friendship centers; veterans groups; cultural centers and other community organizations; and businesses" [33, p.18]. Interestingly the Ministry of Education (Ontario) emphasizes the mutual benefits of community education: Not only the schools but also the communities gain by collaborating in various projects such as skill competitions, career days, or ceremonies. Within these processes and collaborations students are enabled to develop a sense of place by investigating various social spaces: physical, social, cultural [33, p.11].

   From a national perspective, on the other hand, German curricula of the federal state Baden-Württemberg were chosen. Therefore, 61 subjects in total could be collected and analyzed: 17 for Elementary School, 37 for High School, general (without Gymnasium),

7 for Schools of Special Needs Education. Keywords and their lemmata[3] of analysis were community, network, social, social space, space, territory.

A first overview for the distribution of the keywords is given here (Table 1).

Table 1. Overview of curricula (percentages may contain discrepancies due to rounding differences).

	Community	Network	Social	Social space	Space	Territory
Curricula of Elementary School						
In total	134	16	92	0	368	0
In % of ES	22%	3%	15%	0%	60%	0%
Curricula of High School, general (without Gymnasium)						
In total	241	524	757	0	707	23
In % of HS	11%	23%	34%	0%	31%	1%
Curricula of Schools of Special Needs Education						
In total	297	87	367	10	623	7
In % of SN	21%	6%	26%	1%	45%	1%
Curricula of all analyzed school types						
In total	672	627	1.216	10	1.698	30
In % of all	16%	15%	29%	0%	40%	1%

Terms were found that described physical spaces of social interactions – but also in a metaphoric form—e.g., number space, time space,[4] resonance space. Some terms were used in different sectors of their semantic fields, e.g. network is used for social interaction and communication hardware. This makes it sometimes difficult to identify precisely strategies and mindsets the analyses looked for. But at least the intensity of using a specific term and its underlying idea could be determined. So, the numbers do not precisely indicate how often community education was in the mind of the curricula's authors. Still, they obviously had a mindset that constructed reality by these keywords, metaphors, and structures.

Because of such "semantical noise," this analysis cannot define the number of terms that refer to community education in the analyzed curricula. But it makes prominent how pedagogical concepts with relation to community education are presented—and with further semiotic and hermeneutical analyses, more about community relation might be found. Therefore, further research than presented here is needed. But in particular, one

---

[3] In their German version. Terms were translated as Gemeinschaft – community, Netzwerk – network, sozial – social, Sozialraum – social space, Raum – space, Gebiet – territory. The keywords and lemmata were documented as far as they marked not only names but content. To illustrate, "*Gemeinschaft*sschule" was not noted for the term community, but "die *gemeinschaft*lichen Einsichten.".

[4] This term was also recorded because of his structuring effects in social space.

can find definite relations to community education with regard to the term social space. Whenever it is used, a perspective of community relation, education, or work is taken in the curricula.

In view of this theoretical basis (2.1–2.3), the themes (1) "Partners in the communities", (2) "Relationships in Community Education", and (3) "The role of schools and universities" should be discussed in the qualitative study. Additionally, the students' reflections upon the term (4) "Community Education" and their (5) "Experiences with Community Education" can lead to a better understanding of the role of Community Education. These categories are discussed under the perspective of (6) "Digital Environments and Diversity" to focus crucial challenges of today's societies.

3 Community Education and Diversity in Digital Contexts: Description of the Course Design

A project-based Master's course serves as the basis of this study and was conducted at LUE in the summer semester of 2021. The main foci of the course are (a) the discussion of project-based learning, (b) the autonomous implementation of a small-scale project in the field of education, and (c) the presentation and reflection of these projects. The character of the course aims for the interconnection of students with partners in the community and their future work areas through projects in the field of education. The course itself was conducted digitally: approximately 50% of the digital course sessions were conducted in a synchronous way using mostly videoconferences whereas the other 50% of the course were carried out in a digital asynchronous way, using learning management systems such as Moodle. The synchronous session mostly offered guided discussions on the asynchronous content and the planning and implementation of projects.

The videoconferences were based on students' questions and reports offering group discussions and individual tutoring in break-out rooms. The asynchronous sessions offered a blend of learning arrangements: readings, podcasts, audio presentations, interactive forums, videos, and chats were used. To these ends, (a) apt presentations with audio commentary were created; (b) a podcast with experts was offered; (c) common topics were deepened through a selection of pertinent literature; (d) videos of professors and students working in project-based settings were provided; and (e) digital forums were established to offer an ongoing interactive exchange of ideas [34]. The project-based focus of the course aims at helping students to connect with partners in the community and to develop an understanding of the potential of community education and diversity in the context of their studies. Furthermore, students can gain experience in digital settings by actually implementing a project, in digital and face-to-face settings, in the field of education. Students can gain up to three credit points (as defined by the European Credit Transfer and Accumulation System; ECTS) for the course and can use their projects as a basis for their Master's Thesis.

4 Empirical Design of the Study

The study focuses on (1) curriculum perspectives and (2) the reflections and discussions of the participating students of the course. The (1) analysis of the curriculum discusses conceptual and practical impulses, whereas the (2) qualitative interviews focus on individual experiences with community education in the described learning contexts. Therefore, the study uses (1) methods of content analysis [35] to identify crucial factors for community education in the curriculum and (2) methods of qualitative interview research in digital contexts. The complex research focus—community education in digital and diverse environments—asks for an innovative approach observing multiple perspectives through (a) curriculum analysis, (b) synchronous dialogue-based interviews, and (c) individual asynchronous feedback. Semi-structured qualitative interviews were carried out in a (a) synchronous digital way, using video calls and (b) in an asynchronous digital way. The digital asynchronous way of interviewing students is a rather new technique in qualitative research, which encourages respondents to reflect on their answers by allowing them to structure their ideas and responses beforehand [36]. The (a) synchronous interview situations offered the possibility to discuss questions in deep, clarify ambiguities, and develop a constructive dialogue between the interviewer and the participants. In these processes, the study follows established structures of qualitative research and analysis in manifold ways [37].

The (b) asynchronous way offered the possibility to record answers as audio files independently. The participants could take as much time for reflection as they individually considered appropriate. The completed audio files were then sent to the research team via a digital transfer system [36].

Using this combined method, the study looks for data, which offers a comprehensive view through curriculum analysis, and individual, reflected, subjective feedback [38].

5   Findings of the Study

The qualitative study focuses mainly on learners' experiences with Community Education and Diversity in digital contexts. The analysis and discussion of the data show different combined categories, which are based on the deductive categories and new inductive impulses found within the data. The findings are structured and discussed according to these categories. Several answers and reflections show links to more than one category and are therefore discussed in various contexts. Experiences in digital environments and with diversity are discussed in all categories and a special focus is given to these experiences in the category "Digital environments and diversity."[5]

5.1   Reflection of the Term Community Education: "THE Complete Surroundings of an Individual..."

As a first step the term "Community Education" is discussed by the participating students: *"The complete surroundings of an individual. The school and also private surroundings and different people we know. Maybe also hobbies. Basically, the complete life of the individual as a related network."* The interviewed students also discuss social spaces and their physicality, networks of individuals, families, and friends. When some

---

[5] These findings are partly discussed on the basis of the study "Experiential learning in digital contexts" [39].

of these diverse spaces are connected, potential for education arise. Against this background, relationships seem to play a major role: *"Interconnection which exist within local spaces through social encounter and interaction. Interpersonal relationships and the resultant interconnections."* Community education has the potential to bridge gaps between diverse social spaces, as encounters and relationships play a crucial role. Within the project-oriented digital settings, students had various possibilities to experiment with different settings and forms of community education. The students mostly used a mix of digital and face-to-face interaction[6] within their community projects and mostly appreciated the autonomous organization of this mix.

5.2 Experiencing Community Education: "…DIfferent Areas Came Together."

Students' experiences play a major role in all projects as students experience project-based collaboration with different partners in schools and communities. The various collaborations led to experiences of membership when a group of people worked together with a common goal for a certain time. It shows that the students developed a broader understanding of collaboration, diversity, and community while working on their projects: *"The whole social environment, whether it be friends, work or family…diverse areas came together in the projects: such as school, hobbies, and family."* Furthermore, students report intense learning experiences, especially because learning in the context of community education *"…is not one-sided but overreaching."* Students report about intersections of roles and contents while they were working in community contexts: *"I made the experience that one can connect many different spaces."* Additionally, the nature of learning in community education seems to have specific traits: *"…especially social learning, communication and encounter can take place. Learning is expanded."* Students also experienced that their projects developed within the community almost independently: *"Some of the participants kept on meeting after the project. This developed independently and took place without me. Another connection which resulted from the initial project."* Especially the development of relationships within and besides the projects are described by the students in manifold ways. Experiencing that relationships within the community developed on the ground of the students' projects seems to be especially impressive for some students. However, students also experienced setbacks, especially in the early stages of their projects: *"It was difficult to initiate the project. Especially the development of connections within the community was a problem."* In this context, some students discussed the importance of structuring projects beforehand: *"If you want to establish something like this (a project in community education) it has to be well structured and planned. You have to define goals connected to areas of the education organization."* It is noticeable that most collaborations were established digitally, as this seems to be easier than face-to-face. However, students do not discuss this fact in detail. On the contrary, it seems almost conventional to the participating students to use digital ways for establishing collaborations.

---

[6] Face-to-face interaction in this article means analog, physical collaboration.

5.3   Partners in the Community: "...MY Project Affected More and More Social Spaces."

The participating students connected various partners in the community, thus bridging the gap between diverse social spaces. The examples show that students included different individuals, groups, and organizations in their projects: *"...I also realized that it (my project) affected more and more social spaces."* Within these spaces, different types of diversity, such as age, educational background, and belief, played a role. While some students connected for example kindergartens and senior homes, others worked with schools and social institutions. The collaboration with different partners led to a discussion of social competencies of individuals and how people can contribute to and benefit from community education. Networks of individuals developed, bridging the gap between social spaces: *"It can be substantial for personal development to look beyond one's personal spaces and connect with others."* Against this backdrop, education organizations played a major role in most projects. One project resulted not only in the connection of two organizations (senior homes and kindergartens) but also in the connection of groups within these organizations and additional partners: *"...because the parents of the (kindergarten) children wanted to be involved and other seniors came up to me and wanted to participate."* In this project seniors helped other seniors to write letters to their pen pals in the kindergarten—groups within the same organization started to connect as an additional outcome of the project. Digital contexts are not discussed explicitly in this category. However, the interviews indicate that many connections were created in digital contexts. Students report about according experiences during the digital implementation of their projects.

5.4   Community Education as Part of Schools and Universities: "...EXperiences Are Even More Intense When Schools Collaborate with Different Social Spaces."

Students find it important that schools and universities are open to the families of the learners: *"That for example parents and siblings know where the child is learning every day and what is happening there all the time."* At the same time students emphasize that education organizations should also be informed about the diverse contexts of the learners: *"As a teacher, you can structure learning arrangements differently, when you know about the backgrounds of the learner. It also helps to come to a better understanding of certain behavior patterns."* Furthermore, students consider the connection of education organizations and social spaces in general as important: *"It (the connection of education organizations and social spaces) helps to develop social competencies ... and social learning cannot be developed on a theoretical level, it needs practical orientation."* Again, the importance of encounters and relationships is discussed in these contexts. Especially the connection with partners in the community is a crucial factor for the participating student: *"...the experiences are even more intense when schools collaborate with different social spaces."* Students also mention the importance of various experiences which are connected to their projects in community education: *"...for example the empathy children showed (towards other children and seniors), the change of perspectives, the caring—this is very important in the development of individuals."*

Some students make it clear: "*Especially in school contexts, there should be more of these collaborations with social spaces.*" The connection of schools and social spaces are emphasized by the students many times: "*...for example collaboration between young and old people can be very constructive and can definitely be connected with topics in schools.*" Against this backdrop, empathy seems to be an important factor for intensive learning experiences in this study. The interviewed students report many times about experiences of belonging to the project and the group of participants. Additionally, the development of empathy in project-based settings seems to change individual perspectives, when it is connected to encounters. The character of encounters does not seem to be important for the students, as digital and face-to-face encounters are discussed without any further differentiation. Finally, students report about the importance of community education in the higher education sector: "*Every student should implement a project in social spaces. (…) I was very surprised and proud of my group. (…) As future teachers, we can have a positive impact (on social spaces).*"

5.5 Community Education and Relationships: "RELationships Certainly Played a Role."

Students report about relationships as a crucial factor in community education. Within the students' projects, various relationships developed in different dimensions and spaces, digitally and face to face. Participants developed relationships with others in different social spaces, they enhanced their relationships with themselves, they fostered relationships with organizations and groups and society as a whole [40]: "*One could see various relationships which played a role.*" Relationships within the projects were established and developed both, digitally and face-to-face. Students report different digital ways, such as voice messages and videos, which helped foster relationships between various participants in social spaces: "*The relationships between children and seniors developed and created happiness, even though they did not know each other from face-to-face contact.*"

5.6 Digital Environments and Diversity

The interviewed students discuss connections between community education, diversity, and digital environments in all categories and the context of different topics and processes. To implement their projects in digital environments, students used various strategies; some students conducted their projects entirely digitally, whereas others balanced online and face-to-face modes.

Students who chose completely digital ways mostly report about positive effects such as (1) the possibility to establish contact quickly, (2) the opportunity to schedule meetings quickly, (3) the possibility to gather and share information online and simultaneously, (4) the chance to meet without using a car or public transport, and (5) the time-saving aspects of digital communication [34, 39].

In contrast, students who chose face-to-face modes emphasize the necessity of face-to-face interaction as many participants cannot handle the required digital devices independently. Additionally, some students report that they experienced an unobstructed flow of information and communication in face-to-face encounters. The use of certain materials, i.e., pictures or items, may have a better effect.

Digital tools were often used for establishing contacts, organizing schedules, and research: "*...we were looking for pictures and stories online.*" Students explicitly report their experiences with various digital tools: "*I advertised my project digitally. And I created a messenger group for the parents (of the participating children).*" Additionally, students used video calls and messengers to prepare and discuss their projects: "*...to stay connected—especially in Covid times, all stayed in different places. Because of this, we had to connect digitally to communicate constructively.*" However, the use of digital tools in the students' projects also shows limitations: "*...to offer video calls instead of handwritten letters would have been too far from my initial idea of a pen friendship.*" Analog options sometimes seem to be more suitable and constructive: "*There was one child who painted on the back of the letter and the paintings added even more value. I don't think this would have been possible with digital tools.*"

All interviewed students experience diversity and make use of digital environments when planning and implementing their projects in social spaces. Diversity is experienced mostly in the dimensions age, belief, and education.

6   Discussion and Outlook

Documented data show the impact of community education for universities in general and teacher education in particular: Organizations of Higher Education can connect with surrounding actors, organizations, structures, and processes. By this, teacher education organizations prepare a learning space for all involved. In doing so, they also create a social space for them. Community education creates new social spaces that bridge former groups and networks, and that also create new challenges, insights, solutions, and understanding for the future teachers. It prepares an inclusive space for diverse individuals and groups.

The new learning spaces produce awareness for social differences and unique learning content. Therefore, future teachers get an idea of what they might need in a more diverse and complex society. They interact with many different social spaces and, in doing so, find more partners for their didactical and pedagogical practices.

On the other hand, pre-service teachers understand more about the social resources of their tasks and educational plans. In learning this, they create innovative learning processes for their pupils when coming back to those actors from social spaces with whom they have a social relationship. To achieve this, pre-service teachers need to detach from a too intense linkage to the given curricula. It now seems more appropriate to ask for the curricula's goals to achieve them on different didactical approaches.

Against this backdrop, digital environments can play a crucial role in developing and fostering collaborations with partners in diverse social spaces. Students should be prepared to use digital tools to reach out to communities in fast and efficient ways. The development of reliable and sustainable relationships between schools, universities, and social spaces, can be supported by flexible and creative use of digital environments open to all partners.

Finally, this means that concepts of "Engaged Campuses" [27] need to be developed, evaluated, and established to interact in the different given social spaces and the social diversity. Further research should explain how this might work with a particular focus on community education in schools with state-run curricula. Then, a bridged and coherent society might also develop from its educational basis in the community's resources.

References

1. Farros, J.N., Shawler, L.A., Gatzunis, K.S., Weiss, M.J.: The effect of synchronous discussion sessions in an asynchronous course. J. Behav. Educ. 1–13 (2020). https://doi.org/10.1007/s10 864-020-09421-2
2. Yates, A., Starkey, L., Egerton, B., Flueggen, F.: High school students' experience of online learning during Covid-19: the influence of technology and pedagogy. Technol. Pedagog. Educ. **30**(1), 59–73 (2021)
3. Hansen, R.E.: The role of experience in learning: giving meaning and authenticity to the learning process in schools. J. Technol. Educ. **11**(2), 23–32 (2000)
4. Dewey, J.: The middle works. Democracy and Education. Southern Illinois University Press, Carbondale et al. (2008/1916)
5. Llenares, I.I., Deocaris, D.: Measuring the impact of an academe community extension program in the Philippines. Malays. J. Learn. Instr. **15**(1), 35–55 (2018)
6. Castro-Rodríguez, M., Marín-Suelves, D., López-Gómez, S., Rodríguez-Rodríguez, J.: Mapping of scientific production on blended learning in higher education. Educ. Sci. **11**(494). https://doi.org/10.3390/educsci11090494 (2021)
7. Smith, A., Seal, M.: Contested terrain of critical pedagogy and teaching informal education in higher education. Educ. Sci. **11**(476). https://doi.org/10.3390/educsci11090476 (2021)
8. Van Houweling, E.: Decolonising development practice pedagogy: ways forward and persistent challenges in the synchronous online classroom. Int. J. Dev. Educ. Glob. Learn. **13**(2), 136–49. https://doi.org/10.14324/IJDEGL.13.2.06 (2021)
9. Haraway, D.: Situated knowledges: the science question in feminism and the privilege of partial perspective. Fem. Stud. **14**(3), 575–599 (1988)
10. Böhmer, A.: Das Wissen der Situationen. Subjektivität und Objektivitäten in einer Ethnographie der Situation. In: Hitzler, R., Klemm, M., Kreher, S., Poferl, A., Schröer, N. (eds.) Ethnographie der Situation. Erkundungen sinnhaft eingrenzbarer Feldgegebenheiten, pp. 115–126. Essen: Oldib-Verlag (2020)
11. Mossman, A.P. Retrofitting the ivory tower: engaging global sustainability challenges through interdisciplinary problem-oriented education, research, and partnerships in U.S. higher education. J. High. Educ. Outreach Engagem. **22**(1), 35–60 (2018)
12. Weiss, H.H., Norris, K.E.: Community engagement professionals as inquiring practitioners for organizational learning. J. High. Educ. Outreach Engagem. **23**(1), 81–105 (2019)
13. Karakose, T.: The impact of the COVID-19 epidemic on higher education. Opportunities and implications for policy and practice. Educ. Process: Int. J. **10**(1), 7–12. https://doi.org/10.22521/edupij.2021.101.1 (2021)
14. Kara, A.: Covid-19 pandemic and possible trends for the future of higher education: a review. J. Educ. Educ. Dev. **8**(1), 9–26. https://doi.org/10.22555/joeed.v8i1.183 (2021)
15. Funk, J.: Caring in practice, caring for knowledge. J. Interact. Med. Educ. **2021**(1), 11, 1–14 (2021). https://doi.org/10.5334/jime.648
16. Ludlow, A., Armstrong, R., Bartels, L.: Learning Together: localism, collaboration and reflexivity in the development of prison and university learning communities. J. Prison Educ. Reentry **6**(1), 25–45 (2019)
17. Olo, D., Correia, L., Rego, C.: Higher education institutions and development: missions, models, and challenges. J. Soc. Stud. Educ. Res. **12**(2), 1–25 (2020)

18. Böhmer, A.: Management der Vielfalt. Emanzipation und Effizienz in sozialwirtschaftlichen Organisationen. Springer VS, Wiesbaden (2020)
19. Warren-Gordon, K., Hudson, K., Scott, F.: Voices of partnerships within the critical service-learning framework. J. Community Engagem. High. Educ. **12**(2), 17–25 (2020)
20. Solano, G.L.: The end of LIFE: thoughts on the marginalization of powerful service-learning in higher education. Crit. Quest. Educ. **11**(3), 208–229 (2020)
21. Walker, A., Mercer, J., Freeman, L.: The doors of opportunity: how do community partners experience working as co-educators in a service-learning collaboration? J. Univ. Teach. Learn. Pract. **18**(7), 56–70 (2021). https://doi.org/10.53761/1.18.7.05
22. Costley, C.: Definitions of different forms of work and learning in higher education. Work Based Learn. e-J. **10**(2), 53–66 (2021)
23. Carvalho, A.R., Santos, C.: The transformative role of peer learning projects in 21st century schools—achievements from five Portuguese educational institutions. Educ. Sci. **11**(196) (2021). https://doi.org/10.3390/educsci11050196
24. Kimmel, S.C., Burns, E., DiScala, J.: Community at a distance: employing a community of practice framework in online learning for rural students. J. Educ. Libr. Inf. Sci. **60**(4), 265–284 (2019). https://doi.org/10.3138/jelis.2018-0056
25. Tamtik, M.: Informing Canadian innovation policy through a decolonizing lens on indigenous entrepreneurship and innovation. Can. J. High. Educ. Revue canadienne d'enseignement supérieur **50**(3), 63–78 (2020)
26. Sam, C.H., Elder, B.C., Leftwich, S.: Supporting university-community partnerships: a qualitative inquiry with contingent academics to understand their scholarship of engagement. J. High. Educ. Outreach Engagem. **25**(1), 37–50 (2021)
27. Furco, A.: The engaged campus: toward a comprehensive approach to public engagement. Br. J. Educ. Stud. **58**(4), 375–390 (2010). https://doi.org/10.1080/00071005.2010.527656
28. Achilleos, J., Douglas, H., Washbrook, Y.: Educating informal educators on issues of race and inequality: raising critical consciousness, identifying challenges, and implementing change in a youth and community work programme. Educ. Sci. **11**(410). https://doi.org/10.3390/educsci11080410 (2021)
29. Soliman, D., Costa, S., Scardamalia, M.: Knowledge building in online mode: insights and reflections. Educ. Sci. **11**(425) (2021). https://doi.org/10.3390/educsci11080425 (2021)
30. Taylor, A., Butterwick, S., Raykov, M., Glick, S., Peikazadi, N., Mehrabi, S.: Community Service-Learning in Canadian Higher Education. https://www.ualberta.ca/community-service-learning/media-library/documents/reports/ks-report-31-oct-2015-final.pdf (2015). Last accessed 15 March 2022
31. Council of Ontario Universities.: Change agents: Ontario Universities: transforming communities, transforming lives. Toronto, Ontario. www.cou.on.ca/publications/reports/pdfs/community-transformation-final-report (2015). Last access 14 March 2022
32. Ministry of Education.: Ontario. http://www.edu.gov.on.ca/eng/general/elemsec/community/. Last accessed 14 March 2022
33. Ministry of Education.: Ontario. The Ontario Curriculum: Social Studies, History and Geography, vol. 11 (2018)
34. Knoblauch, C.: Digital project-based learning in the higher education sector. In: Guralnick, D., Auer, M.E., Poce, A. (eds.) Innovations in Learning and Technology for the Workplace and Higher Education. Lecture Notes in Networks and Systems (349). Springer, pp. 170–179 (2021)
35. Früh, W.: Inhaltsanalyse. Theorie und Praxis. **8**, 147–157 (2015)
36. Salmons, J.: Qualitative online interviews: strategies. In: Design, and Skills, 2nd edn. SAGE: Los Angeles (2015)
37. Ehlers, U.: Qualitative Onlinebefragungen. In: Mikos, L., Wegener, C. (eds.) Qualitative Medienforschung Ein Handbuch (2), 327–339. UTB, Konstanz, München (2017)

38. Berg, B., Lune, H.: Qualitative research methods for the social sciences (9), 21–30/172–174 (2017)
39. Knoblauch, C.: Experiential learning in digital contexts—a case study. Cham, Springer (2022) (In press)
40. Boschki, R.: Beziehung als Leitbegriff der Religionspädagogik. Stuttgart (2003)

The Use of Indie4All Platform for Visually Impaired Students on the Acquisition of Learning Objects with Computational Thinking Practices in Music, Math and Physics

Sarantos Psycharis[1,2], Paraskevi Theodorou[2,3]($\boxtimes$) [iD], and Pantelis Kydonakis[4] [iD]

[1] School of Pedagogical and Technological Education (ASPETE) Greece, Heraklion, Greece
[2] Hellenic Education Society of S.T.E.M. (E3STEM), Heraklion, Greece
theodoroup@unipi.gr
[3] Department of Digital Systems, University of Piraeus, Piraeus, Greece
[4] MSc in Informatics, Department of Informatics, University of Piraeus, Piraeus, Greece

Abstract. Recently, coding and computational thinking are included in every school curriculum. Within the intelligent tutoring systems community for the sensory disabled, adaptive platforms integrating different forms of inclusive context are emerging as an effective medium for individualized learning in various school grades. To date, relatively few empirical studies have been conducted to assess learning experiences in customized learning environments for the sensory disabled, especially for the blind and visually impaired. In this paper, an authoring tool in the indie4all platform creates accessible digital learning units with new interactive activities for the inclusion of the disabled. The study investigates and presents a scenario for teaching Music, Math and Physics simultaneously via the utilization of the Micro bit programming environment. This is an initial pilot study evaluated by both field experts and a selected group of blind individuals before presenting the platform to the real users, the blind and visually impaired students. Research indicates that the users improved their learning experiences and outcomes with the aid of this educational tool, Indie4all platform and Micro: Bit programming environment.

Keywords: Computational Thinking · Learning Objects · Evaluation · Indie4all Platform

1 Introduction

An important issue in modern society is the ability to ensure inclusive and equal opportunities for education and learning for all citizens, including people with disabilities, within the framework of Universal Design for Learning (UDL). This term includes both those who are blind and those with low vision [1]. According to the World Health Organization, approximately 285 million people worldwide are visually impaired [2]. Although many of them try to socialize and have equal educational opportunities, they still face many difficulties that prevent them from making use of these opportunities. Globally, vision loss is the third most frequent disability. Although there are cost-effective therapies for preventing or treating most types of vision loss, their availability varies greatly between nations and regions. Visual impairment and socioeconomic variables were shown to be tightly linked, which might assist in the identification of nations that need to pay more attention to these concerns. The relationship between vision loss and socioeconomic variables might help with public health planning [3]. According to research, some socio-psychological characteristics among totally blind persons constitute impediments to assistive technology acceptance and use [4]. Furthermore, some socio-psychological factors are likely to continue to inhibit the use of assistive technology [5]. The lack of a gradual transition from one grade to another, particularly among visually impaired pupils, has contributed to many of the challenges they have in receiving and processing information and knowledge. These hurdles are reduced because of these students' and the institute's adoption, usage of assistive technology and the necessary training with it.

Haring and Schiefelbusch [6] dealt with issues related to the education of visually impaired students. They focused mainly on the importance of vision and reading and tried to show how intelligence is manifested in blind and visually impaired people compared to deaf people. Their work highlighted the importance of blindness and information processing and also demonstrated the need to harness the use of all available sensory data during learning, as well as the translation of visual stimuli.

Nowadays, an increasing number of countries have included pre-programming and computational thinking (CT) in education and national curricula [7]. To achieve this goal, many educational tools such as programming languages and learning objects are needed [8].

Rapidly evolving technological development can be helpful in our efforts to provide equal educational opportunities for visually impaired students. However, the standardization of CT highlights the lack of appropriate tools for many children with disabilities. Indeed, although programming is difficult for children, for visually impaired pupils there are some additional barriers that need to be overcome. Interestingly, previous research has examined accessibility [9], bias [10], navigational information [11] and usability [12] of technology for such individuals. However, in several countries, the participation of visually impaired people in educational processes was found to be limited due to insufficient funding, the lack of participation of people with disabilities in decision making and the lack of knowledge about the needs of these people [13]. Therefore, there seems to be a growing need to create accessible learning objects geared towards developing CT practices for visually impaired students, which at the same time is important to better equip them to live and work in an increasingly digital world. This is a challenge, as there is a tendency to rely more and more on visual representations to convey abstract concepts.

CT is originally being referred to as "procedural thinking," coinciding with Piaget's "learning-by-making" principles [14]. Wing [15] proposed the current nomenclature, which stands for Abstraction, Algorithms, Automation, Problem Decomposition, and Generalization determining how well a solution spreads across diverse challenges. Selby and Woollard [16] and Angeli et al. [17] also proposed similar component parts. There is growing interest in how CT should be integrated into the teaching approach, particularly for students who have never used a computer before—specifically, which tools should be used and what would constitute a proper pedagogical structure in order for the learning process to be effective [18].

Computing is related to CT. According to Wing [15], "Computing is the field that encompasses computer science, computer engineering, communications, information science and information technology." Wille, Century and Pike [19] state that we should provide to special education need learners with the same "economic and social mobility opportunities as their peers." And computing should be applied for all students.

According to Bower et al. [20], educators are increasingly interested in CT, with a focus on the most appropriate tools, the most effective pedagogical structure, and the best way to embed the discipline into school curriculum and STEM subjects.

With this rise in the importance of the subject within teaching there has come the realization that CT is difficult for learners with visual disabilities because of its "tendency to convey complicated concepts through visual representations" [21]. As a result, there is a lot of research being conducted on how to effectively deliver an inclusive CT curriculum for visually impaired students.

We should highlight that inclusive programming education for visually impaired students has been under extensive research. There are also many research programs for teaching CT for blind students, one of them being the project Interactive Digital Content Platform for All, INDIe4All that will be more analyzed in a following section. This project is being implemented with the further development of the results of the Erasmus Plus KA201 project INDIe, 2018–1-ES01-KA201–050924, http://indieproject.eu, an authoring tool which implement a layer of generation of accessible content for blind students.

2   Learning Objects

A learning object is a collection of digital data and/or practical elements that work together to achieve a particular educational purpose [22]. A learning object, according to the Institute of Electrical and Electronic Engineering (IEEE), is digital or non-digital accessible, pedagogical material with different guidelines and principles that can be utilized for learning, education, or training [23]. Because the major purpose of the learning approach is educating, a learning object should be developed to take advantage of each student's skills e.g, for a child with vision, a picture or an animation is likely to be more successful than a descriptive one. An auditory or sound-based description of information, on the other hand, is far more beneficial than text for a blind child. From the above, it is evident that a learning object's framework and architecture and are essential [24].

Technological improvements create new accessibility challenges that also affect the educational system.

UDL supports educators in achieving this aim by offering a basis for determining how to design a curriculum that fits the requirements of all students from the start [25], including blind and low-vision students [1]. The framework is founded on ideas like employing many ways of presenting information and thus giving multiple means of assistance and opportunities for multiple methods of action and engagement [25].

However, there are still significant accessibility challenges to overcome when teaching CT to visually challenged pupils. In this research, keeping in mind the various learning styles of visually impaired students, we created differentiated scenarios to teach various subjects and uploaded these scenarios in the platform Indie4all, which is accessible for blind students. Following, in this research, we will focus on the scenarios concerning teaching music and physics through the Micro:bit programming environment. Physical computing through the latter environment has been found to be motivating while also providing chances for cooperation and creativity, and there is increased interest in aiding learning through it [26].

3   Indie4All Platform

This platform is a tool for producing digital learning modules that are accessible for students with vision impairments. These units are adjustable for other categories of special needs also. The project is complementary to the running Erasmus Plus INDIe project 2018–1-ES01-KA201–050924, which has the aim of significantly contributing to the production and use of online rich interactive learning units and has successfully developed an authoring tool INDIeAuthor.

The latter tool (https://github.com/cpcdupct/INDIeAuthor and video demo: http://tv.upct.es/?vim=369260713) was created as part of the Erasmus Plus KA201 project INDIe, 2018–1-ES01-KA201–050924, which ran from September 2018 to August 2021. Its engine, INDIeGenerator, is based on UPCTforma, a Domain Specific Language (DSL) created by the Universidad Politécnica de Cartagena's Digital Content Production Center (CPCD [https://cpcd.upct.es]). The INDIeGenerator engine then converts the syntax of this file into HTML5, CSS3, and Javascript code, producing learning units for students to explore. The modules are then published to INDIeOpen, an OER repository, where other teachers/authors may use, share, and change according to the needs and particularities of every class.

The universal design merits will be implemented into the INDIeGenerator engine, which will ensure that the resulting web programming (HTML5, Javascript, CSS) of the generated digital contents for different levels of education complies with accessibility requirements. Additionally, transcripts will be automatically generated for all videos uploaded to the platform by the authors.

The project's goal is to have a significant impact on the adoption and implementation of inclusive e-learning in the participating regions. It is associated with the priority of "Social Inclusion," because it addresses a new way to "reducing gaps in accessing and interacting with formal and non-formal education."

INDIe4all is precisely aligned with the recently released Digital Education Action Plan (2021–27), which identifies two key goals:

- Educators and education and training workers who are digitally skilled and confident.
- Content of high quality, user-friendly tools, and secure platforms, complying to privacy and ethical norms.

The teachers, after being trained in the platform by specialists and researchers, they produce their own learning objects accessible to blind or visually impaired students, including students with other comorbidities. Additionally, to maintain student interest, INDIeAuthor's rich online learning modules frequently feature several interactive activities. Some activities, such as the drag and drop widget, are changed to make them more accessible and practical with the utilization of the screen reader (Figs. 1, 2).

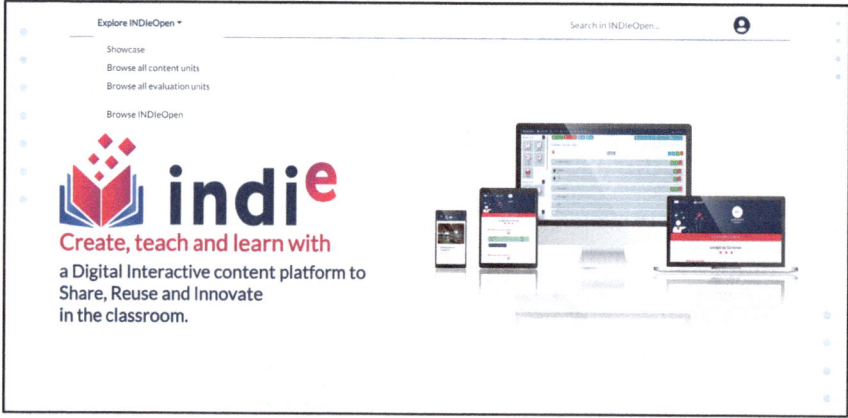

Fig. 1. The basic environment.

Fig. 2. My units.

4 Micro:Bit Programming Environment—an Uploaded Scenario in Indie4All Platform

The Micro:bit is a small programmable and embeddable computer designed, developed, and deployed by the BBC. The Micro:bit Educational Foundation (microbit.org), a non-profit organization founded in 2016, has taken the Micro:bit from a local educational experiment in the United Kingdom to a global effort sponsored by the Micro:bit Educational Foundation [27]. Over four million Micro:bits have been sold in more than 60 countries, with various hardware, content, and education partners involved. All students are being inspired to master essential computing skills in entertaining and creative ways because of this programmable computer which has the size of a credit card [28]. There are other researches like Voštinár in 2020 Slovakia that aims to find out, whether they can motivate normal students to study Computer science by using BBC Micro:bit. The biggest issue in this research was for pupils who had only a basic understanding of music tones. They had to figure out what Middle G, Middle A, and other terms meant [29] (Figs. 3, 4, 5, 6).

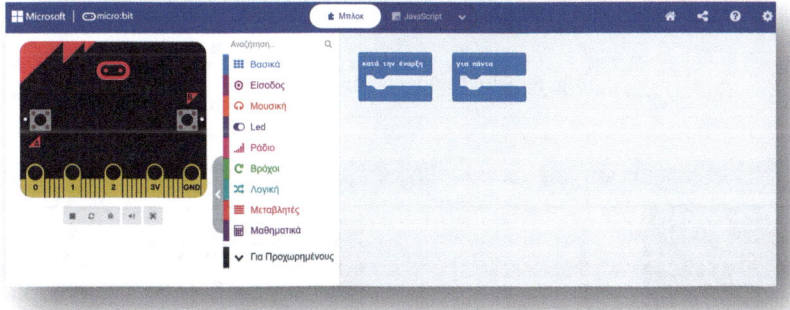

Fig. 3. The programming environment of Micro:bit.

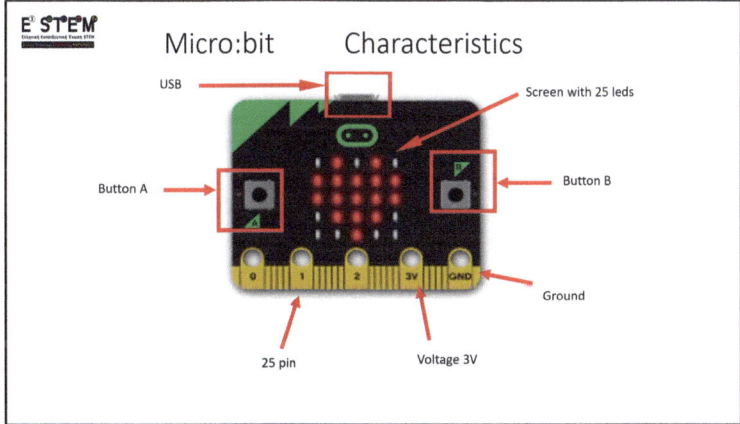

Fig. 4. The buttons of the panel are announced to the blind students.

Students start with creating simple notes (e.g., Mi note), and then continue progressively from the scale of D Major to more complex one. Next, students are reinforced to continue so that other notes of different instruments can be heard (Fig. 7).

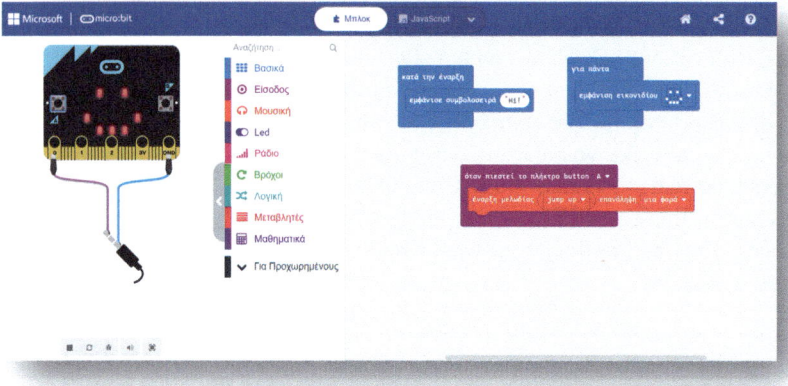

Fig. 5. Blind students' first program.

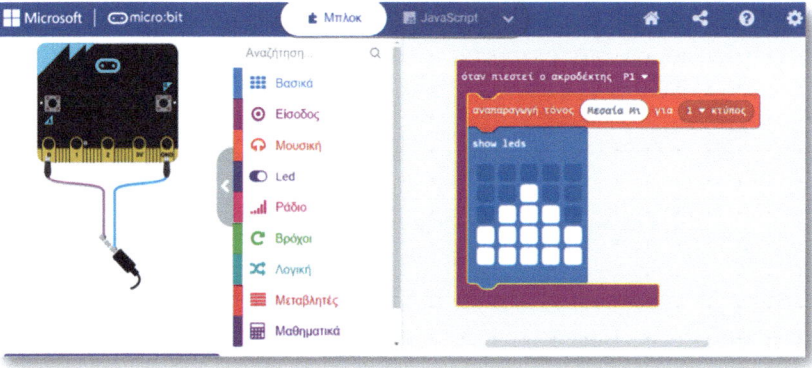

Fig. 6. An example with MI note.

Headphones, Cardboard, crocodile clip leads, and conductive materials like aluminum foil and paper clips are necessary for the creation of projects with the Micro:bit programming environment (Figs. 8, 9).

The above exhibition of the Micro:bit scenario in Indie4all platform has a two-fold goal. The first goal is to teach blind students about basic abstract concepts in Physics via the enjoyable, entertaining, and constructive way of hearing music. The second goal is to teach these students the different sounds of notes, the scales, and different instruments. In the end, they were posed some questions to the students to make them think in depth and search for more information. Some examples are the following: Why doesn't our musical instrument work if we don't touch the two foils? Try putting them together using objects made of different materials like pen, ruler, pencil, and scissors.

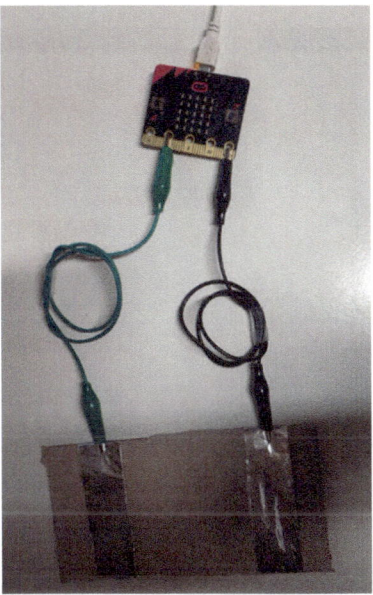

Fig. 7. The construction of a musical instrument.

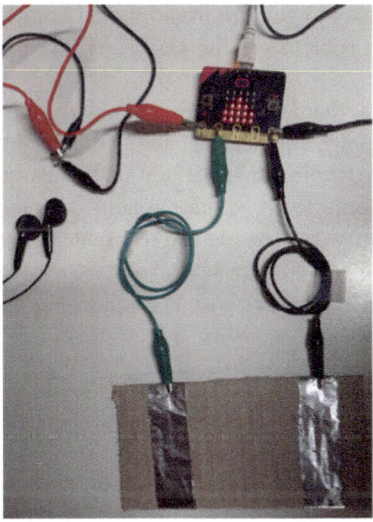

Fig. 8. The connectivity for the headphones to listen to music.

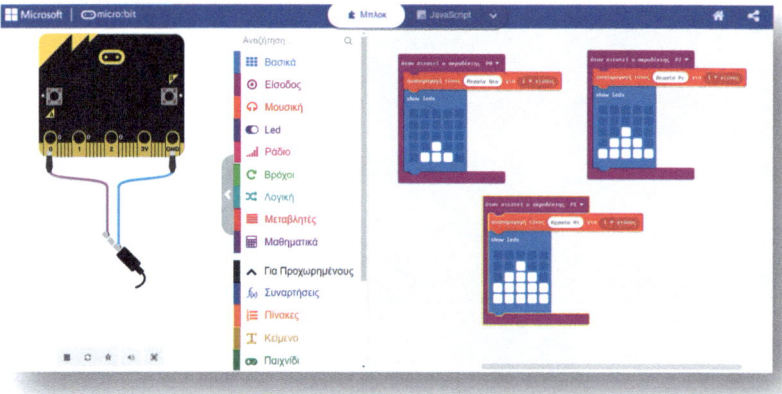

Fig. 9. More notes.

5 Barriers to Computational Thinking Faced by Visually Impaired Students

CT, which includes aspects such as the logical order of executing instructions, has become a significant tool for students to gain information and knowledge via educational platforms. In this way, basic problem-solving skills are developed as part of the learning process [30].

Vision is a key problem that leads to inequalities in education and job possibilities, notably in science, technology, engineering, and mathematics (STEM) [31].

To understand the surrounding environment and construct cognitive maps, a blind and visually impaired student needs additional help from other cues other than vision [32]. People who use cognitive maps build and retain mental representations of distances and routes to areas they can't see. The reception of spatial information from other senses, such as hearing and touch, cooperates with the building of mind maps to represent the world in this way [33, 34].

6 Research Method—Pilot Study

The research focused on the blind and visually impaired students who use the proposed platform and its differentiated scenarios via the provided interface. In fact, the purpose of the pilot study is to gain an understanding of how students interact with the various components of the interface. When the scenarios in Micro:bit programming environment were implemented, three specialists teachers in the school for students with visual problems were invited to use and evaluate the scenarios in the Indie4all platform, its updated versions and determine its practicality, usefulness, and usability. One of the most common themes in the teachers answers in the pilot tests was that the Micro:bit is very simple for pupils to learn and transform concepts and tasks into engaging, motivating and accessible to all. Other opinions about the Micro:bit were that it was good, entertaining, and didn't require any expensive components, and that it could develop as many projects as required.

The aim was to improve the comprehension and the interaction with the platform in order to better perceive and interact with the interface of the Indie4all platform and the learning scenarios for music, math and physics. Considering the comments of the above experts, we developed and gradually improved the learning scenarios adding more drag and drop exercises and creating more interesting interactive videos with video editing explaining the learning concepts with accuracy. After several attempts, the three experts, were very satisfied with the results. We intended to test the software in the field, however because of the COVID-19 pandemic, we did not visit special education classrooms with blind and visually impaired students. This is our next plan when the virus will subside.

7 Application Design

A visually impaired student needs special techniques to receive feedback from the device. Sensitive interfaces such as vibrations, synthesized voice and beep sounds are thus used to provide multisensory information to users [35].

The formation of mental structures based on previously identified things, locations, and memories embedded in a blind student's mind is known as cognitive mapping. According to studies on the acquisition of spatial knowledge, blind students may create mental representations of space in the same way as people with normal eyesight do if they have adequate information [35, 36].

8 Conclusions

Due to the wide internet access and the increased use of mobile devices, the use of CT skills is considered necessary and for that purpose, various tools have been developed for a more effective teaching approach [37]. Even though scientific research has been based on interventions and student engagement in a variety of educational domains, understanding of how to teach CT to children with visual impairments is insufficient.

This study intends to help and stimulate the creation of new technologies to improve the computer learning education of students with visual impairments. Having access to cutting-edge technologies such as text-to-speech and speech recognition, through mobile application development environments such as AppInventor, applications can be easily created, which are accessible to visually impaired people in addition to those without this impairment.

The benefits of accessibility and non-exclusion from digital learning objects provide a starting point by creating a new space that promotes opportunities for people with disabilities to participate in all areas of education.

References

1. Corn, A., Erin, J.: Foundations of low vision: Clinical and functional perspectives. 2nd edn. APH Press (2010)

2. Pascolini, D., Mariotti, S.: Global estimates of visual impairment: 2010. Br. J. Ophthalmol. **96**, 614–618 (2011)

3. Wang, W., Yan, W., Müller, A., Keel, S., He, M.: Association of socioeconomics with prevalence of visual impairment and blindness. JAMA Ophthalmology. **135**, 1295 (2017)

4. Sachdeva, N., Suomi, R.: Assistive Technology for Totally Blind—Barriers to Adoption. In: Bratteteig, T., Anaestad, M., Skorve, E. (ed.) IRIS36. Akademika forlag pp. 182–198. (2013)

5. Bhatt, A., Kumari, A.: Assistive technology for the visually impaired children for their academic excellence. Glob. J. Med. Case Rep. (2015)

6. Haring, N., Schiefelbusch, R.: Methods in special education. McGraw-Hill, New York [etc.] (1967)

7. Grout, V., Houlden, N.: Taking computer science and programming into schools: The Glyndŵr/BCS Turing Project. Procedia. Soc. Behav. Sci. **141**, 680–685 (2014)

8. Maloney, J., Resnick, M., Rusk, N., Silverman, B., Eastmond, E.: The scratch programming language and environment. ACM Trans. Comput. Education. **10**, 1–15 (2010)

9. Giraud, S., Thérouanne, P., Steiner, D.: Web accessibility: Filtering redundant and irrelevant information improves website usability for blind users. Int. J. Hum Comput Stud. **111**, 23–35 (2018)

10. Du, J., Haines, J.: Working with indigenous communities: reflections on ethical information research with ngarrindjeri people in South Australia. Proceedings of the Association for Information Science and Technology. **55**, 794–796 (2018)

11. Guerreiro, J., Ahmetovic, D., Kitani, K., Asakawa, C.: Virtual navigation for blind people. In: Proceedings of the 19th International ACM SIGACCESS Conference on Computers and Accessibility. (2017)

12. Tekli, J., Issa, Y., Chbeir, R.: Evaluating touch-screen vibration modality for blind users to access simple shapes and graphics. Int. J. Hum Comput Stud. **110**, 115–133 (2018)

13. Alotaibi, H., S. Al-Khalifa, H., AlSaeed, D.: Teaching programming to students with vision impairment: impact of tactile teaching strategies on student's achievements and perceptions. sustainability. 12, 5320 (2020)

14. Papert, S.: Mindstorms: Children, computers, and powerful ideas. Basic Books, New York (1980)

15. Wing, J.: Computational thinking. Commun. ACM **49**(3), 33–35 (2006)

16. Selby, C., Woollard, J.: Computational thinking: the developing definition. university of Southampton. (E-prints) (2013)

17. Angeli, C., Fluck, A., Webb, M., Cox, M., Malyn-Smith, J., Zagami, J.: A K-6 computational thinking curriculum framework: Implication for teacher knowledge. Educ. Technol. Soc. **19**(3), 47–57 (2016)

18. Guzdial, M.: Education paving the way for computational thinking. Commun. ACM **51**, 25–27 (2008)

19. Wille, S., Century, J., Pike, M.: Exploratory research to expand opportunities in computer science for students with learning differences. Comput. Sci. & Engineering. **19**, 40–50 (2017)

20. Bower, M., et al.: Improving the computational thinking pedagogical capabilities of school teachers. Aust. J. Teach. Educ. **42**(3), 53–72 (2017)

21. Hadwen-Bennett, A.: computing for learners with visual impairments, https://helloworld.ras pberrypi.org/issues/11, (2019)

22. Basuhail, A.: e-Learning objects designing approach for programming-based problem solving. Int. J. Technol. Educ. (IJTE). **2**(1), 32–41 (2019)

23. IEEE standard for learning object metadata, https://standards.ieee.org/standard/1484_12_1-2002.html, last accessed 12 March 2022

24. De Macedo, C., Ulbricht, V.: Accessibility guidelines for the development of learning objects. Procedia Comput. Sci. **14**, 155–162 (2012)

25. Hastuti, W., Degeng, N., Efendi, M., Praherdiono, H.: Design model mix blended universal design learning in university. Psychol. Educ. **57**(8), 375–393 (2020)
26. Kalelioglu, F., Sentance, S.: Teaching with physical computing in school: the case of the Micro:bit. Educ. Inf. Technol. **25**, 2577–2603 (2019)
27. BBC—Make It Digital - The BBC Micro:bit, https://www.bbc.co.uk/programmes/articles/4hVG2Br1W1LKCmw8nSm9WnQ/the-bbc-micro-bit
28. Austin, J., et al.: The BBC Micro:bit. Commun. ACM **63**, 62–69 (2020)
29. Vostinar, P., Kneznik, J.: Experience with teaching with BBC Micro:bit. In: 2020 IEEE Global Engineering Education Conference (EDUCON). (2020)
30. Catlin, D., Woollard, J.: Educational robots and computational thinking. Teaching Robot-ics Teaching with Robotics (TRTWR) and Robotics in Education (RIE) 2014 Conference (2014)
31. Stehling, V., Plumanns, L., Richert, A., Hees, F., Jeschke, S.: Designing Hands-on robot-ics courses for students with visual impairment or blindness. Causes and coping with visual impairment and blindness. (2018)
32. Long, R., Giudice, N.: Establishing and maintaining orientation for mobility. In Foundations of Orientation and Mobility. 3rd edn. American Foundation for the Blind (2010)
33. Damasio Oliveira, J., de Borba Campos, M., de Morais Amory, A., Manssour, I.H.: Teaching robot programming activities for visually impaired students: a systematic review. In: Antona, M., Stephanidis, C. (eds.) UAHCI 2017. LNCS, vol. 10279, pp. 155–167. Springer, Cham (2017). https://doi.org/10.1007/978-3-319-58700-4_14
34. Lahav, O., Schloerb, D., Kumar, S., Srinivasan, M.: BlindAid: A learning environment for enabling people who are blind to explore and navigate through unknown real spaces. 2008 Virtual Rehabilitation. (2008)
35. Guerrón, N., Cobo, A., Serrano Olmedo, J., Martín, C.: Sensitive interfaces for blind people in virtual visits inside unknown spaces. Int. J. Hum Comput Stud. **133**, 13–25 (2020)
36. Kitchin, R.: Cognitive maps. International Encyclopedia of the Social & Behavioral Scienc-es. 2120–2124 (2001)
37. Bers, M., González-González, C., Armas-Torres, M.: Coding as a playground: Promoting positive learning experiences in childhood classrooms. Comput. Educ. **138**, 130–145 (2019)

Author Index

Printed by Printforce, the Netherlands